WEBER
THE A.W. SCHORGER COLLECTION

ANALYSIS WITH STANDARD CONTAGIOUS DISTRIBUTIONS

STATISTICAL DISTRIBUTIONS IN SCIENTIFIC WORK
Volume 4

ANALYSIS WITH STANDARD CONTAGIOUS DISTRIBUTIONS

J. B. Douglas
Department of Statistics
University of New South Wales
Kensington, New South Wales, Australia

International Co-operative Publishing House
Fairland, Maryland USA

P.O. Box 245, Burtonsville, Maryland 20730

Library of Congress Catalog Card Number: 80-80755
Volume ISBN: 0-89974-012-X
Series ISBN: 0-89974-011-1

Printed in the United States of America

PREFACE

This book is intended both for practical and theoretical investigators - both for those who wish to use known procedures relating to contagious distributions in their work in some field of application, and those who wish to extend the procedures.

The organization of the book reflects this intent. Chapter 1 is introductory, with a principal aim to provide a leisurely and repetitive treatment of generating functions - without these almost all of the distributions discussed are intolerably tedious to manipulate. Chapter 2 is a more formal treatment of discrete distributions so far as is relevant to the following Chapters: most of its detail will not be of interest to those primarily concerned with applications unless they have already a reasonable acquaintance with the kind of theory discussed. Chapter 3 attempts to relate distributions with a "stopped" structure to observational material, with some inconvenience because of the necessity of forward references to the specific distributions discussed in Chapters 4 and 5. These Chapters try to separate applicable procedures from the theory so that working formulae can be identified and used without the necessity of reading all the argument leading to them, but it must be confessed that the burden of symbols is sometimes so great as to make the task almost impossible. Finally, the Appendices represent another effort to cater for differing classes of readers. They do not only contain technical derivations: they include qualitative discussion and qualifications regarding commonly used procedures, partly to discourage the merely mechanical application of such techniques.

For the writing of this book G. P. Patil provided both the initial stimulus to tidy up lecture notes and literature references and subsequent continued goading without which its completion would have been at best uncertain. The hospitality, discussion and computational facilities of The Pennsylvania State University (G. P. Patil, J. B. Bartoo and W. L. Harkness), The University of Georgia (L. R. Shenton), the University of Waterloo (D. A. Sprott), Imperial College London (D. R. Cox) and the Systems Development Institute, I. B. M. Canberra (G. H. Ford and W. N. Holmes) contributed substantially to its progress; but of course the facilities of the University of New South Wales, discussion with A. G. L. Elliott, programming assistance by W. J. Adamson and R. D. Williams, drawings by Miss Jan Keppie and above all the patient and accurate typing by Mrs. Helen Langley provided the essential foundation.

But in thanking those whose assistance is, so to speak, visible in the product it is no less appropriate to express appreciation of the recognition by ones family of the demands such writing makes particularly in deprivation of shared activities. Their not always welcoming reception of these demands is no less essential and equally thanked.

PREFATORY NOTES

Notation

As far as is reasonably practicable a standard notation is used for a number of the distributions which appear repeatedly in this book: it is set out in the following *Brief Table of Distributions* (Table 0).

Absolute uniformity of usage is not practicable, but the following conventions are applied where possible, with as much conformity with tradition as can be provided by sight or sound.

> Greek letters are used for parameters, so that in the binomial distribution the Greek letter pi, π, replaces the traditional p (cf. Kendall and Stuart I (1963)) and ν replaces n; the latter is especially convenient because n is then released for sample size.
>
> Upper case roman letters are used for random variables $(X,Y,\ldots)$ and lower case for their values $(x,y,\ldots)$. The outstanding exceptions to this are probably $\overset{o}{\mu}, \hat{\mu}, \ldots$ for moment, maximum likelihood,... estimators of a parameter μ; these are random variables. Also, in a context rarely likely to be confusing, $F(x), G(x), \ldots$ are used for cumulative distributions functions: these are not random variables.
>
> Underlining with a wavy line is used to denote matrices and vectors.
>
> With much reluctance, and only after repeated encounters with ambiguities between the use of $'$ for transpose, for differentiation and (e.g.) for moments about the origin, raised t, ^{t}, has been used for transpose.

It may also be remarked that *variate* is used synonomously with *random variable*.

Distribution Terminology

When a name for a distribution seems to be very widely used it is also applied here - for brevity in compound names *negative binomial* is often replaced by *Pascal*, though both are used. When required by the context for definiteness, descriptions such as "Poisson distribution with mean λ" are abbreviated to "Poisson(λ) distribution": the exact specificiation is given in the Brief Table of Distributions (Table 0).

Just as one now speaks of the Poisson (not Poisson's) distribution, reference will be made to the Neyman (Type A) distribution; similarly, as with the binomial distribution, so the logarithmic (not logarithmic series) distribution.

References

With the appearance of *A Dictionary and Bibliography of Discrete Distributions*, by G. P. Patil and S. W. Joshi (1968), there is less need to aim at bibliographic completeness. Further, *Distributions in Statistics: Discrete Distributions*, Vol. I, by N. L. Johnson and S. Kotz (1969), gives a very complete coverage of the literature arranged under subject headings with an emphasis on applicable results but without proofs. Mention should also be made of F. A. Haight (1967): *Handbook of the Poisson Distribution*, and J. K. Ord (1972): *Families of Frequency Distributions*.

Appendices

There are various Appendices containing material which different readers will use to widely differing degrees and for different purposes. When a definition is expected to be familiar to most readers it is given, if at all, in an Appendix, to be referred to only in case of need - perhaps especially to refresh ones memory as to precise details. On the other hand, some material in the Appendices is difficult to find in the litearture; or it may be in an Appendix because it is algebraically long and would interrupt the train of argument in the main body of text: the results can be taken on trust by those who wish to do so.

Computer Implementations

Many of the formulae, for example for calculating probabilities and fitting families of distributions, have been implemented on various computer systems. These implementations have been carried out at various times, and with varying purposes in mind: once off calculations, iteractive use by experts, interactive use by the unsophisticated, routine running, etc.; correspondingly some have been hurriedly constructed with inelegance of structure and few cautionary lines, while some (with an increase in the times devoted to their construction and often also with an increase in their line lengths) are more carefully composed and tested. A selection of these, chosen partly for compactness, in APL, are given following the Exercises at the ends of various sections - others are available if requested from the author.

ACKNOWLEDGMENTS

For permission to reproduce materials in this volume, thanks are due to: the Editors of Biometrics (Tables 9 on page 271, 12 on page 286, and 13 on page 299), Biometrika (Figures 25 on page 216, 26 on page 226 and 34 on page 345 and Table 7 on page 211), Contributions from Laboratory of Vertebrate Biology (Figure 10 on page 121), Environment and Planning (Figures 9 on page 120 and 27 on page 229), Journal of Animal Ecology (Figure 5 on page 116); and Charles Griffin and Company Ltd. for Figures 13 on page 162, 18 on page 194, 19 on page 194, 20 on page 195, 21 on page 195, and 24 on page 202.

TABLE O

BRIEF TABLE OF DISTRIBUTIONS

Name	Probability or Density Function	p.g.f.
Poisson(λ)	$e^{-\lambda} \frac{\lambda^x}{x!}$, $\lambda > 0$	$e^{-\lambda+\lambda z}$
binomial(ν,π)	$\binom{\nu}{x}\pi^x\chi^{\nu-x}$, $0 < \pi < 1$, $\pi + \chi = 1$, ν pos. integ. $(x = 0,1,\ldots,\nu)$	$(\chi + \pi z)^\nu$
logarithmic(α)	$\frac{-\alpha^x}{x \ln(1-\alpha)}$, $0 < \alpha < 1$ $(x = 1,2,3,\ldots)$	$\frac{\ln(1-\alpha z)}{\ln(1-\alpha)}$
geometric(π)	$\chi\pi^x$	$\frac{\chi}{1-\pi z}$
negative binomial(κ,ρ) / Pascal(κ ρ)	$\binom{\kappa+x-1}{x}\rho^x \gamma^{-\kappa-x}$ $\rho,\kappa > 0$, $\gamma = 1 + \rho$	$(\gamma - \rho z)^{-\kappa}$
Poisson-binomial(μ,ν,π)	---	$\exp\{-\mu+\mu(\chi+\pi z)^\nu\}$ $\mu > 0$
Poisson-Pascal(μ,κ,ρ)	---	$\exp\{-\mu+\mu(\gamma-\rho z)^{-\kappa}\}$
Polya-Aeppli(μ,ρ)	---	$\exp(-\mu + \frac{\mu}{\gamma-\rho z})$

TABLE O (Continued)

Name	Probability or Density Function	p.g.f.
Neyman(μ,ν) Neyman Type A(μ,ν)	---	$\exp(-\mu+\mu e^{-\nu+\nu z})$ $\mu,\nu > 0$
Thomas(μ,λ)	---	$\exp(-\mu+\mu z e^{-\lambda+\lambda z})$, $\mu,\lambda > 0$
gamma(α,β)	$\dfrac{e^{-x/\beta}x^{\alpha-1}}{\beta^{\alpha}\Gamma(\alpha)}$, $\alpha,\beta > 0$ $(x > 0)$	$(1 - \beta \ln z)^{-\alpha}$
beta(α,β)	$\dfrac{x^{\alpha-1}(1-x)^{\beta-1}}{B(\alpha,\beta)}$ $(0 < x < 1)$	---
log-zero(δ,ρ)	---	$1-\pi+\pi \dfrac{\ln(1-\alpha z)}{\ln(1-\alpha)}$ $= 1-\delta \ln(\gamma-\rho z)$ $\delta = \dfrac{-\pi}{\ln(1-\alpha)}, \quad \rho = \dfrac{\alpha}{1-\alpha}$, $0 < \delta < 1/\ln \gamma$
log-zero-Poisson(δ,ρ,λ)	---	$1-\delta \ln(\gamma-\rho e^{-\lambda+\lambda z})$, $\lambda > 0$

TABLE OF CONTENTS

TABLE OF CONTENTS (Continued)

TABLE OF CONTENTS (Continued)

INDEX OF TABLES

INDEX OF FIGURES

ANALYSIS
WITH STANDARD
CONTAGIOUS DISTRIBUTIONS

Chapter 1
INTRODUCTORY NOTIONS

1.1 INTRODUCTION

Counting is perhaps the earliest formal mathematical process children carry out, and its importance survives at all levels of mathematical sophistication. In experimental work, a scientist often counts as well as or instead of measuring; and the first observations which led to the setting up of a frequentist approach to the notion of probability must have come from counting. Slowly, and perhaps first with respect to games of chance, a degree of stability was recognized about the frequencies of some repetitive events: certain "experiments" led to results predictable only on a long term basis. For example, an apparently symmetrical coin when tossed repeatedly comes down so that after many tosses the number of heads is nearly always about half the total number of tosses.

Gradually, the notion of random variable developed: for such an experiment as the above, with an individual outcome not precisely predictable but in which all controllable conditions are held constant, a numerical valued function, X say, can be defined on the set of outcomes. So in the throw of 2 dice, X might be the total number of spots uppermost; or in the use of a car, X might be the number of kilometers travelled between punctures. Implicit is the idea that probabilities are associated with the set of outcomes on which the random variable is defined (its domain, or sample space).

This set (or a suitable family of subsets of it), which can consist of arbitrary objects, enables one to attach probabilities directly to the random variable (or variate), and when this has been done, it is common also to call the set of values of X its sample space. For a pair of drawing pins or thumb tacks, using

$$X = \text{number of points uppermost after a single toss,}$$

the outcomes are:

Pin No. 1		Pin No. 2	X
Point Up	,	Point Up	2
Point Up	,	Base Up	1
Base Up	,	Point Up	1
Base Up	,	Base Up	0

The sample space of X is 0,1, and 2; to attach probabilities to these points, suppose it is known that for Pin No. 1 the probability of Point Up is π_i (so that of Base Up is $1-\pi_i = \chi_i$, say). Then the probabilities of these compound outcomes are readily found in terms of those of the elementary outcomes on the supposition of independence:

the probability that $X = 2$ is $\pi_1\pi_2$,

the probability that $X = 1$ is $\pi_1\chi_2 + \pi_2\chi_1$,

and the probability that $X = 0$ is $\chi_1\chi_2$.

In an obvious notation, frequently used, these are written

$$\Pr(X=2) = \pi_1\pi_2, \text{ etc;}$$

where $\Pr(X=x)$, $x = 0,1,2$, is called the probability function of X. This probability function (abbreviated to p.f.) is often capable of being summarized in a formula, sometimes rather indirectly, but often advantageously nevertheless. For example, writing

$$\Pr(X=0) = p_0,\ \Pr(X=1) = p_1 \text{ and } \Pr(X=2) = p_2$$

for brevity, the above results are all contained in

$$p_0 + p_1 z + p_2 z^2 = \chi_1\chi_2 + (\pi_1\chi_2 + \pi_2\chi_1)z + \pi_1\pi_2 z^2 = (\chi_1 + \pi_1 z)(\chi_2 + \pi_2 z),$$

where this is interpreted by noticing for the (dummy or carrier) variable z that the coefficient of z^x is precisely the probability that X takes the value x. Such functions of z are called probability generating functions: they arrange in some recognizable order all the probabilities being studied. For three pins, the probabilities of the various outcomes 0,1,2, and 3 points uppermost are arrayed by

$$(\chi_1 + \pi_1 z)(\chi_2 + \pi_2 z)(\chi_3 + \pi_3 z)$$

as above; if the pins have equal probability π of coming point uppermost, the probability generating function (p.g.f.) is

$$(\chi + \pi z)^3 = \chi^3 + 3\chi^2 \pi z + 3\chi\pi^2 z^2 + \pi^3 z^3,$$

so that the probability of 2 points uppermost is $3\chi\pi^2$, the coefficient of z^2. The ease with which such functions are manipulated is the justification for their introduction - in the case of more complicated probability distributions it often would be very difficult to deal directly with explicit formulae for individual probabilities, though of course these can be obtained from generating functions if they are needed.

Finally, in these introductory remarks, a comment is made about "counting" variates versus "measuring" variates. Mostly the discrete variates discussed below run through the integers 0,1,2,3,... or perhaps some part of these, like 1,2,3,.... Sometimes this is not the case - for example if X_i has the former sample space, and one studies the mean

$$\bar{X} = (X_1 + X_2 + \ldots + X_n)/n,$$

it is obvious that $\bar{X}$ has a sample $0, 1/n, 2/n, \ldots$. But the possible values can be arranged in order and are equally spaced: it is this which distinguishes the lattice, or arithmetic, variate from the general discrete variate (though discrete is often used loosely), and even more from the continuous case corresponding to an idealization of measurement in which values in a continuum constitute the sample space. The mathematical apparatus to give a fairly adequate discussion of much discrete variate theory is less than for the general case, and, while continuous variates will be used where necessary, most of the development will need only a slight acquaintance with the details.

The subject matter of this book is the (univariate) contagious distributions, a term introduced by Neyman (1939) and now used generally though rather loosely to describe many of the less simple discrete distributions. (This book excludes a formal treatment of the negative binomial and logarithmic (series) distributions since each will be dealt with in a separate volume.) Originally the notion was that for larvae of low mobility hatched from eggs laid in egg clusters the finding of a larva provided evidence of being in the neighborhood of an egg cluster and thus being rather likely to find another larva: the connection with contagion is clear, although there is no larva serving as a focus of "infection" in this case. Objection has sometimes been taken to the terminology, the circumstances being described as those of "apparent contagion". However, the terminology has been used very widely (and not implausibly), and, in terms of the treatment which follows, a contagious distribution could be defined as that of a variate X whose structure can be represented as

$$X_1 + X_2 + \dots + X_N$$

where each X_i is thought of as independently coming from a specific distribution (e.g., as above, the number of larvae per egg cluster) and where the number N of egg clusters observed also has some specific independent distribution. The description does not correspond to a unique specification in that many distributions with this representation have rather natural alternative representations.

It is difficult to trace the lines of development of the study of these distributions for at least two distinct but often intermingled reasons. For instance, the negative binomial may be regarded as a contagious distribution, first so recognized by Quenouille (1948) (and the details spelt out in Quenouille (1949) though Luders (1934) had exhibited the structure implicitly). However, Greenwood and Yule (1920) in their work on accidents ("accident proneness" (see Kemp (1970)) is a term which seems to lead to even more confusion than "contagious") were considering the distribution as a weighted mixture of other distributions, while the negative binomial as the distribution of the number of tosses of a coin necessary to achieve a fixed number of heads stems at any rate from Montmort (1708). This problem, of which of these dates to consider as primary, has added to it the more practical problem of locating first references: there can be few fields in which so many results have been rediscovered so often in so many different notations with reference to so diverse applications - and published without recognition in so many Journals.

As evidenced by the repeated finding of early gambling devices, notably astralagi (like knuckle bones), and the ancient practice of divination, random events have been the subject of intense interest for a very long period. Perhaps because of the imperfections of these devices, the stability of long run frequencies of specified events does not appear to have been formally recognized until relatively recently - though well made pottery dice of 3000 B.C. are known - and the development of the simplest theory of probability, for an outcome space which was partitioned into sub-sets of equally likely events, was delayed until about 1500 A.D. It is also possible that the imperfections of ordinary shuffling as a randomizing device in card games obscured the appearance of this stability and thus its equally likely analysis.

The theory itself drifted into the mathematical backwater of elaborate combinatorial problems until about this century, when the development of measure theory permitted a formal statement - perhaps most easily signposted by the Kolmogorov axioms in 1933 - of its foundations and thus the modern theories of probability and statistics. Of course neither probability nor statistics waited for this axiomatization, and in particular the other, empirical, face of statistics could be said to have begun to appear about 1660 with John Graunt's Bills of Mortality - however, the connexion between the two faces was not really recognized until the nineteenth century.

The Poisson distribution plays a rather similar role with respect to discrete distributions to that of the normal for absolutely continuous distributions. But its first appearance early in the eighteenth century was purely as a mathematical formula: an approximation to or limit for a binomial probability. It was not until the very end of the nineteenth century that its direct applicability to discrete data was recognized, by von Bortkiewicz. In the early part of the twentieth century the subject of discrete distributions developed strongly, followed by a rather subdued period until the nineteen forties, when there began a revival of interest, perhaps inspired by Neyman's 1939 paper on Contagious Distributions. What might have been a decline in the 1950's was changed by the advent of computers: not only did problems previously labelled too hard become manageable, but numerical results for discrete distributions could often be obtained more easily than for continuous cases, in contrast to the usual experience for analytical results.

The information explosion requires specialization for anything like completeness in an account of a topic currently the subject of investigation, and this book therefore hews to a rather narrow line in order to stay within reasonable bounds - other volumes in the same series will redress the emphasis which appears here on one aspect of the theory and application of discrete distributions.

This book contains an account of some of the most used but not altogether elementary discrete distributions, fitting their discussion into a uniform theoretical framework which is first set up, and then dealing with some of their practical aspects. It does not pursue in an encyclopaediac way the great variety of distributions which can be and have been generated by application of mixing procedures: it rather deals in some depth with those distributions which have in fact been seriously applied, with some emphasis on how recent computer oriented work has changed what was generally believed about them, for example with respect to the distribution of estimators. Of course there are few final answers known, but at the least some questions are now recognized and better posed than in pre-computer days, when the sample size n was almost invariably supposed to be large enough for the first term of an asymptotic expansion to be applicable.

1.2 THE BINOMIAL AND POISSON DISTRIBUTIONS

The reader is supposed to have some acquaintance with the binomial and Poisson distributions; this section is intended partly as a reminder of some important aspects of them, in a rather descriptive fashion and in a manner which will serve also as an introduction to the more complicated distributions which follow.

As before, the probability function $\Pr(X=x)$ of the random variable X is written $p(x)$, or p_x. There are two extreme positions from which one can regard $p(x)$. The first is that it is a mathematically defined formula, like

$$p(x) = 1/6 \text{ for } x = 1,2,3,4,5,6, \text{ or}$$

$$p(x) = e^{-1.469}\frac{(1.469)^x}{x!} \quad \text{for } x = 0,1,2,\ldots$$

$$(\text{where } x! = x(x-1)(x-2)\ldots 3.2.1,\ 0!=1),$$

the properties of which can be studied, and which, it may turn out, happens to describe reasonably accurately the relative frequencies with which the counts of certain natural phenomena appear. The second considers some physically determined system, constructs a mathematical description (or <u>model</u>) of some aspects of it, and - for some models - deduces a formula for the probabilities of occurrence of certain observable events. This formula may turn out to be mathematically tractable and hence useful in verifying the appropriateness of the model constructed, or in making predictions.

However, there is commonly much interaction between these extreme positions; that certain distributions are called "elementary" may be due to their mathematical importance (certainly this is so for the Poisson distribution) or to their simple and immediate relations with physical systems (as, e.g., for the binomial distribution and sampling from a large collection of objects which includes defectives).

The mathematical specifications of these two distributions as used here are as follows.

a) The binomial distribution. If quadrats are randomly placed over a homogeneous field, and the probability of finding a certain plant species within a single quadrat is π (a constant, corresponding to homogeneity), then provided the quadrats are small compared with the field, the numbers of quadrats in which the species are found (or alternatively, not found) will be nearly binomial. A precisely specified model considers a set of objects in which a proportion π is of one kind (called, say, "defective") and the rest, of proportion $1-\pi = \chi$, of another kind ("effective"). If an object is randomly drawn the probability that it is defective is π; if it is replaced and another randomly drawn the probability that it is defective is also π. The probabilities that the objects should be both defective, both effective, or one of each kind, can be obtained by examining in detail the possible constitutions of these outcomes. For later brevity it is convenient to call the particular elementary event in which one is interested a success (e.g. the finding of the species, or a defective) and to score 1 for the occurrence of each success; similarly a failure is scored 0. (Of course these labels are quite arbitrary and could be used in any consistent manner.) Thus if X_i is the outcome score for the ith draw, $X_1 = 1$ means a success is obtained at the first draw; and

$$\Pr(X_1 + X_2 + X_3 = 2)$$

is the probability that in the first 3 draws (in fact, in any 3 draws selected without knowledge of their outcomes) 2 successes, and hence 1 failure, are obtained. An explicit formula for this probability, and the other probabilities associated with 3 draws, is obtained by exhibiting all possible outcomes as follows:

	First Draw X_1 (=0,1)		Second Draw X_2 (=0,1)		Third Draw X_3 (=0,1)	
3 successes $X_1+X_2+X_3=3$	1	+	1	+	1	=3
2 successes $X_1+X_2+X_3=3$	1	+	1	+	0	=2
	1	+	0	+	1	=2
	0	+	1	+	1	=2
1 success $X_1+X_2+X_3=1$	1	+	0	+	0	=1
	0	+	1	+	0	=1
	0	+	0	+	1	=1
0 successes $X_1+X_2+X_3=0$	0	+	0	+	0	=0

Since on each occasion the probability that $X_i = 1$ is π (and for 0 is χ), writing X for $X_1+X_2+X_3$, then

$$\Pr(X=3) = \pi^3,$$

$$\Pr(X=2) = 3\pi^2\chi ,$$

$$\Pr(X=1) = 3\pi\chi^2,$$

and $$\Pr(X=0) = \chi^3.$$

The argument generalizes easily: for ν draws, the probability of any one specific outcome containing x successes (and $\nu - x$ failures) is

$$\pi^x \chi^{\nu-x}$$

irrespective of the order in which these successes and failures occurred; and to obtain the probability that $X_1 + X_2 + \ldots + X_\nu (=X)$ takes the value x whatever the order of the x successes across the ν draws, it is only necessary to find in how many ways these x successes can be distributed. Writing this function as $\binom{\nu}{x}$ ("ν choose x"), it is easy to see that to get to the νth draw with x successes from the $(\nu - 1)^{th}$ draw only two kinds of of path are possible: either $X_1 + X_2 + \ldots + X_{\nu-1} = x$ and then $X_\nu = 0$, or $X_1 + X_2 + \ldots + X_{\nu-1} = x-1$ and then $X_\nu = 1$. So it is clear that

$$\binom{\nu-1}{x} + \binom{\nu-1}{x-1} = \binom{\nu}{x},$$

and by successive applications of this formula (Pascal's Triangle) the number of ways can be written down, both numerically and as a formula:

$$\binom{\nu}{x} = \frac{(\nu-1)(\nu-2)\ldots(\nu-x+1)}{x(x-1)\ldots 3.2.1}$$

$$= \frac{\nu!}{x!(\nu-x)!}, \quad 0 \le x \le \nu,$$

where as usual $0! = 1$.

Hence, generally, for X the number of successes in ν random drawings with replacement after each draw from a set of objects (a <u>population</u>, with probability π of a success at a single draw),

$$\Pr(X=x) = \binom{\nu}{x}\pi^x \chi^{\nu-x}, \quad x = 0,1,2,\ldots,\nu.$$

A slightly more sophisticated way of setting out the above argument involves recollecting the binomial expansion of $(a+b)^n$ to be

$$(a+b)^n = a^n + na^{n-1}b + \frac{n(n-1)}{1.2} a^{n-2}b^2 + \frac{n(n-1)(n-2)}{1.2.3.} a^{n-3}b^3 + \ldots + b^n$$

$$= \binom{n}{0} a^n + \binom{n}{1} a^{n-1}b + \binom{n}{2} a^{n-2}b^2 + \ldots + \binom{n}{n} b^n$$

$$= \sum_{x=0}^{n} \binom{n}{x} a^{n-x} b^x.$$

It follows that

$$(\chi + \pi z)^{\nu} = \sum_{x=0}^{\nu} \binom{\nu}{x} \chi^{\nu-x} \pi^{x} z^{x};$$

i.e., the number of ways x of the π's (and $\nu-x$ of χ's) can arise is as the formula (A) gives; or, in the binomial expansion of $(\chi + \pi z)^{\nu}$, the coefficient of z^x is the probability of exactly x successes. It will be recalled that a function of some variable, like z (which acts as a label only and is often thus called a dummy or carrier variable), is called a generating function: by means of an expansion procedure it can be made to list (or generate) in some systematic way the probabilities of other functions being studied, while holding them all in a compact formula until individual terms are needed. In general, the probability generating function of X will be written as $g(z)$ in what follows: it is a function such that

$$g(z) = p(0) + p(1)z + p(2)z^2 + \ldots + p(x)z^x + \ldots$$
$$= \Sigma\; p(x)z^x.$$

The above argument can now be put more briefly still. For any one draw, the probability generating function is

$$\Pr(X=0)z^0 + \Pr(X=1)z^1$$
$$= \chi + \pi z;$$

for ν independent draws it is thus

$$(\chi + \pi z)(\chi + \pi z)\ldots(\chi + \pi z) \quad \text{to } \nu \text{ terms} \qquad \text{(b)}$$
$$= (\chi + \pi z)^{\nu},$$

and this is the previous expression. In deriving this result, use was made of the "convolution" property that if for the two independent variates X and Y the probability generating functions are $g(z)$ and $h(z)$ respectively, then that of $W = X + Y$ is $g(z)h(z)$. For the probability that $W = X + Y$ takes any specified value w is found by considering the probabilities associated with all the ways $X + Y = w$, and these are $X = 0$, $Y = w$; $X = 1$, $Y = w-1$, $X = 2$, $Y = w-2, \ldots$ to $X = w$, $Y = 0$; which are exactly the terms which come up in successive terms in powers of z in the product $g(z)h(z)$.

In summary, the binomial(ν,π) variate X has the probability function $\Pr(X=x) = \binom{\nu}{x}\pi^x \chi^{\nu-x}$, $x = 0,1,\ldots,\nu$, $0 < \pi < 1$, $\pi+\chi = 1$, the probability generating function $(\chi+\pi z)^\nu$, and its mean and variance are π and $\nu\pi\chi$ respectively.

b) The Poisson distribution. A somewhat similar argument, not here polished, for the numbers X of events per interval of fixed length t (in time or space, e.g.) with suitable assumptions can lead to the Poisson distribution. Divide the interval (0,t)

into n sub-intervals each of length h, so $nh = t$:

$$(0,h)\ ,\ (h,2h)\ ,\ (2h,3h),\ldots,((n-1)h,\ nh).$$

If the probability generating function of X is written $g(z;t)$ and correspondingly those of the numbers of events in each sub-interval as

$$g(z;0,h)\ ,\ g(z;h,2h),\ldots\ ,$$

then if there is independence between the sub-intervals,

$$g(z;t) = g(z;0,h)\ g(z;h,2h)\ldots g(z;(n-1)h,nh).$$

If the (constant) average number of events per unit time is written μ, and h is small, simple but plausible assumptions are that:

Pr(exactly 1 event in a sub-interval) will be approximately jointly proportional to μ and h

$= \mu h +$ higher order terms in h,

and Pr(0 events in a sub-interval) = $1 - \mu h$ + higher order terms; and so

Pr(more than 1 event in a sub-interval) = terms in h of higher order than the first.

(Their plausibility is perhaps enhanced by thinking of what would happen as $h \to 0$.)

For a sub-interval, the probability generating function of the number of events is thus

$$(1-\mu h)z^0 + \mu h z^1 + \text{higher order terms},$$

whence

$$g(z;t) = \{1-\mu h(1-z)\}^n + \text{higher order terms}$$

$$= \{1-\frac{\mu t(1-z)}{n}\}^n + \text{higher order terms}$$

on writing $h = t/n$. Letting $n \to \infty$ and $h \to 0$ in such a way that $nh = t$ remains finite, and recalling the exponential limit

$$\lim_{n\to\infty} (1+\frac{\alpha}{n})^n = e^{\alpha},$$

gives

$$g(z;t) = \exp\{-\mu t(1-z)\}.$$

Introducing $\lambda = \mu t$, the mean number of events per interval, this generating function is

$$e^{-\lambda+\lambda z};$$

using the exponential power series it is

$$e^{-\lambda} e^{\lambda z} = e^{-\lambda}(1+\lambda z+\frac{\lambda^2}{2!}z^2+\ldots+\frac{\lambda^x}{x!}z^x+\ldots)$$

and so the probability function of this Poisson variate X is

$$\Pr(X=x) = e^{-\lambda}\frac{\lambda^x}{x!}, \quad x = 0,1,2,\ldots,\lambda > 0.$$

As can be shown readily, its mean and variance are both λ.

Such a process is exemplified by the numbers of atoms which decay per minute in a sample of radium ore, by the numbers of misprints per page in a type set volume, and by the numbers of cells per cubic centimeter of some recognizable characteristic in a dilute suspension. If insects have no community habits and are sufficiently sparsely distributed so that they interact very little, it would be reasonable to expect the numbers per quadrat in quadrat sampling to be Poisson distributed; there is, however, in spite of the old name "Law of Small Numbers" (Das Gesetz der Kleinen Zahlen) no implication that either the number of insects per quadrat or its expectation should be small.

```
     ∇R←BN PARA;NU;PI;X
[1]   ⍝RETURNS THE PROBABILITY VECTOR FOR THE BINOMIAL(PARA←NU,PI).    PGF:
     [(1-PI)+PI×Z]*NU
[2]   →(((NU-1)=|⌊(NU←PARA[1])-1)∧(PI>0)∧(PI←PARA[2])<0.99999)/LBL
[3]   →ρ0ρ⎕←'NU  MUST BE AN INTEGER ≥ 1 ; AND  0<PI<0.99999 .  START AGAIN.'
[4]  LBL:R←(X!NU)×(PI*X)×(1-PI)*NU-X←0,⍳NU
     ∇


     ∇R←TOL PS PARA;I;SUM;PX
[1]   ⍝RETURNS THE PROBABILITY VECTOR FOR POISSON(PARA←LA), TO AS MANY TERMS
      AS NECESSARY FOR
[2]   ⍝  CUMULATIVE SUM > 1-10*-TOL , WITH A TAIL SUM APPENDED.           PGF
     :   EXP[(-LA)+LA×Z]
[3]   PX←,*-PARA,0ρI←0
[4]  INC:→((SUM←+/PX←PX,PX[I]×PARA÷I←I+1)≤1-10*-TOL)/INC
[5]   R←PX,1-SUM
     ∇
```

FIGURE 1 APL functions for calculating binomial(ν,π) and Poisson(λ) probabilities.

1.3 COMPOUNDING DISTRIBUTIONS

In constructing mathematical models of physical (including biological) phenomena, it is often convenient to use familiar distributions, like the binomial and Poisson, whose properties can be taken as known, and from these to build more complex distributions intended to apply to more complex physical phenomena rather than always to try to return to first principles. Some of the procedures commonly used have already been illustrated with reference to the binomial and Poisson cases; a rather more general treatment is now given, and the next Chapter gives a still more general formulation.

Convolutions (Sums of random terms)

One of the most obvious ways of combining variates is to consider the distribution of their sum, or convolution: if the distributions of X_1 and X_2 are known, what can be said about the distribution of $X_1 + X_2$, where this has the obvious interpretation exemplified by the following? Suppose a randomly chosen egg cluster has X_1 eggs and another randomly chosen cluster X_2 eggs; what is the distribution of the total number of eggs in the two clusters (in terms of the distributions of X_1 and X_2, or, if necessary, their joint distribution)? Exactly this question was discussed in the previous section for special cases.

If X_1 with p.g.f. (probability generating function) $g_1(z)$ is independent of X_2 with p.g.f. $g_2(z)$, a neat formula for the pgf of $X_1 + X_2 = W$, say, can be given. Its p.f., $\Pr(X_1 + X_2 = w)$, say, is obviously given by summing the products of all terms corresponding to the X_1 and X_2 whose sum is w: if the domain of X_1 and of X_2 is 0,1,2,..., then

$$\Pr(X_1 + X_2 = w) = \Pr(X_1 = 0)\cdot\Pr(X_2 = w) + \Pr(X_1 = 1)\cdot\Pr(X_2 = w-1) + \ldots$$

$$\ldots + \Pr(X_1 = w)\cdot\Pr(X_2 = 0)$$

$$= \sum_{x_1+x_2=w}\sum \Pr(X_1 = x_1)\cdot\Pr(X_2 = x_2)$$

$$= \sum_{x_1=0}^{w} \Pr(X_1 = x_1)\cdot\Pr(X_2 = w-x_1) \quad \text{alternatively.}$$

(This is often called the convolution formula for X_1 and X_2; it can also be seen as an application of the Theorem of Total Probability:

$$\Pr(X_1 + X_2 = w) = \sum_{x_1+x_2=w} \Pr(X_1 + X_2 = w \mid X_1 = x_1)\Pr(X_1 = x_1)$$

$$= \sum_{x_1} \Pr(X_2 = w - x_1)\Pr(X_1 = x_1).)$$

But this is precisely the coefficient of z^w in the expression

$$g_1(z)g_2(z) = \Sigma\, z^{x_1} \Pr(X_1 = x_1)\ \Sigma\, z^{x_2} \Pr(X_2 = x)$$

$$= \Sigma\,\Sigma\, z^{x_1+x_2} \Pr(X_1 = x_1)\cdot\Pr(X_2 = x_2)$$

$$= \Sigma\,\Sigma\, z^{w} \Pr(X_1 = x_1)\cdot\Pr(X_2 = x_2)$$

where $w = x_1 + x_2$; in symbols, if $g_W(z)$ is the p.g.f. of $W = X_1 + X_2$,

$$g_W(z) = g_1(z)g_2(z).$$

Of course the process generalizes: if $X_1, X_2, \ldots, X_n$ are independent variates with p.g.f.'s $g_1(z), g_2(z), \ldots, g_n(z)$ respectively, then the distribution of $X_1 + X_2 + \ldots + X_n$ (i.e. of the convolution of $X_1, X_2, \ldots, X_n$) has the p.g.f. $g_1(z)\cdot g_2(z)\ldots g_n(z)$.

Thus if X has a Poisson distribution with mean λ, and Y is independently Poisson with mean μ, so that the p.g.f. of X is $\exp(-\lambda + \lambda z)$, the pgf of $X + Y$ is

$$\exp(-\lambda + \lambda z)\ \exp(-\mu + \mu z)$$

$$= \exp\{-(\lambda + \mu) + (\lambda + \mu)z\};$$

i.e. $X + Y$ also has a Poisson distribution with mean $\lambda + \mu$. A direct argument is quite possible, but longer: see the exercises.

An illustration is provided by photographic film: if X is the number of pin-holes, and Y the number of specks, per square cm of film, each being Poisson distributed, then $X + Y$

is the total number of defects (of these kinds) per square cm; and with the assumptions above (which include independence) the distribution of this total conforms with intuition: it is Poisson, with mean the sum of the means.

Because the sums obtained from a direct approach are often analytically complex, their manipulation is often by-passed by the introduction of generating functions, this of course requiring in turn the recognition of the generating function of the convolution (or the use of an inversion theorem) for the identification of the distribution of the convolution.

Random Number of Terms in a Convolution: Stopped Distributions (Generalized Distributions)

The previous section obtained a formula for the

$$S_n = X_1 + X_2 + \dots + X_n$$

where the p.g.f.s of the independent variates were known and n was a known number. So the probability distribution of the total number of eggs when n egg clusters are observed, the distribution within each cluster being known, can be determined. However, a common situation is that the number of egg clusters observed is itself a random variable N, say, independent of any of the X_i with probability function $\Pr(N=n)$: what is the distribution of

$$S_N = X_1 + X_2 + \dots + X_N$$

where as well all of $X_1, X_2, \dots$ are supposed to have the same distribution? A second important example is in the field of insurance, where S_N is the total sum claimed in a year (say) from a company, corresponding to the individual claims $X_1, X_2, \dots, X_N$.

Since a convenient expression for the former case required the use of p.g.f.s, their introduction here is natural. First, $\Pr(S_N = s)$, the probability function sought, is expressed as far as possible in terms of the previous case. In order to do this, consider a set of events $B_0, B_1, B_2, \dots$ which do not overlap and include the whole of the set of outcomes, the sample space. (In the present example the sample space is 0,1,2,... and the B's are simply 0,1,2,....) Then the probability of an event A (that $S_N = s$) can obviously be obtained by finding the probabilities of both A and B_0, both A and $B_1, \ldots$ (i.e. $S_N = s$ and $N = 0$, $S_N = s$ and $N = 1, \ldots$) and adding. By the definition of conditional probability

$$\Pr(\text{both } A \text{ and } B_n) = \Pr(A \mid B_n) \Pr(B_n)$$

where $\Pr(A \mid B_n)$ is the conditional probability of A given B_n, and hence

$$\Pr(A) = \sum_{n=0}^{\infty} \Pr(A \mid B_n) \Pr(B_n)$$

(often called the Theorem of Total Probability); in the present example

$$\begin{aligned}\Pr(S_N = s) &= \Sigma \Pr(S_n = s \mid N = n) \Pr(N = n) \\ &= \Sigma \Pr(S_n = s) \Pr(N = n) .\end{aligned}$$

The first factor in the terms of the sum has already been dealt with compactly in terms of the p.g.f. So to obtain a simple formula here, introduce the p.g.f. of S_N:

$$\begin{aligned}h(z) &= z^0 \Pr(S_N = 0) + z \Pr(S_N = 1) + z^2 \Pr(S_N = 2) + \ldots \\ &= \sum_s z^s \Pr(S_N = s) \\ &= \sum_s \sum_n z^s \Pr(S_n = s) \Pr(N = n)\end{aligned}$$

But if $g(z)$ is the p.g.f. of N, and $f(z)$ of each X_i (so $g(z) = \Sigma z^n \Pr(N = n)$, and $f(z) = \Sigma z^x \Pr(X_i = x)$), then $\sum_s z^s \Pr(S_n = s)$ is the p.g.f. of $X_1 + X_2 + \ldots + X_n$, i.e., it is $[f(z)]^n$. Thus the p.g.f. of S_N is

$$\sum_n [f(z)]^n \Pr(N = n) \qquad \text{(A)}$$

which is exactly the expression for the p.g.f. for N with z replaced by $f(z)$, and hence $h(z) = g(f(z))$. (A more transparent, but rather tedious, notation would be

$$g_{S_N}(z) = g_N(g_X(z)) \ ;$$

the "ordinary" convolution result would then be

$$g_{S_n}(z) = \{g_X(z)\}^n.)$$

A distribution with the structure of this formula will be called a stopped distribution: the N-stopped X distribution, or the g-stopped f distribution. (It is also called, a generalized, or more rarely a clustered, distribution.)

It is not as easy to anticipate this result as for the previous case: if, for example, insects are laying egg clusters in which the numbers of eggs per cluster are Poisson distributed, and the numbers of egg clusters counted in the area are also Poisson distributed, then the total number of eggs is hardly likely to be guessed to have the Neyman Type A distribution, as in fact it has. In the formula (A) it may be noticed that the previous case is obtained if only one fixed $N = n$ occurs: the sum then reduces to a single term, and as before the p.g.f. is $[f(z)]^n$.

Mixtures of Distributions

A very powerful and versatile method of generating more flexible and complex distributions from simpler ones is to _mix_ them. For example, suppose X has the probability function $\phi_1(x)$ or $\phi_2(x)$, and the distribution (it is obviously a distribution)

$$\alpha_1\phi_1(x) + \alpha_2\phi_2(x), \ \alpha_1, \alpha_2 > 0, \ \alpha_1 + \alpha_2 = 1,$$

is constructed: it is referred to as a mixture of ϕ_1 and ϕ_2, or a ϕ-mixture with weights α_1 and α_2, or a mixed-ϕ distribution with weights α_1 and α_2. Such a case would arise when counting leaves per plant, X, from two different kinds of plant which occur in the proportions α_1 and α_2.

The most useful formulation for the present purpose is to consider the probability function $\phi(x)$ to be a function of some indexing parameter λ, so that the family will be written $\phi(x;\lambda)$, with p.g.f. $f(z;\lambda)$. In general, the mixing process is conveniently thought of probabilistically (the constants α_1 and

α_2 could clearly be taken to be probabilities, of choosing ϕ_1 and ϕ_2 respectively), and so the various weights are themselves thought of as having a distribution. Each value of λ has a certain "weight", $\gamma(\lambda)$, and so the mixture is (again by the Theorem of Total Probability)

$$\sum_i \phi(x;\lambda_i)\,\gamma(\lambda_i), \quad \text{where} \quad \sum_i \gamma(\lambda_i) = 1\ ;$$

even more generally, since continuous weights are often useful in practice, this can be written as

$$\int_\Lambda \phi(x;\lambda)dG(\lambda),$$

where $G(\lambda)$ is the cumulative distribution function of λ defined over the domain Λ and the integral is taken in a generalized sense to include all cases, continuous and otherwise.

Those who prefer to use Bayesian arguments make great use of mixtures: in this context the weight or mixing distribution is called the prior distribution. In the theory of insurance the weight function is called the risk function.

It should be noticed that a convolution is a special case of a mixture, though quite important enough in its own right to merit separate study. For two independent variates X and Y, the convolution formula is

$$\sum_y \phi_1(w-y)\,\phi_2(y)\ ;$$

comparison with $\sum_\lambda \phi(x,\lambda)\,\gamma(\lambda)$, the mixture formula, shows the identity of the two, where $\phi(x,\lambda)$ has the special form $\phi(x-\lambda)$.

A famous example (Greenwood and Yule (1920)) of the mixing process is to consider the number of accidents per individual to be Poisson distributed with mean m:

$$\Pr(X=x \mid M=m) = e^{-m}\frac{m^x}{x!}, \quad x=0,1,2,\ldots,m>0,$$

while across individuals the mean is taken to have a gamma (or chi-squared or Pearson Type III) distribution with probability density function

$$\frac{\rho^{-\kappa}}{\Gamma(\kappa)}\, e^{-m/\rho}\, m^{\kappa-1}, \ m>0,\ \Gamma(\kappa) = (\kappa-1)!\ ,$$

where ρ and κ are disposable positive constants. The mixture is thus

$$\int_0^\infty e^{-m} \frac{m^x}{x!} \frac{\rho^{-\kappa}}{\Gamma(\kappa)} e^{-m/\rho} m^{\kappa-1} \, dm ,$$

which reduces to

$$\frac{(\kappa + x - 1)!}{x! \ (\kappa - 1)} \frac{\rho^x}{(1+\rho)^{\kappa+x}} , \quad x = 0,1,2,\ldots$$

Inspection shows that this is the coefficient of z^x in the expansion of

$$(\gamma - \rho z)^{-\kappa} , \quad \gamma = 1 + \rho \ (\text{note } \frac{(\kappa+x-1)!}{x!(\kappa-1)!} = \binom{\kappa+x-1}{x} = (-1)^x \binom{-\kappa}{x}),$$

and the distribution so obtained is called the negative binomial distribution. (Note the two different uses of γ in this section.)

A briefer treatment is usually to be obtained from the use of generating functions. In general, if the probability generating function of the component X indexed by λ is $\gamma(z;\lambda)$, the p.g.f. of the mixture is

$$\int_\Lambda \gamma(z;\lambda) \, dG(\lambda) .$$

So, for the case already taken,

$$\gamma(z;\lambda) = e^{-\lambda+\lambda z} , \quad dG(\lambda) = \frac{\rho^{-\kappa}}{\Gamma(\kappa)} e^{-\lambda/\rho} \lambda^{\kappa-1} d\lambda,$$

and the p.g.f. of the mixture is

$$\int e^{-\lambda+\lambda z} \, dG(\lambda) = (\gamma - \rho z)^{-\kappa}$$

as before. Similarly, if the components are Poisson distributed with mean $\lambda\nu$, and λ is Poisson distributed with mean μ, the p.g.f. of the mixture is

$$\sum_{\lambda=0}^{\infty} e^{-\lambda\nu+\lambda\nu z} e^{-\mu} \frac{\mu^{\lambda}}{\lambda!}$$

$$= e^{-\mu} \sum_{\lambda=0}^{\infty} \frac{\{\mu e^{-\nu+\nu z}\}^{\lambda}}{\lambda!}$$

$$= e^{-\mu} \exp\{\mu e^{-\nu+\nu z}\}$$

$$= \exp\{-\mu + \mu \exp(-\nu+\nu z)\} ,$$

that of the Neyman Type A distribution.

A more detailed study of the properties of mixtures is given in the next Chapter, and further illustrations occur throughout the book. Reference may also be made to an extended account, with illustrations, of the construction both of mixed and stopped distributions in Douglas (1970).

A Note on Terminology

A terminology relating to the various ways of combining distributions has not yet been stabilized, and some care is necessary to determine what a particular author means. In this book the practice will be:

Mixture or Mixed Distribution

$\int f(z;\lambda)\, dG(\lambda)$ is a mixed-f by G distribution (f-mixture by G weight distribution).

Stopped Distribution

$g(f(z))$ is a g-stopped f distribution, where
$f(z)$ is the p.g.f. of X_i, $i = 1,2,\ldots$, the summed variates,
$g(z)$ is the p.g.f. of N,
and $g(f(z))$ is the p.g.f. of $X_1 + X_2 + \ldots + X_N$.

A somewhat similar labelling is used in Rao, Mitra and Matthai (1966).

Among the alternatives which have been used for these distributions or special cases of them are:

Mixture:

Greenwood and Yule (1920): compound
Satterthwaite (1942): generalized (first development)
Feller (1943): compound
Gurland (1957): compound
Feller II (1966): mixture (but the *process* is called randomizing).

Stopped Distribution:

Satterthwaite (1942): generalized (second development)
Feller (1943): generalized
Gurland (1957): generalized
Feller I (1966): compound.

It might be noticed particularly that Satterthwaite's 1942 paper, which discussed only mixed and stopped Poisson distributions, included both under the title "Generalized Poisson Distribution", the mixture being called "First Development" (of the generalization) and the stopped case "Second Development".

But a great variety of usage is to be found, including 'sum' for a convolution with a random number of terms, a 'composite' distribution for either case, and for a mixture 'superposition'. Gurland's (1957) paper introduced the notation $N \vee X$ for an N-stopped X variate in the present terminology, calling N the generalized variate and X the generalizer; similarly $X \wedge L$ was used for a mixed-X by L variate (with the inclusion of an arbitrary scale factor), and both of these are to be found in the more recent literature. Rao, Mitra and Matthai use variate names, and so, e.g., writing p for a Poisson distribution, $p \overset{+}{\ } p$ denotes a Poisson-stopped Poisson variate, and for n denoting a negative binomial variate, pn denotes a mixed Poisson with negative binomial weights. The label stuttering has also been used with respect to the Poisson distribution - see Johnson and Kotz (1969), p. 112, for reasons it is not used here.

When some, not precisely specified, method of combining variates is being discussed in this book, the process will be called *compounding*, and the variates concerned will be called *components*.

1.4 EXERCISES

1. Find the power generating functions $g(z) = \Sigma\, g_n z^n$, $n = 0,1,2,\ldots$ of the following sequences.

   a) $(1,1,1,1)$ is generated by $(1 - z^4)/(1 - z)$ $(=1 + z + z^2 + z^3)$.

   b) $(1,1,1,\ldots)$ by $1/(1 - z)$.

   c) $(1,-1,1,-1,\ldots)$ by $(1 + z)^{-1}$.

   d) $(1,2,3,4,\ldots)$ by $(1 - z)^{-2}$.

   e) $(0,1,\frac{1}{2},\frac{1}{3}),\ldots)$ by $-\ell n(1 - z)$.

   f) $(1,\frac{1}{2},\frac{1}{4},\frac{1}{8},\ldots)$ by $(1 - \frac{1}{2}z)^{-1}$.

   g) $(1,\frac{1}{1!},\frac{1}{2!},\ldots)$ by e^z.

   h) $\binom{\nu}{x}$ for $x = 0,1,2,\ldots,\nu$ by $(1 + z)^{\nu}$.

   $$\left(\binom{\nu}{x} = \frac{\nu!}{(\nu-x)!\ x!} = \frac{\nu^{(x)}}{x!}\right)$$

   i) for $g(z)$ as above:

   $(g_0, g_0 + g_1,\ g_0 + g_1 + g_2,\ldots)$ by $\frac{g(z)}{1-z}$,

   and

   $(g_1 + g_2 + g_3 + \ldots,\ g_2 + g_3 + \ldots,\ g_3 + \ldots,\ldots)$ by $\frac{1 - g(z)}{1 - z}$.

   j) $(f(a),\ f'(a)/1!,\ f''(a)/2!,\ldots)$ by $f(a + z)$,

   by use of Taylor's Theorem.

   k) $(g_0h_0,\ g_0h_1 + g_1h_0,\ g_0h_2 + g_1h_1 + g_2h_0,\ldots)$ by $g(z),h(z)$ where $g(z) = \Sigma\, g_n z^n$ and $h(z) = \Sigma\, h_n z^n$. Use this result to establish d) from b).

2. Use any convenient set of statistical Tables (e.g. the Biometrika Tables, Vol. I (see Pearson and Hartley (1966), Handbook of Tables for Probability and Statistics (see Beyer (1968)), Owen (1962) or Rao, Mitra and Matthai (1966))

to plot the probability functions of the binomial and Poisson distributions. Mark on each graph the mean, and the mean plus and minus one standard deviation. (You may choose to plot the four binomial cases with $\nu = 10$, $\pi = 0.1, 0.3, 0.5, 0.9$ and the four Poisson cases with $\lambda = 1,3,4,9$.)

3. Obtain the single term recurrence relations, by which probabilities can be calculated in succession, for the

binomial: $$p_{x+1} = \frac{\pi}{\chi} \cdot \frac{\nu - x}{x+1} p_x, \quad x = 0,1,\ldots,\nu-1$$

with $p_0 = \chi^\nu$, and

Poisson: $$p_{x+1} = \frac{\lambda}{x+1} p_x, \quad x = 0,1,2,\ldots$$

with $p_0 = e^{-\lambda}$.

(These can be obtained by direct algebraic manipulation on p_{x+1} and, less obviously, by differentiating the p.g.f. with respect to z, subsequently equating coefficients of z^x.)

4. Suppose numbers of children per family in some community are Poisson distributed. (No such community may exist.) If the mean number of children per family is 1 show that the probabilities of 0 or 1 children are the same, and of 2 children half this 0,1 probability.

5. Use a direct argument to show that the convolution of two independent Poisson variates is also Poisson distributed, with parameter the sum of the parameters of the two components.

$$\left(\Pr(X+Y=w) = \sum_{x+y=w}\sum e^{-\lambda} \frac{\lambda^x}{x!} e^{-\mu} \frac{\mu^y}{y!} \right.$$

$$= e^{-\lambda-\mu} \sum_{y=0}^{w} \frac{\lambda^{w-y}}{(w-y)!} \frac{\mu^y}{y!}$$

$$\left. = e^{-(\lambda+\mu)} \frac{(\lambda+\mu)^w}{w!} . \right)$$

6. X takes the values 0,1 with probabilities 1/3, 2/3 respectively; write down its p.g.f. (E.g. throw a fair die and score 0 for a one or six spot uppermost, 1 for any other

outcome.) Y independently takes the values 0,2 with probabilities 1/2,1/2; write down its p.g.f. (E.g. throw a fair coin and score 0 for a head, 2 for a tail.) Enumerate the values $X+Y$ can take, and by considering the ways these values can be obtained calculate the probability each has associated with it. (Check that these probabilities sum to 1.)

Verify this result by multiplying the p.g.f.'s together (the convolution p.g.f. is $(1+2z+z^2+2z^3)/6$).

7. Identify N and the X's in $X_1+X_2+\ldots+X_N$ where this is to be taken as a model for the variate described in the following situations.

   a) The <u>number of coins</u> in the pockets of the people in a particular bus. (X_i : number of coins in the ith person's pockets, and N: the number of people in the bus.)

   b) The <u>amount</u> claimed from Petty Cash in a week in an office.

   c) The <u>number of dogs</u> in the houses in a specified street.

   d) The <u>number of kittens</u> born in an animal shelter in a week.

   e) The <u>number of individuals</u> of the various species of trees on an island.

8. a) A fair coin, with ONE on the heads side and TWO on the tails side, is tossed. A second coin is tossed this number of times. The second coin has probability 2/3 for the heads side (which bears the digit 0) and 1/3 for the tails side (digit 1).

   Calculate the probabilities of the various possible digit sums both by a complete analysis of the outcomes and by use of p.g.f.'s.

   (The stopped p.g.f. is $(10+7z+z^2)/18$.)

   b) Two unbiased dice are available. The first has two faces marked 0 and the other four marked 1. The second has three faces marked 1 and the other three marked 3. A coin has "first" marked on its head side and "second" on its tail side; it is a biased coin for which the probability of heads is 3/4 and tails 1/4.

The coin is tossed, and according to its outcome one of the dice is tossed. Calculate the probabilities of the various possible scores, of 0,1 and 3; use both a direct exhibition of all outcomes, and p.g.f.'s.

(The mixture pgf is $(2+5z+z^3)/8$.)

9. To investigate income (\$X) distribution, an individual is selected randomly from a list of all residents, 1/3 of whom are rural and 2/3 urban. 99% of rural residents have incomes of \$0, and 1% of \$1,000,000; 25% or urban residents have incomes of \$0, 50% of \$10,000, and 25% of \$20,000. Identify ϕ and G in the resulting mixed-$\phi(x;\lambda)$ distribution with weight function $G(\lambda)$.

(Using $\lambda = r$ for rural and $\lambda = u$ for urban,

$$dG(r) = 1/3,\ dG(u) = 2/3,$$

$$\phi(0;r) = 0.99,\ \phi(10^6;r) = 0.01,$$

$$\phi(0;u) = 0.25,\ \phi(10^4;u) = 0.5,\ \phi(2\times 10^4;u) = 0.25.)$$

Make a similar identification for the following: For certain conditions, the probability that an insect's egg hatches is 1/3. The eggs are laid in clusters, the numbers in the clusters being Poisson distributed with mean 20. For a randomly chosen cluster, what is the p.f. of the number of eggs hatching?

$$\left(\phi(x;n) = \binom{n}{x}\left(\frac{1}{3}\right)^x\left(\frac{2}{3}\right)^{n-3},\ dG(n) = e^{-20}\,20^n/n!\right)$$

Chapter 2
FORMAL DERIVATIONS AND GENERAL PROPERTIES

2.1 FORMAL DEFINITIONS

In this chapter a rather more formal analysis of the mathematical models leading to many commonly used discrete distributions is given. However, the view is primarily directed towards applications; a more precise statement of much of the material in the first section which sets up the definitions may be found in Lukacs (1960), e.g. Hence results such as "If X is a random variable then so is $h(X)$" are taken implicitly (with no reference to measurability). On the other hand it is impossible not to refer to measurability notions in classifying variates: no attempt is made to expand these passing references. The chapter is largely self contained, but presupposes an awareness of the kind of material in the preceding chapter.

The development is confined essentially to the univariate case.

Random variables and distribution functions

A function X is a random variable in some space if for a specified class of sets in the space a probability measure is defined for X : i.e., the probability that X belongs to a set S in the class (written $\Pr(X \in S)$) is well defined. Hence the intuitive description of a random variable as a variable with an associated probability distribution.

Discrete variates are one of three fundamental types: they are described in more detail as follows:

For a random variable X the function $\Pr(X \leq x) = F(x)$, say, is well defined: it is called the (cumulative) distribution function of X, and has the following properties.

1. $F(x+h) \geq F(x)$ for $h > 0$ (monotonic non-decreasing).
2. $F(x+0) = F(x)$ (continuous to the right).
3. As limits, $F(-\infty) = 0$, $F(\infty) = 1$.

For an arbitrary distribution, it can be shown (by the Lebesgue Decomposition Theorem - e.g. Loéve (1963)) that $F(x)$ has a unique representation of the form

$$f(x) = \alpha_1 F_1(x) + \alpha_2 F_2(x) + \alpha_3 F_3(x)$$

$$\text{where } \alpha_i \geq 0, \quad \Sigma\alpha_i = 1,$$

and F_1 corresponds to a discrete variate, F_2 to an absolutely continuous variate (often loosely called continuous), and F_3 to a singular variate. These three basic distribution types may be described in the following way.

A variate X is discrete if there exists a countable set E such that $\Pr(X \in E) = 1$ and if for every countable set $A = \{x_1, x_2, \ldots\}$

$$\Pr(X \in A) = \sum_A \Pr(X = x_i).$$

The very special case in which E consists of a single point corresponds to what is usually called a degenerate or improper variate. In most of this book attention will be focussed on lattice or arithmetic discrete variates: for these the discontinuities of $F(x)$ form a (proper or improper) subset of a sequence of equidistant points for which

$$x_j = \alpha + j\beta$$

where α and β are real and $j = 0, \pm 1, \pm 2, \ldots$.

X is absolutely continuous if, for every set N of Lebesgue measure 0, $\Pr(X \in N) = 0$. It then follows (Radon-Nikodym Theorem: Loéve (1963)) that there exists a function $\phi(x)$, called the (probability) density of X, such that

$$\Pr(X \in B) = \int_B \phi(x)\, dx$$

for every (measurable) B.

Finally, a variate X is <u>singular</u> if there exists a set N of Lebesgue measure 0 such that $\Pr(X \in N) = 1$ and if for every countable set A it is true that $\Pr(X \in A) = 0$. (An example and further discussion may be found in Lukacs (1960), Fisz (1963) and Feller II (1966).)

Because of the variety of usage, it might be pointed out that the strict description <u>continuous</u> is applied to a distribution whose decomposition has $\alpha_1 = 0$ and α_2 and α_3 non-zero: i.e., one which has both absolutely continuous and singular components. However, since standard calculus manipulative procedures can be applied only to the absolutely continuous variate - whose distribution function has an expression as the Riemann integral of a density - only the absolutely continuous variate ordinarily arises or is discussed in many courses in Statistics, and so for brevity the adjective "absolutely" is very often omitted.

On a point of notation rather than theory, since almost no manipulative results are needed, it is convenient to use a generalized integral (Lebesgue - Stieltjes) to include all cases. Whatever the type of variate X, with distribution function F(x), for any set A the probability

$$\Pr(X \in A)$$

will be written as

$$\int_A dF(x).$$

Thus if X is discrete with probability function f(x)

$$\int_A dF(x) = \sum_{x \in A} f(x)$$

$$\left(\text{or if } A = \{x_1, x_2, \ldots, x_n\},\ \int_A dF(x) = \sum_{i=1}^{n} f(x_i)\right);$$

if X is absolutely continuous with probability density function $\phi(x)$

$$\int_A dF(x) = \int_A \phi(x)\, dx$$

$$\left(\text{or if } A = \{a \le x \le b\},\ \int_A dF(x) = \int_a^b \phi(x)dx\right).$$

Similar definitions and results can be given for the multivariate case - for example

$$\Pr(X \leq x,\ Y \leq y) = F_{X,Y}(x,y)$$

defines the (cumulative) distribution in the bivariate case.

One of the commonly occurring formulae is that for a convolution. If X and Y have this joint distribution function, the distribution of the convolution $X + Y = W$, say, is given by

$$\Pr(X + Y \leq w) = \iint_{x+y \leq w} dF_{X,Y}(x,y),$$

the integral being taken over the shaded region A.

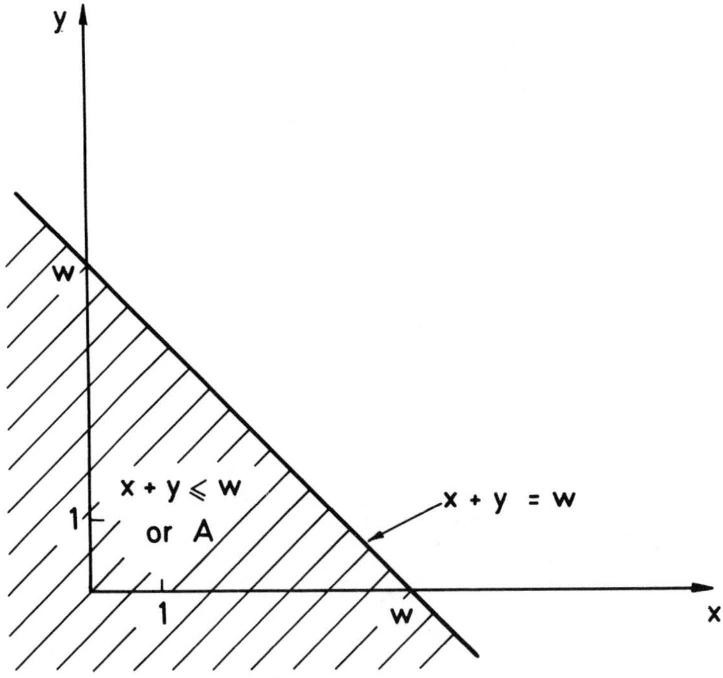

FIGURE 2 Convolution of X and Y.

In the most common application, where X and Y are independent and so $F_{X,Y}(x,y) = F_X(x)F_Y(y)$, the product of the marginal distributions,

$$\Pr(X+Y \leq w) = \iint_{x+y \leq w} d\{F_X(x)\ F_Y(y)\},$$

and the use of a conditional expectation argument (or the reduction of a multiple integral to repeated integrals) enables this to be written as

$$\int_{-\infty}^{\infty} \left\{ \int_{-\infty}^{w-x} dF_Y(y) \right\} dF_X(x)$$

$$= \int_{-\infty}^{\infty} F_Y(w-x)\ dF_X(x).$$

(Where the variates are positive, so the region A is wholly in the positive quadrant bounded by $x+y=w$ and the two axes, the explicit formula becomes

$$\int_{0}^{w} F_Y(w-x)\ dF_X(x),$$

as can be seen by examining the range of variation of x.)

However, no further formal definitions will be given at this point - they will be introduced as they are subsequently needed.

2.2 GENERATING FUNCTIONS

A generating function (g.f.) for a sequence of elements $\alpha_0, \alpha_1, \alpha_2, \ldots$ is a function which lists in some systematic, and hence identifiable, way these elements; the single function and hence the sequence as a whole can thus be manipulated until it is convenient to look at the separate elements.

There are two especially common g.f.'s, one being the power g.f.

$$\alpha_0 + \alpha_1 z + \alpha_2 z^2 + \ldots ,$$

and the other the exponential g.f.

$$\alpha_0 + \alpha_1 \frac{z}{1!} + \alpha_2 \frac{z^2}{2!} + \ldots,$$

z being a dummy variable used only for labelling: the ith element is the coefficient of z^i, or of $z^i/i!$. An example of a power g.f. is the probability g.f., and of an exponential g.f. is the moment g.f.; both forms are useful, in different circumstances, and hence both are widely used.

Most generating functions can be defined as readily for the general as for the discrete case, though their usefulness may be restricted to certain classes. Definitions which apply in general are therefore given, and particular applications follow, little attention being paid to existence questions. (See Niven (1969) for a formulation in abstract terms.)

If X has the distribution F(x), the expectation of any function m(X) is defined (or derived) to be

$$\int m(x)\,dF(x), \quad \text{and written } \mathcal{E}\{m(X)\},$$

over the domain of definition of X. For the discrete case, where the countable set on which

$$\int dF(x) = 1 \quad \text{is} \quad (\ldots, x_{-1}, x_0, x_1, \ldots) \quad \text{and}$$

$$\Pr(X = x_i) = f(x_i),$$

these two integrals are

$$\sum_i m(x_i)\,f(x_i) = \mathcal{E}\{m(X)\} \quad \text{and} \quad \sum_i f(x_i) = 1.$$

Examples of the above and the following definitions are given after section (d).

(a) Probability generating function (p.g.f.). Although this is not the most general type of generating function, it is the most used for discrete variates. Formally, writing it as g(z), the definition is

$$g(z) = \mathcal{E}(z^X)$$

with of course the implication that existence is to be the subject of investigation: if the right side exists, it is the p.g.f.

For the common case where $X = 0,1,2,\ldots$ only,

$$g(z) = p_0 + zp_1 + z^2p_2 + \ldots + z^xp_x + \ldots$$

where as before $p_x = \Pr(X = x)$. Since this is a power series in z, the usual techniques are available: p_x may be determined from $g(z)$ by equating coefficients of z^x, or by using a Taylor series expansion - so $p_x = g^{(x)}(0)/x!$ - for example. Unlike most of the g.f.'s used here it is of power, not exponential type.

Of great importance in applications is the result relating to convolutions: if X and Y are independent variates, the p.g.f. of $X+Y$ is the product of those of X and Y. For that of $X+Y$ is

$$\mathcal{E}(z^{X+Y})$$
$$=\mathcal{E}(z^X z^Y)$$
$$=\mathcal{E}(z^X)\mathcal{E}(z^Y) \quad \text{from the independence assumption.}$$

In some cases, the p.g.f. of some function of X may be of interest: the p.g.f. of $h(X)$ is

$$\mathcal{E}\{z^{h(X)}\}.$$

The p.g.f. of a continuous variate is sometimes useful (Seal (1949) attributes its first use to Lagrange in 1773): the formal definition is unchanged, though no power series expansion need exist.

(b) Characteristic function. Of major importance from a mathematical point of view is the characteristic function $\phi(t)$, defined as $\mathcal{E}(e^{itX})$ where t is real. Its importance chiefly lies in the facts that $\phi(t)$ always exists for any variate, and that it is a kind of integral transform for which many general theorems are available - for example, there is an inversion theorem: given a $\phi(t)$ known to be a characteristic function, there is a formula for the unique distribution function corresponding to it. (Perhaps naturally, the formula is rather complex and not necessarily easy to use: see e.g. Lukacs (1960).)

With reference to the discrete variates here studied, for which $x_j = \alpha + j\beta$, $j = 0, \pm 1, \pm 2, \ldots$ certain results are of interest.

i) The characteristic function is

$$\phi(t) = e^{it\alpha} \sum_j p_{x_j} e^{itj\beta}, \quad p_{x_j} = \Pr(X = x_j),$$

ii) Hence

$$\begin{aligned} |\phi(\tfrac{2\pi}{\beta})| &= |e^{2\pi i\alpha/\beta} \sum p_{x_j} e^{2\pi ij}| \\ &= |e^{2\pi i\alpha/\beta}| \sum p_{x_j} \\ &= 1; \end{aligned}$$

i.e., for an arithmetic distribution there exists a (not unique) $t_0 \neq 0$ such that

$$|\phi(t_0)| = 1.$$

The converse is also true (whence necessity and sufficiency). For suppose there exists a $t_0 \neq 0$ such that $|\phi(t_0)| = 1$. Then there is a real ξ such that $\phi(t_0) = e^{it_0\xi}$,
i.e. $\int e^{it_0X} dF(x) = e^{it_0\xi}$,
or $\int e^{it_0(x-\xi)} dF(x) = 1$.

Equating real parts,

$$\int \{1 - \cos [t_0(x-\xi)]\} dF(x) = 0.$$

But $1 - \cos [t_0(x-\xi)]$ is non-negative, and since $\int dF(x) = 1$ discontinuities of $F(x)$ must be at the zeroes of $1 - \cos [t_0(x-\xi)]$,

$$\text{i.e., at } x = \xi + 2n\pi/t_0, \ n = 0, \pm 1, \ldots;$$

hence X must be an arithmetic variate.

iii) Obviously any arithmetic variate must have a periodic characteristic function - the converse is also true (see Lukacs (1960) for a proof). In fact, if the period is $2\pi/|\beta|$, the discontinuities are at

$$\alpha + j\beta, \ j = 0, \pm 1, \ldots$$

(Without proofs, the following statements might be of interest as a means of identifying variate types through their characteristic functions. A discrete variate has an almost periodic characteristic function; and conversely. (A function $\phi(t)$ is periodic if for all t a number τ can be found such that $\phi(t+\tau) - \phi(t) = 0$. A function $\phi(t)$ is almost periodic if, given an arbitrary $\varepsilon > 0$, for all intervals of t of length $\ell(\varepsilon) > 0$ a number τ can be found such that

$$|\phi(t+\tau) - \phi(t)| < \varepsilon .$$

Clearly a periodic function is almost periodic.) For an absolutely continuous variate,

$$\lim_{|t| \to \infty} \phi(t) = 0.$$

For a singular variate,

$$\limsup_{|t| \to \infty} |\phi(t)| \text{ may take any value in } (0,1.)$$

iv) For the special, but very common and important, case of an integral valued variate, the Inversion Theorem takes a particularly simple form. Suppose $X = \ldots,-1,0,1,\ldots$ (or $0,1,2,\ldots$ etc.) and write $p_x = \Pr(X = x)$. Then if the characteristic function of X is $\phi(t)$, the distribution of X is given by

$$p_x = \frac{1}{2\pi} \int_{-\pi}^{\pi} e^{-itx} \phi(t)\, dt.$$

For $\phi(t) = \sum_x e^{itx} p_x$; taking any fixed $x = \xi$ say,

$$e^{-it\xi} \phi(t) = \sum_{x \neq \xi} e^{it(x-\xi)} p_x + p_\xi .$$

Hence, integrating with respect to t from $-\pi$ to π, and using

$$\int_{-\pi}^{\pi} e^{it(x-\xi)}\, dt = 0, \quad x \neq \xi,$$

gives the formula immediately.

As a simple illustration, if it is known that the characteristic function $\exp(-\lambda + \lambda e^{it})$ corresponds to a discrete variate, its probability function is given by

$$\frac{1}{2\pi} \int_{-\pi}^{\pi} e^{-\lambda+\lambda e^{it}} e^{-itx}\, dt$$

$$= \frac{e^{-\lambda}}{2\pi} \int_{-\pi}^{\pi} \sum_{0}^{\infty} \frac{\lambda^y}{y!} e^{ity} e^{-itx}\, dt$$

$$= \frac{e^{-\lambda}}{2\pi} \sum_{0}^{\infty} \frac{\lambda^y}{y!} \int_{-\pi}^{\pi} e^{it(y-x)} dt$$

$$= e^{-\lambda} \frac{\lambda^x}{x!}$$ since the integral vanishes unless $y = 0$ and is 2π if $y = x$.

The result is of course familiar as the probability function of a Poisson variate.

A more serious example is given by Neyman's (1939) original derivation of the Neyman Type A distribution (see also Exercise 2).

(c) Moment generating functions. The rth (power) moment about the origin of X is defined as $\mathscr{E}(X^r)$; the rth factorial moment is

$$\mathscr{E}\{X(X-1)(X-2)\ldots(X-r+1)\} = \mathscr{E}\{X^{(r)}\}.$$

Because the set of variate values so often consists of the integers, factorial moments are very often simpler and more useful for discrete distributions than are the power moments.

Since

$$e^{tX} = 1 + tX + \frac{t^r}{r!} X^2 + \ldots + \frac{t^r}{r!} X^r + \ldots = \sum_{r=0}^{\infty} \frac{t^r}{r!} X^r,$$

$$\mathscr{E}(e^{tX}) = \Sigma \frac{t^r}{r!} \mathscr{E}(X^r)$$

(given that the operations indicated are valid, as always: the mere writing down of the symbols does not guarantee this), so that the coefficient of $t^r/r!$ is the rth power moment, and

$$\mathcal{E}(e^{tX}) = m(t), \quad \text{say},$$

is a moment generating function, of exponential type. Since

$$(\frac{d}{dt})^r\ m(t) = m^{(r)}(t) = \mathcal{E}(X^r e^{tX}),$$

it follows that $m^{(r)}(0) = \mathcal{E}(X^r)$, and so the rth moment can be obtained by differentiation, for example. Similarly,

$$\mathcal{E}\{(1+t)^X\} = \mathcal{E}\Sigma\binom{X}{r}t^r = \mathcal{E}\Sigma\ X^{(r)}\ \frac{t^r}{r!}$$

$$= \Sigma\ \frac{t^r}{r!}\ \mathcal{E}(X^{(r)})$$

is a generating function, also of exponential type, for the factorial moments.

Writing $\mu'_r = \mathcal{E}(X^r)$ and $\mu'_{(r)} = \mathcal{E}[X(X-1)\ldots(X-r+1)]$, relations between these are obtainable in a variety of ways, by direct expansion of the expectations, or by manipulation of the expansions, replacing t by $\ell n(1+t)$ in the expansion for μ'_r, for example. (These relations may be systematized by the use of Stirling numbers: see the Appendix.) So

$$\mu'_{(1)} = \mu'_1 = \mu,$$

$$\mu'_{(2)} = \mu'_2 - \mu, \text{ etc.}$$

Moments about the mean, $\mathcal{E}[(X-\mu)^r] = \mu_r$, are also often used. If the generating function for moments about the origin is $m_X(t)$ and that for moments about the mean is $m_{X-\mu}(t)$, then

$$m_X(t) = \mathcal{E}(e^{tX}),$$

and

$$m_{X-\mu}(t) = \mathcal{E}\{e^{t(X-\mu)}\} = e^{-\mu t}\mathcal{E}(e^{tX})$$

$$= e^{-\mu t} m_X(t)\ ;$$

and similarly for factorials. Again, these provide relations between moments about the mean and the origin. (See also Kendall and Stuart , I. (1963), and the Appendix on Moments and Cumulants.)

Further, the p.g.f. can be used to generate moments. For

$$g(z) = \mathscr{E}(z^X)$$

and so

$$g^{(r)}(z) = \mathscr{E}\{X^{(r)} z^{X-r}\},$$

whence

$$g^{(r)}(1) = \mathscr{E}(X^{(r)}) = \mu'_{(r)}.$$

There is a sometimes useful Inversion Theorem which gives the probabilities p_x associated with a variate $X = 0,1,2,\ldots$ explicitly in terms of its factorial moments. Writing the p.g.f. as

$$g(z) = \sum_{r=0}^{\infty} z^r p_r,$$

and supposing (as always) the existence of the expansion, the factorial moment g.f. is

$$m_{(\ )}(t) = \Sigma(1+t)^r \; p_r.$$

Hence

$$g(z) = m_{(\ )}(z-1) = \Sigma \; \mu'_{(r)} \frac{(z-1)^r}{r!} \quad ,$$

and

$$g^{(x)}(z) = m_{(\ )}^{(x)}(z-1)$$

$$= \sum_{r=x}^{\infty} \mu'_{(r)} \; r(r-1)\ldots(r-x+1) \frac{(z-1)^{r-x}}{r!}$$

$$= \sum_{r=x}^{\infty} \mu'_{(r)} \frac{(z-1)^{r-x}}{(r-x)!} \;;$$

writing $z = 0$ gives the Theorem:

$$x! \; p_x = \sum_{r=x}^{\infty} \frac{(-1)^{r-x}}{(r-x)!} \mu'_{(r)},$$

corresponding to

$$\mu'_{(x)} = \Sigma\, r^{(x)}\, p_r.$$

And since

$$\mu'_{(m)} = \Sigma\, S_m^{(n)} (-1)^{m-n}\, \mu'_{(m-n)},$$

where $S_m^{(n)}$ is a Stirling Number of the First Kind, the result can also be used to give p_x in terms of power moments. (See the Exercises for an alternative proof.)

(d) Cumulant generating functions. If a power series expansion in t of the logarithm of the moment generating function $\mathcal{E}(e^{tX})$ can be carried out in the form

$$\ln\{\mathcal{E}(e^{tX})\} = \kappa_1\, t + \kappa_2 \frac{t^2}{2!} + \ldots + \kappa_r \frac{t^r}{r!} + \ldots,$$

the coefficient of $t^r/r!$ is the rth cumulant. Similarly, if

$$\ln[\mathcal{E}\{(1+t)^X\}] = \kappa_{(1)} t + \kappa_{(2)} \frac{t^2}{2!} + \ldots + \kappa_{(r)} \frac{t^r}{r!} + \ldots$$

the coefficient of $t^r/r!$ is the rth factorial cumulant; and the functions from which these expansions follow are appropriate generating functions. Again, the factorial cumulants have advantages for use with discrete distributions.

It is clear that all these generating functions are closely related; for example, the characteristic function is a generating function for moments, the coefficient of $(it)^r/r!$ being the rth moment; or if $g(z)$ is the p.g.f. then $g(1+t)$ is the cumulant generating function. It is thus not really necessary to remember all the details for the various cases, since they can be obtained from one another.

Furthermore, the devices most conveniently used to recover more explicit expressions for (say) the probabilities p_x from the p.g.f. $g(z)$ depend on the particular analytical form of $g(z)$ and the kind of expressions sought. One device often appropriate is the Leibnitz formula for the x-fold derivative of a product: if

$$g(z) = u(z)\ v(z),$$
$$g'(z) = u'v + uv',$$
$$g''(z) = u''v + 2u'v' + uv'',$$

and

$$g^{(x)}(z) = \sum_{r=0}^{x} \binom{x}{r} u^{(x-r)} v^{(r)}.$$

For example, if

$$g(z) = \exp\{-\mu + \mu \exp(-\nu + \nu z)\} = \Sigma\ p_x z^x,$$

then

$$g'(z) = \mu\nu e^{-\nu} \cdot e^{\nu z} g(z)$$

on recognizing the appearance of $g(z)$ again. Hence, differentiating x times,

$$g^{(x+1)}(z) = \mu\nu e^{-\nu}\ \Sigma \binom{x}{r} \nu^{x-r}\ e^{\nu z}\ g^{(r)}(z)$$

and, writing $z = 0$ $(r!\ p_r = g^{(r)}(0))$,

$$(x+1)!\ p_{x+1} = \mu\nu e^{-\nu}\ \Sigma \frac{x!}{(x-r)!r!} \nu^{x-r}\ r!\ p_r$$

and

$$p_{x+1} = \frac{\mu\nu e^{-\nu}}{x+1} \sum_{r=0}^{x} \frac{\nu^{x-r}}{(x-r)!} p_r,\quad x = 0,1,2,\ldots$$

(e) Examples.

Binomial distribution.

$$\Pr(X = x) = \binom{\nu}{x} \pi^x \chi^{\nu-x},\quad x = 0,1,\ldots,\nu,\quad 0 < \pi = 1 - \chi < 1.$$

The p.g.f. is $\mathscr{E}(z^X) = \sum_{x=0}^{\nu} z^x \Pr(X = x)$

$$= (\chi + \pi z)^{\nu}.$$

Characteristic function: $(\chi + \pi e^{it})^{\nu}$.

Moment generating function: $(\chi + \pi e^{t})^{\nu}$.

Factorial moment generating function: $(\chi + \pi(1+t))^{\nu} = (1+\pi t)^{\nu}$.

Cumulant generating function: $\nu \ln(\chi + \pi e^{t})$.

Factorial cumulant generating function: $\nu \ln(1+\pi t)$.

Of these, the factorial moment and cumulant generating functions are mathematically the most attractive, because of their simple expansions: thus the rth factorial moment, $\mu'_{(r)}$, is given by

$$\mu'_{(r)} = \nu(\nu-1)(\nu-2)\ldots(\nu-r+1)\pi^{r}$$
$$= \nu^{(r)}\pi^{r},$$

and the rth factorial cumulant by

$$\kappa_{(r)} = (-)^{r+1}\ \nu(r-1)!\ \pi^{r}.$$

From these results, any of the usual moments or cumulants may readily be found (for explicit formulae, see Kendall and Stuart, I (1963)) - e.g.

mean: $\mu'_{(1)} = \nu\pi$,

variance: $\nu\pi\chi$.

(See Figure 1, p. 13.)

Poisson Distribution

$$\Pr(X=x) = e^{-\lambda}\frac{\lambda^{x}}{x!}\ ,\ x = 0,1,2,\ldots,\lambda > 0.$$

Then the p.g.f. is $e^{-\lambda+\lambda z}$; not to write down all the generating functions already listed, notice that the factorial cumulant generating function is simply

$$\lambda t,$$

i.e., the first factorial cumulant is λ, and all higher factorial cumulants are 0. (Compare the normal distribution in which in the usual notation the first cumulant is μ, the second σ^2, and all others zero: each of these properties of the cumulants is a characterization property of the distribution concerned.)

Thus the mean is λ, and so is the variance. (See Figure 1, p. 13.)

(f) Multivariate case. All these definitions can be extended to the multivariate case. By way of illustration, consider the bivariate X, Y, for which the expectation of a function $m(X,Y)$ is

$$\mathcal{E}[m(X,Y)] = \int\int m(x,y)\ dF(x,y)$$

where $F(x,y)$ is the distribution function of X, Y.

Thus the p.g.f. of X, Y with dummy or carrier variables u, v is defined as

$$\mathcal{E}(u^X v^Y) = \int\int u^x v^y\ dF(x,y);$$

and the moment g.f. as

$$\mathcal{E}(e^{uX+vY}).$$

The moment g.f. about the means $\mu_X = \mathcal{E}(X)$, $\mu_Y = \mathcal{E}(Y)$ (where now

$$\mathcal{E}(X) = \int\int_{\text{all } x,y} x\ dF(x,y) = \int_{\text{all } X} x\ dF_X(x),$$

$F_X(x)$ being the marginal distribution of X with

$$F_X(x) = \int_{\text{all } y} dF(x,y)\ )$$

is likewise

$$\mathcal{E}[e^{u(X-\mu_X)+v(Y-\mu_Y)}].$$

Expanding in powers of $X-\mu_X$ and $Y-\mu_Y$ leads not only to the univariate moments but also to mixed moments, of which the most important practically is the covariance $\mathcal{C}(X,Y)$, the coefficient of uv:

$$\mathcal{C}(X,Y) = \mathcal{E}[(X-\mu_X)(Y-\mu_Y)].$$

In the general m-variate case, confining attention to the pgf, if the (joint) p.f. of $X_1,\ldots,X_m$ is $\phi_{X_1,\ldots,X_m}(x_1,\ldots,x_m)$, i.e. $\Pr(X_1 = x_1,\ldots,X_m = x_m) = \phi_{X_1,\ldots,X_m}(x_1,\ldots,x_m) = \phi_{\underset{\sim}{X}}(\underset{\sim}{x})$ for brevity, then the p.g.f. is

$$g_{\underset{\sim}{X}}(\underset{\sim}{z}) = \mathcal{E}(z_1^{X_1} z_2^{X_2} \cdots z_m^{X_m}) = \sum_{\text{all } x_i} \cdots \sum \phi_{\underset{\sim}{X}}(\underset{\sim}{x}) z_1^{x_1} z_2^{x_2} \cdots z_m^{x_m}.$$

(If the X_i are independent, the obvious factorization follows:

$$\phi_{\underset{\sim}{X}}(\underset{\sim}{x}) = \phi_{X_1}(x_1)\, \phi_{X_2}(x_2) \cdots \phi_{X_m}(x_m).)$$

Moments, etc., follow as usual:

e.g. $$\mathcal{E}(X_i) = \frac{\partial}{\partial z_i} g_{\underset{\sim}{X}}(\underset{\sim}{z})\Big|_{\text{all } z_k = 1},\quad k = 1,2,\ldots,m,$$

and $$\mathcal{E}(X_i\, X_j) = \frac{\partial^2}{\partial z_i\, \partial z_j} g_{\underset{\sim}{X}}(\underset{\sim}{z})\Big|_{\text{all } z_k = 1}.$$

The convolution result follows as before: if

$$W = \Sigma X_i \quad \text{and} \quad w = \Sigma x_i$$

with p.f. $\phi_W(w)$ and p.g.f. $g_W(z)$, then

$$g_W(z) = \mathcal{E}(z^W) = \mathcal{E}(z^{X_1} z^{X_2} \cdots z^{X_m})$$

$$= g_{\underset{\sim}{X}}(z,z,\ldots,z)$$

(and for independence, $g_W(z) = g_{X_1}(z) \cdots g_{X_m}(z)$ where $g_{X_i}(z)$ is the p.g.f. of X_i).

Generally, results for transformed variates are not simple, although marginal distributions are obtained immediately by writing $z_i = 1$ for appropriate values of i. For example, the marginal p.g.f. of X_1 is

$$\mathcal{E}(z_1^{X_1}) = g_{\underset{\sim}{X}}(\underset{\sim}{z})\Big|_{z_2 = z_3 = \cdots = z_m = 1}$$

$$= g_{\underset{\sim}{X}}(z_1, 1, \ldots, 1).$$

The classical multinomial result concerning independent Poisson(λ_i) variates X_i constrained to have a constant sum w follows by noting

$$\phi_{\underset{\sim}{X}}(\underset{\sim}{x}) = e^{-\Sigma\lambda_i} \frac{\lambda_1^{x_1}\cdots\lambda_m^{x_m}}{x_1!\cdots x_m!},$$

and for $W = \Sigma X_i$,

$$\phi_W(w) = e^{-\Sigma\lambda_i} \frac{(\Sigma\lambda_i)^w}{w!}.$$

Hence the conditional p.g.f. of $X_1,\cdots,X_m \mid \Sigma X_i = w$

is $\displaystyle \frac{w!}{(\Sigma\lambda_i)^w} \sum_{\Sigma x_i = w}\cdots\sum \frac{(\lambda_1 z_1)^{x_1}\cdots(\lambda_m z_m)^{x_m}}{x_1!\cdots x_m!}$

$$= \sum_{\Sigma x_i = w}\cdots\sum \frac{w!}{x_1!\cdots x_m!}\left(\frac{\lambda_1 z_1}{\Sigma\lambda_i}\right)^{x_1}\cdots\left(\frac{\lambda_m z_m}{\Sigma\lambda_i}\right)^{x_m}$$

$$= \left(\frac{\lambda_1}{\Sigma\lambda_i} z_1 + \cdots + \frac{\lambda_m}{\Sigma\lambda_i} z_m\right)^w ,$$

a multinomial p.g.f. Unfortunately this result cannot be obtained directly from $g_{\underset{\sim}{X}}(\underset{\sim}{z})$ and $g_W(z)$, essentially because

$$\sum_{\text{all } x_i}\cdots\sum \psi(x_1,\cdots,x_m) \quad \text{and} \quad \sum_{\Sigma x_i = w}\cdots\sum \psi(x_1,\cdots,x_m)$$

need bear no simple relation.

2.2a EXERCISES

1. i) If desirable, attempt Exercise 1 from the preceding chapter.

   ii) The following are probability generating functions of

integral valued variates. Find explicit expressions for the corresponding probability functions.

a) $\frac{1}{2}(1+z^2)$;

b) $\frac{z}{4} \cdot \frac{1-z^4}{1-z}$;

c) $\frac{\chi}{1-\pi z}$ and $\frac{\chi z}{1-\pi z}$ where $\pi+\chi=1,\ 0<\pi<1$;

d) e^{z-1};

e) $\frac{\ell n(1-\alpha z)}{\ell n(1-\alpha)}$, $0<\alpha<1$.

2. If the following two functions are characteristic functions of discrete variates, obtain their probability functions.

$$(\gamma-\rho e^{it})^{-\kappa}\ ,\ \rho>0,\ \gamma=1+\rho,\ \kappa>0,$$

$$\exp\{-\mu+\mu\ \exp(-\nu+\nu e^{it})\},\ \mu,\ \nu>0.$$

(Assume discreteness, use the Inversion Formula, and verify the result.)

3. Give examples to show that if $g(z)$ and $h(z)$ are p.g.f.'s then $g(z)\cdot h(z)$ is also a p.g.f. (of the convolution), but that $g(z)/h(z)$ need not be (and "usually" is not) a p.g.f.

Show that if

$$K_{(g)}(t) = \ell n\ g(1+t) = \sum_{r=1}^{\infty} \kappa_{(r):g}\ t^r/r!$$

is the factorial cumulant g.f. for g, and similarly for h, then the factorial cumulants of the convolution are additive:

$$\kappa_{(r):\text{convolution}} = \kappa_{(r):g} + \kappa_{(r):h}\ .$$

4. Show that $\mathcal{E}\{1/(1-tX)\}$ is a (power) generating function for the moments $\mathcal{E}(X^r)$ - but observe that since

$$\mathcal{E}\left(\frac{1}{1-t(X+Y)}\right) \neq \mathcal{E}\left(\frac{1}{1-tX}\right)\mathcal{E}\left(\frac{1}{1-tY}\right)$$

for independent variates X, Y the familiar convolution formula does not apply.

5. Suppose X is the number of points in a fixed interval, with $\Pr(X = x) = p_x$, $x = 0, 1, 2, \ldots$. For each point suppose z is the (constant) probability that it is a "success", this labelling being independent of X. Then for a given number of points, x, the probability that none is a "success" is

$$p_x z^x ,$$

and hence the unconditional probability over all numbers of points that none is a "success" is

$$\sum_{x=0}^{\infty} p_x z^x ,$$

the probability g.f. of X.

(For a detailed examination of this idea, due to van Dantzig, see Rade (1972).)

6. For the p.g.f. $p_0 + p_1 + p_2 z^2 + \cdots + p_x z^x + \ldots = g(z)$, define the 'tail' probabilities $q_x = \Pr(X > x) = p_{x+1} + p_{x+2} + \ldots, x = 0, 1, \ldots$ Show that the p.g.f. of $q_0, q_1, \ldots$ is

$$\frac{1 - g(z)}{1 - z}$$

and that the p.g.f. of $\Pr(X \leq x)$ is

$$\frac{g(z)}{1 - z} .$$

In particular, if the probability of a head at the single toss of a coin is π , show that the p.g.f. of at least $\nu - x$ heads in ν tosses is

$$\frac{(\pi + \chi z)^{\nu}}{1 - z} .$$

7. Consider the discrete distribution

$$\Pr(X = -\tfrac{1}{2}) = \Pr(X = \tfrac{1}{2}) = \tfrac{1}{2}.$$

By use of the p.g.f., show that for independent identically

distributed such variates $X_1,\ldots,X_n$ with $\bar{X} = \Sigma X_i/n$, then $\Pr(\bar{X} = r - \frac{n}{2}) = \frac{1}{2^n}\binom{n}{r}$, $r = 0,1,\ldots,n$.

8. Suppose $X_1, X_2, \ldots, X_n$ are independent variates with corresponding p.g.f.'s $\exp(-\lambda_1 + \lambda_1 z)$, $\exp(-\lambda_2 + \lambda_2 z^2) \ldots, \exp(-\lambda_n + \lambda_n z^n)$ and that the p.g.f. of their n-convolution is

$$\Sigma p_x z^x .$$

Show that $p_0 = \exp(-\Sigma\lambda_i)$,

$$p_1 = p_0\lambda_1$$

$$p_2 = p_0\left(\frac{\lambda_1^2}{2!} + \lambda_2\right)$$

$$p_3 = p_0\left(\frac{\lambda_1^3}{3!} + \lambda_1\lambda_2 + \lambda_3\right)$$

$$p_4 = p_0\left(\frac{\lambda_1^4}{4!} + \frac{\lambda_1^2\lambda_2}{2!} + \frac{\lambda_2^2}{2!} + \lambda_1\lambda_3 + \lambda_4\right)$$

$$\ldots$$

and $$\lambda_1 = \frac{p_1}{p_0}$$

$$\lambda_2 = \frac{p_2}{p_0} - \frac{1}{2}\left(\frac{p_1}{p_0}\right)^2$$

$$\lambda_3 = \frac{p_3}{p_0} - \frac{p_2}{p_0}\frac{p_1}{p_0} + \frac{1}{3}\left(\frac{p_1}{p_0}\right)^3$$

$$\lambda_4 = \frac{p_4}{p_0} - \frac{p_3}{p_0}\frac{p_1}{p_0} - \frac{1}{2}\left(\frac{p_2}{p_0}\right)^2 + \frac{p_2}{p_0}\left(\frac{p_1}{p_0}\right)^2 - \frac{1}{4}\left(\frac{p_1}{p_0}\right)^2$$

$$\ldots$$

whence estimates of $\lambda_1, \lambda_2, \ldots$ may be obtained. (Maritz (1952), Kemp and Kemp (1965)).

9. Obtain the Inversion Theorem

$$p_x = \frac{1}{x!} \sum_{r=x}^{\infty} (-1)^{x+r} \frac{\mu'_{(r)}}{(r-x)!}$$

by differentiation of the p.g.f. $g(z) = \Sigma z^r p_r$ and the factorial moment g.f. $m_{(\,)}(t) = \Sigma(1+t)^r p_r = g(1+t)$.

(For $g(z) = m_{(\,)}(z-1) = \Sigma \mu'_{(r)} \frac{(z-1)^r}{r!}$;

$$g^{(x)}(z) = m_{(\,)}^{(x)}(z-1) = \Sigma \mu'_{(r)} \frac{r^{(x)}(z-1)^{r-x}}{r!}$$

$$= \Sigma \mu'_{(r)} \frac{(z-1)^{r-x}}{(r-x)!},$$

and writing $z = 0$ gives the result.)

10. For $S_N = X_1 + X_2 + \ldots + X_N$,
where $X_1, X_2, \ldots$ are independent identically distributed variates with common moment g.f. $m_X(t)$ and N is an independent variate with p.g.f. $g_N(t)$, show that the moment g.f. of S_N is

$$g_N(m_X(t)) .$$

By differentiating twice with respect to t, establish the results

$$\mathcal{E}(S_N) = \mathcal{E}(N)\,\mathcal{E}(X)$$

and $$\mathcal{V}(S_N) = \mathcal{V}(N)\mathcal{E}^2(X) + \mathcal{E}(N)\mathcal{V}(X).$$

11. Verify the following results for the combinatorial generating functions exhibited.

a) $(1+\alpha z)(1+\beta z)(1+\gamma z)$ generates the combinations (i.e. sub-sets) of the objects α, β, γ taken without repetition;

b) $(1+\alpha z+\alpha^2 z^2)(1+\beta z)(1+\gamma z)$ generates the combinations of α,β,γ where α may be chosen twice ($\alpha^2\beta$ means two α's and one β);

c) $(1+\alpha z)(1+\alpha^2 z)(1+\beta z)(1+\gamma z)$ generates the results of choosing bags, one with α, another with two α's, another with β and another with γ.

12. <u>Multivariate probability generating functions</u>

i) Bivariate binomial

a) Consider the four fold table classifed as shown:

	A	$\overline{A}$	
B	p_{11}	p_{01}	$p_{.1}$
$\overline{B}$	p_{10}	p_{00}	$p_{.0}$
	$p_{1.}$	$p_{0.}$	1

where $U=1$ for A, 0 for $\overline{A}$, $V=1$ for B, 0 for $\overline{B}$, $p_{uv}=\Pr(U=u,\ V=v)$ and the margins are totals. Write down the p.g.f. $g(x,t)$ of U,V. Express this p.g.f. in terms of $p_{.1}$, $p_{1.}$ and p_{11}; and write down the condition for independence of the A and B classifications in terms of p_{11} and $p_{1.}, p_{.1}$, and the p.g.f. for this case. If ν trials are made with the above structure, the resulting p.g.f. $\{g(s,t)\}^\nu$ is that of a bivariate binomial. Write down its factorial moment g.f.

$(p_{00}+p_{01}s+p_{10}t+p_{22}st = 1+p_{.1}(s-1)+p_{.7}(t-1)+p_{11}(1-s)(1-t)$,

$p_{11}=p_{.1}p_{1.}$, $(p_{.1}s+p_{.0})(p_{1.}t+p_{0.})$.

Replacing s by $s+1$ etc.,

$$\{(1+p_{.1}s)(1+p_{1.}t)+(p_{11}-p_{.1}p_{1.})st\}^\nu.)$$

b) If (X_1,X_2,X_3) is multinomial with p.g.f.

$$(1-p_1-p_2-p_3+p_1z_1+p_2z_2+p_3z_3)^\nu,$$

then for $Y_1 = X_1 + X_3$, $Y_2 = X_2 + X_3$

show that (Y_1, Y_2) has the bivariate binomial distribution with

$$p_{10} = 0_1,\ p_{01} = p_2,\ p_{11} = p_3$$

$$(\text{and } p_{00} = \Sigma\, p_i).$$

(Kendall and Stuart I)

ii) Bivariate Poisson (Aitken (1947), Ahmed (1961), Holgate (1964), Kemp and Kemp (1965))

If U,V,W are independent Poisson variates with parameters λ, μ, ν show that

$$X = U + W,\quad Y = V + W$$

has the bivariate Poisson p.g.f.

$$g(x,t) = \exp\{\lambda(s-1) + \mu(t-1) + \nu(st-1)\}$$

with p.f.

$$\phi(x,y) = \Pr(X=x,\ Y=y) = e^{-(\lambda+\mu+\nu)} \sum_{z=0}^{\min(x,y)} \frac{\lambda^{x-z}\,\mu^{y-z}\,\nu^{z}}{(x-z)!\,(y-z)!\,z!}$$

Show that

$$x\phi(x,y) = \lambda\phi(x-1,y) + \nu\phi(x-1,y-1)$$

$$y\phi(x,y) = \mu\phi(x,y-1) + \nu\phi(x-1,y-1)\ ,$$

recursive relations for computational purposes.

Deduce from the factorial cumulant g.f.

$$(\lambda+\nu)s' + (\mu+\nu)t' + \nu\, s't',\quad s = s'+1,\quad t = t'+1,$$

that the distribution reduces to the obvious Poisson marginal distributions, with

$$\mathscr{E}(X) = \lambda+\nu,\quad \mathscr{E}(Y) = \mu+\nu \quad \text{and} \quad \mathscr{C}(X,y) = \nu.$$

If U,V,W are independent variates with joint p.g.f. $g(z_1, z_2, z_3)$, $X = U+W$, $Y = V+W$ with p.g.f. $h(\zeta_1,\zeta_2)$ say, show that

$$h(\zeta_1\zeta_2) = g(\zeta_1,\zeta_2,\zeta_1\zeta_2)$$

and that the p.g.f. of $X+Y$ is $g(\zeta,\zeta,\zeta^2)$. Adding the above Poisson assumptions for U,V,W deduce that $X+Y$ is Poisson-binomial with parameter $\nu=2$ (or Hermite).

iii) Bivariate Neyman (Holgate (1966))

a) Obtain the p.g.f.

$$\exp[-\mu+\mu \exp\{\nu_1(s-1)+\nu_2(t-1)+\nu_{12}(st-1)\}]$$

as a Poisson-stopped bivariate Poisson distribution.

b) As an alternative distribution with the label "bivariate Neyman", obtain the p.g.f.

$$\exp(-\mu_1+\mu_1 e^{-\nu+\nu s}-\mu_2+\mu_2 e^{-\nu+\nu t}-\mu_3+\mu_3 e^{-\nu+\nu st})$$

for X,Y where $X=U+W$, $Y=V+W$ and U,V,W are independent Neyman variates with parameters (μ_1,ν), (μ_2,ν), (μ_3,ν) respectively.

iv) Multivariate Negative Binomial (Neyman (1965))
For the p.g.f.

$$g(z_1,z_2,\ldots,z_s) = \{1+\sum_{i=1}^{s} p_i(1-z_i)\}^{-\kappa}, \text{ all } p_i, \kappa > 0,$$

any marginal joint distribution has a p.g.f. of the same form. And if $X_1,\ldots,X_s$ has this distribution, then given

$$X_i = x_i \quad \text{for} \quad i=m+1,m+2,\ldots,x$$

the conditional distribution of $X_1,\ldots,X_m$ has a p.g.f. of the same form depending on

$$\sum_{i=m+1}^{s} x_i$$

but not on the individual $x_{m+1},\ldots,x_s$.

13. Prove the Inversion Theorem for probabilities p_x in terms of factorial moments $\mu'_{(r)}$ by expanding

$$\sum_{r=0}^{\infty} (-1)^r \frac{\mu'_{(r)}}{r!} (1-z)^r$$

as
$$\sum_{r=0}^{\infty} (-1)^r \frac{\mu'_{(r)}}{r!} \sum_{x=0}^{r} (-1)^x \binom{r}{x} z^x$$

and interchanging the order of summation.

2.3 MIXTURES OF DISTRIBUTIONS

Consider the family of discrete variates X_λ, $\lambda \varepsilon \Lambda$, indexed by a parameter λ with $\Pr(X_\lambda = x) = p_x(\lambda)$, say. Then if $G(\lambda)$ is a weight function on Λ such that

$$G(\lambda) \geq 0,$$

for no λ is $G(\lambda+0) - G(\lambda-0) = 1$ (no unit jump),

and
$$\int_\Lambda dG(\lambda) = 1,$$

then $\int_\Lambda p_x(\lambda)\, dG(\lambda)$, with variate X, is called a $p_x(\lambda)$ mixture on λ induced by the weight function $G(\lambda)$; or $p_x(\lambda)$ is said to be mixed on λ by $G(\lambda)$. Since a "mixed-Poisson" or "Poisson mixture" means the compounded distribution

$$\int_0^\infty e^{-\lambda} \frac{\lambda^x}{x!}\, dG(\lambda)$$

and not the Poisson distribution $e^{-\lambda} \frac{\lambda^x}{x!}$ which is mixed, $p_x(\lambda)$ will be called a component of the mixture; similarly $G(\lambda)$ will always be called the weight function (or distribution), and not the mixing distribution, so that one can speak of mixing Poisson distributions (as above). Gurland's (1957) notation writes a family of such mixed variates as $X_\lambda \wedge \lambda$, or $p_x(\lambda) \wedge dG(\lambda)$. or an equivalent in words, with the inclusion of an arbitrary constant c:

$$X_\lambda \wedge \lambda : \int p_x(\lambda)\, dG(\lambda/c).$$

From a statistical point of view, a common interpretation (or derivation of a particular distribution) is that there is a random variable L such that $\Pr(L \leq \lambda) = G(\lambda)$, i.e. $G(\lambda)$ is a cumulative distribution function. Thus a first experiment leads to the random selection of a value ℓ of L from the distribution $G(\cdot)$, and so to a specific $p_x(\ell)$ from which in turn a random selection is made: the variate X, or X_L, from this sequence of experiments is an X_λ mixed by $G(\lambda)$ variate, or more vaguely an X_λ mixed variate. It will be recognized that the Lebesgue Decomposition Theorem writes the most general distribution as a mixture of the three fundamental types, discrete, absolutely continuous, and singular, with a discrete weight function.

Of course the definition may be rephrased in terms of generating functions: for the purposes of discrete variates, if $f(z;\lambda)$ is the p.g.f. corresponding to the family $p_x(\lambda)$, then the p.g.f. of the mixture induced by G is

$$\int f(z;\lambda)\, dG(\lambda).$$

If, for example, a binomial variate with p.g.f. $(\chi + \pi z)^\nu$ is mixed on π with a uniform distribution for π the p.g.f. of the mixture is

$$\int_0^1 (\chi + \pi z)^\nu \, d\pi = \int_0^1 \{1 - \pi(1 - z)\}^\nu \, d\pi = \frac{1}{\nu+1}\,\frac{1-z^{\nu+1}}{1-z},$$

i.e. $\Pr(X=x) = \dfrac{1}{\nu+1}$ for $x = 0,1,\ldots,\nu$ (and zero otherwise): a uniform discrete distribution. On the other hand, if $(\chi+\pi z)^\nu$ is mixed on ν with Poisson(μ) weights, the p.g.f. of the mixture is

$$\begin{aligned}\sum_{\nu=0}^{\infty} (\chi+\pi z)^\nu \, e^{-\mu} \frac{\mu^\nu}{\nu!} &= e^{-\mu} \sum \frac{\{\mu(\chi+\pi z)\}^\nu}{\nu!} \\ &= \exp\{-\mu+\mu(\chi+\pi z)\} \\ &= e^{-\mu\pi+\mu\pi z},\end{aligned}$$

corresponding to a Poisson($\mu\pi$) distribution in which only the product $\mu\pi$ appears.

It is often convenient to argue directly in terms of conditional expectations (see the Appendix). If

$$f(z;\lambda) = \mathscr{E}_X(z^{X_\lambda})$$

is the p.g.f. already introduced, with the additional subscript for reasons which will be apparent in a moment, the the p.g.f. of the mixture induced by $G(\lambda)$ is

$$\int f(z;\lambda)\ dG(\lambda)$$

$$= \mathscr{E}_L[f(z;L)] = \mathscr{E}_L[\mathscr{E}_X(z^{X_L})] = \mathscr{E}(z^{X_L}).$$

(This is very commonly written as $\mathscr{E}(z^X)$ when the context supplies the background.) Taking the second of the above examples, if X_λ is a standard binomial variate,

$$X_\lambda : (\chi + \pi z)^\lambda,$$

with the probability that $N = \lambda$ being given by the Poisson(μ) distribution,

i.e., $$N : e^{-\mu+\mu z} = \mathscr{E}(z^N),$$

then the p.g.f. of the mixture of X with the Poisson weight function stated is

$$\mathscr{E}_N[(\chi + \pi z)^N]$$

$$= \exp[-\mu + \mu(\chi + \pi z)]$$

$$= \exp(-\mu\pi + \mu\pi z)$$

without the necessity of an explicit summation.

A mixture is again a discrete distribution if the discontinuity points of p_x are independent of those of G; but this condition is merely sufficient. That it is not necessary is obvious from the case where a binomial family is mixed on ν by a Poisson distribution: this leads to a discrete distribution (the Poisson) even though the points depend on ν. However, some conditions are necessary to ensure discreteness, as is shown by considering

$$p_x(\lambda) = \begin{cases} 1, & x = \lambda \\ 0 & \text{otherwise} \end{cases}$$

when the mixture is

$$\int p_x(\lambda)\ dG(\lambda) = G(x),$$

which is not discrete unless $G(\cdot)$ is discrete. If $G(\cdot)$ is discrete, the independence condition leads to the conclusion that the points of positive probability of a mixture are at the common points of discontinuity.

Comparisons may be made between, e.g., the expectation of a mixed-variate, $\mathcal{E}(X)$, and the expectation over the weight distribution of the expectation of the component variate X_λ. Thus, writing

$$\mu_\lambda = \Sigma x p_x(\lambda) \quad \text{and} \quad \sigma_\lambda^2 = \Sigma(x-\mu_\lambda)^2\ p_x(\lambda),$$

a direct application of conditional expectations gives

$$\mathcal{E}(X) = \mathcal{E}(\mathcal{E}(x|L)) = \mathcal{E}(\mu_L),$$

and

$$\mathcal{V}(X) = \mathcal{V}(\mathcal{E}(X|L)) + \mathcal{E}(\mathcal{V}(X|L)) = \mathcal{V}(\mu_L) + \mathcal{E}(\sigma_L^2)$$

$$\geq \mathcal{V}(\mu_L) \quad \text{with equality iff} \quad \sigma_\lambda^2 \equiv 0.$$

An alternative derivation is to replace z by $1+t$ in $h(z) = \int f(z;\lambda)\ dG(\lambda)$ and take logarithms, giving the factorial cumulant g.f. Expansion in powers of t then yields

$$\mu_h = \int \mu_{f,\lambda}\ dG(\lambda), = \int \mu_f\ dG \quad \text{for brevity}$$

(or $\mathcal{E}(X) = \mathcal{E}(\mu_L))$,

and

$$\kappa_{(2):h} = \int \kappa_{(2):f}\ dG + \int \mu_f^2\ dG - \mu_h^2$$

$$= \int \kappa_{(2):f}\ dG + \int \mu_f^2\ dG - \{\int \mu_f dG\}^2$$

$$= \int \kappa_{(2):f}\ dG + \mathcal{V}(\mu_L).$$

Thus the Index of Cluster Size, defined as $\iota_{CS} = \kappa_{(2)}/\mu$ (see the next Chapter), is given by

$$\iota_{CS:h} = \frac{\int \iota_{CS:f}\, \mu_f\, dG}{\int \mu_f\, dG} + \frac{\mathcal{V}(\mu_L)}{\mathcal{E}(\mu_L)} ,$$

and the Index of Cluster Frequency $\iota_{CF} = \mu^2/\kappa_{(2)}$ by

$$\iota_{CF:h} = \frac{\mu_h^2}{\int \iota_{CS:f}\, \mu_f\, dG + \mathcal{V}(\mu_L)} .$$

However, the literature on mixtures is enormous, and attention here will be confined almost entirely to Poisson mixtures. A general survey of discrete mixtures is given by Blischke (1965), and Johnson and Kotz (1969) give many specific examples; see also Molenaar (1965).

Mixed-Poisson Distributions (Poisson Mixtures).

Of families of mixed discrete distributions, those derived by mixing Poisson distributions have received most attention.

These mixed-Poisson distributions can be written as

$$\int_0^\infty e^{-\lambda+\lambda z}\, dG(\lambda) = \Sigma\, z^x\, p_x^*$$

where $e^{-\lambda+\lambda z}$ is the p.g.f. of the Poisson family, and the whole of the parametric specification of p_x^* is carried implicitly by $G(\lambda)$. Alternatively and often more usefully, writing the Poisson family as

$$e^{-u\nu+u\nu z} ,$$

where u is the realization of a variate with cumulative distribution $G(u)$ and p.g.f. $g(z)$, the p.g.f. of the mixture is

$$\int_0^\infty e^{-u\nu+u\nu z}\, dG(u) = \int_0^\infty \{e^{-\nu+\nu z}\}^u\, dG(u)$$

$$= g(e^{-\nu+\nu z})$$

since $$g(z) = \int_0^\infty z^u \, dG(u) \; .$$

For example, if $g(z) = e^{-\mu+\mu z}$, the mixture obtained by mixing Poisson($u\nu$) on u by Poisson(μ) thus has p.g.f.

$$\exp\{-\mu + \mu e^{-\nu+\nu z}\}, \quad \text{the Neyman Type A}(\mu,\nu);$$

mixing Poisson($u\rho$) on u by gamma(κ,1) (so the gamma p.g.f. is

$$1/(1 - \ln z)^{\kappa})$$

gives the negative binomial with p.g.f.

$$(1 + \rho - \rho z)^{-\kappa} \; .$$

Any number of mixed-Poisson distributions can be written down at sight in a similar fashion, but only a few have been studied, at least in sufficient depth to appear in the literature. An early example, not followed up to any extent, was given by Fisher (1931) where $G(\lambda)$ is the three parameter distribution specified by

$$\frac{\lambda^q}{\sigma^{q+1} I_q(a) q! \sqrt{(2\pi)}} \exp\left\{-\frac{1}{2}\left(\frac{\lambda + \sigma^a}{\sigma}\right)^2\right\}, \quad 0 \le \lambda < \infty,$$

$$I_q(a) = \frac{1}{\sqrt{(2\pi)}} \int_a^\infty \frac{(t-1)^q}{q!} e^{-\frac{1}{2}t^2} \, dt:$$

see Kemp and Kemp (1967). For an account of cases of the form

$$\sum_{i=1}^m e^{-\lambda_i+\lambda_i z} p_i \; , \; \Sigma \, p_i = 1 \text{ and no } p_i < 0,$$

especially for $m = 2$, see Johnson and Kotz (1969), Chapter 4, Section 10.3.

Properties of Mixed-Poisson Distributions

$$h(z) = g(e^{-\nu+\nu z}) = \Sigma \, h_x z^x$$

i) Probabilities of the mixture and weight distributions

If $g(z) = \Sigma g_r z^r$, a direct expansion gives

$$h(z) = \Sigma g_r e^{-\nu r+\nu rz}$$

$$= \Sigma g_r e^{-\nu r} \Sigma \frac{\nu^s r^s}{s!} z^s$$

$$= \Sigma z^s \frac{\nu^s}{s!} \Sigma g_r e^{-\nu r} r^s$$

whence $h_x = \frac{\nu^x}{x!} \Sigma g_r e^{-r\nu} r^x$.

Alternatively, by the application of the Faa di Bruno Formula (Appendix on Stirling Numbers, section (c)) to h(z),

$$h_x = \frac{\nu^x}{x!} \Sigma g^{(n)}(e^{-\nu}) e^{-n\nu} \mathcal{S}_x^{(n)},$$

$\mathcal{S}_x^{(n)}$ being a Stirling number of the Second Kind.

ii) Moments of the mixture and weight distributions

The factorial moment g.f. is $h(1+t)$;

$$h(1+t) = g(e^{\nu t}),$$

and, since $g(e^t)$ is the moment g.f. of the weight variate,

$$\mu'_{(r):h} = \nu^r \mu'_{r:g},$$

where $\mu'_{(r):h}$ is the factorial moment about the origin of the mixture.
In an exactly similar way,

$$\kappa_{(r):h} = \nu^r \kappa_{r:g}.$$

Thus the moments of a (perhaps unobserved but postulated) weight distribution may be obtained if those of the (perhaps observable) mixture are known: cf. Tucker (1963), or more informally, Newbold (1927).

A converse also exists: a distribution is a mixture of Poisson distributions iff its factorial moment g.f. is moment g.f. of a non-negative variate (Maceda (1948)); see also the section dealing with Stopped Distributions.

The Indices of Cluster Size and Cluster Frequency (see the next Chapter) can be expressed in terms of those of the weight distribution as follows:

Index of Cluster Size $\dfrac{\kappa_{(2):h}}{\mu_h} = \nu \dfrac{\kappa_{2:g}}{\mu_g}$,

or $\iota_{CS:h} = \nu + \nu\ \iota_{CS:g}$.

Index of Cluster Frequency $\dfrac{\mu_h^2}{\kappa_{(2):h}} = \dfrac{\mu_g^2}{\kappa_{2:g}}$,

or $_{CF:h} = \dfrac{\mu_g\ _{CF}}{\mu_g + \iota_{CF:g}} = \dfrac{\mu_g}{1 + \iota_{CS:g}}$, which is independent of ν. Further, it follows from

$$\kappa_{2:h} = \mu_h \left\{ 1 + \nu \frac{\kappa_{2:g}}{\mu_g} \right\}$$

that, in an alternative notation, writing X for the mixture and L for the weight variate,

$$\mathcal{V}(X) = \mathcal{E}(X) \left\{ 1 + \nu \frac{\mathcal{V}(L)}{\mathcal{E}(L)} \right\}.$$

A mixed-Poisson variate is therefore "overdispersed" - i.e. its variance exceeds its mean-unless $\mathcal{V}(L) = 0$. Since the inequality is strict (if $\mathcal{V}(L) \neq 0$), a mixed-Poisson variate cannot itself be Poisson.

iii) Identifiability

That not more than one weight distribution should correspond to the one mixture of a Poisson distribution is fundamental to the estimation of weight distributions of mixtures. (Generally, a mixture is called <u>identifiable</u> iff

$$\int f(z;\lambda)\, dG_1(\lambda) = \int f(z;\lambda)\, dG_2(\lambda)$$

implies $G_1 \equiv G_2$ (for some class of weight functions).)

Since $g_1(e^{-\nu+\nu z}) = g_2(e^{-\nu+\nu z})$ implies $g_1 \equiv g_2$,

the mixed-Poisson is identifiable. This also follows from the use of Laplace Transform theory (which is how a mathematician would view the p.g.f. of a Poisson mixture), and it can be shown thus that no mixed-Poisson is itself Poisson. (See also the section on comparisons of Poisson and mixed-Poisson distributions.)

iv) Reproductive Property.

If a Poisson distribution is mixed by two independent weight functions, then the convolution of these two mixed-Poisson distributions is itself a mixed-Poisson distribution with weight function the convolution of the weight functions.

For the mixtures have p.g.f.'s $h_i(z) = g_i(e^{-\nu+\nu z})$, $i = 1,2$, and the p.g.f. of their convolution is

$$h_1(z)\, h_2(z) = g_1(e^{-\nu+\nu z})\, g_2(e^{-\nu+\nu z});$$

the p.g.f. of the convolution of the weight functions is $g_1(z)\, g_2(z)$, whence the result.

Teicher (1960) has extended this theorem to additively closed families.

It is to be noticed that the Poisson distributions being mixed are supposed the same. Consider, as an example of this case, egg clusters of two kinds where the numbers of eggs per cluster are Poisson distributed about means of u_j where the u_j are realizations of random variables U_j with gamma densities

$$\frac{1}{\Gamma(\kappa_j)}\, e^{-u}\, u^{\kappa_j - 1},$$

$j = 1$ and $j = 2$ corresponding to the two kinds. (That is, there are different proportions of cluster size u in the two kinds.) Then for a cluster randomly selected from each kind the total number of eggs in the two clusters has distribution given by the p.g.f.

$$1/(1+\rho-\rho z)^{\kappa_1+\kappa_2}$$

since cluster size for each has the distribution

$$\int_0^\infty e^{-u\rho+u\rho z}\ \frac{1}{\Gamma(\kappa_j)}\ e^{-u}u^{\kappa_j-1}\ du$$

$$= (1+\rho-\rho z)^{-\kappa_j},$$

a negative binomial (as was previously shown), verifying immediately the general theorem. However, if the numbers of eggs per cluster are Poisson distributed about means $u_j\rho_j$ with u_j as above, the convolution distribution is

$$(1+\rho_1-\rho_1 z)^{-\kappa_1}\ (1+\rho_2-\rho_2 z)^{-\kappa_2}$$

and this is not a negative binomial; if the numbers of eggs per cluster are Poisson($u\rho_j$) with the distribution of U the same for both kinds, the p.g.f. of the convolution is

$$\{(1+\rho_1)(1+\rho_2)+(\rho_1+\rho_2+2\rho_1\rho_2)z+\rho_1\rho_2 z^2\}^{-\kappa}\ .$$

A further example, leading to one extension of the Neyman family, is to take the cluster size to be Poisson($u_j\nu$) where U_j is Poisson(μ_j), $j = 1,2$. The convolution distribution has p.g.f.

$$\exp(-\ \overline{\mu_1+\mu_2}+\overline{\mu_1+\mu_2}\ e^{-\nu+\nu z}),$$

a Neyman Type A($\mu_1+\mu_2,\nu$) distribution corresponding to the general theorem. But if the cluster sizes are Poisson($u\nu_1$) and Poisson($u\nu_2$) with U: Poisson($\tfrac{1}{2}\mu$), the p.g.f. of the convolution is

$$\exp\{-\mu+\tfrac{1}{2}\mu(e^{-\nu_1+\nu_1 z}+e^{-\nu_2+\nu_2 z})\},$$

a <u>three parameter</u> Neyman Type A distribution. Again, allowing variation of the proportions of cluster sizes in the two kinds gives the convolution p.g.f.

$$\exp\{-\mu_1-\mu_2+\mu_1 e^{-\nu_1+\nu_2 z}+\mu_2 e^{-\nu_2+\nu_2 z}\},$$

a four parameter distribution.

v) Infinite Divisibility

A variate is said to be infinitely divisible if its characteristic function $\phi(t)$ is such that

$$\{\phi(t)\}^{1/n} = \chi(t)$$

is also a characteristic function for every positive integer n (choosing the branch for which $\chi(0) = 1$ and $\chi(t)$ is continuous). Thus an infinitely divisible variate can be written as the convolution of an arbitrary number of identical independent variates - obviously Poisson and normal variates are infinitely divisible, less obviously so is a translated logarithmic variate with p.g.f. $\{\ell n(1-\alpha z)\}/z\ \ell n(1-\alpha)$, and binomial and uniform variates are not.

Since the argument concerning the reproductive property in v) applies without change to characteristic functions, it is easy to see that if the weight distribution of a mixed-Poisson variate is infinitely divisible so is the mixture.

Katti (1967) gives necessary and sufficient conditions for an integer valued discrete variate to be infinitely divisible.

For further related discussion of infinite divisibility see Lukacs (1960), Linnik (1964), Feller (1966) and the following section on Stopped Distributions.

vi) Comparisons of mixed-Poisson and Poisson distributions

It is natural to compare a mixed-Poisson variate (X, with probabilities h_x) with a Poisson variate (Y, with probabilities p_y), where in order to make the comparison meaningful the means are adjusted to be equal. Then for $\mathcal{E}(X) = \mathcal{E}(Y)$ by the above standardization (this is the only connection Y has with X, for Y has nothing to do with the mixing process, and is introduced purely for comparative reasons):

a) $\mathcal{V}(X) > \mathcal{V}(Y) = \mathcal{E}(Y)$

(the variance of the mixture exceeds that for the corresponding Poisson),

b) $h_0 > p_0$

(there is more in the zero class of the mixture),

c) $h_1/h_0 < p_1/p_0 = \mathcal{E}(Y)$

(relative to the zero class the mixture has less in the first class);

all these inequalities are strict for a non degenerate weight distribution (Feller (1943)).

That the variance of a mixed-Poisson exceeds its mean except for a weight function with zero variance has been already proved: this establishes a). The proofs of b) and c) are similar; with $h(z) = \int \exp(-\lambda+\lambda z)\, dG(\lambda)$ and taking c), consider

$$h_1 - \bar{\lambda}\, h_0 = \int \lambda e^{-\lambda}\, dG - \bar{\lambda} \int e^{-\lambda}\, dG, \text{ where}$$

$$\bar{\lambda} = \int \lambda\, dG = h'(1) = \mathcal{E}(X),$$

$$= \int (\lambda - \bar{\lambda}) e^{-\lambda}\, dG$$

$$= e^{-\bar{\lambda}} \int (\lambda - \bar{\lambda}) e^{\bar{\lambda}-\lambda}\, dG$$

$$= e^{-\bar{\lambda}} \int (\lambda - \bar{\lambda}) \left\{ 1 - (\lambda - \bar{\lambda}) e^{\theta(\bar{\lambda}-\lambda)} \right\} dG$$

by the Mean Value Theorem, $0 < \theta < 1$,

$$= e^{-\bar{\lambda}} \int \left\{ \lambda - \bar{\lambda} - (\lambda - \bar{\lambda})^2\, e^{\theta(\bar{\lambda}-\lambda)} \right\} dG$$

$$< e^{-\bar{\lambda}} \int (\lambda - \bar{\lambda})\, dG = 0.$$

Hence the result follows; a generalization is given by Molenaar and Van Zwet (1966).

Pictorially, the graphs of typical p.f.'s are shown in Figure 3.

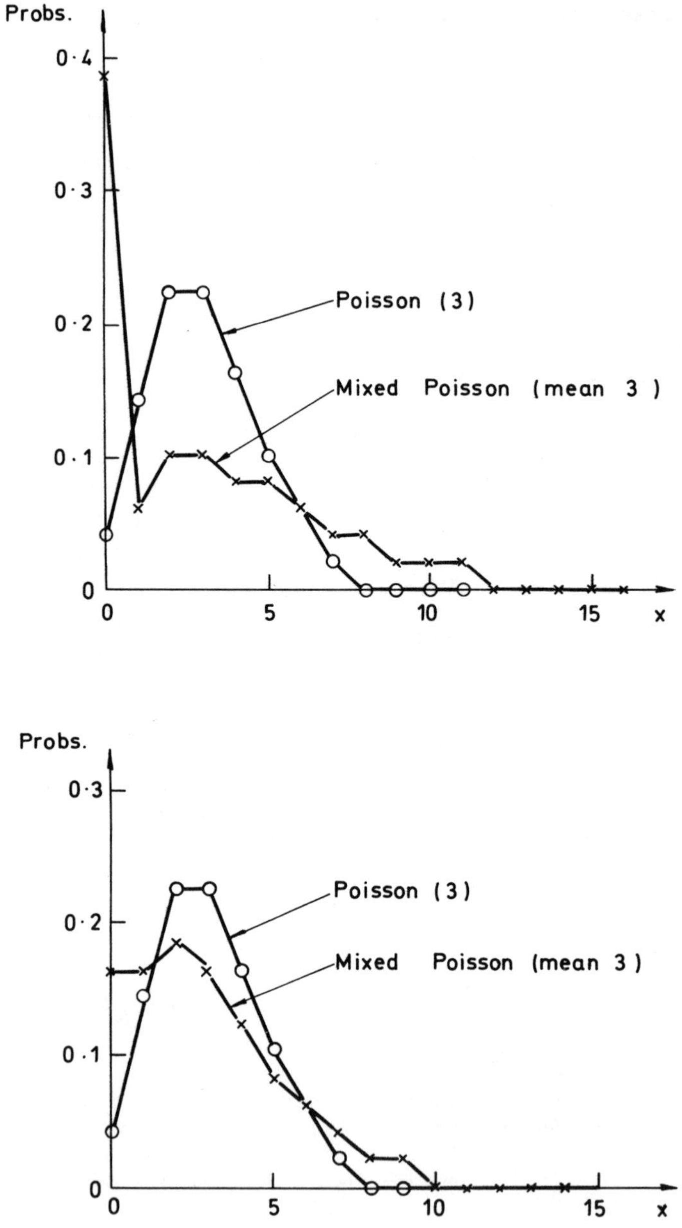

FIGURE 3 Poisson and mixed-Poisson distributions.

vii) An important family of distributions is the exponential, or Koopman-Pitman, class, with p.f. or p.d.f.

$$\exp\{p(\theta)k(x)+s(x)+q(\theta)\} :$$

for various desirable properties, regularity conditions must be placed on the otherwise arbitrary functions p, k, s, and q and on the domain of X. The linear exponential family is the special case

$$\exp\{\theta x+s(x)+q(\theta)\}; \qquad \text{(A)}$$

suppose this is the form of the weight distribution, with p.g.f.

$$g(z;\mu) = \int_0^\infty z^x e^{\mu x+s(x)+q(\mu)}\,dx$$

and moment g.f. $m(t;\mu) = g(e^t;\mu)$. Then the Poisson(u,ν) distribution mixed by this weight distribution on u has p.g.f.

$$\int_0^\infty e^{-u\nu+u\nu z}\,e^{\mu u+s(u)+q(\mu)}\,du,$$

and this can be written in two interesting forms (Sankaran (1970)).

First, it is

$$e^{q(\mu)-q(\mu+\nu z-\nu)} \int_0^\infty e^{(\mu+\nu z-\nu)u+s(u)+q(\mu+\nu z-\nu)}\,du$$

$$= \exp\{q(\mu) - q(\mu+\nu z-\nu)\} ;$$

and second, with $z = e^t$, it is

$$e^{q(\mu)-q(\mu-\nu)}\, m(t^\nu;\mu-\nu).$$

This second result shows that the probabilities in the mixture are proportional to the corresponding moments about the origin of the linear exponential weight distribution - precisely, if X is the mixture variate and $\mu'_x(\theta)$ the xth moment about the origin corresponding to (A), then

$$\Pr(X=x) = \frac{\nu^x}{x!}\,\mu'_x(\mu-\nu)\; e^{q(\mu)-q(\nu)} .$$

Since the factorial moment g.f. of the mixture is

$$\int_0^\infty e^{u\nu t+u\mu+s(u)+q(\mu)}\,du$$

$$= m(e^{\nu t};\mu),$$

this verifies the result of the general theory: the rth factorial moment of the mixture is ν^r times the rth power moment of the weight distribution.

viii) Modality

Holgate (1970) shows that if the weight function $G(\lambda)$ is absolutely continuous with a unimodal density - where this means the mode need not be unique but the set of modal values must constitute an interval - then the corresponding mixed-Poisson is unimodal (with the same kind of qualification on "mode"). A corresponding result does not hold for a discrete weight function - cf. the case of the Neyman distribution; see also Keilson and Gerber (1971).

2.3a EXERCISES

1. Show that the mixture of the negative binomial with p.g.f. $(\gamma-\rho z)^{-\kappa}$, $\gamma=1+\rho$, $\kappa=1,2,3,\cdots$, on κ with weights given by the geometric distribution with p.g.f. $\chi z/(1-\tilde{\omega}z)$ is itself geometric with p.g.f. $(1-\delta)/(1-\delta z)$, $\delta=\rho/(\gamma-\tilde{\omega})$. (Magistad (1961)).

2. Mix the geometric p.g.f. $\chi z/(1-\tilde{\omega}z)$ on $\tilde{\omega}$ with the $B(\alpha,\beta)$ weight distribution, showing its p.f. to be

$$\Pr(X=x) = \frac{B(\alpha+x-1,\beta+1)}{B(\alpha,\beta)}$$

with $\mathscr{E}(X) = \dfrac{\alpha+\beta-1}{\beta-1}$, $\mathscr{E}(X^2) = \dfrac{(2\alpha+\beta-2)(\alpha+\beta-1)}{(\beta-1)(\beta-2)}$

and $\Pr(X=x+1) = \dfrac{\alpha+x-1}{\alpha+\beta+x}\cdot\Pr(x=x)$. (Pielou (1962)).

3. If the Poisson(λ) distribution is mixed with weights given by the density

$$\frac{\theta}{\theta+1} \cdot (\lambda+1)e^{-\lambda\theta}, \ \theta > 0,$$

show that the mixed variate has p.f.

$$\Pr(X=x) = \theta^2 \cdot \frac{\theta+2+x}{(\theta+1)^{x+3}}, \ x = 0,1,2,\ldots .$$

Show further that its p.g.f. is

$$\frac{\theta}{\theta+1} \cdot \frac{\theta+2-z}{(\theta+1-z)^2},$$

that $\Pr(X=x+1) = \dfrac{\theta+3+x}{(\theta+1)(\theta+2+x)} \cdot \Pr(X=x)$,

and that the mean and variance are

$$\frac{\theta+2}{\theta(\theta+1)} \quad \text{and} \quad \frac{\theta^3+4\theta^2+6\theta+2}{\theta^2(\theta+1)^2}$$

respectively. (Sankaran (1970))

4. Obtain the following reductions of the indicated mixtures.

a) {Poisson($u\nu$) mixed on u by Poisson(μ)} = Neyman Type A(μ,ν).

b) {Poisson(λ) mixed on λ by gamma(κ,ρ)} = Pascal(κ,ρ).

c) {Pascal($k\kappa,\rho$) mixed on k by Poisson(μ)} = Poisson Pascal(μ,κ,ρ).

d) {Binomial($k\nu,\pi$) mixed on k by Poisson(μ)} = Poisson binomial(μ,ν,π). (Note the special case: {binomial(ν,π) mixed on ν by Poisson(λ)} = Poisson($\lambda\pi$).)

e) {Binomial(ν,π) mixed on π by uniform(0,1)} = discrete uniform.

f) {Gamma($\kappa+k,\beta$) mixed on k by Pascal(κ,ρ)} = gamma(κ,β,γ).

g) {Poisson($\mu\pi$) mixed on π by Beta(α,β)}

= {binomial(ν,π) mixed on π by Beta(α,β)} mixed on ν by Poisson(μ).

5. For the Poisson mixture $h(z) = g(e^{-\nu+\nu z})$, where $g(z)$ is the p.g.f. of the weight distribution, use the relation $\mu'_{(r):h} = h^{(r)}(1)$ to show that

$$\mu'_{(r):h} = \nu^r \mu'_{r:g}$$

$$= \nu^r \sum \mathcal{S}_r^{(n)} \mu'_{(n):g} ,$$

$\mathcal{S}_r^{(n)}$ being a Stirling Number of the Second Kind.

6. Suppose U is a weight variate with cumulative distribution function $G(u)$, for which the corresponding mixed-Poisson variate X has p.g.f.

$$h(z) = \int_0^\infty e^{-\nu u+\nu uz} \, dG(u) . \qquad (*)$$

Writing $\mu_g = \int_0^\infty u dG(u)$, the mean of the weight distribution, and introducing Y:Poisson$(\nu\mu_g)$, show by twice differentiating the f.m.g.f. obtained from (*) that

$$\mathcal{E}(X) = \nu\mu_g = \mathcal{E}(Y),$$

and $\mathcal{V}(X) = \mathcal{V}(Y) + \nu^2 \mathcal{V}(U)$.

(Thus the variance of a mixed Poisson variate exceeds that of a Poisson variate with the same mean unless $\mathcal{V}(U) = 0$.)

7. For an insect which lays eggs in clusters it is often reasonable to suppose that by the end of some fixed time the probability that a single egg hatches and the larva survives is a constant π ; and hence that for a given egg cluster size n the number of larvae may be binomial(n,π).

Suppose further that the numbers N of eggs per cluster have the following distributions:

a) N:Poisson(α).
Then the unconditional probability distribution of the number of larvae, X, from a randomly selected egg cluster is Poisson$(\alpha\pi)$. Observe that from observations on X only the product $\alpha\pi$ (and neither α nor π separately) can be estimated.

b) N:binomial(ν,π') leads to X:binomial$(\nu,\pi\pi')$.

c) N:negative binomial(κ,ρ), to X:negative binomial($\kappa,\rho\pi$)

d) N:logarithmic(α), to

$$\Pr(X=0) = \frac{\ln(1-\alpha\chi)}{\ln(1-\alpha)} ,$$

$$\Pr(X=x) = \frac{-\beta^x}{x\ln(1-\alpha)} \quad \text{for} \quad \beta = \frac{\alpha}{1-\alpha\chi} , \; x = 1,2,\ldots .$$

But if $X=0$ is rejected (i.e., zero truncation),

$$\Pr(X=x) = \frac{-\beta^x}{x\ln(1-\beta)} , \; x = 1,2,\ldots .$$

In this second case α and π are not separately estimable; in the first case they are.

8. Continuing the previous exercise, if N has the p.f. f_n and p.g.f. $f(z)$, show that for a randomly chosen cluster the unconditional number of larvae has the p.g.f. $f(\chi+\pi z)$.

Suppose, however, larvae and not clusters are randomly selected, so that the probability of selecting a cluster is proportional to its size. Show that the observed distribution of the cluster size M will be given by

$$\Pr(M=m) = \frac{m f_m}{\sum_{\text{all } n} n f_n} , \; m = 0,1,\ldots,$$

with p.g.f. $zf'(z)/f'(1)$. In particular, show that an observed geometric distribution for cluster size when sampling has been carried out thus provides some evidence that the "genuine" distribution of cluster size is logarithmic.

(Observe that this might apply, e.g., to the distribution of family sizes when families are selected, e.g., by choosing children at random from school records.) (Rao (1965), Patil and Rao (1978)).

9. Suppose in a probability distribution of X , for which $p_x = \Pr(X=x)$, the method of sampling leads to a weight w_x being attached to $X=x$. (For example, the larger x, the more likely it may be to be observed - in this case w_x will

increase with x.) Then the probabilities associated with the sampled variate Y are

$$\frac{p_y \; w_y}{\sum\limits_{\text{all } x} p_x \; w_x}, \quad w_x \geq 0.$$

Consider the simple case $w_x = x$ for $x = 0,1,2,\ldots$ and obtain the p.g.f. for the sample variate in terms of the p.g.f. of the original variate. Verify the following explicit results for this case.

i) Binomial: $p_x = \binom{\nu}{x}\pi^x \chi^{\nu-x} = b(x;\nu,\pi)$ say. Then the sampled variate has p.f. $b(y-1;\nu-1,\pi)$.

ii) Poisson: $p_x = p(x;\lambda)$ leads to $p(y-1;\lambda)$.

iii) Negative binomial: $p_x = nb(x;\kappa,\rho)$ leads to $nb(y-1;\kappa+1,\rho)$.

iv) Logarithmic: $-\alpha^x/x \,\ell n(1-\alpha)$, $x = 1,2,3,\ldots$ leads to geometric $(1-\alpha)\alpha^{y-1}$, $y = 1,2,3,\ldots$

(If X has p.g.f. $g(z) = \Sigma\, z^x\, p_x$, then the sampled variate has p.g.f. $z\, g'(z)/g'(1)$.) (Rao (1965), Patil and Rao (1978))

10. Consider families of f children, where "exceptional" children are being studied, and the probability that such a child is selected is $\tilde{\omega}$ (irrespective of family), the composition of the population being such that the proportion of exceptional children is π. Repeated selections are made (with repetition of families permitted): show that the probability that x exceptional children are found from a family of size f, the family being found t times, is

$$\Pr(X{=}x,T{=}t) = \binom{x}{t}\tilde{\omega}^t(1-\tilde{\omega})^{x-t}\binom{f}{x}\pi^x(1-\pi)^{f-x}$$

(so that given the frequencies with which families of size f turn up t times with x exceptional children the parameters could be estimated).

Show also that the probability of a family with x exceptional children being selected is

$$\Pr(X{=}x) = \{1-(1-\tilde{\omega})^x\}\binom{f}{x}\pi^x(1-\pi)^{n-x},$$

so that given the frequencies of families with x exceptional children (and not the number of times such families have been selected) estimates of parameters can be made. (Rao (1965), Patil and Rao (1978))

11. If $\phi(x;\beta) = \dfrac{\beta^{\alpha} x^{\alpha-1} e^{-\beta x}}{\Gamma(\alpha)}$ for $x > 0,\ \alpha,\beta > 0$

and $\psi(\beta) = \dfrac{\delta^{\gamma}\beta^{\gamma-}\theta^{-\delta\beta}}{\Gamma(\gamma)}$ for $\beta > 0,\ \gamma,\delta > 0$

then the mixed ϕ distribution with weight function ψ has probability density function

$$\frac{\delta^{\gamma} x^{\alpha-a}}{B(\alpha,\gamma)(\delta+x)^{\alpha+\gamma}} \quad .$$

Show that its moments are

$$\mathcal{E}(X^m) = \frac{\delta^m}{B(\alpha,\gamma)} B(\gamma-m,\alpha+m),\ \gamma > m,$$

and thus that the moment estimators for α and γ are

$$\overset{o}{\gamma} = 2 + \frac{\bar{X}(\bar{X}+\delta)}{M_2},$$

$$\overset{o}{\alpha} = \frac{\bar{X}}{\delta}(\overset{o}{\gamma}-1),$$

where δ is known. (Dubey (1970))

12. The Hermite (or Poisson-binomial with $\nu=2$) distribution has p.g.f.

$$\exp(-\theta + \theta z - \phi + \phi z^2),\ \theta,\ \phi > 0.$$

a) Show that this can be obtained formally as a mixed-Poisson distribution with normal weights, or as a normal-stopped Poisson distribution. More precisely,

$$\int_{-\infty}^{\infty} e^{-\lambda+\lambda z} \frac{1}{\sqrt{(2\pi\sigma^2)}} \exp\left\{-\frac{(-\lambda-\mu)}{2\sigma^2}\right\} d\lambda$$

$$= \exp(-\rho+\rho z-\tfrac{1}{2}\sigma^2 + \tfrac{1}{2}\sigma^2 z^2) = g(z),\ \text{say},$$

for $\rho = \mu - \sigma^2 > 0.$

b) Show that its mean is $\theta + 2\phi$ and variance $\theta + 4\phi$, and that for $g(z) = \Sigma z^x p_x$,

$$p_0 = \exp(-\rho - \tfrac{1}{2}\sigma^2)$$

$$(x+1)p_{x+1} = \rho\, p_x + \sigma^2 p_{x-1}, \quad x \geq 0 \quad \text{and} \quad p_{-1} = 0.$$

c) When the normal weight distribution is truncated so that $0 \leq \lambda < \infty$, show that the p.g.f. is

$$\frac{\Phi\left(\frac{\rho+\sigma^2 z}{\sigma}\right)}{\Phi\left(\frac{\rho+\sigma^2}{\sigma}\right)} g(z)$$

where $\Phi(u) = \frac{1}{\sqrt{(2\pi)}} \int_{-\infty}^{u} e^{-\frac{1}{2}t^2} dt.$ (Kemp and Kemp (1966))

13. Examples of distributions with various relations between their means and variances

$$X : p_1(x;\lambda) = e^{-\lambda} \lambda^x / x!$$

for which $\mathcal{E}(X) = \mathcal{V}(X)$.

$$Y : p_2(y;\lambda) = \binom{\nu}{y}\left(\frac{\lambda}{\nu}\right)^y \left(1 - \frac{\lambda}{\nu}\right)^{\nu-y}$$

for which $\mathcal{E}(Y) > \mathcal{V}(Y)$.

$$Z : p_3(z;\lambda) = \binom{\mu+z-1}{z} \frac{\mu^\mu \lambda^z}{(\mu+\lambda)^{\mu+z}}$$

for which $\mathcal{E}(Y) < \mathcal{V}(Y)$.

Also, for

$$dG_1(\ell) = \frac{\mu^\mu}{\lambda^\mu \Gamma(\mu)} \ell^{\mu-1} e^{-\mu\ell/\lambda} d\ell,$$

$$dG_2 = \frac{1}{B\left(\frac{\lambda\mu}{\nu}, \left(1-\frac{\lambda}{\nu}\right)\mu\right)\nu^{\lambda\mu/\nu}} \ell^{\frac{\lambda\mu}{\nu}-1} \left(1-\frac{\ell}{\nu}\right)^{\left(1-\frac{\lambda}{\nu}\right)\mu-1} d\ell,$$

then

$$U : \int_0^{\nu} p_2(u;\ell)\ dG_2(\ell) \quad \text{has } \mathscr{E}(U) < \mathscr{V}(U),$$

while

$$V : \int_0^{\infty} p_1(u;\ell)\ dG_1(\ell) \quad \text{has } \mathscr{E}(V) \lesseqgtr \mathscr{V}(V)$$

$$\text{with equality iff} \quad \lambda = \frac{\nu(\nu-1)}{\mu+\nu} .$$

Note also the log-zero-Poisson distribution. (F.E. Binet)

14. A coin with probability π of heads at a single toss is tossed ν times. If X heads are obtained, toss the coin again X times, obtaining Y heads this time.

Show that a common sense and maximum likelihood estimator of π is

$$\hat{\pi} = \frac{X+Y}{\nu+X} ,$$

and obtain expressions for $\mathscr{E}(\hat{\pi})$ and $\mathscr{V}(\hat{\pi})$. (Bartlett (1952)).

(Writing $\ell n\ L = \ell n\{L(\text{expt. } 1).\ L(\text{expt. } 2|\text{expt. } 1)\}$,

show $\dfrac{\partial\ \ell n\ L}{\partial\pi} = \dfrac{1}{\pi\chi}\ (X-\nu\pi+Y-X\pi)$,

whence the estimator. Also,

$$\mathscr{E}(\hat{\pi}|X=x) = x\ \frac{\pi+1}{\nu+x} ,$$

whence introducing $u = x-\nu\pi$ and expanding

$$= \pi \left\{ 1 + \frac{u}{\nu\pi(1+\pi)} - \frac{u^2}{\nu^2\pi(1+\pi)^2} + 0(1/\nu^3) \right\} .$$

Replacing x by X and u by U,

$$\mathscr{E}(\hat{\pi}) = \mathscr{E}_X\{\mathscr{E}(\hat{\pi}|X)\} = \pi - \frac{\pi\chi}{\nu(1+\pi)^2} + 0(1/\nu^2),$$

$$\mathscr{V}(\hat{\pi}) = \frac{\pi\chi}{(1+\pi)} + 0(1/\nu^2).)$$

15. Using the notations of the text for a mixed Poisson variate (p.g.f. $h(z) = \Sigma\, z^x h_x$) and a Poisson variate (p.g.f. $\Sigma\, z^y p_y$) with the same mean, obtain the following extensions of the results there given, by retaining further terms of the Taylor expansions:

$$h_0 \le p_0(1 + \frac{1}{2}\mu_{2:\lambda} - \frac{1}{6}\mu_{3:\lambda}),$$

$$\frac{h_1}{h_0} \le \frac{p_1}{p_0} - \frac{p_0}{h_0}(\mu_{2:\lambda} - \frac{1}{2}\mu_{3:\lambda}),$$

$\mu_{r:\lambda}$ being the rth moment of the weight function. (Feller (1943)).

2.4 RANDOMLY STOPPED DISTRIBUTIONS (Random convolutions; generalized or clustered distributions)

It is often necessary to consider sums of the form

$$X_1 + X_2 + \ldots + X_N = S_N, \quad \text{say},$$

where $X_1, X_2, \ldots$ are identical independent variates - say binomial(ν,π) - and N is an independent non-negative integer valued variate - say Poisson(λ). (For $N = 0$, S_N is taken to be 0, so $\Pr(S_N = 0|N = 0) = 1$.) The distribution of S_N is then conveniently called a (randomly) stopped distribution, since the number of terms in the sum is determined by a probability distribution - in the example above it would be called more specifically a Poisson-stopped summed-binomial distribution. In general, let $f(z)$ be the p.g.f. of the components of the sum S_N, and $g(z)$ the p.g.f. of N. Then the p.g.f. of this g-stopped summed-f distribution is given by a conditional expectation argument:

$$\begin{aligned}\mathcal{E}\left(z^{S_N}\right) &= \mathcal{E}_N\left[\mathcal{E}_X\left(z^{S_N}|N\right)\right]\\ &= \mathcal{E}_N\left[\left\{f(z)\right\}^N\right]\\ &= g(f(z)),\end{aligned}$$

where the 'inner' function corresponds to the components X of the sum, called the summed variates, and the 'outer' to N, the stopping variate. However, if $f(z)$ is a p.g.f., whether or not this interpretation as a convolution is intended, $g(f(z))$ will be referred to as a g-stopped f p.g.f.; in particular, it is not necessarily the case that $g(z)$ refers to an integral valued variate. (The formula, in a somewhat different context, was given by H. W. Watson (1873); see also Heyde and Seneta (1972) for an earlier attribution to Bienaymé.)

So, for example, if a coin for which the probability of a head at a single toss is π (whence the p.g.f. for the number of heads is $\chi+\pi z$, $\chi = 1-\pi$) is tossed a number of times N given by the number of atomic disintegrations of a certain type for a specific sample of radio-active material in a specified time interval, the mean of N being μ (whence p.g.f. $\exp\{-\mu+\mu z\}$), the number of heads observed will have the p.g.f.

$$\exp\{-\mu+\mu(\chi+\pi z)\} = \exp(-\mu\pi+\mu\pi z),$$

i.e., it will also be Poisson distributed, but with mean $\mu\pi$. If on the other hand each single toss is replaced by the toss of ν identical coins, so that the p.g.f. of the number of heads for each of the N "sub-experiments" is $(\chi+\pi z)^{\nu}$, then the p.g.f. for the total number of heads will be

$$\exp\{-\mu+\mu(\chi+\pi z)^{\nu}\},$$

that of a Poisson-stopped binomial, or simply a Poisson-binomial distribution. And if in a certain experimental plot the number of plants of a certain species of weed can be supposed Poisson(μ) distributed, while for each plant the number of seeds which mature can be supposed to have a negative binomial (or Pascal) distribution with parameters κ and ρ, then the total number of seeds yielded by the plot will have a distribution with p.g.f.

$$\exp\{-\mu+\mu(\gamma-\rho z)^{-\kappa}\}, \quad \gamma = 1+\rho ,$$

that of a Poisson-stopped Pascal, or Poisson-Pascal, distribution.

Comments on terminology are to be found at the end of Chapter I, and some brief remarks may be added here. The most common terminology for a stopped variate derives from Satterthwaite (1942), who was seeking distributions which in some sense were generalizations of the Poisson distribution, whence generalized variate (although he equally applied this name to a mixed variate). There is no particular reason to attach generalized to this specific "generalization"; stopped describes the structure of the convolution initially considered, is hardly likely to conflict with "stopping rules" in different contexts, and abbreviates to a convenient verbal description (e.g., a Poisson-stopped variate, or a Poisson-stopped binomial variate), and so will be adopted here. A caution about an apparently natural terminology, also applicable to "generalized", might be given. It is easy but confusing to speak of the variate with p.g.f. $f(z)$, whose convolution is "stopped", as a "stopped variate" - however, the almost inevitable abbreviation, already used, for the variate corresponding to $g(f(z))$ is stopped variate; so the variate corresponding to $f(z)$ will here invariably be called a summed variate. Gurland's (1957) notation would write the g-stopped f variate as $g \vee f$.

Because of the apparent generality to be obtained thus, it is natural to consider characteristic functions in place of p.g.f.s (Molenaar (1965)). Suppose $F(\cdot)$ has the characteristic function $\phi(t)$, and is such that $\{\phi(t)\}^y$ is a uniquely defined characteristic function for all y in the set of points of increase of the distribution $G(y)$. Then the corresponding definition for the characteristic function of the stopped distribution is $\int \{\phi(t)\}^y \, dG(y)$. But although this new definition permits negative variates, for example, the question of when there is a uniquely defined characteristic function is not settled. (Clearly, there will be no problems in the common case in which $F(\cdot)$ is infinitely divisible.)

It should be noticed that a stopped distribution is a special kind of mixture, as is made obvious by writing

$$\mathscr{E}(z^{S_N}) = \Sigma \mathscr{E}_X(z^{S_n}) \cdot \Pr(N=n) \quad \text{for discrete } n$$
$$= \int \{f(z)\}^n \, dG(n) ,$$

$G(\cdot)$ being the cumulative distribution function of N. Since it is a special mixture it of course has properties not possessed by mixtures in general.

An interpretation of this result is given by Gurland's (1957) 'equivalence' theorem (see also Molenaar (1965)): if a variate X has a p.g.f. of the form $\{f(z)\}^u$ where u is the realization of a variate U, then

$$X \wedge U = U \vee X:$$

i.e., the mixture of X_U on a multiple of U is equivalent to a U-stopped X variate. (Recall the arbitrary constant in the definition of $\wedge$; the notation implies that this can be chosen so that the distributions of the left and right hand variates are the same.) The result of Maceda (1948), that there is an infinite class of distributions which can be generated both as mixed-Poisson and Poisson-stopped distributions, has been completed by Godambe and Patil (1969), who showed that iff the (non-negative) weight function of a mixed-Poisson is infinitely divisible is there also a construction as a Poisson-stopped distribution; i.e., $\int e^{-\lambda+\lambda z} dG(\lambda)$, in which λ is positive and $G(\lambda)$ is a cumulative distribution function, can be written as $\exp(-\mu+\mu\, f(z))$, in which μ is positive and $f(z)$ is a p.g.f. iff G is infinitely divisible.

Before exhibiting a series of specific examples, consider simple physical (or biological) sampling procedures which lead to the distributions being studied. Suppose there is a heterogeneous population of insects which lay eggs in clusters where the sizes of clusters vary across insects. If for a given "kind" of insect, indexed by u with cumulative distribution function $G(u)$, the distribution of numbers of eggs per cluster has p.g.f. $f(z;u)$, then for a "randomly selected" insect the number of eggs laid in a cluster has the p.g.f.

$$\int f(z;u)\, dG(u).$$

(This is the mixture formulation: perhaps the insect has been "randomly selected" by choosing the leaf on which has been laid the egg cluster in which the individuals are counted.) Alternatively, suppose there is an insect (or a homogeneous population of insects) randomly laying eggs in clusters across a region. Consider a "quadrat" (a sampling sub-area) chosen at random in the region, and suppose the number of egg clusters laid in the quadrat to have p.g.f. $g(z)$, with $f(z)$ the p.g.f. of the number of eggs per cluster. Then the number of eggs in the quadrat has the p.g.f.

$$g(f(z)).$$

(This is the stopped formulation: perhaps the "quadrat" is a plant which has been "randomly selected" by the insects.) The two sampling procedures are quite different: think of repetitions to build up frequency distributions in the two cases. The equivalence result above, however, asserts that if $f(z;u)$ has the special form $\{f(z)\}^u$, corresponding to a convolution, so that the heterogeneity in the population corresponds precisely to additivity, then the two sampling procedures lead to the same compound distribution.

Writing the equivalence of two distributions (or variates) with names n_1 and n_2 as $n_1 = n_2$, several examples are as follows (see also the Exercises). Since their verification is usually quite straightforward, though sometimes tedious, the details will be omitted, and only comments on particular aspects will be included.

1. Poisson($u\nu$) mixed on u by Poisson(μ)

   = Poisson(μ)-stopped Poisson(ν)

   Both lead directly to the Neyman(μ,ν) distribution, a familiar result. It may be worth pointing out that the mixture formulation specifies a set of Poisson distributions with means $0,\nu,2\nu,3\nu,\ldots,$ and such a set of populations would be artificial in many applications (for which a continuously distributed set of means might be more plausible). On the other hand, replacing ν by ν/c and μ by $c\mu$ leads to the Neyman($c\mu,\nu/c$) distribution and an appropriate choice of c will lead to approximating distributions which might be more acceptable.

2. The negative binomial (or Pascal) distribution is an historically well known example with a number of interesting aspects, some of which are mentioned below.
   Poisson($u\rho$) mixed on u by gamma(κ,1)

   = gamma(κ,1)-stopped Poisson(ρ)

   = Pascal(κ,ρ)

   Accident data in certain populations have frequency distributions which more resemble Pascal than Poisson probability distributions. They have therefore sometimes been interpreted through the first line above: for such a population the propensity of individuals to incur accidents is gamma distributed, with the numbers of accidents to an individual at any level Poisson distributed. Such an interpretation might be valid - what is <u>not</u> valid is the belief that even if the overall distribution were known to be Pascal (instead of mere speculation that it might be) this would necessarily lead to the mixture interpretation quoted. (Note that the mixture is equally described as

   Poisson(u) mixed on u by gamma (κ,ρ) = gamma (κ,ρ)-stopped Poisson (1).)

   The stopping decomposition of the second line has a continuous stopping distribution, with some loss of simplicity in interpretation. However, another stopping decomposition is

$$\text{Pascal}(\kappa,\rho) = \text{Poisson}(\kappa \cdot \ell n\ \gamma)\text{-stopped logarithmic}(\rho/\gamma),$$
$$\gamma = 1+\rho\ ,$$

with a natural interpretation in the accident context if records contain only those to whom at least one accident occurs. But this has an inconvenience, further discussed in the next Chapter, in that <u>both</u> parameters of the two parameter distribution occur in one of the single parameter component distributions; an alternative specification is (see Figure 1 at the end of this Section):

$$\text{Poisson}(\mu)\text{-stopped logarithmic}(\alpha)$$
$$= \text{Pascal}\left(\frac{\mu}{\ell n \frac{1}{1-\alpha}}, \frac{\alpha}{1-\alpha}\right).$$

Moreover, if the summed variate has a logarithmic distribution with added zeroes, with p.g.f.

$$1-\pi+\pi \frac{\ell n(1-\alpha z)}{\ell n(1-\alpha)}, \quad 0 < \pi \leq 1\ ,$$

then the Poisson(μ)-stopped distribution with this summed distribution leads to the

$$\text{Pascal}\left(\frac{\mu\pi}{\ell n \frac{1}{1-\alpha}}, \frac{\alpha}{1-\alpha}\right)$$

distribution: again the negative binomial, perhaps unexpectedly.

It is surely very clear that, without separate information, no inference regarding the structure of a family of distributions can be drawn from the identification of a distribution as a member of the family - indeed the identification of a distribution (see in various places, under Estimation and the headings of particular distributions) is itself a generally uncertain affair.

3. Pascal($u\kappa,\rho$) mixed on u by Poisson(μ)

$$= \text{Poisson}(\mu)\text{-stopped \{geometric}(\tfrac{\rho}{\gamma}) \text{ summed to } \kappa \text{ terms\}}$$

$$= \text{Poisson}(\mu)\text{-stopped Pascal}(\kappa,\rho)$$

It is not uncommon to state results such as these in the form:

Pascal mixed on its exponent by Poisson
= Poisson-stopped Pascal,

where the notation is to be disentangled by the reader.

4. Pascal($u\kappa,\rho$) mixed on u by gamma(α,β)
= gamma(α,β)-stopped Pascal(κ,ρ)
= Pascal($\alpha,\kappa\beta \ln \gamma$)-stopped logarithmic($\rho/\gamma$).

Relations between stopped variates and their components

The important special case of

$$h(z) = g(f(z))$$

when $g(z)$ corresponds to a Poisson variate is taken up separately in a later section; the general case is here discussed.

Relations between probabilities

The expansion of $h(z)$ as a power series in z uses the Faa di Bruno Formula for the derivative of a "function of a function" (see the Appendix on Stirling Numbers). Thus, writing

$$h(z) = \Sigma h_x z^x , \text{ so } h^{(x)}(0) = x!\, h_x ,$$

with similar definitions for g_x and f_x, it follows that

$$h_x = \sum_{n=1}^{x} n!\, g_n \Sigma \frac{f_1^{n_1} f_2^{n_2} \cdots f_x^{n_x}}{n_1!\, n_2! \cdots n_x!}$$

where the second sum is over the non-negative integers $n_1, n_2, \ldots$ such that $n_1+n_2+\ldots+n_x=n$, $1.n_1+2.n_2+\ldots+x.n_x=x$.

Relations between factorial moments (and cumulants)

In a corresponding notation, the rth factorial moment is

$$\mu'_{(r):h} = \mathscr{E}[X(X-1)\ldots(X-r+1)] = h^{(r)}(1)$$

(similarly for $\mu'_{(r):g}$, $\mu'_{(r):f}$). Using the same expansion but writing $z=1$ gives

$$\mu'_{(r):h} = \sum_{n=1}^{r} \mu'_{(n):g} \sum \frac{r!}{n_1!\ldots n_r!} \{\mu'_{(1):f}\}^{n_1} \times$$

$$\left\{\frac{\mu'_{(2):f}}{2!}\right\}^{n_2} \cdots \left\{\frac{\mu'_{(r):f}}{r!}\right\}^{n_r}$$

where the second sum ranges over the same non-negative integers $n_1, n_2, \ldots, n_r$ as before.

Correspondingly, the factorial cumulants are given by

$$\kappa_{(r):h} = \sum_n \kappa_{(n):g} \sum \frac{r!}{n_1!\ldots n_r!} \{\mu'_{(1):f}\}^{n_1} \cdots \left\{\frac{\mu'_{(r):f}}{r!}\right\}^{n_r}.$$

The coefficients in the first six terms are given explicitly in the Appendix on Stirling Numbers, under the heading of the Faa di Bruno result; they are the same as those which appear in the general formulae which connect moments and cumulants for the one variate, and are given up to order 10 in Kendall and Stuart (1963) and to order 12 in David, Kendall and Barton (1966).

There is particular interest attaching to the second order factorial cumulants. From the formulae just quoted,

$$\frac{\kappa_{(2):h}}{\mu_h} = \frac{\mu'_{(2):f}}{\mu_f} + \mu_f \frac{\kappa_{(2):g}}{\mu_g},$$

and, introducing the Index of Cluster Size $\iota_{CS} = \kappa_{(2)}/\mu$, (see the next Chapter) the Index of h can be written as

$$\iota_{CS} = \mu_f + \iota_{CS:f} + \mu_f \iota_{CS:g}.$$

Neater results are obtained by using the variance to mean ratio $\tau = \mathcal{V}(X)/\mathcal{E}(X)$ and the Index of Mean Crowding $\iota_C = \mu'_{(2)}/\mu$:

$$\tau_h = \tau_f + \tau_g \mu_f, \quad \text{and} \quad \iota_{C:h} = \iota_{C:f} + \iota_{C:g} \mu_f.$$

A similar expression for the Index of Cluster Frequency $\iota_{CF} = \mu^2/\kappa_{(2)}$ is given by

$$\iota_{CF} = \frac{\mu_g}{\dfrac{1+\iota_{CF:f}}{\iota_{CF:f}} + \iota_{CS:g}} = \frac{\mu_g}{\iota_{P:f} + \iota_{CS:g}}$$

on introducing the <u>Index of Patchiness</u> $\iota_P = (1+\iota_{CF})/\iota_{CF}$.
For Poisson-stopped variates, $\iota_{CS:g} = 0$, $\iota_{CF:g} = \infty$,
and (see section ii) under Properties of Poisson-Stopped Variates) hence

$$\iota_{CS:h} = \mu_f + \iota_{CS:f} \text{ , not a function of } g \text{ ,}$$

and $$\iota_{CF:h} = \mu_g \frac{\iota_{CF:f}}{1+\iota_{CF:f}}$$

where the second factor is not a function of g.

It follows that the Index of Cluster Size depends only on f if $\kappa_{(2):g} = 0$ or $\iota_{CS:g} = 0$ (as it therefore does if $g(\cdot)$ corresponds to a Poisson variate or a gamma(κ,1) variate). Similarly, the Index of Cluster Frequency depends only on g if $\mu'_{(2):f} = 0$ or $\iota_{CF:f} = -1$.

A special case of some interest is given by

$$h(z) = g(1-\theta+\theta z) \text{ , } 0 < \theta < 1.$$

This corresponds to a g-stopped summed-binomial(1,θ) variate, but is perhaps better thought of directly as a modification of the g variate corresponding to the circumstance in which every individual in the g distribution has a probability $1-\theta$ of not appearing in the sample (mortality, migration, malfunction of counting equipment, etc.). (In terms of the cluster description already used, the clusters are of size 0 (with probability $1-\theta$) or 1 (with probability θ).)
If as well $g(z) = e^{-\lambda+\lambda z}$, then $h(z) = e^{-\theta\lambda+\theta\lambda z}$,
i.e. a Poisson distribution is unaltered in form by such a process but the parameter is modified, in such a way that unless information beyond that from the simple sample is available neither θ nor λ but merely their product can be estimated.

<u>Probabilities computed from factorial moments</u>

$$\text{Since } h_x = \sum_{r=x}^{\infty} (-1)^{r+x} \binom{r}{x} \frac{\mu'_{(r):h}}{r!} = \sum (-1)^{r-x} \frac{\mu'_{(r):h}}{(r-x)!}$$

and factorial moments corresponding to $h(z)$ can be expressed in terms of those corresponding to $g(z)$ and $f(z)$, it may turn out that, if these latter are sufficiently simple, the above inversion formula can be used to compute numerically the probabilities of a complex stopped distribution. For computation the formula will necessarily be truncated: whether the technique is worthwhile will depend on the rapidity with which the $\mu'_{(r):h}$ tend to 0 with r and the simplicity of their expression in terms of the components. Katti (1966) has applied the process with varying success to a number of specific examples.

(Questions relating to the extent to which a distribution is (uniquely) determined by its moments are not discussed here - see, e.g., Moran (1968).)

<u>Poisson-Stopped Distributions</u> (Generalized Poisson)

Just as the mixed-Poisson distribution has proved important, so has the Poisson-stopped distribution. In the general formula

$$g(f(z)),$$

taking $$g(z) = e^{-\mu+\mu z}$$

for the Poisson gives the Poisson-stopped p.g.f. to be

$$\exp\{-\mu + \mu\, f(z)\}$$

where $f(z)$ is the p.g.f. of the summed variate.

Examples are:

a) Poisson(μ)-stopped Poisson(ν):
the p.g.f. is $\exp(-\mu + \mu e^{-\nu+\nu z})$,
that of a Neyman Type A variate.

b) Poisson(μ)-stopped {"Poisson(ν) + parent"}.
This is the formulation of the Thomas (1949) distribution, where the number of offspring X from a parent is supposed Poisson(ν) and also the parent is included in the count. For this case the p.g.f. of the count of a parent plus offspring is

$$\mathscr{E}(z^{X+1}) = z\mathscr{E}(z^X) = ze^{-\nu+\nu z},$$

and hence the p.g.f. of the compounded variate is

$$\exp\{-\mu + z\mu e^{-\nu+\nu z}\}.$$

c) Poisson($\kappa \ln \gamma$)-stopped logarithmic($\frac{\rho}{\gamma}$), with $\gamma = 1+\rho$, is

$$\exp\left(-\kappa \ln \gamma + \kappa \ln \gamma \frac{\ln \frac{\gamma - \rho z}{\gamma}}{-\ln \gamma}\right)$$

$$= (\gamma - \rho z)^{-\kappa} ,$$

the negative binomial (Pascal) distribution, as already exhibited.

d) Poisson-stopped binomial:

$$\exp(-\mu + \mu(\chi + \pi z)^{\nu}) ,$$

mostly called the "Poisson binomial" distribution in the literature. (But also notice that this name has been and is used for "Poisson's generalization of the binomial", with p.g.f.

$$(\chi_1 + \pi_1 z)(\chi_2 + \pi_2 x) \ldots (\chi_\nu + \pi_\nu z)$$

where each term is binomial with variable π_i - see e.g. Aitken (1947) and Hodges and LeCam (1960)).

e) Poisson-stopped Pascal ("Poisson-Pascal"):

$$\exp(-\mu + \mu(\gamma - \rho z)^{-\kappa}),$$

sometimes also called the "generalized Polya-Aeppli", since the case with $\kappa = 1$ is attributed to Aeppli (1924).

f) Two parameter Neyman Types A, B, C,... .

Here the p.g.f. is

$$\exp(-\mu + \mu\ \beta! \sum_0^\infty \frac{\nu^s (z-1)^s}{(\beta+s)!})$$

where $\beta = 0$ corresponds to the Neyman type A,
$\beta = 1$ corresponds to the Neyman type B,
$\beta = 2$ corresponds to the Neyman type C
etc.,
a generalization given by Beall and Rescia (1953).

g) A further generalization of f) using the confluent hypergeometric function

$$M(\alpha, \beta, x) = 1 + \frac{\alpha}{\beta} x + \frac{\alpha(\alpha+1)}{\beta(\beta+1)} \frac{x^2}{2!} + \ldots .$$

as a p.g.f., so the compounded p.g.f. (Gurland (1958)) is

$$\exp[-\mu + \mu\ M\{\alpha,\alpha+\beta,\nu(z-1)\}]\ .$$

(It is hard not to mention also (Subrahmaniam (1966)) the p.g.f. $M\{\alpha;\alpha+\beta;-\mu+\mu\ \exp(-\nu+\nu z)\}$, irrevelant though it is to the present sequence.)

There is obviously no end to the increasing variety and complexity of distributions which can be constructed in this fashion; Philipson (1960) attempts to obtain a degree of generality through the use of the confluent hypergeometric function.

Properties of Poisson-Stopped Distributions

It is to be noticed that the roles of the Poisson distribution in the mixed-Poisson and in the Poisson-stopped distributions are quite different. The mixed-Poisson distribution has a p.g.f. of the form

$$f(e^{-\nu+\nu z}) = \int_0^\infty e^{-u\nu+u\nu z}\ dG(u) = \int_0^\infty \{e^{-\nu+\nu z}\}^u\ dG(u).$$

That of the Poisson-stopped distribution is

$$\exp\{-\mu+\mu\ f(z)\} = \sum_{n=0}^\infty \{f(z)\}^n\ e^{-\mu}\frac{\mu^n}{n!} = \int_0^\infty \{f(z)\}^n\ dF(n)$$

where $dF(n) = e^{-\mu}\ \mu^n/n!$. In the mixture case the Poisson distribution is the family which is mixed with weight $G(\cdot)$, while in the stopped case the Poisson constitutes the weight function $F(\cdot)$; as a consequence the results for the mixed-Poisson distribution have no special relation with those for Poisson-stopped distributions. (But of course results which apply to general mixtures apply to stopped variates.)

i) Probabilities of stopped and summed variates

From $h(z) = \exp\{-\mu+\mu f(z)\}, = \sum h_x z^x$,

and similarly for f_x, direct application of the Faa di Bruno Formula leads to

$$h_x = e^{-\mu+\mu f_0} \sum_{n=1}^{x} \mu^x \sum \frac{f_1^{n_1}\ f_2^{n_2}\ldots f_x^{n_x}}{n_1!\ n_2!\ldots n_x!}$$

where the n_i are non-negative integers such that

$$n_1+\ldots+n_x = n \ , \quad 1.n_1+\ldots+x.n_x = x \ .$$

This can also be obtained by a direct rearrangement of the p.g.f. (Ord (1972)):

$$h(z) = \{\exp(-\mu+\mu f_0)\} \cdot \exp\{\mu z\phi(z)\}$$

$$\text{where} \quad \phi(z) = z\{f(z) - f_0\}$$

$$= z \sum_{x=0}^{\infty} f_{x+1} \ z^x$$

$$= \{\exp(-\mu+\mu f_0)\} \sum_{r=0}^{\infty} \frac{\{\mu\phi(z)\}^r}{r!} \ z^r .$$

In the latter sum the coefficient of z^x can be identified, to give an expansion for h_x with a finite number of terms, involving only $f_0, f_1, \ldots, f_x$, e.g.

$$h_0 = e^{-\mu+\mu f_0} ,$$

$$h_1 = h_0 \mu f_1 ,$$

$$h_2 = h_0 \mu (f_2 + \tfrac{1}{2} \mu f_1^2) ,$$

$$h_3 = h_0 \mu (f_3 + \mu f_1 f_2 + \tfrac{1}{6} \mu^2 f_1^3) .$$

But perhaps the most generally useful formulae are the recurrence relations obtained by direct differentiation of the definition

$$h(z) = \exp\{-\mu+\mu \ f(z)\}$$

(e.g. Khatri and Patel (1961)). Differentiating this expression,

$$h'(z) = \mu \ h(z) \ f'(z) ,$$

i.e., $\Sigma(x+1) \ h_{x+1} z^x = \mu \ \Sigma \ h_p z^p \ \Sigma \ (q+1) \ f_{q+1} z^q$.

Equating coefficients of z^x , with $p+q=x$ on the right, gives

$$(x+1)\ h_{x+1} = \mu \sum_{p=0}^{x} (x-p+1) f_{x-p+1}\ h_p$$

$$= \mu \sum_{q=0}^{x} (q+1) f_{q+1} h_{x-q} \ .$$

Thus $h_0 = \exp(-\mu+\mu\ f_0)$,

$h_1 = \mu\ h_0 f_1$,

$h_2 = \frac{1}{2}\ \mu(2h_0 f_2 + h_1 f_1)$,

$h_3 = \frac{1}{3}\ \mu(3h_0 f_3 + 2h_1 f_2 + h_2 f_1)$, etc.,

a very simple set of recurrence relations.

ii) Moments of the Poisson-stopped and summed variates

The cumulants of a Poisson-stopped distribution are directly proportional to the corresponding moments about the origin of the summed variate. For writing $z = e^t$ in the p.g.f. and taking logarithms,

$$\ln h(e^t) = -\mu+\mu\ f(e^t) \ ,$$

where the left side is the cumulant g.f. of the stopped variate and the right side the moment g.f. of the summed variate (modified to remove the term in t^0) . In an obvious notation,

$$\kappa_{r:h} = \mu \cdot \mu'_{r:f} \ , \quad r = 1,2,\ldots$$

The same relation holds for factorial functions:

$$\kappa_{(r):h} = \mu \cdot \mu'_{(r):f} \ .$$

It follows if X is a Poisson(μ)-stopped summed-Y variate that

$$\mathscr{E}(X) = \mu\ \mathscr{E}(Y),\ \mathscr{V}(X) = \mu[\mathscr{V}(Y) + \{\mathscr{E}(Y)\}^2] \ ,$$

and so

$$\mathscr{V}(X) = \mathscr{E}(X) \cdot \frac{\mu'_{2:Y}}{\mu'_{1:Y}} \ .$$

(Cf. the Poisson variate, for which the variance is equal to the mean.) For a discrete distribution,

$$\frac{\mu'_{2:Y}}{\mu'_{1:Y}} = \frac{\mathscr{E}(Y^2)}{\mathscr{E}(Y)} = \frac{\Sigma y_i^2 p_i}{\Sigma y_i p_i} \quad \text{with} \quad p_i = \Pr(Y = y_i),$$

and if $y_i = 0,1,2,\ldots,$ then $y_i^2 \geq y_i$ with equality iff $y_i = 0,1$. Hence $\frac{\mathscr{E}(Y^2)}{\mathscr{E}(Y)} > 1$ if $p_i > 0$ for any $i \geq 2$, and so for any distribution on $0,1,2,\ldots$

$$\mathscr{V}(X) \geq \mathscr{E}(X)$$

with equality iff the summed variate is concentrated on 0,1. It follows that a Poisson-stopped distribution cannot be Poisson unless the summed variate is concentrated on 0,1 (is binomial $(1,\pi)$ or degenerate on 0 or 1), i.e., except in those cases it is "overdispersed".

The Indices of Cluster Size and Frequency take attractively simple forms:

<u>Index of Cluster Size</u> $\kappa_{(2):h}/\mu_h = \iota_{CS:h}$

$$= \mu_f + \iota_{CS:f},$$

which is independent of μ (Skellam (1952)), and is exactly μ_f if $\iota_{CF:f} = 0$;

<u>Index of Cluster Frequency</u> $\mu_h^2/\kappa_{(2):h} = \iota_{CF:h}$

$$= \mu \frac{\iota_{CF:f}}{1 + \iota_{CF:f}}$$

which is nearly μ if $\iota_{CF:f}$ is large.

In each case $\mu_h = h'(1)$, the mean of X, etc.

iii) Identifiability

Unlike the mixture formulation, where the question referred to the uniqueness of the weight function for a given component, here the weight function is given and the question relates to the uniqueness of the summed variate. Since

$$\exp\{-\mu+\mu f_1(z)\} = \exp\{-\mu+\mu f_2(z)\}$$

implies $\quad f_1(z) = f_2(z)$,

for given Poisson-stopped distribution <u>and</u> given Poisson stopping distribution there is a unique summed distribution.

However, it is not sufficient merely to know that the distribution is Poisson-stopped (Skellam (1952)). For

$$\exp\{-\mu+\mu\ f(z)\} \equiv \exp\{-M+M\ F(z)\}$$

if $M = \mu(\alpha+1)$ and $F(z) = \dfrac{\alpha + f(z)}{\alpha + 1}$

for arbitrary $\alpha(> - f_0$ for a positive probability in the zero class of $F(z))$. Thus, for example, the Neyman Type A (6,5) distribution with p.g.f.

$$\exp\{-6+6\ e^{-5+5z}\}$$

$$= \exp\{-12+12\ \frac{1+e^{-5+5z}}{2}\}\ ,\ \alpha = 1,$$

and so it can be interpreted as

a Poisson(6)-stopped Poisson(5) distribution

or as

a Poisson(12)-stopped{½Poisson(5) with extra probability of ½ in the zero class} distribution.

Hopefully, the more "natural" representation will be generally appropriate - i.e., of $F(z)$ and $f(z)$ one will often be regarded as more fundamental. Skellam suggests that in the absence of other information one may prefer to take $f_0 = 0$ (i.e. the probability of the zero class in the summed distribution to be zero), when the probability of the zero class in the stopped distribution is $e^{-\mu}$ and μ is the density of clusters of size 1 or more (i.e., of "actual" clusters, not including those unobservable clusters of size 0).

iv) Reproductive Properties

a) Since a Poisson stopped variate is a mixture with a Poisson weight function, the general theory for mixtures shows that the Poisson-stopped variate is reproductive in the Poisson distribution. Precisely, and directly from the p.g.f.s, if X_i is a Poisson(μ_i)-stopped f variate, where the X_i are independent (and the summed distribution the same), then ΣX_i is a Poisson($\Sigma\mu_i$)-stopped f variate. For the p.g.f. of each X_i is

$$\exp\{\mu_i + \mu_i\ f(z)\}$$

and hence that of ΣX_i is

$$\exp(-\Sigma\mu_i + f(z)\ \Sigma\mu_i).$$

Thus if X and Y are independent Neyman(μ_X,ν) and Neyman(μ_Y,ν) variates, then X+Y is Neyman($\mu_X+\mu_Y,\nu$); if X,Y are independent Pascal(κ_X,ρ), Pascal(κ_Y,ρ), then X+Y is Pascal($\kappa_X+\kappa_Y,\rho$) , familiar results.

b) These results can also be examined from the point of view of decomposition (see the next Section v) for infinite divisibility).

1) If λ is the rate of occurrence per unit time in a Poisson distribution, so that in a time interval t the mean rate is $\mu = \lambda t$, write

$$h(z;t) = \exp\{-\lambda t + \lambda t\ f(z)\}\ .$$

Then $h(z;t_1+t_2) = h(z;t_1)\ h(z;t_2)$;
i.e. if $X(t_i)$ has the p.g.f. $h(z;t_i)$ with $X(t_1)$ and $X(t_2)$ independent, and if X has the p.g.f. $h(z;t_1+t_2)$, then $X = X(t_1) + X(t_2)$ and X has been 'decomposed' into a sum of independent random variables.

2) Suppose one thinks of events like numbers of egg clusters laid per plant, say, consisting of clusters in which one egg is laid (a singleton event), two eggs (a doublet),..., the numbers of these events, written $Y_1,Y_2,\ldots$, being Poisson(λf_1), Poisson(λf_2),... distributed. Then the total number of eggs laid on the plant is

$$Y_1 + 2Y_2 + 3Y_3 + \ldots = X, \quad \text{say,}$$

and on the supposition of independence of the Y_p the p.g.f. of X is

$$\mathscr{E}(z^X) = \mathscr{E}\{z^{Y_1}\}\,\mathscr{E}\{(z^2)^{y_2}\}\,\mathscr{E}\{(z^3)^{y_3}\}\ldots$$

$$= e^{-\lambda f_1+\lambda f_1 z} \cdot e^{-\lambda f_2+\lambda f_2 z^2} \ldots$$

$$= e^{-\lambda\Sigma f_p+\lambda\Sigma f_p z^p} .$$

Writing $f_0 + \sum_1^\infty f_p = 1$

and $f(z) = \sum_0^\infty f_x z^x$, $0 \leq f_x < 1$,

$$\mathscr{E}(z^X) = e^{-\lambda+\lambda f(z)} ,$$

the general Poisson-stopped variate can thus be decomposed into an infinite convolution of "Poisson-like" variates $Y_1, 2Y_2, \ldots$ corresponding to singleton events, doublet events,...

If (Luders (1934))

$$\lambda f_p = \frac{\kappa}{p}\left(\frac{\rho}{1+\rho}\right)^p ,$$

the p.g.f. $\exp\{-\lambda\Sigma f_p + \lambda\Sigma_p z^p\}$

$$= \frac{1}{(1+\rho-\rho z)^\kappa} ,$$

that of the negative binomial; if

$$\lambda f_p = \mu e^{-\nu}\frac{\nu^p}{p!} ,$$

the p.g.f. becomes

$$\exp\{-\mu+\mu e^{-\nu+\nu z}\} ,$$

that of the Neyman distribution. The former of these may be thought of as the first derivation of the negative binomial as a Poisson-stopped logarithmic distribution - see Quenouille (1949) for a more explicit statement.

Kemp and Kemp (1965) (see also Kemp (1967a) and Patel (1976)) offer an interesting analysis, in terms of this decomposition, of the circumstance that for a particular set of data it may happen that there is very little to choose between the goodness of fit of a number of different families of Poisson-stopped distributions. For several of the distributions examined in detail in later sections the decompositions are, explicitly, as follows.

Poisson(λ)

$$\exp(-\lambda+\lambda z)$$

Neyman(μ,ν)

$$\exp\left[\mu e^{-\nu}\{\nu(-1+z) + \frac{\nu^2}{2!}(-1+z^2) + \frac{\nu^3}{3!}(-1+z^3)+\ldots\}\right]$$

Pascal(κ,ρ)

$$\exp\left\{\frac{\kappa\rho}{\gamma}(-1+z) + \frac{\kappa\rho^2}{2\gamma^2}(-1+z^2) + \frac{\kappa\rho^3}{3\gamma^3}(-1+z^3) + \ldots\right\}$$

Poisson-Pascal(μ,κ,ρ)

$$\exp\left[\frac{\mu}{\gamma^\kappa}\left\{\frac{\kappa\rho}{\gamma}(-1+z) + \frac{\kappa(\kappa+1)}{2!}\frac{\rho^2}{\gamma^2}(-1+z^2) + \ldots\right\}\right]$$

Poisson-binomial(μ,ν,π)

$$\exp\left[\mu\,\chi^\nu\left\{\frac{\nu\pi}{\chi}(-1+z) + \frac{\nu(\nu-1)}{2!}\frac{\pi^2}{\chi^2}(-1+z^2) + \ldots\right\}\right]$$

For Student's 1907 data on yeast cells, good and nearly identical fits are given by a variety of distributions, and for these Kemp and Kemp have calculated the values of $f_1, f_2, \ldots$ in the notation of a previous paragraph. These are:

	f_1	f_2	f_3	f_4
Poisson (0.682)	0.682	0	0	0
Neyman ($3.6_+, 0.19_-$)	0.565	0.053	0.004	0.000
Pascal ($3.6_-, 0.19_+$)	0.573	0.046	0.005	0.000
Poisson-Pascal ($7.2_+, 1, 0.094_+$) or Polya-Aeppli	0.570	0.049	0.004	0.000
Poisson-binomial ($2.2_-, 2, 0.16_-$) or Hermite	0.576	0.054	0	0

The fit in the Poisson case is less good than for the others - in terms of the present explanation, f_2, though small, is not negligible. For the other cases f_1 and f_2 are very similar and f_3 and f_4 negligible - hence the agreement between the distributions and, in particular, an explanation of the goodness of fit for the Hermite case (for which $f_3 = f_4 = \ldots = 0$).

v) Infinite divisibility.

Poisson stopped distributions are infinitely divisible, since

$$\exp\left\{-\frac{\mu}{n} + \frac{\mu}{n} f(z)\right\}$$

is obviously also a p.g.f. (of a Poisson-stopped variate, in fact) for every $n > 0$.

Remarkably, a converse is true : if $h(z)$ is the p.g.f. of an <u>integral valued</u> infinitely divisible variate, then $h(z)$ has the form $\exp\{-\mu+\mu' f(z)\}$ where $f(z)$ is a p.g.f.; i.e. every integral valued infintely divisible variate has a representation as a Poisson-stopped variate. There is an accessible proof in Feller I (1967), which also establishes the equivalent conditions that $h(z)$ is the p.g.f. of an integral valued infinitely divisible variate iff

$$h(1) = 1$$

and $$\ell n \frac{h(z)}{h(0)} = \sum_1^\infty a_k z^k$$

where no a_k is negative

and Σa_k converges.

Perhaps even more remarkably, <u>all</u> infinitely divisible distributions are limits of Poisson-stopped distributions (de Finetti; Lukacs (1960)).

vi) Comparison of Poisson-stopped and Poisson distributions.

Choosing a Poisson variate Y with the same mean as the Poisson(μ)-stopped f variate X (with p.g.f. $\exp\{-\mu+\mu f(z)\} = \Sigma h_x z^x$):

a) $\mathcal{V}(X) \geq \mathcal{V}(Y)$,

b) $\Pr(X=0) \geq \Pr(Y=0)$,

c) $\dfrac{\Pr(X=0)}{\Pr(X=1)} \geq \dfrac{\Pr(Y=0)}{\Pr(Y=1)}$,

with equality iff the f distribution is concentrated on 0,1.

The result a) was established above. For b) and c), noting

$$X: h(z) = \exp\{-\mu+\mu f(z)\},$$

$$h'(z) = \mu \cdot h(z) f'(z) ,$$

whence

$$\mathcal{E}(X) = h'(1) = \mu \cdot f'(1) = \mathcal{E}(Y).$$

Hence, writing $p_y = \Pr(Y=y)$,

$$p_0 = e^{-\mu \cdot f'(1)} ,$$

so that

$$p_1 = \mu \cdot f'(1) p_0,$$

$$\begin{aligned}\Pr(X=0) &= h_0 \\ &= \exp(-\mu+\mu f_0) \\ &= p_0 \exp[\mu\{f(1) - 1 - f'(1)\}].\end{aligned}$$

A Taylor expansion gives

$$f(z) = f(1) + (z-1)f'(1) + \frac{1}{2}(z-1)^2\ f''\{1+\theta(z-1)\},$$

$$\text{where}\quad 0 < \theta < 1,$$

i.e.

$$f(0) = f(1) - f'(1) + \frac{1}{2}\ f''(\phi),\quad 0 < \phi = 1 - \theta < 1,$$

whence

$$\frac{h_0}{P_0} = \exp\{\mu \cdot \frac{1}{2} \cdot f''(\phi)\}.$$

But $f''(\phi) \geq 0$, with the equality iff the f-distribution is concentrated on 0,1, and hence

$$h_0 \geq p_0.$$

Similarly, for c),

$$h_1 = h'(0) = \mu \cdot h(0) \cdot f'(0) = \mu\ h_0\ f_1;$$

i.e.

$$\frac{h_1}{h_0} = \mu\ f_1 \leq \mu(f_1 + 2f_2 + 3f_3 + \ldots)$$

or

$$\frac{h_1}{h_0} \leq \mu\ f'(1) = \frac{\mu\ f'(1) \cdot e^{-\mu f'(1)}}{e^{-\mu f'(1)}}$$

$$= \frac{p_1}{p_0}$$

with the equality in the same circumstances.

These conclusions, which are similar to those for mixed-Poisson variates, accord well with commonsense: if individuals are heavily clustered (i.e., many in a small number of clusters), many quadrats will be empty. But their utility with respect to a single sample is limited because of random variability, though over replicated samples the effect may by quite conspicuous.

vii) Maximum Likelihood Estimation in Stopped Distributions

The application of the method of maximum likelihood to the estimation of parameters in contagious distributions rather often produces an equation of the form

$$\Sigma F_x (x+1) \frac{\hat{p}_{x+1}}{\hat{p}_x} = \Sigma x\, F_x \ ,$$

where F_x is the observed frequency and p_x the probability of the xth class, the circumflex denoting the replacement of parameter values by their maximum likelihood estimators.

Following Sprott (1965), the structure which leads to this equation can be explained in terms of stopped distributions. The random variable is

$$S_N = X_1 + X_2 + \ldots + X_N$$

where $X_1, X_2, \ldots$ each has p.g.f. $f(z) = \Sigma\, f_x z^x$ depending on a single parameter ν, and N has p.g.f. $g(z) = \Sigma\, g_n z^n$ with parameter μ, so that S_N has the p.g.f.

$$h(z) = g(f(z)) = \Sigma\, z^x p_x \ .$$

(A more explicit notation, used below when appropriate, is $f(z;\nu)$, $g(z;\mu)$ and $h(z) = g(f(z;\nu),\mu) = G(z;\mu,\nu)$.) Then the log-likelihood is

$$\ell n\, L = \Sigma\, F_x\, \ell n\, p_x \ ,$$

and
$$\frac{\partial\, \ell n\, L}{\partial \nu} = \Sigma \frac{F_x}{p_x} \frac{\partial p_x}{\partial \nu} \ .$$

Equating to zero, and omitting the circumflexes for convenience,

$$\Sigma \frac{F_x}{p_x} \frac{\partial\, p_x}{\partial \nu} = 0.$$

But the equation at the head of this section is

$$\Sigma \frac{F_x}{p_x} \left\{ (x+1)\; p_{x+1} - x\; p_x \right\} = 0,$$

and for these two to be identical for all F_x ,

$$\frac{\partial p_x}{\partial \nu} = \{(x+1)\ p_{x+1} - x\ p_x\}\ u(\mu,\nu)$$

where $u(\mu,\nu)$ is arbitrary.

However,

$$\Sigma\ (x+1)\ p_{x+1}\ z^x = \frac{\partial h}{\partial z}$$

and $\quad \Sigma\ x\ p_x\ z^x = z\ \frac{\partial h}{\partial z}$,

whence $\quad \frac{\partial h}{\partial \nu} = (1-z)\ \frac{\partial h}{\partial z}\ u(\mu,\nu)$.

Since $\quad dh = \frac{\partial g}{\partial f}\ (\frac{\partial f}{\partial z}\ dz + \frac{\partial f}{\partial \nu}\ d\nu) + \frac{\partial g}{\partial \mu}\ d\mu$

$$= \frac{\partial G}{\partial z}\ dz + \frac{\partial G}{\partial \mu}\ d\mu + \frac{\partial G}{\partial \nu}\ d\nu\ ,$$

equating coefficients of the differentials leads to

$$\frac{\partial h}{\partial \nu} = \frac{\partial g}{\partial f} \cdot \frac{\partial f}{\partial \nu} = (1-z)\ \frac{\partial g}{\partial f} \cdot \frac{\partial f}{\partial z}\ u(\mu,\nu)\ ,$$

and as f is not a function of μ ,

$$\frac{\partial f}{\partial \nu} = (1-z)\ \frac{\partial f}{\partial z}\ v(\nu)\ , \quad \text{say}\ ,$$

or $\quad 0 = \frac{\partial f}{\partial \nu}\ \frac{1}{v(\nu)} - (1-z)\ \frac{\partial f}{\partial z}$.

Comparing this with

$$df = \frac{\partial f}{\partial \nu}\ d\nu + \frac{\partial f}{\partial z}\ \partial z\ ,$$

$$\frac{df}{0} = v(\nu)\ d\nu = -\ \frac{dz}{1-z}$$

which is satisfied by

$$f = \alpha, \quad \text{a constant,}$$

and $\quad \int v(\nu) d\nu = \ell n(1-z) - \ell n\ \beta$ where β is constant

$$\text{i.e.} \quad (1-z)\ \exp\{-\int v(\nu) d\nu\} = \beta\ ,$$

so that an expression of the form

$$f = f^*\{(1-z)\ \exp(-\int v(\nu)\ d\nu)\} = f^*(\beta)$$

satisfies the differential equation for f. (Differentiation provides an immediate verification, since

$$\frac{\partial f}{\partial z} = \frac{df}{d\beta}\ \frac{\partial \beta}{\partial z} = \frac{df}{d\beta}\ \{-\exp(-\int v(\nu)d\nu)\}\ ,$$

and $$\frac{\partial f}{\partial \nu} = \frac{df}{d\beta}\ (1-z)\exp(-\int v(\nu)d\nu)\{-v(\nu)\} = (1-z)v(\nu)\ \frac{\partial f}{\partial z}\ .)$$

If therefore the p.g.f. of the components of the g-stopped f distribution has the form

$$f(z;\nu) = f^*\{(1-z)\ v(\nu)\}$$

where $v(\nu)$ is arbitrary, so that the stopped distribution has p.g.f.

$$g(f^*\{(1-z)\ v(\nu)\},\mu)\ ,$$

then the maximum likelihood equation obtained by differentiating with respect to ν is

$$\Sigma\ F_x(x+1)\ \cdot\ \frac{\hat{p}_{x+1}}{\hat{p}_x} = \Sigma\ x\ F_x\ .$$

In fact $v(\nu) = \nu$ is a very commonly occurring specialization.

Since the Neyman Type A distribution is of this form, with

$$g(z;\mu) = e^{-\mu+\mu z}\ ,\ f^*(\cdot) = \exp(\cdot)\ ,\ v(\nu) = \nu\ ,$$

it has a maximum likelihood equation of the above type; so has the Poisson-Pascal distribution, for which g is as above and $f^*(x) = (1+x)^{-\kappa}$, $v(p) = p$. Trivially, since $(x+1)p_{x+1}/p_x = \lambda$ in this case, the Poisson distribution leads to an equation of the specified form; the structure above is also realized, however, for g as above and $f^*(x) = 1 + x$ with $v(\pi) = \pi$, since a Poisson(μ)-stopped binomial($1,\pi$) variate is Poisson($\mu\pi$). On the other hand, the negative binomial, though it can be generated as a Poisson-stopped logarithmic variate, has neither the

the appropriate form nor a maximum likelihood equation of of the form discussed.

It is also common to find that, as well as the above maximum likelihood equation, the second equation (for a two parameter distribution) - the "equation of the mean" - is of the form

$$\hat{\mu}\,\hat{\nu}\,\Sigma F_x = \alpha\,\Sigma\, xF_x ,$$

where α is some known constant. For an analysis of the conditions in which both these equations are satisfied, see Sprott (1965): in brief, the distribution must be expressible as a mixed-Poisson with generalized power series distribution weights, or equivalently a mixed-binomial with g.p.s.d. weights.

2.4a EXERCISES

1. Show that the following equivalences between distributions can be established.

   a) Poisson($u\nu$) mixed on u by Poisson(μ)

      = Poisson(μ)-stopped Poisson(ν)

      = Poisson(μ)-stopped [binomial(μ,π) mixed on u by Poisson(ν/π)]

      = Neyman(μ,ν).

   b) Poisson mixed by gamma

      = Poisson-stopped logarithmic = gamma-stopped Poisson

      = Pascal.

   (N.B.: Poisson(μ)-stopped logarithmic(α)

   $$= \text{Pascal}\left(\frac{\mu}{-\ell n(1-\alpha)}, \frac{\alpha}{1-\alpha}\right),$$

   and Poisson(μ)-stopped $\left\{1-\theta+\theta\,\dfrac{\ell n(1-\alpha z)}{\ell n(1-\alpha)}\right\}$

   $$= \text{Pascal}\left(\frac{\mu\theta}{-\ell n(1-\alpha)}, \frac{\alpha}{1-\alpha}\right), \quad 0 < \theta \le 1.$$

c) Pascal($u\kappa,\rho$) mixed on u by Poisson

= Poisson-stopped Pascal
= Poisson-Pascal.

d) Binomial($u\nu,\pi$) mixed on u by Poisson

= Poisson-stopped binomial
= Poisson-binomial.

e) Pascal($u\kappa,\rho$) mixed on u by gamma

= gamma-stopped Pascal
= Pascal-stopped logarithmic

f) Binomial(ν,π) mixed on ν by Poisson

= Poisson-stopped binomial($1,\pi$).

2. Taking the geometric distribution with the p.g.f. $\chi z/(1-\pi z)$ show that this geometric-stopped Poisson(λ) distribution has the first and second moments

$$\lambda/\chi \quad \text{and} \quad \lambda^2(2-\chi)/\chi^2+\lambda/\chi.$$

(Pielou (1962)).

3. Write $S_N = X_1 + X_2 + \ldots + X_N$, as usual, where each X_i has the p.g.f. $f(v)$ and N the p.g.f. $g(u)$. Show that the joint p.g.f. of N, S_N is $g(uf(v))$. Obtain the familiar marginal p.g.f. of S_N. Verify that $\mathscr{C}(N,S_N = \mathscr{E}(X)\,\mathscr{V}(N)$.

4. Consider a model for accidents which supposes accidents occur only in 'spells', where the numbers of spells per period of of observation are Poisson(μ) and the numbers of accidents per spell are Poisson(ν). Then it is well known that the numbers of accidents per period are Neyman(μ,ν), thought of as a Poisson(μ)-stopped Poisson(ν) distribution (Cresswell and Froggatt's 'Long' distribution (1963)).

Suppose instead the numbers of accidents per spell are log-with-zeroes, with p.g.f.

$$\theta + (1-\theta)\,\frac{\ell n(1-\alpha z)}{\ell n(1-\alpha)} \quad .$$

As in 1.b) show that the corresponding Poisson-stopped distribution of accidents per period negative binomial $\left(\frac{\mu(1-\theta)}{-\ell n(1-\alpha)}, \frac{\alpha}{1-\alpha}\right)$.

Returning to the first paragraph, suppose in addition there can be accidents outside spells, and that these are Poisson(λ) distributed. Show that the p.g.f. of the numbers of accidents per period has p.g.f.

$$\exp(-\lambda+\lambda z-\mu+\mu e^{-\nu+\nu z})$$

(Cresswell and Froggatt's 'Short' distribution).

5. Cresswell and Froggatt (1963) define their 'Short' distribution as corresponding to the convolution of a Neyman and and an independent Poisson distribution.

a) Show that its p.g.f. has the form

$$\exp\{-\lambda+\lambda z-\mu+\mu \exp(-\nu+\nu z)\}, \quad \lambda,\mu,\nu > 0 \cdot$$

$$= \Sigma\, z^x\, p_x, \text{ say.}$$

b) Obtain the recurrence relation

$$(x+1)\, p_{x+1} = \lambda\, p_x + \mu\nu e^{-\nu} \sum_{i=0}^{x} p_{x-i} \frac{\nu^i}{i!}$$

with $p_0 = \exp(-\lambda-\mu+\mu e^{-\nu})$.

c) Obtain the factorial cumulants:

$$\kappa_{(r)} = \mu\nu^r \quad \text{with} \quad \kappa_{(1)} = \mu\nu+\lambda$$

whence the variance $\mu\nu(1+\nu)+\lambda$. (Kemp (1967)).

6. In the Poisson-stopped distribution

$$h(z) = \exp\{-\mu+\mu\, f(z)\},$$

show that

$$\kappa_{r+1:h} = \mu \cdot \kappa_{r+1:f} + \sum_{s=1}^{r} \binom{r}{s} \kappa_{s:h}\, \kappa_{r-s+1:f}$$

where the suffixes h and f have their obvious meanings.

7. In Gurland's (1958) generalized Poisson-stopped distribution

$$\exp\{-\mu+\mu f(z)\} = \sum p_x z^x ,$$

in which $f(z) = M(\alpha,\alpha+\beta,\nu(z-1)) = \sum f_x z^x$ is the confluent hypergeometric function, show that the standard recurrence relation

$$p_{x+1} = \frac{\mu}{x+1} \sum_{r=0}^{x} f_r p_{x-r} , \quad p_0 = e^{-\mu+\mu f_0}$$

can be used with

$$f_x = \frac{\nu^{x+1}}{x!} \cdot \frac{\alpha(\alpha+1)\ldots(\alpha+x)}{(\alpha+\beta)\ldots(\alpha+\beta+x)} \cdot M(\alpha+x+1,\ \alpha+\beta+x+1,-\nu)$$

$$= \frac{\nu+\alpha+\beta+x-1}{x} \cdot f_{x-1} - \nu\,\frac{\alpha+x-1}{x(x-1)} \cdot f_{x-2} .$$

8. Consider the binomial(ν,π)-stopped Poisson(μ) distribution, with p.g.f. $(\chi+\pi e^{-\mu+\mu z})^{\nu}$.

Obtain an explicit formula in terms of Stirling Numbers of the Second Kind for the probability function, and an expression for the general factorial cumulant.

Show that the distribution can equally be regarded as a mixed Poisson$(K\mu)$ distribution, mixed on K with binomial(ν,π) weights, and obtain the mean and variance by a conditional expectation argument based on this formulation.

9. Show that the "generalized Polya-Aeppli distribution", with p.g.f. $\exp\{-\mu+\mu z^{\kappa}(\gamma-\rho z)^{-\kappa}\}$, κ a positive integer, is a Poisson-stopped distribution where the summed variate is a Pascal(κ,ρ) variate plus κ (κ being a positive integer). Interpret the summed variate as the number of tosses required to achieve a specified number κ of heads in a sequence of independent Bernoulli trials in each of which the probability of a head is $1/\gamma$.

(This distinction, between a Poisson-Pascal and a generalized Polya-Aeppli distribution, is not maintained consistently in the literature, any more than is the distinction between a negative binomial and Pascal (κ a positive integer) distribution.)

10. Show that the following distributions with p.g.f.s of the form $g[f^*\{(1-z)\nu\}, \mu]$ have a maximum likelihood equation of the form

$$\Sigma\, F_x(x+1)\,\frac{\hat{p}_{x+1}}{\hat{p}_x} = \Sigma\, x\, F_x .$$

Poisson-hypergeometric:

$$f^*\{(1-z)\nu\} = \sum_{r=0}^{\infty} \binom{k}{r}(-1)^r\, \nu^r\, (1-z)^r .$$

General binomial-general binomial:

$$f^*\{(1-z)\nu\} = (1-\nu+\nu z)^k .$$

General binomial-Poisson:

$$f^*\{(1-z)\nu\} = e^{-\nu+\nu z} .$$

(Sprott (1965)).

11. For the Poisson distribution with $p_x = e^{-\lambda}\lambda^x/x!$ show that

$$\Pi_x \equiv (x+1)p_{x+1}/p_x = \lambda\ ,\ x = 0,1,2,\ldots;$$

and that $\Pi_x = \lambda$, a positive constant, together with non-negative p_x such that $\Sigma p_x = 1$ defines a Poisson(λ) distribution.

Note that, if $F_0, F_1, \ldots, F_x, \ldots$ are sample frequencies,

$$\tilde{\Pi}_x = (x+1)\, F_{x+1}/F_x\ ,\ x = 0,1,2,\ldots$$

are estimators of λ, and that

$$\Sigma\, F_x\, \tilde{\Pi}_x/\Sigma\, F_x = \overline{X}$$

is the sufficient unbiased estimator of λ.

Show that for the truncated distribution with

$$p'_x = \frac{\lambda^x/x!}{\sum_{y=\alpha}^{\beta} \lambda^y/y!}\ ,\ x = \alpha,\ \alpha+1, \ldots, \beta\ ,$$

the analogous $\Pi'_x = \Pi_x$ and $\tilde{\Pi}'_x = \tilde{\Pi}_x$ lead to the weighted estimator

$$\text{for } \lambda \text{ of } \frac{\sum_{x=\alpha}^{\beta-1} F_x \tilde{\Pi}_x}{\sum_{x=\alpha}^{\beta-1} F_x} = \frac{\overline{X}' - \alpha F_\alpha}{n' - F_\beta}$$

$\left(n' = \sum_\alpha^\beta F_x,\ \overline{X}' = \sum_\alpha^\beta x\, F_x / n'\right)$. Note the special cases with $\beta = \infty$, when the estimator is $\overline{X}' - \alpha F_\alpha / n'$, and with $\alpha = 0$, with estimator

$$\frac{\overline{X}'}{1 - F_\beta / n}.$$

(Gart (1970)).

12. Show that the associative law holds for mixing and stopping:

h-mixed with weight (g-mixed with weight f)
= (h-mixed with weight g)-mixed with weight f ;

h-stopped summed-(g-stopped summed-f)
= (h-stopped summed-g)-stopped summed f.

(Molenaar (1965)).

13. Considering the degenerate distribution with p.g.f. $g(z) = \sqrt{z}$ and the Bernoulli p.g.f. $f(z) = \chi + \pi z$ show that even if $g(z)$ and $f(z)$ are p.g.f.'s it is not necessarily true that $g(f(z))$ is a p.g.f.

14. i) Consider the bivariate (X,Y) with p.g.f.

$$\mathcal{E}(s^X t^Y) = g_{X,Y}(s,t),$$

and the independent N with p.g.f. $\mathcal{E}(r^N) = g_N(r)$.
Show that the p.g.f. of

$$W = X_1 + \ldots + X_N,\quad Z = Y_1 + \ldots + Y_N$$

(where X_i and Y_i are the random variables described above) is $g_N\{g_{X,Y}(s,t)\}$, an N-stopped (X,Y) p.g.f. Thus if (X,Y) has the bivariate Poisson p.g.f.

$$g_{X,Y}(s,t = \exp\{\alpha+\beta+\gamma(st-1)\}$$

and N is Poisson(λ), the joint p.g.f. of W,Z

$$\exp\{-\lambda+\lambda_{X,Y}(s,t)\} .$$

ii) Consider the bivariate M,N with p.g.f. $g_{M,N}(s,t)$ and X,Y independent variates with p.g.f.'s $g_X(r), g_Y(r)$. Then the p.g.f. of

$$W = X_1 + \ldots + X_N, \; Z = Y_1 + \ldots + Y_N$$

is

$$g_{M,N}\{g_X(s),\; g_Y(t)\}.$$

So for $g_{M,N}(s,t = g_{X,Y}(s,t)$ above and X,Y : Poisson(λ), Poisson(μ) the p.g.f. of W,Z is

$$\exp\{(e^{-\lambda+\lambda s} + \beta e^{-\mu+\mu t} + \gamma e^{-\lambda\mu+\lambda s+\mu t}\};$$

this could be called a (M,N)-stopped (X,Y) p.g.f.

15. A parasite lays an egg (with probability one) on a host at first encounter; on each "subsequent" encounter it lays a further egg with probability $\pi < 1$. Supposing the number of encounters in each fixed interval is Poisson(λ) distributed, show that the number of "subsequent" encounters has p.g.f.

$$e^{-\lambda}(e^{\lambda z} + z - 1)/z .$$

Hence the p.g.f. of the number of super-parasite eggs (i.e. eggs beyond the first) has p.g.f.

$$h(z) = e^{-\lambda}(e^{\lambda(\chi+\pi z)} - \pi + \pi z)/\chi+\pi z$$

and the total number of eggs has the p.g.f.

$$zh(z) + e^{-\lambda}/(1-z)$$

$$= \frac{ze^{-\lambda\pi+\lambda\pi z} + \chi e^{-\lambda}(1-z)}{\chi+\pi z}$$

$$= \Sigma\, p_x z^x \text{ , say.}$$

Deduce that

```
      ∇R←TOL LG AL;SUM;PX;I
[1]    ⍝RETURNS THE PROBABILITY VECTOR FOR LOGARITHMIC[PARA←AL],
       TO AS MANY TERMS AS NECESSARY FOR
[2]    ⍝  CUMULATIVE SUM > 1-10*-TOL , WITH A TAIL SUM APPENDED.
          PGF:    [LN(1-AL×Z)]÷LN[1-AL]
[3]    →((AL>0)∧AL<1)/LBL1
[4]    →ρ0ρ⎕←'AL  MUST BE BETWEEN 0 AND 1.  START AGAIN.'
[5]   LBL1:PX←,-AL÷⍟1-AL+I←0
[6]   LBL2:→((SUM←+/PX←PX,PX[I]×AL×I÷1+I←I+1)≤1-10*-TOL)/LBL2
[7]    R←PX,1-SUM
      ∇


      ∇R←TOL NB PARA;I;KA;RH;SUM;PX
[1]    ⍝RETURNS THE PROBABILITY VECTOR FOR NEGATIVE-BINOMIAL(PARA←KA,RH), TO
       AS MANY TERMS AS NECESSARY FOR
[2]    ⍝  CUMULATIVE SUM > 1-10*-TOL , WITH A TAIL SUM APPENDED.       PGF:
       [(1+RH)-RH×Z]*-KA
[3]    PX←,(1+RH←PARA[2])*-KA←PARA[1]+I←0
[4]   INC:→((SUM←+/PX←PX,PX[I]×(KA+I-1)×RH÷(I←I+1)×1+RH)≤1-10*-TOL)/INC
[5]    R←PX,1-SUM
      ∇
```

FIGURE 4 APL programs for calculating logarithmic(α) and negative-binomial(κ,ρ) probabilities.

$$p_0 = e^{-\lambda},$$

$$p_x = \frac{e^{-\lambda\pi}\pi^x}{(-\chi)^x}\left\{e^{-\lambda\chi} - \sum_{i=0}^{x-1}\frac{\lambda^i(-\chi)^i}{i!}\right\}, \quad x \geq 1,$$

and $$\mathscr{E}(X) = e^{-\lambda}\left\{\chi + \frac{\lambda\pi}{1-e^{-\lambda}}\right\}.$$

(Griffiths (1977).)

2.5 GENERALIZED POWER SERIES DISTRIBUTIONS

There is rather a dearth of useful general formulations relating to discrete distributions; perhaps the nearest approach to this generality as yet has been studied by Noack (1950), Patil (1961, and a series of later papers) and many subsequent authors.

If a variate X has as its sample space a countable set A of real numbers x for which the sequence $\alpha_x \geq 0$ has the g.f.

$$f(\theta) = \sum_{x \varepsilon A} \alpha_x \theta^x$$

and for which $\Pr(X=x) = \dfrac{\alpha_x \theta^x}{f(\theta)}$ for $x \varepsilon A$, X is said to have a generalized power series distribution (g.p.s.d.) with series function $f(\theta)$. Rather obviously only the case $f(\theta) > 0$ is of interest; the p.g.f. of X, also obviously, is

$$\frac{f(\theta z)}{f(\theta)}.$$

It is also clear that a truncated g.p.s.d. is also a g.p.s.d.

An example is the Poisson distribution with p.g.f.

$$e^{-\lambda+\lambda z}$$

$$= \frac{e^{\lambda z}}{e^{\lambda}}$$

so that for $\lambda = \theta$ the series function is e^{θ}. For the logarithmic distribution, $f(\theta) = -\ell n(1-\theta)$, $\theta = \alpha$.

Moments and cumulants are simply expressed in terms of the series function and its derivatives - for example, since the factorial moment g.f. is

$$\frac{f(\theta(1+t))}{f(\theta)},$$

the rth factorial moment is

$$\theta^r \frac{f^{(r)}(\theta)}{f(\theta)} .$$

There is a very simple recurrence relation for cumulants:

$$\kappa_{r+1} = \theta \frac{d\kappa_r}{d\theta} ,$$

which follows immediately from the Qvale relation given in the Appendix on Moments and Cumulants. For if $k_\theta(t)$ is the cumulant g.f.

$$k_\theta(t) = \ell n\ f(\theta e^t) - \ell n\ f(\theta)$$

and

$$\frac{\partial k}{\partial \theta} - \beta \frac{\partial k}{\partial t} = e^t(1 - \beta\theta) \frac{f'(\theta e^t)}{f(\theta e^t)} - \frac{f'(\theta)}{f(\theta)}$$

is independent of t if $\beta = 1/\theta$, the result quoted.

Some results on uniformly minimum variance unbiased estimators (u.m.v.u.e.) due to Roy and Mitra (1957) exhibit the attractive nature of the power series family

$$\Pr(X=x) = \frac{\alpha_x \theta^x}{f(\theta)} , \ x = 0,1,2,\ldots .$$

First, $\mathscr{E}\left(\frac{\alpha_{x-r}}{\alpha_x}\right) = \theta^r$ (where the ratio is taken to be 0 if $X < r$), and second, $T = \sum_{i=1}^{n} X_i$ is a complete sufficient estimator for θ. But T also has a power series distribution, with p.g.f.

$$\left\{\frac{f(\theta z)}{f(\theta)}\right\}^n = \frac{1}{\{f(\theta)\}} \sum_{t=0}^{\infty} c(t;n)\theta^t z^t , \text{ say,}$$

so $\Pr(T=t) = \dfrac{c(t;n)\theta^t}{\{f(\theta)\}^n}$ (where $c(t;n) = \Sigma \prod_{i=1}^{n} \alpha_{x_i}$ summed over $x_1+\ldots+x_n=t$). It then follows that

$$\mathscr{E}\left\{\frac{c(t-r;n)}{c(T;n)}\right\} = \theta^r \quad \text{(again, 0 if } T<r)$$

and from the Rao-Blackwell Theorem (e.g. Rao (1965)) this ratio is a u.m.v.u.e. of θ^r. In particular,

$$\begin{cases} \dfrac{c(T-1);n)}{c(T;n)} & , \quad T \geq 1 \\ 0 & , \quad T=0 \end{cases}$$

is the u.m.v.u.e. of θ, with the u.m.v.u.e. of its variance given by

$$\frac{c(T-1;n)}{c(T;n)}\left\{\frac{c(T-1;n)}{c(T;n)} - \frac{c(T-2;n)}{c(T-1;n)}\right\} .$$

For the negative binomial $(\gamma-\rho z)^{-\kappa}$ with κ known and $\theta=\rho=\gamma-1$, $f(\theta)=(1-\theta)^{-\kappa}$, and

$$c(t;n) = \frac{(\kappa n+t-1)!}{(\kappa n-1)!t!} ,$$

Hence the u.m.v.u.e. of p is

$$\frac{T}{\kappa n+T-1} ,$$

and the u.m.v.u.e. of its variance is

$$\frac{(\kappa n-1)T}{(\kappa n+T-1)^2(\kappa n+T-2)} .$$

These results also apply to the truncated case - thus, e.g., for the Poisson(λ) distribution truncated at $x=0$, $c(t;n) = \mathcal{S}_t^{(n)}$, and the u.m.v.u.e. of λ is $\mathbf{u}(T) = T\,\mathcal{S}_{T-1}^{(n)}/\mathcal{S}_T^{(n)}$, with the u.m.v.u.e. of its variance also given by

$$u(T)\{u(T) - u(T-1)\} .$$

It might also be noticed that for the family the Index of Cluster Size, $\kappa_{(2)}/\kappa_1$,

$$= \theta \cdot \frac{d^2 \ln f(\theta)}{d\theta^2}$$

and the Index of Cluster Frequency, $\kappa_1^2/\kappa_{(2)}$,

$$= \frac{1}{\frac{d}{d\theta}\left[\ln\left\{\frac{d\ln f(\theta)}{d\theta}\right\}\right]} .$$

Further results and references for the family appear in Johnson and Kotz (1969) and Gupta (1975).

As formulated, the g.p.s.d. is a one parameter family. It can be extended by considering the p.f.

$$\frac{\alpha_x(\phi)\ \theta^x}{f(\theta,\phi)} , \quad f(\theta,\phi) = \sum_A \alpha_x(\phi)\ \theta^x ,$$

where ϕ is a second parameter, and the p.g.f. is thus $f(\theta z,\phi)/f(\theta,\phi)$. (See also Khatri (1959).)

Examples are:

<u>Binomial</u> $$(\chi + \pi z)^\nu = \chi^\nu \left(1 + \frac{\pi}{\chi} z\right)^\nu$$

$$= \frac{(1+\theta z)^\phi}{(1+\theta)^\phi} \quad \text{for} \quad \theta = \frac{\pi}{\chi} , \ \phi = \nu$$

whence $f(\theta,\phi) = (1 + \theta)^\phi$.

<u>Negative binomial</u>

$$f(\theta,\phi) = (1 - \theta)^{-\phi} , \ \theta = \rho, \ \phi = \kappa.$$

(The 'inverse binomial', corresponding to

$\chi = \nu, \nu+1, \nu+2, \ldots$ has

$$f(\theta) = (\frac{\theta}{1-\theta})^{\nu}.)$$

Neyman

$$f(\theta,\phi) = \exp(-\phi + \phi e^{\theta})$$

where $\phi = \mu e^{-\nu}$,

$\theta = \nu$.

Stirling Distributions

First Kind $f(\theta,\phi) = \{-\ln(1-\theta)\}^{\phi}$, $\theta = \alpha$, $\phi = n$.

Second Kind $f(\theta,\phi) = (e^{\theta}-1)^{\phi}$, $\theta = \lambda$, $\phi = n$.

Polya Aeppli

$$f(\theta,\phi) = \exp \frac{\phi\theta}{1-\theta} , \quad \theta = \frac{\rho}{1+\rho} , \quad \phi = \frac{\mu}{1+\rho} .$$

Hyper-Poisson with parameters (θ,ϕ)

(See Bardwell and Crow (1964) for details)

$$f(\theta,\phi) = M(1,\theta;\phi) ,$$

M being the confluent hypergeometric function.

For these families, the earlier recurrence relation quoted for cumulants gives

$$\mathscr{E}(X) = \theta \frac{\partial \ln f(\theta,\phi)}{\partial\theta} \quad \text{and} \quad \mathscr{V}(X) = \theta \frac{\partial \mathscr{E}(X)}{\partial\theta} ,$$

and various results on estimation are available - see papers by G. P. Patil from 1961 onwards.

2.5a EXERCISES

1. If $f(\theta)$ is the series function of the generalized power series variate X for which the p.g.f. is $f(\theta z)/f(\theta)$,

show that

$$\mathcal{E}(X) = \frac{d \ln f(\theta)}{d \ln \theta} = \theta \frac{d}{d\theta} \ln f(\theta),$$

and $$\mathcal{V}(X) = \frac{d^2 \ln f(\theta)}{d(\ln \theta)^2} = \theta \frac{d}{d\theta} \mathcal{E}(X) .$$

By obtaining the p.g.f. of the convolution of n independent such variates $X_1, X_2, \ldots, X_n$, show that $T + X_1 + X_2 + \ldots + X_n$ also has a g.p.s.d. with series function $\{f(\theta)\}^n$, whence the conditional probability of $X_1, X_2, \ldots, X_n$ given that $X_1 + X_2 + \ldots X_n = t$ is independent of θ. Deduce that T is sufficient for θ, and that the moment and maximum likelihood estimators of θ are the same.

2. For the logarithmic distribution

$$\Pr(x=x) = \beta\alpha^x/x \ , \ x = 1,2,\ldots$$

where $0 < \alpha < 1$, $\beta = -1/\ln(1-\alpha)$, show that the m.l. estimator of α from a random sample $X_1, X_2, \ldots, X_n$ is given by the solution $\hat{\alpha}$ of

$$\bar{X} = \frac{\hat{\beta}\hat{\alpha}}{1-\hat{\alpha}} ,$$

with asymptotic variance

$$\frac{\alpha}{n \frac{d\mu}{d\alpha}} = \frac{\alpha^2}{n \ \mu_2(\alpha)} ,$$

where $\mu = \frac{\beta\alpha}{1-\alpha}$, and $\mu_2 = \mu[\frac{1}{1-\alpha} - \mu]$, the variance of the logarithmic distribution.

Using the methods set out in Patil (1962c) show that the bias $b(\hat{\alpha})$ is given to order $1/n$ by

$$b(\hat{\alpha}) = -\frac{1}{n} \cdot \frac{\alpha}{2} \cdot \frac{\mu_3 - \mu_2}{\mu_1^2} ,$$

where μ_3 is the third central moment of the logarithmic distribution.

3. For the generalized power series distribution

$$\Pr(X=x) = \frac{\alpha_x \theta^x}{f(\theta)}, \quad f(\theta) = \sum_{x \in A} \alpha_x \theta^x ,$$

show that for $\mu'_{(r)}$, the rth factorial moment,

$$\theta \frac{\partial \mu'_{(r)}}{\partial \theta} = \mu'_{(r+1)} + (r-\mu)\, \mu'_{(r)}, \quad \mu = \mu'_{(1)} .$$

Show by an induction that

$$\theta^s \frac{\partial^s \mu'_{(r)}}{\partial \theta^s} = \mathscr{E}\left\{ X^{(r)}\, g_s^{(X)} \right\}, \quad s = 0,1,2,\ldots,$$

where $g_s(x)$ is a polynomial of degree s in x.

(Hence orthogonal polynomials on $\Pr(X=x)$ may be found.)
(van Heerden and Gonin (1966)).

Chapter 3
CLUSTERING AND AGGREGATION

3.1 INTRODUCTION

When animals, plants, shops or individuals of some kind are dispersed over an area, one of the aspects of interest is the occurrence of "communities" (here called clusters); these may be formed because of social instincts, mode of egg laying, propagation of root systems, competition for nutriment or customers (perhaps 'negative' clustering), etc. Obviously a complete discussion of any specific application must make substantial reference to details of the individuals concerned, but a somewhat formal analysis in terms of general distributions may serve to expose common features more clearly than specific discussion with its special features. Neyman's (1939) analysis of larvae hatching from egg clusters began by pointing out that, if clustering is present, the detection of a larva suggests being in the neighborhood of a cluster, whence an increased likelihood of finding at least one more larva - thus the name contagious for distributions for which a derivation includes such clustering.

Much observational material for analysis comes from counts on quadrats. These are sub-areas (or volumes,...) of an experimental region on which observations of a specified character are made. The quadrats may be selected in a field, e.g., by an experimenter casting a hoop in the field, the sub-area included within the hoop being the quadrat; or the experimental region may be covered by a grid which defines the quadrat size, a random quadrat being selected by a coordinate system and a table of random numbers. But in some applications a quadrat may be constituted by a single plant or animal (e.g. the number of fleas on a rat) -

the term can be interpreted very widely. A distinction which often needs to be drawn is between quadrat sampling, when the quadrats (hopefully) can be thought of as random samples from those available from the experimental region, and quadrat censusing, when the quadrats cover the whole of the experimental region and contiguous quadrats are thus unlikely to be independent. Standard statistical inference procedures, as set out in the following sections, require randomness for their strict applicability and probabilistic interpretations. It is the responsibility of the user of such methods to ensure that the data to which the procedures are applied have been obtained in a way which at least appears not to have done violence to the assumptions - notably independence - built into the techniques.

Several examples of the kind of material giving rise to quadrat counts are provided below. Figure 5 is of a beech forest area in England in which leaf mould insects were studied (Lloyd (1967)); because of the practical difficulty of observation, the location within a quadrat of an individual is not possible, and indeed for mobile insects it is obvious that sampling must be highly time dependent. The second example is of plants in inland Australia, where Figures 6, 7 and 8 show an experimental area and analytical maps of it (Williams (1972)). It is clear that no small amount of observational effort must be invested to obtain frequency distributions of counts of individuals per quadrat, and that questions of judgement (not merely classification, unambiguous and entirely repeatable) are involved. The third example (Figure 9) is of the spatial distribution of certain types of shops in Ljublijana, Yugoslavia, showing how these differ with respect to type of goods sold, and other factors (Rogers (1969)). Finally, Figure 10 shows plant distributions in Michigan (Cain and Evans (1952)).

Some general discussion, with references, of sampling designed to detect clustering, and particularly against a physical background, will be found in Greig-Smith (1964), Lloyd (1967), Pielou (1969) and Rogers (1969, 1972). Although quadrat sampling is not the only sampling method appropriate it is the most widely used and will generally be assumed in what follows. Such sampling gives information rather well about the mean density of individuals - what else can be inferred? Commonsense suggests that an appropriate quadrat area might depend on the area of a cluster, and various suggestions have been made concerning this (e.g. Greig-Smith (1964)), an interesting recent one (Rogers and Gomar (1969)) being the maximization of the power of the classical chi-squared goodness of fit test of the distribution being considered (and therefore aginst some specified alternative distribution); there are many relevant papers in Statistical Ecology Vol. 1, The Pennsylvania State University, University Park (1971).

FIGURE 5 Beech grove in Wytham Woods near Oxford from which samples of leaf litter animals were collected from contiguous equilateral triangular quadrats with 15 foot sides. The stakes at the corners of the quadrats can be seen. (Lloyd, 1967)

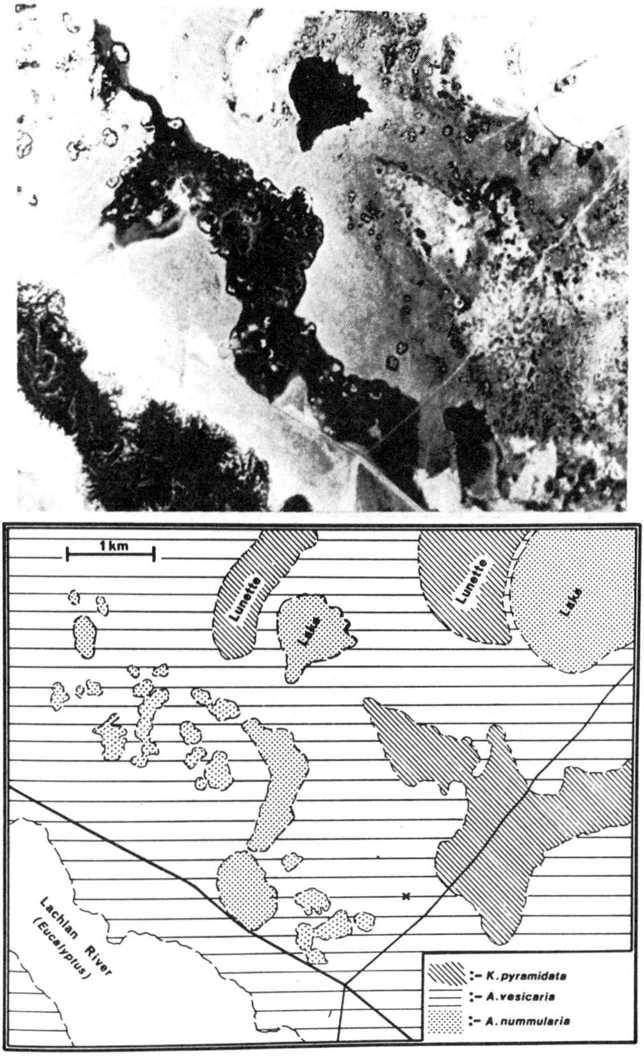

FIGURE 6 (*upper*) Aerial photograph of a section of the Riverina showing the sparse vegetation and irregularly flooded land. (*lower*) Map of the same area, with identification of various features, and × marking the location of the following two figures.
(Williams, 1972)

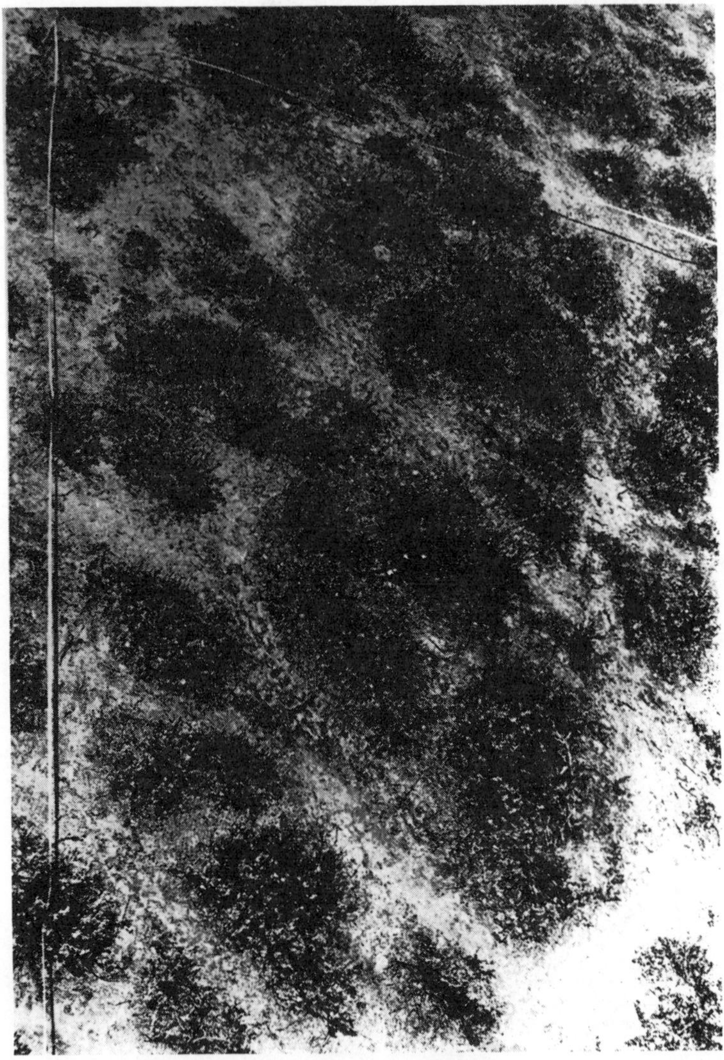

FIGURE 7 Oblique aerial photograph of part of a permanent study area for the examination of salt bush (*Atriplix vesicaria*). (Williams, 1972)

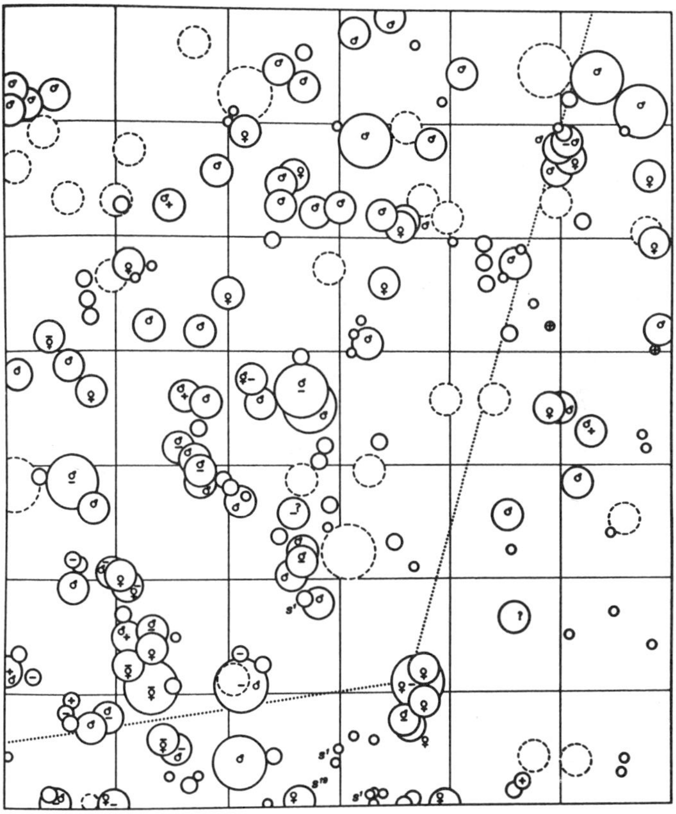

FIGURE 8 Map of part of the permanent study area - the lower and right boundaries of the photograph correspond to the dotted lines. The sizes of the circles correspond to classes of maturity of the plants, with dead plants shown by broken circles. Other symbols correspond to foliage rank and sex. (Williams, 1972)

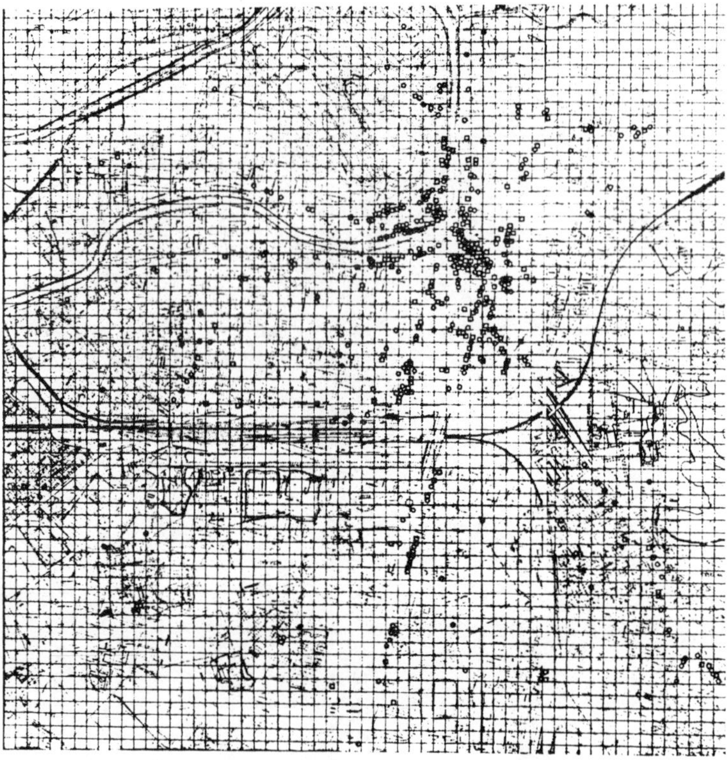

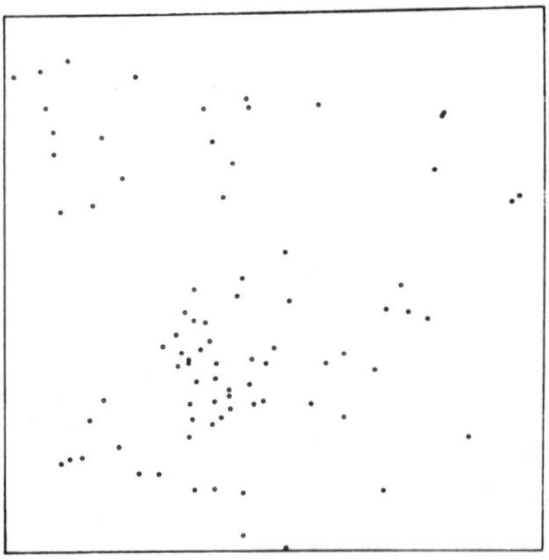

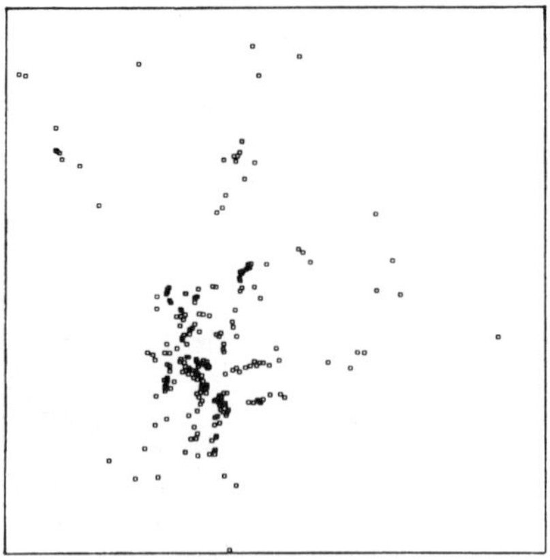

FIGURE 9 Maps of Ljubljana, Yugoslavia, showing the overall distribution of retail stores (*top*), the distribution of grocery stores (*bottom left*), and the distribution of non-food stores (*bottom right*). (Rogers, 1969)

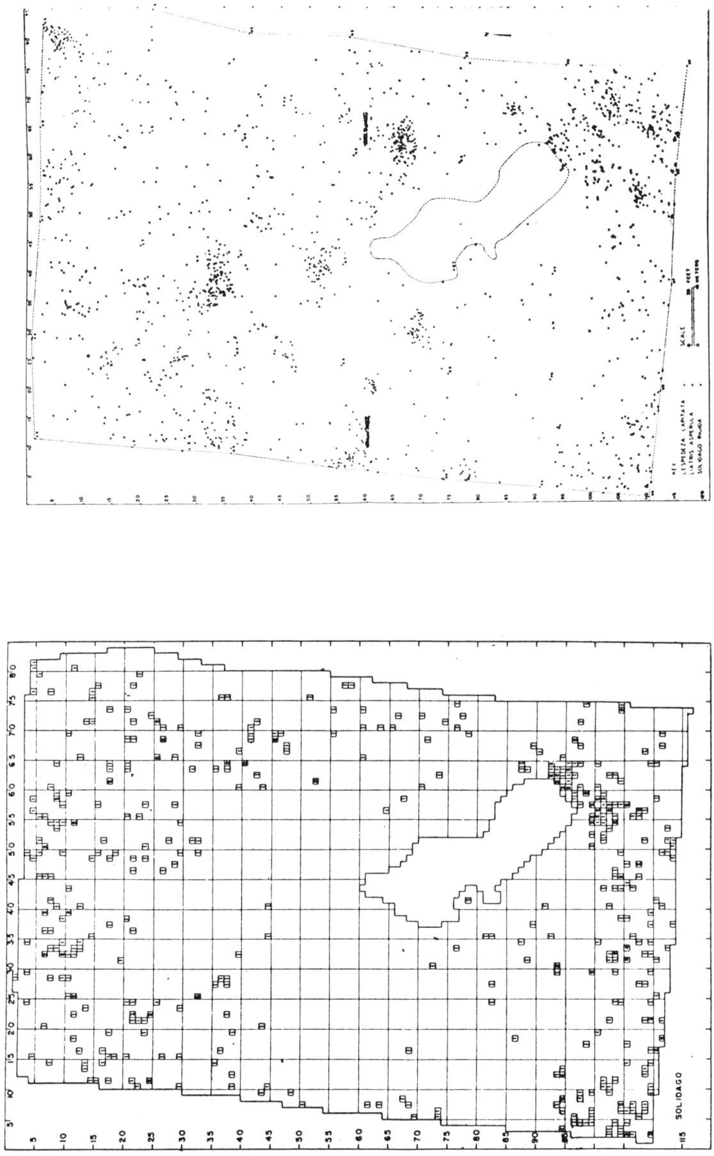

FIGURE 10 Maps of two acres in Michigan showing (*at left*) quadrats 1 m square and the numbers of golden rod (*Solidago rigida*) per quadrat and (*at right*) the distribution of three plant species. (Cain and Evans, 1952)

The following discussion deals with some special aspects of clustering against the general theme of contagious distributions, and so is much less than a comprehensive account. In particular, it is put in terms of means, and hence is necessarily a 'summary' account: a mean does not usually characterize a distribution.

Two measures associated with clustering when random quadrat sampling is supposed will be particularly discussed: the Index of Cluster Size and the Index of Cluster Frequency. If the random variable is the number of counts in a quadrat with $\kappa_{(2)}$ its second factorial cumulant and μ its mean, then:

$$\text{Index of Cluster Size} \quad \iota_{CS} = \frac{\kappa_{(2)}}{\mu} ,$$

$$\text{Index of Cluster Frequency} \quad \iota_{CF} = \frac{\mu^2}{\kappa_{(2)}} .$$

The purpose of introducing such measures is primarily analytic: how do they behave under modifications of environment? From the experimental aspect, how are they affected as the area of a quadrat is varied? Or as the density is increased of a predator which attacks only larvae (or eggs) in the case of insects laying eggs? Mathematically, how can they be related to specific families of distributions? What are the effects of modifying a distribution in a way intended to correspond to mortality? And how are such measures to be estimated from data? In this Chapter attention is primarily focussed on the mathematical questions, with some discussion of applications; but since many of the answers turn out to be highly distribution dependent they are dealt with further in the Chapters which relate to the appropriate distributions, and there are aspects already discussed in Chapter II.

3.2 INDEX OF CLUSTER SIZE

In the analysis of counts of individuals, "random" is very often equated to "Poisson", the primary reason perhaps being the kind of probabilistic derivation of the Poisson distribution given in Chapter I, in which each individual is thought of as "behaving" independently of all other individuals. For a Poisson distribution (though not uniquely so) the variance is equal to the mean, and so a possible measure of aggregation is the quotient

$$\frac{\text{variance}}{\text{mean}} ,$$

or equivalently (so that the reference standard, the Poisson, has a zero measure)

$$\frac{\text{variance}}{\text{mean}} - 1 = \frac{\kappa_{(2)}}{\kappa_1} = \iota_{CS} ,$$

often referred to as the over-dispersion, called the Index of Clumping by David and Moore (1954) and here called the Index of Cluster Size.

However, it is necessary to consider whether the Index performs in a satisfactory fashion when the Poisson distribution does not apply. An immediate interpretation for contagious distributions follows from the frequent appearance of the variance in such cases in the form

$$\text{mean}(1 + \text{positive term}):$$

this "positive term" is the Index already defined. In a more precise fashion, can the Index be related to the mean number of individuals in a cluster? (The term size is meant to refer to the number in, and not area of, a cluster.) For some common distributions the Index turns out to be as follows in Table 1.

If these are, where appropriate, interpreted as stopped distributions (with a convenient concrete background of random quadrat sampling of individuals in clusters with areas small compared with the quadrat area) Table 2 is obtained. It can be seen that the Index generally turns out to be closely related to but, except for appropriate structuring of the important cases of the Neyman and negative binomial distributions, not equal to the mean which was to have been hoped for, that of the number of individuals per cluster. However, as one would hope, since the Index is generally not a function of the mean of the stopping variate, the number of clusters per quadrat, it should be useful first as a descriptive measure of clustering and second for comparisons between sets of data because of its elimination of explicit reference to the stopping variate.

A warning might be given against interpreting (within this structure) a Poisson variate as an observation from a clustered distribution in which either the cluster size is 1 and the numbers of clusters Poisson distributed, or the cluster sizes are Poisson distributed and there is 1 cluster; a valid interpretation in terms of limiting results is suggested in the section on the Index of Cluster Frequency.

TABLE 1

Distribution	Index of Cluster Size $\iota_{CS} = \kappa_{(2)}/\kappa_1$
Poisson(λ)	0
binomial(ν,π)	$-\pi$
logarithmic(α)	$(1-\beta)\frac{\alpha}{1-\alpha}$, $\beta = -1/\ell n\ (1-\alpha)$
geometric(π)	π/χ
negative binomial(κ,ρ)	ρ
Poisson-binomial(μ,ν,π)	$(\nu-1)\pi$
Poisson-Pascal(μ,κ,ρ)	$(\kappa+1)\rho$
Polya-Aeppli(μ,ρ)	2ρ
Neyman(μ,ν)	ν
Thomas(μ,λ)	$\frac{\lambda(\lambda+2)}{\lambda+1}$
gamma(α,β)	$\beta - 1$
log-zero Poisson(δ,ρ,λ)	$\lambda(\gamma-\delta\rho)$

TABLE 2

Distribution	Index of Cluster Size	Mean of Summed Variate
negative binomial	ρ	-
Poisson($\kappa\ \ell n(1+\rho)$)-stopped logarithmic($\frac{\rho}{1+\rho}$)	$=\rho$	$\frac{\rho}{\ell n(1+\rho)}$
gamma(κ,1)-stopped Poisson(ρ)	$=\rho$	ρ
Poisson-binomial(μ,ν,π)	$(\nu-1)\pi$	-
Poisson(μ)-stopped binomial(ν,π)	$=\nu\pi-\pi$	$\nu\pi$
Poisson-Pascal(μ,κ,ρ)	$(\kappa+1)\rho$	-
Poisson(μ)-stopped Pascal(κ,ρ)	$=\kappa\rho+\rho$	$\kappa\rho$
Neyman(μ,ν)	ν	-
Poisson(μ)-stopped Poisson(ν)	$=\nu$	ν
Thomas(μ,λ)	$\lambda\,\frac{\lambda+2}{\lambda+1}$	-
Poisson(μ)-stopped {1 + Poisson(λ)}	$= \lambda+1 - \frac{1}{\lambda+1}$	$\lambda+1$
Poisson(μ)-stopped Poisson-Pascal(α,κ,ρ)	$\alpha\kappa\rho + (\kappa+1)\rho$	$\alpha\kappa\rho$
Log-zero-stopped Poisson(δ,ρ,λ)	$\lambda(\gamma-\delta\rho)$	λ

As has already been shown in the section on Stopped Distributions, for the general stopped variate with p.g.f. $h(z) = g(f(z))$,

$$\iota_{CS:h} = \mu_f + \iota_{CS:f} + \iota_{CS:g}\mu_f \; ,$$

or in terms of the variance-to-mean quotient $\tau = \kappa_2/\kappa_1$,

$$\tau_h = \tau_f + \mu_f \tau_g \; .$$

It follows that the Index will be a function only of f, corresponding to the distribution describing the numbers in clusters, if $\iota_{CS:g}$ (or $\kappa_{(2):g}$) is zero; further the Index will be exactly the mean of f if also $\iota_{CS:f} = 0$. Since this is only exceptionally so (though it can be for two popular compound distributions, the Neyman and negative binomial), it is perhaps more relevant to see in what circumstances the Index is close to μ_f. This is answered by the preceding Table for the cases listed - e.g. for the Thomas(μ,λ) distribution if λ is large (compared to 1). Of course for the special case of a Poisson-stopped distribution, for which $\iota_{CS:g} = 0$ (since all factorial cumulants except $\kappa_{(1)} = \kappa_1$ of a Poisson variate are zero),

$$\iota_{CS:h} = \mu_f + \iota_{CS:f} \; .$$

Thus the Index of Cluster Size for a Poisson-stopped variate exceeds the mean number in a cluster by the Index of Cluster Size for the cluster size distribution, a rather unexpected relation.

Alternatively, following a route suggested by

$$S_N = X_1 + X_2 + \ldots + X_N$$

in the notation of a previous Chapter, what has been sought is an expression for $\mathscr{E}(X)$ in terms of $\mathscr{E}(S_N)$ alone. (From the point of view of a data analyst, observations on S_N but neither on N nor X separately are available.) The familiar expression

$$\mathscr{E}(S_N) = \mathscr{E}(N)\,\mathscr{E}(X)$$

is available but is not enough because of the appearance of $\mathscr{E}(N)$. The next simplest relation is

$$\mathscr{V}(S_N) = \mathscr{E}(N)\,\mathscr{V}(X) + \mathscr{V}(N)\,\mathscr{E}^2(X) \; ,$$

but as well as involving $\mathcal{E}(N)$ this also introduces both $\mathcal{V}(X)$ and $\mathcal{V}(N)$ - consequently no universally applicable expression for $\mathcal{E}(X)$ can be obtained along these lines;

$$\frac{\mathcal{V}(S_N)}{\mathcal{E}(S_N)} = \mathcal{E}(X)\,\frac{\mathcal{V}(N)}{\mathcal{E}(N)} + \frac{\mathcal{V}(X)}{\mathcal{E}(X)}$$

is the same expression as has just been given in terms of Indexes of Cluster Size, and the same conclusions follow.

An interesting application, communicated to me by Dr. G. B. Hill, McMaster University, refers to the numbers of people killed per year by lightning strikes in England and Wales averaged over successive decades this century. A single strike may kill several people in a group: this corresponds to the cluster. It is probably reasonable to take the number of strikes per year ("quadrat") to be Poisson distributed with constant parameter μ, though it is possible this may have altered due to changes in atmospheric ionization, for example. Then for $h(z) = \exp\{-\mu+\mu f(z)\}$ describing the numbers killed where $f(z)$ describes the numbers in clusters,

$$\iota_{CS:h} = \mu_f + \iota_{CS:f}\ .$$

The observed values of $\iota_{CS:h}$, estimated by the sample values for the ratio (variance-mean)/mean, are as shown on the following graph. These suggest that after about 1940 there was a decrease in μ_f or $\iota_{CS:f}$ (or both), relating to the numbers of individuals killed per strike, and a possible explanation is that groups outdoors are on the average now smaller, for instance because of the introduction of television.

Returning to the general argument, for an "appropriate" choice of the stopping structure the Index of Cluster Size is simply related (though not necessarily equal) to the mean cluster size in the stopping structure. That the choice of the structure must be important has already become obvious - because a structure is not unique - and three examples are as follows.

i) Poisson(λ) = Poisson($\frac{\lambda}{\alpha}$)-stopped binomial(1,α). Here the Index is always 0, but the mean of the summed variate is the arbitrary α, $0 < \alpha < 1$.

ii) (a) Pascal(κ,ρ) = Poisson($\kappa\ \ell n(1+\rho)$)-stopped logarithmic($\frac{\rho}{1+\rho}$). This Index is ρ, but the mean of the

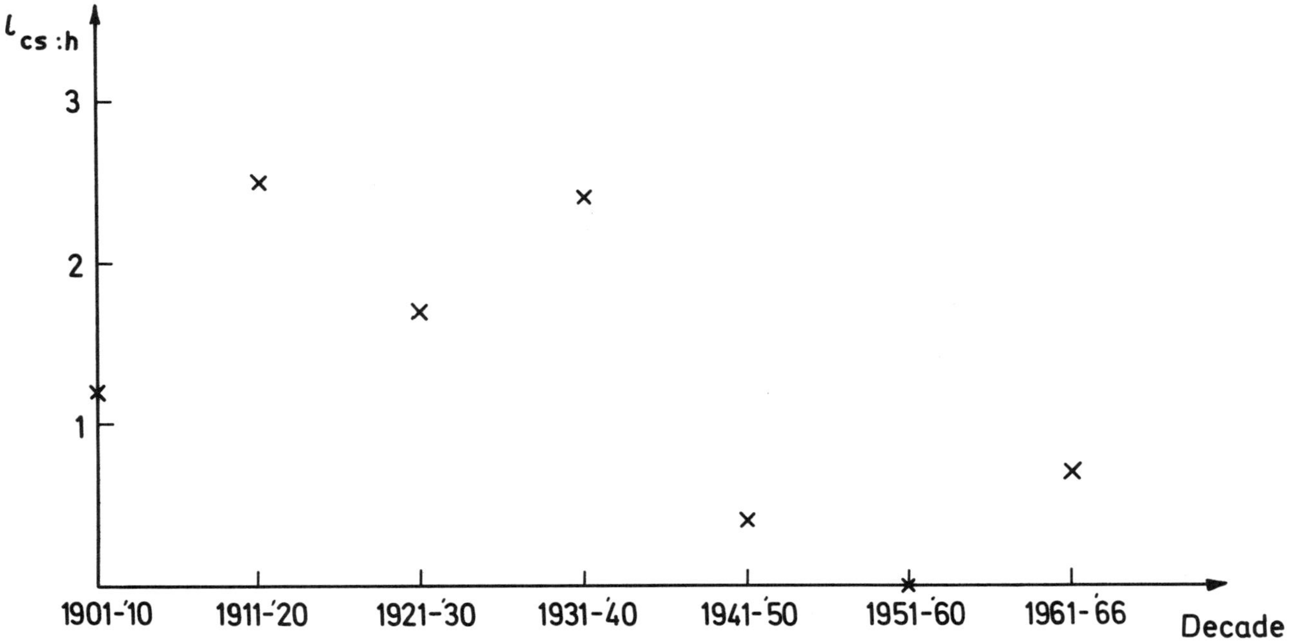

FIGURE 11 Deaths per year due to lightning strikes, England and Wales. (G. B. Hill)

summed variate is $\rho/\ln(1+\rho)$ - it is at least a function of ρ only. Put inversely,

Poisson(μ)-stopped logarithmic(α)

$$= \text{Pascal}\left(\frac{\mu}{-\ln(1-\alpha)}, \frac{\alpha}{1-\alpha}\right).$$

(b) Pascal(κ,ρ) = gamma(κ,1)-stopped Poisson(ρ). Both the Index and the mean of the summed variate are ρ. Put inversely,

$$\text{gamma}(\kappa,\beta)\text{-stopped Poisson}(\lambda) = \text{Pascal}(\kappa,\beta\lambda);$$

The Index is $\beta\lambda$ and the mean of the summed variate is λ, for arbitrary $\beta > 0$.

iii) (a) Neyman(μ,ν) = Poisson($\beta\mu$)-stopped "Poisson-zero($\frac{1}{\beta},\nu$)" where "Poisson-zero(θ,ν)" has p.g.f. $1-\theta+\theta e^{-\nu+\nu z}$. The Index is ν, but the mean of the summed variate is ν/β.

(b) Neyman(μ,ν) = Poisson(μ)-stopped Poisson(ν). Both Index and mean are ν.

In the anomalous examples, the parameters of the two component variates are either not separately specified or do not appear separately in the parameters of the stopped variate - this leads to fortuitous cancellations in the combination of the distributions and no straightforward general interpretation will hold. Unless the parametric specifications of the stopping and summed variates are (functionally) independent and lead to independent parameters (if more than one is involved) in the stopped distribution, no meaningful interpretation along the above lines can be given.

It is obviously of interest to examine the behaviour of any proposed Index of Cluster Size under simple modifications of the structure of a distribution. If the overall density were uniformly reduced, for example, two extreme kinds of behaviour are possible, one being that the Index is reduced in the same (absolute) way - because the density in the clusters has been so reduced - and the other that the Index does not alter - because relative to the overall density the density in the clusters is unaltered. Both kinds of behavior may be important in particular problems - they simply reflect different properties. (As well, of course, a different Index would refer to the density of clusters irrespective of their sizes.)

First, then, the p.g.f. of a variate when individuals are removed at random is investigated. To fix ideas, suppose Y with p.g.f. $f(z) = \mathscr{E}(z^Y)$ is the basic variate but that each individual is recognized as belonging to the population not with certainty but with probability θ. If X is the variate corresponding to this "recognized" population, an examination of the events $X = 0,1,2,\ldots$ in turn leads to

$$\Pr(X = x) = \sum_{y=0}^{\infty} \Pr(Y=y)\binom{y}{x}\theta^x(1-\theta)^{y-x}$$

and so to the p.g.f. of X as

$$\sum \Pr(Y=y)(1-\theta+\theta z)^y$$

$$= f(1-\theta+\theta z).$$

(A more sophisticated, and briefer, argument recognizes directly that if $\mathscr{E}(z^Y)$ is the p.g.f. of the original variate, then that of X is $\mathscr{E}\{(1-\theta+\theta z)^Y\}$. The modified variate is an f-mixture with a binomial weight function.)

This result shows immediately that the factorial moment g.f., obtained by writing $z = 1+t$, is

$$\mathscr{E}\{(1+\theta t)^Y\},$$

and this is the factorial moment g.f. of Y with t replaced by θt (Skellam (1952)). The same is therefore true for the factorial cumulant g.f., and hence the Index of Cluster Size $(\kappa_{(2)}/\kappa_1)$ for the modified variate is θ times that for the original variate - i.e., this Index is directly proportional to the overall density of individuals.

An application may be made to successive observations on an initially given population of insects say, where for observations at times $t = 0,t_1,t_2,\ldots$ the overall mean density is $\mu(t)$ and the Index of Cluster Size $\iota_{CS}(t)$. Supposing across the interval t_i,t_{i+1} the factor which gives the new density is θ_i, so

$$\mu(t_{i+1}) = \theta_i\,\mu(t_i), \quad i = 0,1,\ldots, \quad \text{with} \quad t_0 = 0,$$

and therefore also

$$\iota_{CS}(t_{i+1}) = \theta_i \ \iota_{CS}(t_i) \ .$$

Then $\iota_{CS}(t)$ may be plotted against $\mu(t)$, where t is a parameter in the mathematical rather than the statistical sense, and because of these relations the graph will be a straight line of unit gradient extending towards the origin if $\theta < 1$ from $(\mu(0), \iota_{CS}(0))$ as t increases (if $\theta > 1$, it extends away from the origin). On the other hand, if for example food supplies became exhausted in dense clusters leading to increased mortality or emigration from clusters out of the counting area, then the Index would decline more rapidly than for the random removal case; conversely, if isolated individuals are more subject to high mortality it will decline less slowly. This behaviour is illustrated below. (Perhaps these results are the basis for the "regression" procedure (Iwao and Kuno (1971)) briefly referred to in Figure 12 after the Index of Crowding is introduced.)

If a clustered structure $h(z) = g(f(z))$ is supposed, this of course leaves unaffected the result just obtained for removing individuals at random: it can also be regarded as removing individuals at random from clusters, with a new p.g.f. of $g(f(1-\theta+\theta z))$. However, with the clustered structure one may also considering removing <u>clusters</u> at random. By the previous argument, if θ is the probability that a cluster remains, the new p.g.f. is

$$h^{\dagger}(z) = g(1-\theta+\theta f(z)) \ ,$$

and hence

$$\iota_{CS:h^{\dagger}} = \iota_{CS:f} + \mu_f + \theta\mu_f \ \iota_{CS:g}$$

$$= \theta \ \iota_{CS:h} + (1-\theta)(\iota_{CS:f} + \mu_f)$$

The mean density for $h^{\dagger}$ is θ times the original mean density for h, but the Index of Cluster Size changes in the rather more complicated way shown as a result of its dependence on mean density. (See some further discussion under Index of Cluster Frequency.)

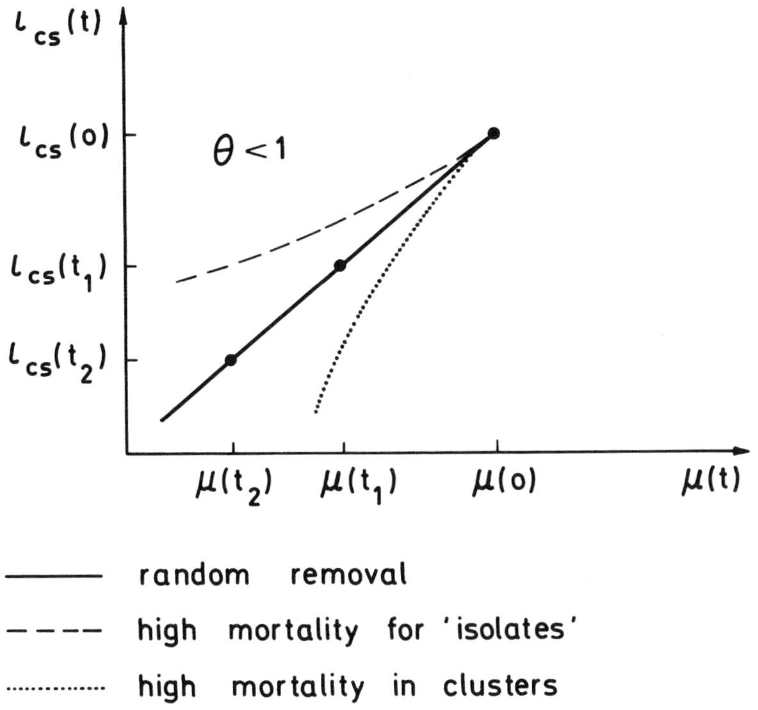

FIGURE 12 Index of Cluster Size and Mean Density.

3.3 INDEX OF CLUSTER FREQUENCY

Within the stopping structure, the second mean of importance is that of the stopping variate (of the numbers of clusters per quadrat with the quadrat sampling model already considered): it is this, formulated to apply more widely, which will be called the Index of Cluster Frequency. It is clearly no less a characteristic of the clustering than the Index of Cluster Size. Most directly,

(mean individuals per cluster) × (mean clusters per quadrat)

= mean individuals per quadrat,

i.e. $\iota_{CS} \times \iota_{CF}$ = overall mean. Introducing the earlier definition for ι_{CS},

$$\iota_{CF} = \frac{\kappa_1^2}{\kappa_{(2)}} = \frac{(\text{mean})^2}{\text{variance-mean}} ,$$

and it is immediate that this is μ for the Neyman(μ,ν), and κ for the negative binomial(κ,ρ), distribution, just as one would like.

If individuals are removed at random, this does not affect the numbers of clusters per quadrat (clusters of size zero are counted under the model); and the analysis carried out at the end of the last section has shown that as one would wish $\kappa_1^2/\kappa_{(2)}$ does not alter under such removal.

A rather more round about approach to the Index, which illuminates further the idea of aggregation and its quantitative expression, can be given by following an interpretation by Lloyd (1967). He proceeds by defining an Index of (Mean) Crowding in terms of the number of an individual's "neighbours in a quadrat": the mean number of neighbours in a quadrat standardized by division by the mean number of individuals in a quadrat (cf. "nearest neighbour" distance analyses). Since in a quadrat with X individuals each individual has $X-1$ neighbours, this is clearly

$$\text{Index of Crowding} = \frac{\mathscr{E}\{X(X-1)\}}{\mathscr{E}(X)} = \frac{\mu'_{(2)}}{\mu'_1} ,$$

and since $\mu'_{(2)} = \kappa_{(2)} + \mu'^2_1$ it follows that

Index of Crowding = Index of Cluster Size + Overall Mean.

Iwao and Kuno (1971) remark that, over a wide range of empirical and theoretical distributions, it is found that the Index of Crowding varies linearly with overall density:

$$\iota_C = \alpha + \beta\mu \ , \ \alpha, \beta \ \text{constants.}$$

Indeed, for a g-stopped f variate given by the p.g.f.s.

$$h(z) = g(f(z)),$$

the variance to mean ratio result is, in this notation,

$$\iota_{C:h} = \iota_{C:f} + \iota_{C:g}\,\mu_f$$

$$= \iota_{C:f} + \iota_{P:g}\,\mu_h \ ,$$

and so, for conditions in which the Index of Crowding ι_C for f and the Index of Patchiness ι_P for g are constant with varying overall density, such a linear relation holds. For empirical data where such assumptions hold, one might presumably estimate $\iota_{C:f}$ and $\iota_{P:g}$ by Least Squares ("regression"), perhaps on the grounds that the overall mean will be well, and the overall Index of Crowding poorly, determined; and Iwao and Kuno exhibit sets of data for which linearity appears to be a quite reasonable assumption in spite of theoretical qualms on the above grounds. (See also Taylor (1971), especially p. 370 and the discussion.)

Division of the Index Crowding by the overall mean produces a density independent Index, called by Lloyd the Index of Patchiness. (Some of his discussion of patchiness (e.g., p. 20) may underestimate the extent to which genuine randomness appears patchy - see, e.g., Feller I, Chap. III and Chap. VI. 7 Ex. (b), and Lloyd himself, p. 13.) For the distributions already listed these Indices are as shown in the following Table. The appearance of 1 in so many of the expressions for the Index of Patchiness suggests the alternative definition of an Index as

$$\frac{\kappa_{(2)}}{\kappa_1^2} \quad (= \text{Index of Patchiness} - 1),$$

which in turn suggests its reciprocal as being more meaningful dimensionally, this reciprocal being recognized as the Index of Cluster Frequency and listed as such in the Table.

TABLE 3

	ι_C Index of Crowding	ι_P Index of Patchiness	ι_{CF} Index of Cluster Frequency
Poisson(λ)	λ	1	-
binomial(ν,π)	$(\nu-1)\pi$	$1-\frac{1}{\nu}$	$-\nu$
logarithmic(α)	$\frac{\alpha}{1-\alpha}$	$-\ell n(1-\alpha)=\frac{1}{\beta}$	$\frac{\beta}{1-\beta}$
geometric(π)	$2\pi/\chi$	2	1
negative binomial(κ,ρ)	$(\kappa+1)\rho$	$1+\frac{1}{\kappa}$	κ
Poisson-binomial(μ,ν,π)	$(\mu\nu+\nu-1)\pi$	$1+\frac{\nu-1}{\mu\nu}$	$\mu\,\frac{\nu}{\nu-1}$
Poisson-Pascal(μ,κ,ρ)	$(\mu\kappa+\kappa+1)\rho$	$1+\frac{\kappa+1}{\kappa\mu}$	$\mu\,\frac{\kappa}{\kappa+1}$
Polya-Aeppli(μ,ρ)	$(\mu+2)\rho$	$1+\frac{2}{\mu}$	$\frac{1}{2}\mu$
Neyman(μ,ν)	$(\mu+1)\nu$	$1+\frac{1}{\mu}$	μ
Thomas(μ,λ)	$\frac{\lambda(\lambda+2)}{\lambda+1}+\mu(1+\lambda)$	$1+\frac{\lambda(\lambda+2)}{\mu(\lambda+1)^2}$	$\mu\,\frac{(\lambda+1)^2}{\lambda(\lambda+2)}=\frac{\mu}{1-1/(\lambda+1)^2}$
gamma(α,β)	$\alpha\beta+\beta-1$	$1+\frac{\beta-1}{\alpha\beta}$	$\alpha\,\frac{\beta}{\beta-1}$
log-zero Poisson(δ,ρ,λ)	$\gamma\lambda$	$\frac{1+\rho}{\delta\rho}$	$\frac{\delta\rho}{\gamma-\delta\rho}$

For the special case of the Poisson distribution the value of the Index is not defined (or is infinite): this corresponds to its introduction from $\iota_{CS}\,\iota_{CF} = \kappa_1$. A better appreciation of this follows from considering, for example, the Neyman(μ,ν) distribution, which is well approximated by the Poisson($\mu\nu$) distribution when ν is small: if the mean (number of individuals per quadrat) $\mu\nu$ is fixed, then for the mean (number of individuals per cluster) ν to tend to zero the mean (number of clusters per quadrat) μ must increase indefinitely.

For the g-stopped summed-f (or clustered) variate defined by $h(z) = g(f(z))$, the Index of Cluster Frequency $\iota_{CF:h}$ can be expressed in terms of its components as

$$\iota_{CF:h} = \mu_g \frac{1}{\dfrac{1 + \iota_{CF:f}}{\iota_{CF:f}} + \iota_{CS:g}} ,$$

or, since the Index of Patchiness $\iota_P = \dfrac{1 + \iota_{CF}}{\iota_{CF}}$,

$$\iota_{CF:h} = \mu_g \frac{1}{\iota_{P:f} + \iota_{CS:g}} .$$

For the Index of Crowding $\iota_C = \mu'_{(2)}/\mu$ the relation is rather simple, and is exactly the variance to mean ratio result

$$\tau_h = \tau_f + \tau_g\,\mu_f :$$

$$\iota_{C:h} = \iota_{C:f} + \iota_{C:g}\,\mu_f .$$

In the special case of a Poisson-stopped variate, $\iota_{CS:g} = 0$, and hence

$$\iota_{CF:h} = \frac{\mu_g}{\iota_{P:f}} = \mu_g \frac{\iota_{CF:f}}{1 + \iota_{CF:f}} ;$$

similarly $\iota_{C:h} = \mu_h + \iota_{C:f}$.

Returning to the general clustered variate, the earlier modifications of removing individuals and clusters at random can be examined. Removing an individual with probability $1-\theta$ gives the p.g.f. $h^*(z) = g(f(1-\theta+\theta z))$ and so, corresponding to the density invariant Index of Cluster Frequency,

$$\iota_{CF:h^*} = \iota_{CF:h} ;$$

while for the Index of Crowding,

$$\iota_{C:h^*} = \theta\, \iota_{C:h} .$$

Removing clusters at random with probability $1-\theta$ gives the p.g.f. $h^\dagger(z) = g(1-\theta+\theta f(z))$ whence

$$\iota_{C:h^\dagger} = \theta\, \iota_{C:h} + (1-\theta)\, \iota_{C:f}$$

(Iwao and Kuno (1971)), and

$$\iota_{CF:h^\dagger} = \frac{\theta\mu_g}{\theta\, \iota_{CS:g} + \iota_{P:f}} ,$$

compared with

$$\iota_{CF:h} = \frac{\mu_g}{\iota_{CS:g} + \iota_{P:f}} .$$

For completeness, though it is foreign to the line of argument in this Chapter, some interest attaches to the form taken by the Indices for the class of mixtures in general. Writing $h(z) = \int f(z;u\nu)dG(u)$ corresponding to the mixture variate X and weight variate U (so $G(u)$ is the cumulative distribution function of U),

$$\mu_h = \int \mu_{u\nu}\, dG(u) , \quad \text{where} \quad \mu_{u\nu} = f'(1,u\nu) ,$$

or $\mathscr{E}(X) = \mathscr{E}(\mu_{U\nu})$, and

$$\kappa_{(2):h} = \int \kappa_{(2):u\nu}\, dG + \int \mu_{u\nu}^2\, dG - \mu_h^2$$
$$= \int \iota_{CS:u\nu}\, \mu_{u\nu}\, dG + \mathscr{V}(\mu_{U\nu}) .$$

Hence

$$\iota_{CS:h} = \frac{\int \iota_{CS:u\nu}\, \mu_{u\nu}\, dG}{\int \mu_{u\nu}\, dG} + \frac{\mathscr{V}(\mu_{U\nu})}{\mathscr{E}(\mu_{U\nu})} ,$$

and $$\iota_{CF:h} = \frac{\mu_h^2}{\int \iota_{CS:u\nu}\mu_{u\nu}\, dG + \mathcal{V}(\mu_{U\nu})} .$$

If h corresponds to a mixed-Poisson distribution (so $h(z) = g(e^{-\nu+\nu z}))$ these results reduce to

$$\iota_{CS:h} = \nu + \nu\ \iota_{CS:g}$$

and $$\iota_{CF:h} = \frac{\mu_g}{1 + \iota_{CS:g}} = \frac{\mu_g\ \iota_{CF:g}}{\mu_g + \iota_{CF:g}} ,$$

the latter being independent of ν. The results are of course consistent with those obtained for stopped distributions (as special mixtures).

3.4 ESTIMATION AND THE INTERPRETATION OF ESTIMATED INDICES

Too literal an interpretation of the Indices of Cluster Size and Frequency ought not to be made; each has a descriptive background which corresponds to its name but, just as for other technical terms, the properties each possesses follow from its definition and not from its descriptive interpretation. Lloyd's (1967) extensive numerical analysis of the notion of Cluster Frequency (expressed in terms of Patchiness) depended heavily on the assumption of a negative binomial distribution (together with the appropriate estimation of its associated κ) and this simply does not apply if the negative binomial is not the correct model; as the above Tables show, even for Poisson-stopped distributions interpretations are not as direct as one might like. Further general discussion along similar lines is to be found in Pielou (1969).

In addition to this, as the remark above implies, the discussion of the previous pages concerning aggregation referred to distributions, or populations. Given samples, as is always the case in practice, the goodness of an estimation procedure depends vitally on the specific model (negative binomial, Neyman, etc.) which holds. It should also be recalled that $\iota_{CS}\ \iota_{CF} = \mu$ in general; while this is obviously necessary behavior in terms of the Indices' interpretations as means of Cluster Sizes and Frequencies it usually leads to rather inconvenient behaviour of estimators in that they tend to be very highly correlated. For the two common cases of the negative binomial and Neyman distributions it may be repeated that the interpretation in terms of parameters is precisely as one would wish - but the estimation problem remains.

Since in practice estimates have to be used for parametric values, what little is known from sampling experiments is worth examining, though a general discussion of estimation procedures as such will be avoided as far as possible (see the Appendix on Estimation, and sections on individual distributions): only intuitive ideas will be needed at this point.

An investigation by Pielou (1957), a brief report by Westman (1971), and some studies of Gleeson and Douglas (1975), are not encouraging for the direct interpretation of sample Indices, particularly in the cases of quadrat sampling and censusing. (Quadrat censusing is "sampling" with contiguous quadrats which cover the whole area being studied, in contrast to quadrat sampling in which the quadrats are selected at random from the area being studied: in the census case adjacent quadrats are clearly dependent, and the theory for estimation from random samples is obviously not exactly applicable. For direct sampling from theoretical distributions, such as the Neyman, see the appropriate sections in the chapters which follow.)

Pielou by hand, and Gleeson and Douglas by computer, placed cluster centres at "random" (with pairs of independent rectangular random numbers defining their Cartesian coordinates) over a rectangular area, with specified mean numbers of clusters per unit area, the actual "random" numbers being Poisson distributed with the specified mean μ (over a range of values). They then associated with each cluster a "random" number of individuals, "random" here meaning either with a Poisson distribution of mean ν (leading to a Neyman(μ,ν) distribution of the total number of individuals) or with "1 + a Poisson(ν-1) number of individuals" (leading to a Thomas(μ,ν-1) distribution for a total), with ν = 3 or 6. The individuals in a cluster were either concentrated at the cluster centre ("point clusters"), or placed randomly (again coordinate-wise) about the cluster centre with an average area of 0.12 square units ("compact clusters") or 0.48 square units ("diffuse clusters") per individual, all this with respect to a total rectangular area 24 x 24 = 576 square units. (These descriptions apply precisely to Gleeson and Douglas' work, but also correspond quite closely to Pielou's.) Seven different quadrat areas were used, all being square, ranging in size from 0.25 of a square unit to 36 square units; they were used both to carry out quadrat censusing and sampling of the area, the "censusing" covering the whole experimental area, and 16 randomly chosen (coordinate-wise) quadrats being used for the "sampling".

The results are complex, and only a brief summary can reasonably be given here. There was little difference, qualitatively at least, between results for the Neyman and Thomas cases. Nor was there, for the range of investigation, much difference between the "censusing" and "sampling" cases, one of the few encouraging outcomes. μ and ν were estimated by the two methods most commonly used by biologists (at least): moments, and mean and zero-class-frequency. (Work reported later indicates that the use of maximum-likelihood estimation would be unlikely to alter the qualitative conclusions.) The particularly unfortunate part of the picture is that μ was over-, and ν under-, estimated, frequently grossly. Rather more precisely, to illustrate the results consider a set of circumstances corresponding to increasing quadrat areas. The mean number of clusters μ in a quadrat is then proportional to the quadrat area, while of course ν, the mean number of individuals per cluster is fixed; it is therefore more convenient to think of μ/(quadrat area) = μ', say, over such variation since this is constant. But in simulations neither the estimates of μ' nor of ν were constant (allowing for random fluctuation): μ' was overestimated, and ν underestimated, in both cases rather less as the quadrat area increased, but in neither case did the effect become unimportant even when the quadrat area was as large as 50 times the expected cluster area (for compact clusters). Since one natural explanation of this under and over estimation is, for the same total number of individuals, the inclusion of part clusters - i.e. more clusters with fewer individuals per cluster - and this effect should certainly decline sharply at high ratios of quadrat to cluster area, it is clear that other factors are present. (Again, forward reference is made to the discussion of methods of estimation for specific distributions.)

It is possible that satisfactory empirical corrections might be put in terms of this observed under and over estimation by taking the estimates to be not of μ and ν but of $\phi\mu$ and $\theta\nu$ where $\phi \geq 1$ and $\theta \leq 1$. Both these factors will be functions of the ratio of quadrat/mean cluster area and of sample size (and of the method of estimation), but for repeated use in clearly defined situations simple recommendations might be practicable. Since in almost any likely situation the mean of the total number of observed individuals will be (well) estimated by the product of the above separate estimates, it is not necessary to introduce both "inflation" and "deflation" constants: their product must be unity and it will be enough to consider the estimates to be of

$$\mu/\theta \quad \text{and} \quad \theta\nu \ .$$

Useful analyses of data in terms of these may well be possible even though θ is unknown: over conditions in which an assumption of constancy of θ is plausible, comparisons of importance may be able to be made.

A practical illustration can be found in Beall (1940) who reported observations on European cornborers which included data on egg clusters, and on the larvae which later hatched from these clusters when of course their originating egg clusters were not identified but the larvae were unlikely to have changed plants. A simple model is the assumption of a Poisson(μ) distribution of egg clusters per plant (quadrat), a constant probability ψ of survival of a larva on the plant, and a Poisson(λ) distribution of eggs per egg cluster. This gives the p.g.f. of the numbers of egg clusters per plant as

$$e^{-\mu+\mu z}$$

and the numbers of larvae per quadrat as

$$\exp[-\mu + \mu \exp\{-\lambda + \lambda(1 - \psi + \psi z)\}]$$
$$= \exp(-\mu + \mu e^{-\nu+\nu z}) \quad \text{for} \quad \nu = \lambda\psi \ .$$

Beall found good agreement (measured by chi-squared goodness of fit) for the frequencies of egg clusters with the Poisson(μ) - but for the larvae with the Neyman Type C (or B) rather than with the Type A(μ,ν) distribution given by the above argument. Nevertheless, using the data for 1936 (in his Tables VII and VIII), μ, the mean number of clusters per plant common to the above models, is estimated from the egg cluster data as 0.86 and from the larvae data (by moments) as 0.73 - especially considering the relatively poor fit of the Neyman Type A distribution the agreement is almost surprisingly good. In his Table II (not III: the use of the smoothing described has obvious dangers), another set of data, for Treatment 2, gave moment estimates for μ,ν of 2.18, 1.45. On the above model this is interpreted as meaning there are a little more than 2 egg clusters laid per plant (on the average) and that about 1½ larvae per egg cluster are to be found on each plant about 4 months after hatching - if the text figure of approximately 20 eggs laid per egg cluster is accepted the mortality or migration rate is clearly (but not surprisingly) high: the "effective" cluster number, of Index of Cluster Size, is 1.45. But there are no internal or external checks available of these figures.

Corresponding work on retail shops by Rogers (1969) and Rogers and Gomar (1969)) has to be interpreted in the light of these sketchy results and their own useful remarks. Here censusing of an area is carried out, distributions fitted (e.g. Neyman(μ,ν)), and, using estimates of the parameters comparisons made between different areas and types of shop with respect to clustering (e.g. California versus Poland, and food versus non-food shops). This analytical use is certainly very important and likely to lead to more precision than a merely descriptive treatment, but it equally needs more support along its longer chain of mathematical modelling. It should be remarked, however, that dismissing such models (as Poisson(μ)-stopped Poisson(ν)) as being oversimplified - though they may be - on abstract grounds, such as "people's behaviour is more complex than this", is not defensible: real materials do not bend exactly in accordance with Hooké's Law, though it is not only useful but almost essential to suppose that within certain limits they do in order to design many engineering structures. However, the behaviour of estimators of μ and ν (or μ/θ and $\theta\nu$) with changing quadrat areas is unfortunately complex - thus, e.g., in Rogers and Gomar's (1969) Table 2 (where the number given for the quadrat size is its area in m^2: e.g. 50 m^2 means a square quadrat of area 2500 m^2) shows that the overall density of food stores is very nearly directly proportional to the quadrat area - as it should be. But the maximum likelihood estimators of the density of clusters on either a Poisson-stopped Poisson (=Neyman) or gamma-stopped Poisson (=Pascal) model do not increase linearly with quadrat area over the range of areas examined, nor consequently do those of mean numbers per cluster remain constant. However, it is to be recalled that censusing, not sampling, was carried out; that estimators of the two densities, of clusters and individuals per cluster, are usually very highly correlated; and that what is being estimated for cluster density might be thought of as μ/θ (for the Neyman model) or κ/θ (for the Pascal model), and $\theta\nu$ or $\theta\rho$ for mean individuals per cluster where θ also varies with quadrat area. (Curiously, both the estimates of μ/θ and $\theta\nu$, e.g., appear to be proportional to the side rather than the area of the quadrats. But the simple supposition that shop densities are related to street length rather than area is not supported by the overall density variation.)

It will be recalled that, in the simulation work already reported on, good estimation of "actual" densities of clusters and individuals per cluster with Neyman and Thomas models was not usually achieved with commonly used sample sizes and quadrat areas. As well, especially in instances such as above, interaction between clusters might assume substantial importance: whereas the simulation clusters which overlapped were merely taken thus, it is more than likely that at densities at which this happened frequently the interaction of such clusters would need to be taken explicitly into account.

It is clear that if a specific model, as the Neyman(μ,ν) distribution above, is assumed then the use of estimation procedures appropriate to it will give, e.g., estimated standard errors for estimators of the Index of Cluster Frequency μ/θ and of Cluster Size $\theta\nu$, the details depending on the procedure (moments, etc.) used. However, even if no specific model is assumed, consistent (but not generally unbiased) estimation can still be carried out from the sample moments or sample cumulants. (There is no obvious advantage in using sample k-statistics.) Use of the methods given in the Appendix under Approximate Means, Variances and Covariances will lead to approximate standard errors - unfortunately the use of these will require the employment of fourth order sample moments and so is relatively unattractive because of high sampling variability.

However, even if it is the case that only very large sample sizes and quadrat areas yield direct estimators of μ and ν, it is possible that such sampling may be practicable for automated procedures, though ordinary visual and mechanical quadrat counts are not normally effective at very high densities: large counts as a practical matter, tend to be unreliable. Anticipating the discussion of the Neyman distribution in the next Chapter it is interesting to notice that for this distribution (and others) the complexities of behaviour examplified by its multimodal character tend to be realized only for large parameter values - and these have been relatively uncommon in practical applications. Perhaps the more complex contagious distributions have for this reason additional areas of application beyond those immediately recognized.

Taking indications afforded by the simulation results, it may be that μ/θ and $\nu\theta$ are <u>really</u> the densities of interest: if one is concerned with the mean number of clusters (whether complete or not) which are represented on a quadrat, and the mean number of individuals per cluster (whether complete or not) represented on the quadrat, then this is so. It is then not the case that one compares observed frequencies of empty quadrats, quadrats with one representative,... with fitted frequencies using estimates of μ and ν: the estimates to be used are those of μ/θ and $\theta\nu$, and of course, in the absence of knowledge of θ, as a matter of fact this is what has always been done in the numerous applications in the literature.

Finally, in this section, formal moment - analogue estimators and their estimated covariance matrix will be briefly set out together with some numerical illustrations. For the observations $X_1,X_2,\ldots,X_n$, these being the total counts of individuals on n randomly chosen quadrats (or for $x = 0,1,2,\ldots$ the observations may be in the form of frequencies $F_0,F_1,F_2,\ldots$), these estimators are:

$$\text{for } \iota_{CS} : I_{CS} = \frac{K_{(2)}}{\overline{X}}$$

$$\text{for } \iota_{CF} : I_{CF} = \frac{\overline{X}^2}{K_{(2)}}$$

where $K_{(2)} = M_2' - \overline{X}^2 - \overline{X}$, $M_2' = \Sigma\, x^2\, F_x/n$.
($K_{(2)}$ is the sample factorial cumulant, and not the factorial k-statistic which is an unbiased estimator of $K_{(2)}$: it is unimportant which is used here since the sample Indices are not in any case unbiased for the corresponding population expressions in terms of cumulants.) Their asymptotic covariance matrix is obtained from Taylor expansions; e.g., for

$$Y = \overline{X}\ ,\ Z = M_2' \quad \text{and} \quad I_{CS} = f(Y,Z)$$

$$\mathcal{V}(I_{CS}) = f_Y^2\, \mathcal{V}(Y) + 2f_Y\, f_Z\, \mathcal{C}(Y,Z) + f_Z^2\, \mathcal{V}(Z)$$

to order $1/n$,

where $f_Y = \partial f/\partial Y$ is evaluated at $Y = \mathcal{E}(Y)$, $Z = \mathcal{E}(Z)$, etc.,
and $\mathcal{V}(\overline{X}) = \dfrac{\mathcal{V}(X)}{n}$, $\mathcal{V}(M_2') = \dfrac{\mu_4' - \mu_2'^2}{n}$, $\mathcal{C}(\overline{X},M_2') = \dfrac{\mu_3' - \mu_1' \cdot \mu_2'}{n}$.

(A matrix formulation is given in the following exercises; and there is a more detailed account of these procedures in the Appendix on Estimation.)

For two parameter distributions such as the negative binomial and Neyman, the moment-analogue estimators and those obtained by first estimating the parameters by moments and then substituting their values in the sample analogues of the distributional equivalents are obviously the same. Even here, however, the estimated standard errors will be different according to whether one continues this distributional assumption in order to substitute (e.g.) for μ_4' the expressions in terms of parameter estimates or whether one prefers to dispense with the specific distributional assumption and estimate μ_4' directly by the sample fourth moment M_4' . If a 3-parameter distribution such as the Poisson-Pascal is assumed, then even the moment-analogue, and moment, estimators are different.

Three "biological" numerical illustrations, with comments, are given of the estimation of the Indices: the comments mainly concern robustness (stability) across the theoretical distributions. For uniformity, maximum likelihood estimation is used throughout.

i) Beet web worm larvae counts on 325 quadrats: Beall (1940) p. 462 Table 1, Treatment 4, with fitting mostly by Katti and Rao (1965), Distribution 5.

The fit by maximum likelihood of each of the Neyman Type A, negative binomial, Poisson binomial with $\nu=2$ and 3, and log-zero-Poisson distributions is excellent, and the m.l. estimates of the Indices of Cluster Frequency and Size are very uniform.

	Neyman	Neg. bin.	P. bin. $\nu=2$	P. bin. $\nu=3$	1-0-P.
$\hat{\iota}_{CF}$	1.463	1.423	1.662	1.539	1.56
$\hat{\iota}_{CF}$	0.281	0.289	0.248	0.268	0.264

The moment-analogue estimates are also very close: $\overset{o}{\iota}_{CF} = 1.590$, $\overset{o}{\iota}_{CS} = 0.259$. (So good is the agreement that the caution against a direct interpretation of the Indices as mean numbers may bear repetition. Further, the error structure is worth noticing, since the estimated standard errors are not small and the correlations high. Thus, for example:

Moment-analogue	ι_{CF}	s.e.	0.61
		correl.	-0.97
	ι_{CS}	s.e.	0.10
Neyman m.l.	ι_{CF}	s.e.	0.56
		correl.	-0.97
	ι_{CS}	s.e.	0.11

ii) Plant counts, on Lespedeza Capitata, from 7640 quadrats, said to be heavily clumped (Cain and Evans (1952)). See Katti and Rao (1965) Distribution 25 for details of fitting.

In this case there are two good fits (marked *), with highly concordant m.l. estimates of the Indices (which also agree quite well with the moment estimates), but for the three poorly fitting distributions the estimates are widely discrepant among themselves. To save space only values of chi-squared and the corresponding degrees of freedom from goodness of fit tests are given here, the differences in the goodness of fit being very substantial.

	χ^2(df)	CF	CS
Moment-analogue		0.0608	1.756
M.L. estimates			
Neyman	91(4)	0.093	1.147
Neg. bin.	2.0(4)	0.062*	1.703*
P.-bin.(ν=2)	4257(4)	0.188	0.570
P.-bin.(ν=3)	824(4)	1.305	0.818
1.-0-P	2.2(3)	0.063*	1.68*

Perhaps one will not always wish to pay the price often exacted by a highly distribution-dependent method of estimation like maximum likelihood, and may wish instead to use a moment-analogue method in the common situation where the choice of a specific distribution is largely a matter of taste or prejudice. (The estimated error structure here is exemplified by:

Moment-analogue	ι_{CF}	s.e.	0.0062
		correl.	-0.87
	ι_{CS}	s.e.	0.21
Neyman m.l.	ι_{CF}	s.e.	0.0052
		correl.	-0.56
	ι_{CS}	s.e.	0.062

The estimation appears more reliable in this case - for which there is a total of 7640 observations, compared with 325 for the proceeding illustration.)

iii) One day industrial absences for 195 males over a year: Froggatt (1970) p. 35, Table 76.

Again, excellent fits of the Neyman Type A, negative binomial, Poisson-binomial with $\nu=2$ (Hermite) and "short" (convolution of Poisson and Neyman Type A) distributions were found, together with stable m.l. estimators of the Indices.

	Moment-Analogue	Maximum Likelihood			
		Neyman	Neg. bin.	P. bin. $\nu=2$	Short
ι_{CF}	4.577	4.080	4.014	5.24	4.31
ι_{CS}	0.611	0.685	0.696	0.533	0.649

(The estimated errors are, however, fairly high - about 1/4 of the estimates - with the usual high negative correlation.)

Mention is repeated of the lack here of detailed interpretation of these Indices with respect to the data and physical circumstances from which the estimates have been derived - this requires a more specific knowledge of the background of these data than I have. What is aimed at in the exhibition of these instances, not specially selected except for diversity of application, is to suggest that the reproducibility of the Indices under a variety of assumptions makes worthwhile their further study in such detail.

3.5 USE OF 'CONTAGIOUS DISTRIBUTION'

Popular usage has adopted "contagious" distribution in the general circumstances associated with clustering, in spite of some advocacy for the restriction of the term to a concept of contagion which requires a 'carrier'. For clustering, the simplest model is provided by

$$X_1 + X_2 + \ldots + X_N = S_N$$

where each independently identically distributed X_i has a distribution (within a cluster) and N has a further independent distribution (of the number of clusters in the sampling unit). Its p.g.f. has the form

$$g_N(g_X(z)),$$

in a self evident notation, and any distribution whose p.g.f. can be written in this form will here be called a contagious distribution. The description is, in these terms, too general since any p.g.f. can be obtained by taking $g_X(z) = z$ (corresponding to the degenerate distribution $\Pr(X=1) = 1$), and also too restrictive since it implies only distributions with p.g.f.'s can be contagious. However no refinement of the definition will be sought for the purpose of this work, in which the semi-popular usage of "more complex than the binomial or Poisson distributions" will be approached, partly because of the variety of structures, clustering being only one, which leads to the same distribution, so that although the label 'contagious' might be appropriate for the distribution it could be quite inappropriate for the physical structure for which the distribution may be an excellent model.

The original usage of Neyman (1939) was slightly different. As Feller (1943) pointed out, Neyman's structure, though expressed in terms of egg clusters, was essentially that of mixing, as was that of Greenwood and Yule (1920), and the so-called "apparent" contagion resulted from variability (or "inhomogeneity") of the cluster sizes. (For a fuller account of Neyman's argument, see the section in the Chapter on the Neyman Distribution, on the Neyman Family of Contagious Distributions.) This has sometimes been described as "merely an artifact of sampling", though indeed it is quite obvious that by suitably adjusting the method of sampling one can produce almost any sampling distribution from any given distribution: the sampling distribution (e.g. the Neyman distribution) is what it is under specified random sampling conditions. There is no inconsistency and nothing misleading in having the dual interpretation of the overall number of individuals T in a quadrat firstly as

$$T = X_1 + X_2 + \ldots + X_N \; : \; g_N(g_X(z)) \; ,$$

coming as above from clusters $1,2,\ldots,N$, and secondly as

$$T : \int_{\text{all } u} g_{T|U}(z|u) \; dG_U(u)$$

where this implies the selection of a cluster in a quadrat (with T individuals characterized by $U = u$ with probabilities specified by the cumulative distribution $G_U(u)$. The "contagion" can be interpreted in (not less than) these two ways, in each case in terms of probably being near a group when an individual is found (whence an increased chance of finding another individual), so that the appropriateness of the term is enhanced.

On the whole, only "simple" stopped distributions (and mixtures) will be discussed here, meaning those in which not more than one or two of these operations have been considered. (Obviously they may be applied repeatedly, and combined with other operations.) Partly this is because no new points of principle seem to be involved, partly to keep the complexity low (while great enough to be interesting), and partly because of the lack of uniqueness inherent in the models so that one becomes more uncertain of the appropriateness of the model as its complexity increases. The literature of discrete distributions seems especially full of rediscoveries of particular distributions, not always recognized as such, by different routes and with different parameterizations: some of these represent genuinely differing contributions, but many amount merely to a lack of recognition of earlier discussions in a different notation of the same problem.

If an exact correspondence is sought between the formal assumptions of a probability model, as for example independence of certain events, and some physical process, it is very rare indeed for these to be "exactly" satisfied (whatever this may mean): there is at the least usually some doubt as to how well they correspond, and verification is normally sought as to whether they correspond "nearly enough" by examining deductions from the model (i.e. predictions) and comparing these with observations. A backwards and forwards process is thus used if the predictions are not nearly enough verified, with a modification of the formal assumptions and perhaps further observations. (A somewhat fuller discussion may to be found in Chapter I of Koerts and Abrahamse (1969).) Since indeed a return to first principles in the formulation of the model very commonly refers to unobservables (like independence, or behaviour on a sub-atomic scale), there is something to be said for expressing the model in terms of distributions which are, or may be made, observable; and it is this rather than a strictly probabilistic approach which is used here. So, for example, the kind of treatment in Roach (1968) while undoubtedly applicable to many important problems is not particularly appropriate to the kind of complex aggregate (such as of grocery stores, or insect communities) which involves environmental and social influences; and some of the complexities which arise even in the simpler models - the cluster-satellite concept - when treated from this point of view may be found in Skellam (1958) and Warren (1971).

In the Chapters which follow, emphasis is placed on the estimation of parameters which are interpretable, generally in terms of means, against some biological background which in principle makes it possible to consider how meaningful particular estimates are, and how they change with sampling conditions. Unfortunately, authors only rarely record sufficient detail in the published literature to enable a subsequent reader to reconstruct the experimental circumstances with much completeness, so that the applications exhibited all too often do not achieve this aim.

3.6 EXERCISES

1. Where $S_n = X_1 + X_2 + \dots + X_N$, use the formulae for $\mathscr{E}(X_N)$ and $\mathscr{V}(X_N)$ in terms of means and variances of X_i and N to obtain

$$\iota_{CS:S_N} = \iota_{CS:X} + \mu_X + \iota_{CS:N}\,\mu_X .$$

2. For the sample Index of Cluster Size

$$I_{CS} = \frac{S^2 - \bar{X}}{\bar{X}}, \quad \text{where} \quad S^2 = n\,M_2/(n-1),$$

show that if X is Poisson(λ) then the Index is unbiased for 0, and if X is binomial(ν,π) the Index is nearly unbiased for χ.

(E.g. write $\varepsilon(\lambda) = \mathscr{E}\left(I_{CS}\right) = \sum_{\bar{x}} \frac{s^2 - \bar{x}}{\bar{x}} \prod_{i=1}^{n} p_{x_i}$.

If X is Poisson(λ)

$$\prod p_{x_i} = e^{-n\lambda} \frac{\lambda^{n\bar{x}}}{\prod x_i!}$$

and $\varepsilon(\lambda) = e^{-n\lambda} \sum \frac{s^2 - \bar{x}}{\bar{x}} \lambda^{n\bar{x}} / \prod x_i!$

so $\frac{d\varepsilon(\lambda)}{d\lambda} = -n\,\varepsilon(\lambda) + \frac{n}{\lambda} \mathscr{E}(S^2 - \bar{X}) = -n\,\varepsilon(\lambda)$.

Hence $\varepsilon(\lambda) = (\text{const.})\, e^{-n\lambda}$, and using $\iota_{CS} = 0$ for all λ, including $0+$, to support $\varepsilon(0+) = 0$, Const. $= 0$, whence $\varepsilon(\lambda) \equiv 0$. Ord (1970).)

3. If $\iota_{C:h} = \mu'_{(2):h}/\mu_h$ is the Index of Crowding for a distribution with p.g.f. $h(z) = g(f(z))$, $g(z)$ and $f(z)$ also

being p.g.f.s, prove that

$$\iota_{C:h} = \iota_{C:f} + \mu_f \iota_{C:g} .$$

4. For the variate obtained by adding 1 to a Poisson(λ) variate, derive the factorial cumulants:

$$\kappa_1 = \lambda+1, \kappa_{(r)} = (-1)^{r-1} (r-1)! \quad \text{for} \quad r > 1 .$$

Hence obtain its Index of Cluster Size as $-1/(\lambda+1)$, and so deduce from the general formula for a Poisson-stopped variate that the Index of Cluster Size for a Thomas(μ,λ) variate is $\lambda(\lambda+2)/(\lambda+1)$, a result independent of μ . Obtain the Index of Cluster Frequency (by division).

Observe that both Thomas Indices behave as "one would like" (i.e. are consistent with the interpretation as a stopped distribution) for large λ , but not for small λ.

5. Suppose the number of individuals $X = 0,1,2,...$ in a quadrat has the p.g.f. $f(z) = \Sigma f_x z^x$. If the probability of declaring an individual in the quadrat to be present (of recognizing it) is θ , and of declaring it to be absent (failing to recognize it) is $1-\theta$, it has been shown in the text that the p.g.f. is $f(1-\theta+\theta z)$.

If the probability of declaring an individual in the quadrat to be present is θ , and of declaring it to be 2 individuals is $1-\theta$, show that the p.g.f. is $f\{z(\theta+(1-\theta)z)\}$.

Obtain the means and factorial second moments for the two cases.

6. In most cases where the clustering structure

$$X = Y_1 + Y_2 + ... + Y_N ,$$

can be applied (with the Y_j identically independently distributed, and N independently distributed), a sample consists of $X_1,X_2,...,X_n$, observations on X but not on Y (nor N).

Suppose that observations can be made on each Y as well, so the data consist of

$$Y_{ij}, \; i = 1,...,n, \; j = 1,...,N_j ,$$

where $X_i = Y_{i1} + \ldots + Y_{iN_i}$, $i = 1,\ldots,n$.

Introducing the notations

$$\mu_Y = \mathscr{E}(Y_{ij}) \ , \quad \mu_N = \mathscr{E}(N_i) \ , \quad \mathscr{E}(X_i) = \mu_N \mu_Y$$

and σ_Y^2 , σ_N^2 , show that obvious estimators are:

$$\text{for } \mu_N : \bar{N} = \frac{\Sigma N_i}{n} \ ,$$

$$\text{for } \mu_Y : \tilde{Y} = \frac{\sum_{i=1}^{n} \sum_{j=1}^{N_i} Y_{ij}}{\Sigma N_i} = \frac{\Sigma X_i}{\Sigma N_i} \ .$$

Derive the following, where $A = \Sigma X_i$, $B = \Sigma N_i$:

$$\sigma_B^2 = \mathscr{V}(B) = n\,\sigma_N^2 \quad \text{so} \quad \mathscr{V}(\bar{N}) = \sigma_N^2/n \ ,$$

$$\sigma_A^2 = \mathscr{V}(A) = n\,\sigma_X^2 = n(\mu_N \sigma_Y^2 + \mu_Y^2\,\sigma_N^2) \ ,$$

$$\mathscr{C}(A,B) = n\,\mathscr{C}(N_i,X_i) = n\,\mu_Y \sigma_N^2 \ ,$$

$$\mathscr{V}\left(\frac{A}{B}\right) = \frac{\sigma_A^2}{\mu_B^2} - 2\,\frac{\mu_A\,\mathscr{C}(A,B)}{\mu_B^2} + \frac{\mu_A^2\,\sigma_B^2}{\mu_B^2} + O(1/n^2) \ ,$$

$$\text{so } \mathscr{V}(\tilde{Y}) = \frac{1}{n} \cdot \frac{\sigma_Y^2}{\mu_N} + O(1/n^2) \ , \quad \mathscr{C}(\bar{N},\tilde{Y}) = 0 + O(1/n^2) \ .$$

Given postulated models for the distribution of Y and N, it is often possible to estimate μ_Y and μ_N from observations on X only - see the exercises on the Neyman distribution for a comparison with the above.

7. The following is a matrix formulation of the covariance matrix for the moment-analogue estimators

$$I_{CS} = K_{(2)}/\bar{X} \quad \text{and} \quad I_{CF} = \bar{X}^2/K_{(2)}$$

of the Indices of Cluster Size and Frequency.

Writing M_1' , M_2' , M_3' , M_4' for the sample moments about the origin, if

$$\text{est. } \underset{\sim}{\mathcal{V}}(\underset{\sim}{M}') = \frac{1}{n} \begin{pmatrix} M_2' - {M_1'}^2 & M_3' - M_1' M_2' \\ M_3' - M_1' M_2' & M_4' - {M_2'}^2 \end{pmatrix},$$

$$\text{and } \underset{\sim}{J} = \begin{bmatrix} -\dfrac{1 + M_2'}{M_1'}, & \dfrac{1}{M_1'} \\ -\dfrac{I_{CF}^2 (M_1' - 2M_2')}{{M_1'}^3}, & -\dfrac{I_{CF}^2}{{M_1'}^2} \end{bmatrix},$$

$$\text{then est. } \underset{\sim}{\mathcal{V}}\begin{pmatrix} I_{CS} \\ I_{CF} \end{pmatrix} = \underset{\sim}{J} \, (\text{est. } \underset{\sim}{\mathcal{V}}(\underset{\sim}{M}')) \, \underset{\sim}{J}^t .$$

Chapter 4

THE NEYMAN DISTRIBUTION

4.1 INTRODUCTION

In a fundamental paper, Neyman (1939) set out an argument based on the dispersion of recently hatched larvae which led to a family of two parameter distributions called by him Type A, Type B, and Type C, together with a multiparameter generalization. (Some further detail is given in the section The Neyman Family of Contagious Distributions.) It was this paper, too, which introduced the term "contagious". However, in practice the two parameter Type A distribution, subsequently called almost universally the Neyman Type A distribution, has been used far more than any others of the family, and so throughout this book the description has been abbreviated to Neyman distribution except where other members of the families appear in the same context. The Thomas (1949) distribution, subsequently introduced, has been referred to as a Double Poisson distribution: if the term is to be used it ought surely to be applied to the Neyman distribution, however.

It has already been shown that the Neyman distribution with pgf $\exp\{-\mu+\mu \exp(-\nu+\nu z)\}$, $\mu, \nu > 0$, can arise either as:

(a) the sum of a random number of independent Poisson(ν) variates, the random number being independently Poisson(μ) , i.e. as a Poisson(μ)-stopped Poisson(ν) distribution;

or as

(b) the mixture of Poisson($k\nu$) distributions in the proportions given by a Poisson(μ) variate across the values $k = 0,1,2,\ldots$, i.e. as a mixed-Poisson($k\nu$) on k by Poisson(μ) distribution.

Of course more complicated chains of combination may lead to the same result: e.g. if a variate is binomial(ν,π) and ν is taken to be Poisson(α), the mixture thus obtained is Poisson($\alpha\pi$), and <u>this</u> Poisson distribution may itself be used as in (a) or (b).

One of the distribution's more striking properties is its possible multimodality, remarked on by Neyman (1939) as characteristic of much biological material: any number of modes may occur (Barton (1957), and Bowman and Shenton (1967), though if there are many they are necessarily not pronounced. With the models in mind, it is natural to expect these modes to occur approximately at multiples of ν.

The distribution is of course "reproductive" (from its derivation as either a mixed or stopped distribution): if $X_1, X_2, \ldots, X_m$ are independently Neyman distributed with parameters $(\mu_1,\nu), (\mu_2,\nu), \ldots, (\mu_m,\nu)$, then their convolution $X_1 + X_2 + \ldots + X_m$ is also Neyman distributed with parameters $(\Sigma\mu_i,\nu)$; this follows directly from the p.g.f.

4.2 FUNDAMENTAL PROPERTIES

a) The probability function. Either by expansion of the p.g.f., or perhaps more simply by thinking of the distribution as a mixture of Poisson($k\nu$) variates with weights on k given by a Poisson(μ) distribution,

$$\sum_{k=0}^{\infty} \left\{ e^{-k\nu} \frac{(k\nu)^x}{x!} \right\} e^{-\mu} \frac{\mu^k}{k!} ,$$

an explicit expression for the probability p_x is obtained as

$$p_x = \frac{e^{-\mu}\,\nu^x}{x!} \sum_{k=0}^{\infty} \frac{(\mu e^{-\nu})^k}{k} k^x , \quad x = 0,1,2,\ldots$$

$$\text{where } 0^x = \begin{cases} 1 \text{ for } x = 0 \\ 0 \text{ otherwise} . \end{cases}$$

Hence

$$p_0 = e^{-\mu+\lambda} , \quad \lambda = \mu e^{-\nu} ,$$

$$p_1 = e^{-\mu+\lambda} \lambda\nu ,$$

$$p_2 = e^{-\mu+\lambda} \frac{\nu^2}{2!} (\lambda+\lambda^2) ,$$

$$p_3 = e^{-\mu+\lambda} \frac{\nu^3}{3!} (\lambda+3\lambda^2+\lambda^3) \ , \quad \text{etc.,}$$

and in fact

$$p_x = e^{-\mu+\lambda} \frac{\nu^x}{x!} \sum_{n=1}^{x} \mathcal{S}_x^{(n)} \lambda^n \ ,$$

where $\mathcal{S}_x^{(n)}$ are Stirling Numbers of the Second Kind, a result first given by Cernuschi and Castagnetto (1946). The general result follows either from the formula given for general Poisson-stopped distributions in Chapter 3 or by using the exponential generating function for Stirling Numbers of the Second Kind, as follows.

$$\begin{aligned}
\exp(-\mu+\mu\, e^{-\nu+\nu z}) &= e^{-\mu+\lambda e^{\nu z}} \qquad \text{for} \quad \lambda = \mu e^{-\nu} \ , \\
&= e^{-\mu+\lambda} e^{\lambda(e^{\nu z}-1)} \\
&= e^{-\mu+\lambda} \sum_{n=0}^{\infty} \frac{\{\lambda(e^{\nu z}-1)\}^n}{n!} \\
&= e^{-\mu+\lambda} \sum_{n} \lambda^n \sum_{x=n}^{\infty} \mathcal{S}_x^{(n)} \frac{\nu^x z^x}{x!} \\
&= e^{-\mu+\lambda} \sum_{x=0}^{\infty} \frac{\nu^x}{x!} z^x \sum_{n=0}^{x} \mathcal{S}_x^{(n)} \lambda^n \ ,
\end{aligned}$$

and hence the coefficient of z^x is as stated.

However, since $\Sigma\, \mathcal{S}_x^{(n)} \lambda^n$ is the xth power moment about the origin, $\mu'_{x:\lambda}$, of a Poisson(λ) variate, p_x is a simple function of $\mu'_{x:\lambda}$:

$$p_x = p_0 \frac{\nu^x}{x!} \mu'_{x:\lambda} \ , \qquad x = 1,2,\ldots;$$

and a direct derivation of this result, avoiding the introduction of Stirling Numbers, is given a little below.

Recurrence formulae for p_{x+1} can be obtained: the common routine is to differentiate the p.g.f. $g(z) = \exp(-\mu+\mu e^{-\nu+\nu z})$ once and rearrange, and then to write $z=0$ after differentiating x further times, though a simpler procedure is to equate coefficients after a single differentiation. (Of course the general result for a Poisson-stopped variate may be applied directly.) Thus

$$g'(z) = \mu\nu e^{-\nu} e^{\nu z} g(z) ,$$

i.e. $\Sigma(x+1)p_{x+1}z^x = \mu\nu e^{-\nu} \Sigma \frac{\nu^n}{n!} z^n \Sigma p_q z^q$,

and equating coefficients of z^x (on the right $n+q = x$)

$$(x+1)p_{x+1} = \mu\nu e^{-\nu} \sum_{r=0}^{x} \frac{\nu^r}{r!} p_{x-r} , \quad x = 0,1,2,\ldots,$$
$$p_0 = \exp(-\mu+\mu e^{-\nu}) .$$

For numerical computation this recurrence relation is commonly used and is very convenient for computer application, though its length varies directly as x. (For hand computation, the preparation of a slip of paper carrying the numerical values of $1,\nu,\nu^2/2!,\nu^3/3!,\ldots$ is worthwhile: this can be moved so that in turn

1 is against p_0 ;

1 is against p_1 and ν against p_0 ;

1 is against p_2, ν against p_1 and $\nu^2/2!$ against p_0; etc.

Alternatively, if tables of the Poisson(ν) distribution are conveniently available the same procedure can be applied to

$$(x+1)\ p_{x+1} = \mu\nu\ \Sigma\ \pi_r p_{x-r} ,$$

π_r being the rth Poisson probability.) However, putting $\lambda = \mu e^{-\nu}$ in the p.g.f. enables it to be written as

$$p_0 \cdot e^{-\lambda+\lambda e^{\zeta}} , \quad \text{with } \zeta = \nu z .$$

But this second factor is the moment g.f. (the coefficients of $\zeta^x/x!$ are the moments about the origin) of a Poisson variate with parameter λ : writing the xth moment about the origin of this Poisson variate as $\mu'_{x:\lambda}$, then

$$p_x = p_0 \frac{\nu^x}{x!} \mu'_{x:\lambda} , \quad x = 1,2,\ldots,$$

expressing the probabilities in terms of moments of an associated Poisson variate. Tables of these - or more precisely of $p_x^* = \mu'_{x+1:\lambda}/\mu'_{x:\lambda}$ for which

$$p_{x+1} = \nu \frac{p_x}{x+1} p_x^*, \quad x = 0,1,2,\ldots$$

- are available (Douglas (1955) where p_x^* is written p_x), but are of course only worth using for hand computation (see under Maximum Likelihood Estimation for the ranges of tabulation). Graphs are exhibited in the section on Modality.

b) Moments and cumulants. An obvious simplification is obtained by replacing z in the p.g.f. by $1+t$ and taking logarithms: the factorial cumulant g.f. is thus

$$-\mu + \mu e^{\nu t}$$

and expanding in powers of t gives the rth factorial cumulant as

$$\kappa_{(r)} = \mu\nu^r, \quad r = 1,2,3,\ldots .$$

The mean and variance are therefore

$$\mathcal{E}(X) = \mu\nu$$

and $$\mathcal{V}(X) = \mu\nu(1+\nu)$$

respectively; the third and fourth moments about the mean are $\mu\nu(1+3\nu+\nu^2)$ and $\mu\nu(1+7\nu+6\nu^2+3\mu\nu(1+\nu)^2)$ respectively.

However, the same argument as was used on the p.g.f. for explicit formulae for p_x also gives the cumulants, and the factorial moments. For cumulants, the g.f. is

$$-\mu + \mu e^{-\nu+\nu e^t} = \mu e^{\nu(e^t-1)} - \mu$$

$$= \mu \sum \frac{\nu^n(e^t-1)^n}{n!} - \mu$$

$$= \mu\left(\sum_{r=0}^{\infty} \frac{t^r}{r!} \sum_{n=0}^{r} \mathcal{S}_r^{(n)} \nu^n - 1\right)$$

$$= \mu \sum_{r=1}^{\infty} \frac{t^r}{r!} \sum_{n=1}^{r} \mathcal{S}_r^{(n)} \nu^n ;$$

i.e. the rth cumulant is $\kappa_r = \mu\Sigma \mathfrak{S}_r^{(n)} \nu^n$, $r = 1,2,3,\ldots$.

A similar argument gives for the rth factorial moment about the origin $\mu'_{(r)} = \nu^r \Sigma \mathfrak{S}_r^{(n)} \mu^n$. (These two results of course essentially follow from the known expression of Poisson moments in this form, and the general theorem for mixed Poissons relating power moments (cumulants) of the mixing distributions and factorial moments (cumulants) of the mixture.)

Johnson and Kotz (1969) point out that, where $\gamma_4 = \kappa_4/\kappa_2^2$ and $\gamma_3 = \kappa_3/\kappa_2^{3/2}$, the ratio

$$\frac{\gamma_4}{\gamma_3^2} = \frac{(1+7\nu+6\nu^2+\nu^3)(1+\nu)}{(1+3\nu+\nu^2)^2}$$

is not a function of μ and is in addition remarkably stable for variations in ν , tending to 1 as ν tends to 0 or ∞ with a maximum almost 1.2149 near $\nu = 0.5099$. To the extent that γ_4 is related to kurtosis and γ_3 to skewness this implies a restriction on possible shapes of the distribution. Working in terms of factorial rather than power cumulants would lead to

$$\gamma_{(4)}/\gamma_{(3)}^2 = 1$$

($\gamma_{(4)} = \kappa_{(4)}/\kappa_{(2)}^2$, $\gamma_{(3)} = \kappa_{(3)}/\kappa_{(2)}^{3/2}$), which is a function of neither μ nor ν . However, a previous analysis (Anscombe (1950)) noted that if one writes the first factorial cumulant (or mean) in the form $\alpha\beta$ and the second factorial cumulant in the form $\alpha\beta$ (or the variance as $\alpha(\beta+1)$) then $\kappa_{(3)}/\alpha\beta^3$ may be called the "skewness" and $\kappa_{(4)}/\alpha\beta^4$ the "kurtosis". Then for the Neyman(μ,ν) distribution the "skewness" and "kurtosis" are both unity. It is difficult to interpret any of these combinations in an absolute sense, though comparisons across distributions can be illuminating.

The relative length of tail of a distribution is of interest when considering the appropriateness of its use to represent data, and the possible multimodality of the Neyman distribution might suggest that its tail is very long: that the probabilities fall off very slowly for large x. Although this is certainly true in comparison with the Poisson distribution, Anscombe (1950) suggested that in comparison with the negative binomial this was not so, from examining standardized third and fourth order factorial cumulants for a number of two parameter discrete distributions.

Writing the mean in the form $\alpha\beta$ and the variance as $\alpha\beta(1+\beta)$, with third and fourth factorial cumulants $\kappa_{(3)}$ and $\kappa_{(4)}$, the following Table 4 is found in which some additional distributions are included for later reference (see also Katti and Gurland (1961)). Bowman and Shenton (1967) confirmed some of the conjectures suggested by this table by computing successive probabilities p_x until $p_x < 10^{-20}$ (approximately): e.g., with $\alpha = 4$, $\beta = 5$, for the Poisson (with mean $\alpha\beta$) this occurred at $x = 75$, for the Neyman Type A at $x = 207$, and for the negative binomial at $x = 298$; for $\alpha = 2$, $\beta = 1$ at $x = 27$, 54 and 71 respectively.

A family of distributions with extremely long tails is given by Irwin (1965).

Two recurrence relations for cumulants, due to Shenton (1949), can be obtained by direct differentiation of the cumulant g.f.

i) Thus, $k'(t) = \nu\{\mu + k(t)\}e^t$.

Differentiating r times further (or equating coefficients of $t^r/r!$),

$$k^{(r+1)}(t) = \nu \sum_{s=0}^{r} \binom{r}{s} k^{(s)}(t)\, e^t + \mu\nu e^t$$

and writing $t = 0$, when $k^{(r)}(0) = \kappa_r$,

$$\kappa_{r+1} = \nu\left(\sum_{s=1}^{r} \binom{r}{s} \kappa_s + \mu\right), \quad r = 1,2,\ldots, \kappa_1 = \mu\nu .$$

ii) $$\frac{\partial k(t)}{\partial\nu} = \mu \exp(-\nu + \nu e^t)(-1 + e^t)$$

$$= -(k(t) + \mu) + \frac{k'(t)}{\nu} .$$

Hence, equating coefficients of $t^r/r!$,

$$\frac{\partial\kappa_r}{\partial\nu} = -\kappa_r + \frac{1}{\nu}\kappa_{r+1}$$

i.e. $$\kappa_{r+1} = \nu\kappa_r + \frac{\partial\kappa_r}{\partial\nu} , \quad r = 1,2,\ldots,\kappa_1 = \mu\nu .$$

TABLE 4

Distribution	"Skewness" $\kappa_{(3)}/\alpha\beta^3$	"Kurtosis" $\kappa_{(4)}/\alpha\beta^4$
Poisson	$\kappa_{(3)} = 0$	$\kappa_{(4)} = 0$
Neyman Type A	1	1
Negative Binomial	2	6
Thomas	$\frac{3}{4} + \frac{1}{8\beta} + 0(\beta)$	$\frac{1}{2} + 0(\beta)$
Neyman Type B	9/8	27/20
Neyman Type C	6/5	8/5
Poisson-Binomial	(0,1)	(0,1)
Poisson-Pascal	(1,2)	(1,6)
Log-zero-Poisson	$(-\infty, \infty)$	$(-3, \infty)$

Of course these results also follow by direct manipulation of the Stirling Number expressions; see also the Appendix on Moments and Cumulants.

c) Modality. In 1939 Neyman pointed out the far from unusual occurrence of multimodality in observed frequency distributions of biological origin, and that the Neyman distribution was capable of multimodality.

The occurrence of modes is naturally determined by the values of μ and ν, but it has not proved easy to obtain a specification of the parameter plane in which there are 1,2,3,... modes. Barton's (1957) investigation remains the most complete although its arguments are rather piecemeal. There is no limit to the number of modes which may occur, very roughly at integral multiples of ν (though perhaps with gaps - e.g. for $\nu = 16$,

$\mu = 2$	gives modes at	0, 16, 32, 46
5		0, 16, 33, 51, 67
6		0, 16, 34, 54
7		0, 16, 102
20		0, 311).

Further calculations on the problem, from which the above figures are taken, are given by Bowman and Shenton (1967); both Barton and Bowman and Shenton, give diagrams of the parameter plane showing the approximate boundaries of multimodal regions, the former being in terms of a transformation of the parameters and the latter not distinguishing between a mode at the origin (a half mode) and elsewhere.

It can be seen that the occurrence of a half mode is rather common, and an inspection of the examples of the graphs of the p.f. shows that this can be very striking: a casual interpretation of data corresponding, say, to $\mu = 2$, $\nu = 12\frac{1}{2}$ would lead to a fallacious belief in a distribution with "added zeroes" (or of counting equipment which mal-functions at zero counts). Thinking e.g. of the stopped model

$$X_1 + X_2 + \ldots + X_N$$

with N : Poisson(μ) and X_i : Poisson(ν) leads plausibly to the observed pattern of modes approximately at multiples of ν (as an

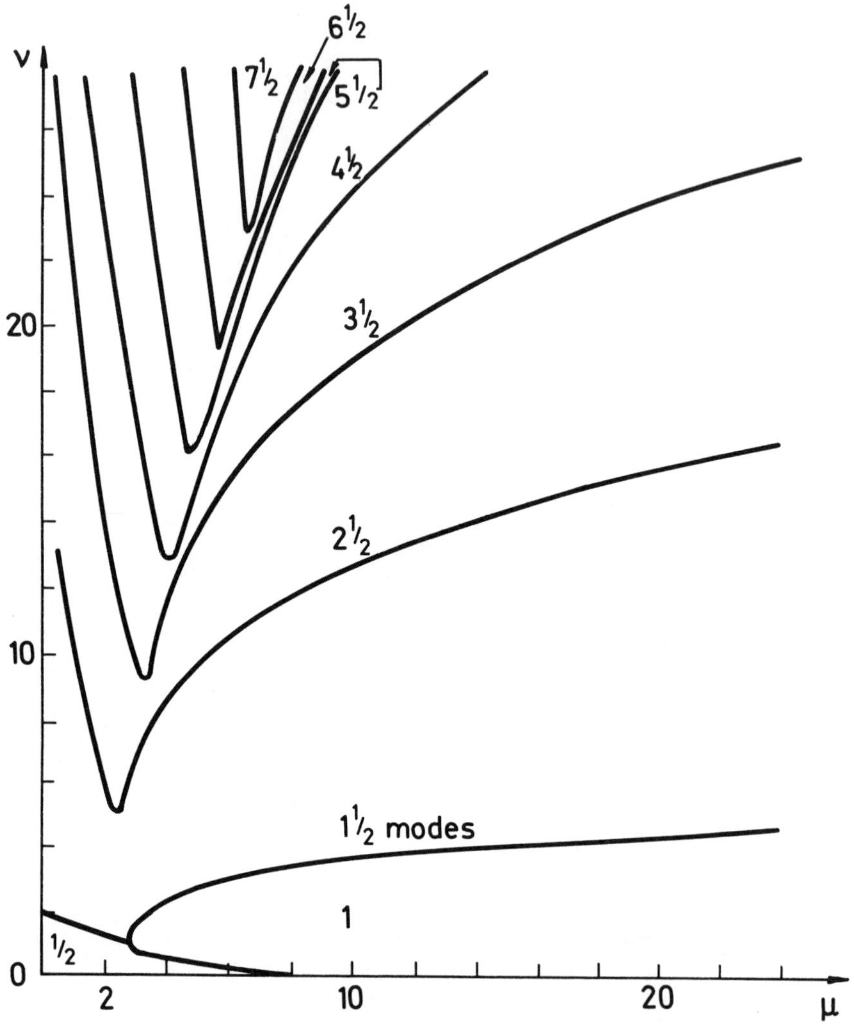

FIGURE 13 Approximate modality contours for the Neyman distribution. (Bowman and Shenton, 1967)

additional X is added its modal value is roughly ν). The half-mode at zero occurs because of a relatively high probability at $N = 0$, when the second Poisson(ν) distribution is not relevant:

$$\Pr(S_N = 0 \mid N = 0) = 1 .$$

Similarly, "missing" modes are sometimes thus because the local change in probabilities occurs during a global increase or decrease: what would have been a not very pronounced mode had the global rate of change been small enough may fail to appear as such - e.g. $\mu = 2$, $\nu = 12\frac{1}{2}$ near $x = 40$.

Since relatively small changes in the parameter values may change the number of modes quite sharply, this behaviour is rather inconvenient in estimation (where errors in the value used for the parameters are certain) however attractive it may be from the view point of flexibility; fortunately the most complex behaviour occurs for 'large' values of μ and ν, not often encountered in data.

Illustrations of the possible shapes of the p.f. are given, where for graphs which require "large" values of x, say $x > 25$, instead of the separate points corresponding to the lattice points $x = 0,1,2,...$ a continuous curve is drawn. The appearance of a half-mode at $x = 0$ is easily overlooked, and should be specially noticed. Brief comments are as follows, with the members of a group having an equal mean. (It is also worth following through sequences such as $\mu = 1$ and $\nu = 1,2,4$.)

μ,ν	
1,1	½ mode
1,2	1½ modes
2,1	½ mode
1,4	1½ modes; the half-mode is very conspicuous.
4,1	1 mode
2,12½	2½ modes; conspicuous half-mode
5,5	1½ modes
12½,2	1 mode, with a marked tendency to "normality"
5,17	5½ modes.

In the case of the last illustration, e.g., if such a case arose in data fitting it is hard to believe that the fit would have any real significance unless the occurrence of the modes (at 0, 17, 35, 53, 70 and 82, approximate multiples of 17)

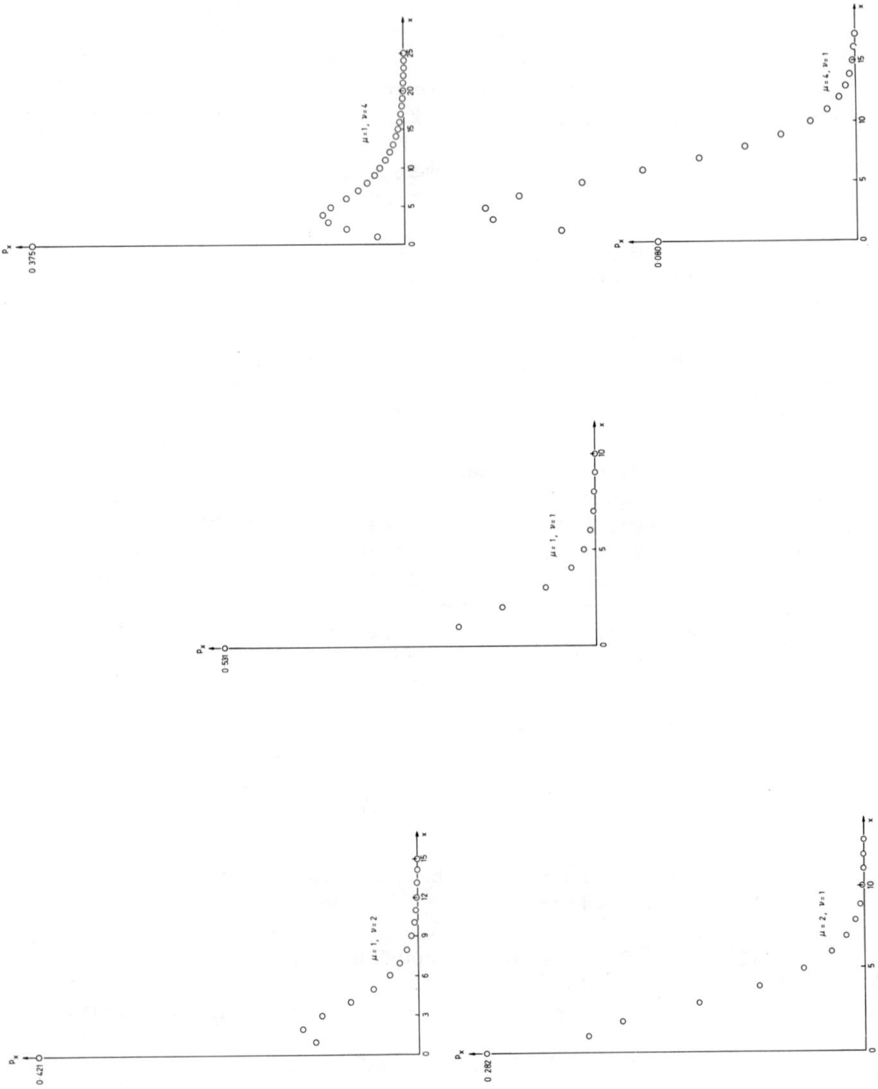

$\mu = 1,\ \nu = 4$
$\mu = 4,\ \nu = 1$
$\mu = 1,\ \nu = 1$
$\mu = 1,\ \nu = 2$
$\mu = 2,\ \nu = 1$
0 375
0 080
0 531
0 421
0 282

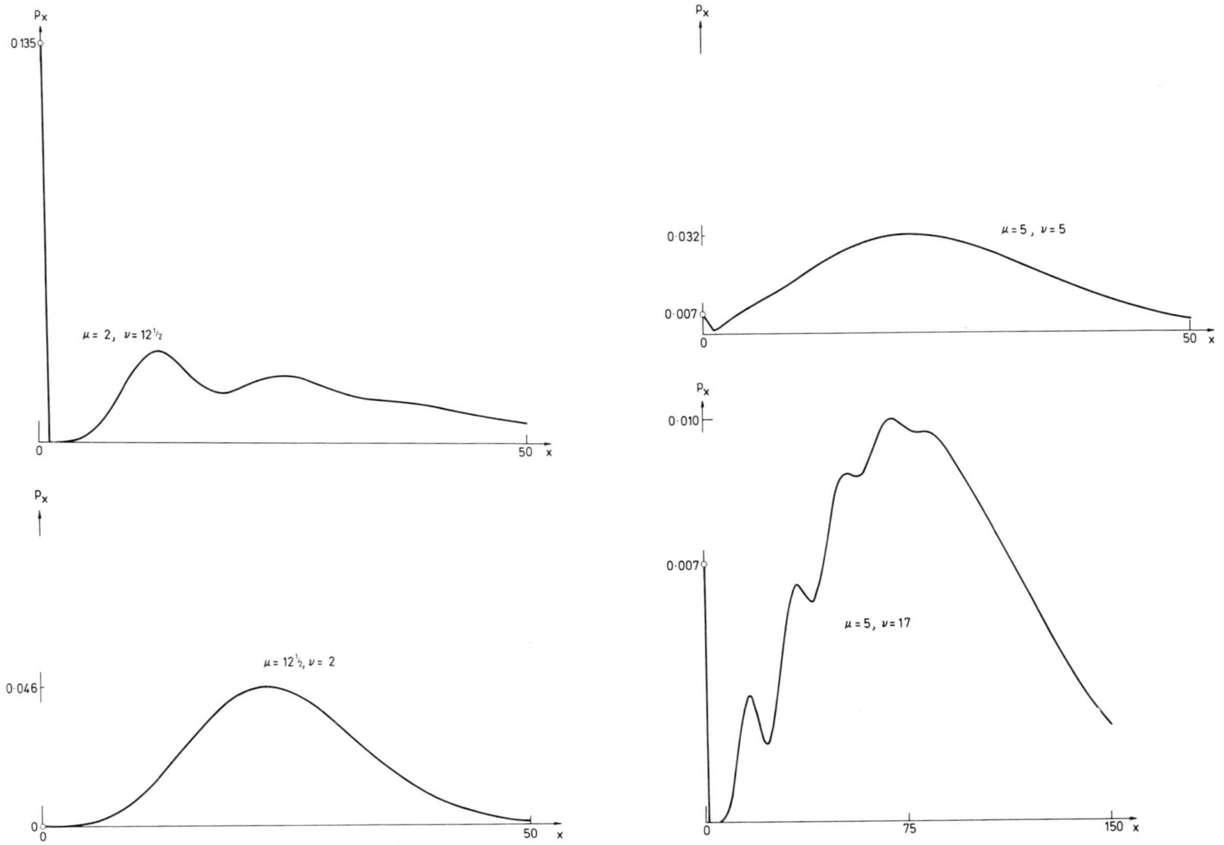

FIGURE 14 Graphs of some probability functions for the Neyman(μ,ν) distribution.

correspond to a cluster size of 17, or some equivalent biological or other physical phenomenon.

d) Approximations. Since the mean is $\mu\nu$ and the variance $\mu\nu(1+\nu)$, there is some reason to look for a Poisson approximation for small ν. Expanding the p.g.f. in powers of ν,

$$\begin{aligned}\exp(-\mu+\mu e^{-\nu+\nu z}) &= \exp\{-\mu+\mu(1-\nu(1-z)+\frac{\nu^2}{2}(1-z)^2-\dots)\}\\ &= \exp\{-\mu\nu(1-z)\}+0(\nu^2),\end{aligned}$$

and the conjecture is verified (in the sense that if $\nu \to 0$ and $\mu \to \infty$ in such a manner that $\mu\nu$ is finite, the variate will have in the limit a Poisson distribution with mean $\mu\nu$).

A similar analysis for small μ leads immediately to the p.g.f.

$$\begin{aligned}&1+\mu(e^{-\nu+\nu z}-1)+0(\mu^2)\\ &= 1-\mu+\mu e^{-\nu+\nu z}+0(\mu^2),\end{aligned}$$

and this is the p.g.f. of a "Poisson with zeros" - i.e. of a variate X with probabilities $1-\mu+\mu e^{-\nu}$ at $X=0$ and $\mu e^{-\nu}\frac{\nu^x}{x!}$ for $X = 1,2,\dots$.

In order to examine the cases where μ and ν are large without restriction, it is convenient both to standardize (since otherwise the mean and variance also are large, and the region of relatively high probability moves away from the origin and is very diffuse) and, since a continuous limiting distribution would naturally be expected, to transform to the cumulant g.f. Hence the moment g.f. of the standardized variate is

$$\mathscr{E}\left[\exp\left(t\,\frac{X-\mu\nu}{\surd\{\mu\nu(1+\nu)\}}\right)\right],$$

whence introducing $\alpha = 1/\surd\{\mu\nu(1+\nu)\}$, and taking logarithms to give the cumulant g.f., leads to

$$-\alpha\mu\nu t-\mu+\mu\exp(-\nu+\nu e^{\alpha t})$$

(recalling that the cumulant g.f. of the Neyman(μ,ν) variate is

$-\mu+\mu\exp(-\nu+\nu e^t))$. Re-arranging gives

$$-\alpha\mu\nu t-\mu+\mu\exp\{\nu(e^{\alpha t}-1)\} = -\alpha\mu\nu t-\mu+\mu\sum_{s=0}^{\infty}\nu^s\frac{(e^{\alpha t}-1)^s}{s!}$$

$$= -\alpha\mu\nu t+\mu\sum_{s=1}^{\infty}\nu^s\sum_{r=s}^{\infty}\mathcal{S}_r^{(s)}\alpha^r\frac{t^r}{r!}$$

$$= -\alpha\mu\nu t+\mu\sum_{r=1}^{\infty}\frac{t^r}{r!}\alpha^r\sum_{s=1}^{r}\mathcal{S}_r^{(s)}\nu^s .$$

So the various coefficients of $t^r/r!$ (i.e. κ_r) are:

t : $-\alpha\mu\nu+\alpha\mu\nu = 0$.

$\frac{t^2}{2!}$: $\mu\,\alpha^2(\nu+\nu^2) = 1$

$\frac{t^3}{3!}$: $\mu\,\alpha^3(\nu+3\nu^2+\nu^3) = \frac{1+(3/\nu)+1/\nu^2}{\mu^{\frac{1}{2}}(1+1/\nu)^{3/2}} = \frac{1}{\mu^{\frac{1}{2}}}\{1+0(1/\nu)\}$.

$\frac{t^r}{r!}$: $\frac{1}{\mu^{\frac{1}{2}r-1}}\{1+0(1/\nu)\}$, $r \geq 2$.

Hence for large μ the cumulant g.f. behaves like that of a standardized normal variate; alternatively, since the factorial cumulant g.f. of the general Neyman variate is

$$-\mu+\mu e^{\nu t}$$

and that of a Poisson(λ) variate is $e^{\lambda t}$ there is a certain correspondence between these for large ν . (See Martin and Katti (1962) for further details.)

Two asymptotic formulae may be quoted. The first, due to Cernuschi and Castagnetto (1946) (see also Douglas (1965)), is that obtained by the method of steepest descent:

$$p_x \sim \frac{e^{-\mu}}{\sqrt{(2\pi)}}\cdot\frac{\nu^x\exp(x/z_x)}{z_x^x\sqrt{\{x(1+z_x)\}}} ,\quad x = 1,2,3,\ldots,$$

where z_x is the root of $z_x e^{z_x} = x/\lambda$, $\lambda = \mu e^{-\nu}$, and

$p_0 = e^{\lambda-\mu}$. The second, due to Philpot (unpublished: quoted in Bowman and Shenton (1967)) is:

$$p_x \sim e^{\lambda-\mu} \frac{\nu^x}{x!} m_x ,$$

where

$$m_x = \frac{1}{\sqrt{L(y_x)}} \exp(f(y_x) - \lambda) ,$$

$$L(y_x) = \frac{x-\frac{1}{2}}{y_x^2} + \frac{1}{y_x} ,$$

y_x is a root of $\dfrac{x-\frac{1}{2}}{y_x} = \ln \dfrac{y_x}{\lambda}$,

and $f(y_x) = y_x \; (x-\frac{1}{2}) \ln (\lambda + \frac{x-\frac{1}{2}}{y_x} - 1)$.

This implies that m_x is an approximation to the xth moment of a Poisson variate with parameter λ , i.e. that m_x approximates

$$\sum_{n=0}^{x} \mathcal{S}_x^{(n)} \lambda^n .$$

An approximation of a different type seeks to represent the p.f. p_x as the sum of a finite number of Poisson terms (Bowman and Shenton (1967)):

$$p_x \sim \alpha_1 e^{-m_1} \frac{m_1^x}{x!} + \alpha_2 e^{-m_2} \frac{m_2^x}{x!} + \ldots + \alpha_s e^{-m_s} \frac{m_s^x}{x!}$$

with all $m_i, \alpha_i > 0$. The use of factorial moments leads to manageable equations; a particular conclusion, confirming the earlier results, is that if μ is small compared with ν , then approximately

$$m_1 = \nu , \; m_2 = 2\nu, \ldots, m_s = (s-1)\nu .$$

It may also be noticed that the Neyman distribution can be obtained as the limit of more complex distributions, the most obvious perhaps being the Poisson-binomial. Here the p.g.f. is

$$\exp\{-\mu + \mu(\chi + \pi z)^\nu\} ;$$

by letting the binomial tend to a Poisson, a Neyman distribution is obviously obtained.

4.3 TABLES AND COMPUTING FORMULAE

Tables of the cumulative probability function of the Neyman distribution have been given by Grimm (1964), to 5 decimal places for

$$\mu\nu = 0,1(0.1)\ 1.0(0.2)4.0,\ 6,\ 10$$

$$\nu = 0.2,\ 0.5,\ 1,\ 2,\ 3,\ 4$$

up to and including the term for which the cumulative probability first exceeds 0.999900. The APL function TOL NY PARA given in Figure 15 computes Neyman probabilities until their cumulative sum contains at least TOL 9's, for PARA = (μ,ν). For example 5 NY 2 3 will return as many probabilities as necessary, beginning with $p_0(2,3)$, $p_1(2,3),\ldots,$ until their sum is at least 0.99999.

Tables of Stirling Numbers of the Second Kind can obviously be used for individual probabilities, and these can be found in Abramowitz and Stegun (1965) up to order 25, and in Fisher and Yates (1967) under the title Initial Differences of Powers of the Natural Numbers. The APL function X NYI PARA, also in Figure 15, returns individual probabilities: e.g., 3 4 7 NYI 2 3 will return $p_3(2,3)$, $p_4(2,3)$ and $p_7(2,3)$.

The Tables given in Douglas (1955), referred to in the section on Maximum Likelihood Estimation, can also be used in the form

$$p_{x+1} = \frac{\nu}{x+1}\, p_x^* \; p_x$$

(where in the 1955 Tables P_x is used for p_x and p_x for p_x^*).

```
     ∇ R←TOL NY PARA;PRO;I;A;MU;NU;SUM;PX
[1]    PX←PX,(PX←*MU×¯1+*-NU)×PRO←(MU←PARA[1])×NU×*-NU←PARA[2],0ρA
       ←,1+I←0
[2]   LBL:→((SUM←+/PX←PX,PRO×(÷I+1)×+/PX×⌽A←A,A[I]×NU÷I←I+1)≤1-10*
       -TOL)/LBL
[3]    R←PX,1-SUM
     ∇

     ∇ R←X NYI PARA;PRO;I;A;MU;NU;SUM;PX
[1]    PX←PX,(PX←*MU×¯1+*-NU)×PRO←(MU←PARA[1])×NU×*-NU←PARA[2],0ρA
       ←,1+I←0
[2]   LBL:→((ρPX←PX,PRO×(÷I+1)×+/PX×⌽A←A,A[I]×NU÷I←I+1)≤⌈/X)/LBL
[3]    R←PX[1+X]
     ∇
```

FIGURE 15 APL functions for calculating Neyman(μ,ν) probabilities.

4.4 ESTIMATION

Some general remarks and theory on Estimation are given in the Appendix with that title, with emphasis on the three procedures most used: moment, maximum likelihood and minimum chi-squared. This Appendix should be referred to if the following account appears too brief. Broadly speaking, the point of view taken is that the sampling distribution of any proposed estimators is the basis for recommending or otherwise their use, although in practice a more or less adequate approximation to this distribution is all that is available.

The basis for the estimation of μ and ν is a random sample of fixed size n, conveniently considered as F_0 observations corresponding to $X = 0$, F_1 to $X = 1,\ldots,$ these frequencies $F_0, F_1, \ldots, F_x, \ldots$ being the random variables, with $n = \Sigma F_x$. There are no (two component) sufficient estimators for μ and ν; contagious distributions commonly do not have sufficient estimators.

Anticipating the analysis which follows, it turns out that estimators of the parameters μ and ν are usually very highly correlated. Since this tends to complicate their interpretation (especially with respect to standard errors) as well as being mathematically inconvenient, it is tempting to look for orthogonal parameters (Jeffreys (1961), Huzurbazar (1956-7)). In an application to the Neyman(μ,ν) distribution, Philpot (1964) found that a parametric specification in terms of

$$\alpha = \mu\nu \ , \ \lambda = \mu e^{-\nu}$$

led to such orthogonality (not uniquely), but although α has a simple interpretation as the overall mean, λ is neither convenient from this point of view nor with respect to moments or cumulants in spite of its appearance in the probability function, and little use of this specification seems to have been made. (With this specification the estimation of α turns out to be relatively accurate, but that of λ is very poor.) It may also be remarked here that the usual high correlation between estimators of μ and ν implies a degree of intrinsic ill-conditioning of the model, and high sampling variability is, unfortunately, to be anticipated, as is very generally the case over most contagious distributions. There are further remarks in the section which discusses Comparisons between Moment and Maximum Likelihood Estimation Procedures.

a) Moment estimation. The moment estimators $\overset{\circ}{\mu}, \overset{\circ}{\nu}$ of μ, ν are given by the moment equations

$$\left.\begin{aligned} \overset{\circ}{\mu}\overset{\circ}{\nu} &= \bar{X} \\ \overset{\circ}{\mu}\overset{\circ}{\nu}(1+\overset{\circ}{\nu}) &= M_2' - \bar{X}^2 = M_2 \end{aligned}\right\} ,$$

where M_2 is the sample second moment about the mean. (An unbiased estimator of the variance can be used instead of the biased M_2 ; but it makes no difference to the large sample properties, and these alone are examined here.)

These equations can be solved explicitly:

$$\left.\begin{aligned} \overset{\circ}{\mu} &= \frac{\bar{X}^2}{M_2 - \bar{X}} = \frac{K_{(1)}^2}{K_{(2)}} \\ \overset{\circ}{\nu} &= \frac{M_2 - \bar{X}}{\bar{X}} = \frac{K_{(2)}}{K_{(1)}} \end{aligned}\right\} .$$

(For hand calculation, introducing $S_r = \Sigma x^r F_x$ leads to the convenient expressions

$$\overset{\circ}{\nu} = \frac{S_2}{S_1} - \frac{S_1}{S_0} - 1, \ \overset{\circ}{\mu} = \frac{S_1}{S_0} \cdot \frac{1}{\overset{\circ}{\nu}} \ .)$$

As will be seen shortly, although the estimated standard errors and correlation of these estimators need careful interpretation, it is at the least desirable to calculate them to see whether it is even remotely plausible that the accuracy aimed at has been achieved. (An important function of statistical data analysis is to warn when the techniques used are inherently incapable of delivering the kind of results being sought.) Since the "large sample", or "asymptotic", sampling properties of sample moments are well known (see the Appendix on Moments and Cumulants), those of the moment estimators can also be derived by standard techniques.

Using the Taylor expansions (see the Appendix: Approximate Means, Variances and Covariances) for these estimators, e.g., for

$$\overset{\circ}{\nu} = \frac{M_2'}{M_1'} - M_1' - 1$$

with random variables M_1' and M_2',

$$\mathscr{E}(\overset{\circ}{\nu}) = \nu + 0(1/n) ,$$

and

$$\mathscr{V}(\overset{\circ}{\nu}) = (-\frac{\mu_2'}{\mu_1'^2} - 1)^2 \frac{\mu_2}{n} + (\frac{1}{\mu_1'})^2 \frac{\mu_4' - \mu_2'^2}{n}$$

$$+ 2(-\frac{\mu_2'}{\mu_1'^2} - 1) \frac{1}{\mu_1'}(\frac{\mu_3' - \mu_1' \mu_2'}{n}) + 0(1/n^2)$$

$$= \frac{1}{n} \cdot \frac{2 + \nu + 2\mu(1+\nu)^2}{\mu} + 0(1/n^2) .$$

Similarly,

$$\mathscr{V}(\overset{\circ}{\mu}) = \frac{1}{n} \cdot \frac{\mu(2+\nu^2+2\mu(1+\nu)^2)}{\nu^2} + 0(1/n^2) ,$$

$$\mathscr{C}(\overset{\circ}{\mu},\overset{\circ}{\nu}) = \frac{-1}{n} \cdot \frac{2(1+\mu(1+\nu)^2)}{\nu} + 0(1/n^2) .$$

The "first order generalized variance", the determinant of the asymptotic first order covariance matrix of $\overset{\circ}{\mu},\overset{\circ}{\nu}$, is thus

$$\frac{1}{n^2} \cdot \frac{1+(1+\nu)^2 + 2\mu(1+\nu)^3}{\nu} .$$

(See the Exercises for a matrix formulation of these results).

These formulae may be applied directly, and one of Neyman's (1939) examples of European cornborers, provides an illustration (Table 5). The fit is visually good (a graph is shown in the section on maximum likelihood estimation), with small relative log-likelihood and chi-squared, but the standard errors of μ and ν are uncomfortably large and the correlation of -0.94

TABLE 5

EUROPEAN CORNBORERS: NEYMAN (1939)

Moment and Maximum Likelihood Fits

```
      NYFIT 24 16 16 18 15 9 6 5 3 4 3 0 1

P.G.F.:   EXP[-MU+MU×EXP(-NU+NU×Z)]
MEAN = MU×NU ,  VARIANCE = MU×NU×(1+NU) .



MOMENT FIT:  [MUO,NUO]  FOR  [MU,NU]
------------------------------------


M1' = 3.166666667 ,  N    = 120 ,  N×M1'  = 380
M2  = 7.705555556


MUO  =  2.2093, WITH S.E. =   .5204
     CORRELATION =   ¯.9417,    (GEN.VAR.)*1÷4  = .2418
NUO  =  1.4333, WITH S.E. =   .3341

'2×LN RELATIVE LIKELIHOOD' =  7.27 PERHAPS WITH 9 DF



MAXIMUM LIKELIHOOD FIT:  [MUL,NUL]  FOR  [MU,NU]
-------------------------------------------------


1
2.05195179 1.543246134
2
2.061343413 1.536214998
3
2.061381718 1.536186452
4
2.061381718 1.536186451

MUL  =  2.0614, WITH S.E. =   .3457
     CORRELATION =   ¯.8786,    (GEN.VAR.)*1÷4  = .2036
NUL  =  1.5362, WITH S.E. =   .2511


'2×LN RELATIVE LIKELIHOOD' =  7.10 PERHAPS WITH 9 DF
```

TABLE 5
(continued)

	OBSERVED	*FITTED[1]*	*FITTED[2]*
0	24	22.3	23.8
1	16	16.9	16.2
2	16	18.4	18.0
3	18	16.5	16.1
4	15	13.4	13.2
5	9	10.3	10.2
6	6	7.5	7.5
7	5	5.2	5.3
8	3	3.5	3.6
9	4	2.3	2.4
10	3	1.4	1.5
11	0	.9	1.0
12	1	.5	.6
13	0	.7	.8
	120	120.0	120.0

LN-LIKELIHOOD[1] = ¯269.782175
CHI-SQUARED[1] = 4.94 *WITH* 9 *DF*

LN-LIKELIHOOD[2] = ¯269.6961791
CHI-SQUARED[2] = 4.58 *WITH* 9 *DF*

implies very high dependence so that statements about μ carry strong implications about ν.

In fact, however, the APL functions used to carry out the fitting are shown in Figure 16; they do not correspond to the algebraic formula above which are of course specific to the Neyman distribution. Instead, the fitting functions were written so that with minor modifications they can be used for a variety of contagious (and other) distributions - e.g. FREQSAMM calculates moments of any specified order from a sample arranged as a frequency distribution, PMOΔFC converts a set of factorial cumulants to the corresponding set of power moments about the origin, COVMO computes the covariance matrix of a vector of moments about the origin: these, and others, are used by the Neyman moment fitting function NYMO.

Ignoring the correlation, the common large sample approach would now give approximate 95% confidence intervals for each parameter as

(estimate $\pm$ twice the estimated standard error);

i.e. for μ : (1.17, 3.25)

for ν : (0.76, 2.10).

In spite of the not small sample size $(n = 120)$, the results suggest that μ and ν are not very precisely estimated. (There are many further comments about the interpretation of these standard errors in the sequel.)

Overall, the mean number of cornborer larvae per sampling unit is 3.17, the overall mean of the frequency distribution. Using the second moment of the distribution, or the moment estimates, the approximate 95% confidence interval for the distribution mean $\mu\nu$ is thus

$$\left(3.17 \pm 2 \times \sqrt{\frac{7.7}{120}}\right) = (2.7,\ 3.7),$$

also rather long; it will be recalled that Neyman distributions "often" have long tails.

Returning to the previous (Chapter 3) discussion of the Indices of Cluster Frequency and Size, the moment estimate of the Index of Cluster Frequency is 2.21 clusters per sampling unit (with s.e. 0.51) and of the Index of Cluster Size 1.43 larvae per cluster (with s.e. 0.33). Moment-like estimates without the assumption of a Neyman distribution are the same for the Indices but with standard errors estimated as 0.45 and 0.27, with correlation -0.92.

More detailed analyses are possible - on the basis of more (and more or less plausible) assumptions.

Consider first the Poisson(μ)-stopped Poisson(ν) model - i.e. suppose the sampling is such that in each sampling unit (quadrat - in the present case probably a plant) the number of clusters is Poisson(μ) distributed. (For the purpose of brevity in this discussion take heavily rounded figures from the above.) Then the mean number of clusters per plant is estimated to be between 1.7 and 2.8; and the mean number of larvae per cluster on the plant at the time of sampling is estimated to be between 1.1 and 1.8, where 120 plants were randomly chosen to generate the frequency distribution. (Three cautionary comments might be inserted: the original clusters were of eggs, not larvae; all larvae from a cluster might not be counted - in terms of Chapter 3, μ/θ and $\theta\nu$ rather than μ and ν might be relevant; and there is further discussion later regarding approximations and asymptotic expansions.)

Secondly, for a mixed-Poisson($U\nu$) on U by Poisson(ν) model, two somewhat different interpretations might be given - neither is very plausible in these particular circumstances. In the one case, suppose the sampling area is divided into sub-areas of varying "attractiveness" to the individuals being sampled, in terms of dryness, windiness, ground cover, etc. For these sub-areas, suppose the mean numbers of individuals per quadrat are $0, 1.4, 2 \times 1.4, \ldots, 1.4u, \ldots,$ with quadrats selected so that the probability of a quadrat coming from sub-area u is

$$e^{-2.2}\frac{(2.2)^u}{u!}, \quad u = 0,1,2,\ldots$$

and where for given sub-area u the probability of x individuals in a quadrat is

$$e^{-1.4u}\frac{(1.4u)^x}{x!}, \quad x = 0,1,2,\ldots;$$

the mixed distribution is thus Neyman(2.2,1.4). In the second case, the variation could be supposed to be of parental reproductive powers (rather than of the attractiveness of areas) - the rest is the same. For these cases, the random selection of 120 sub-areas, or parents, with a subsequent selection generates the frequency distribution.

No amount of study of a single frequency distribution will distinguish between these (or other) models - a model will be assumed of knowledge regarding the experimental circumstances, and, hopefully, its appropriateness investigated. For example, with

```
      ∇ R←NYMO FX;I;M;MJP;MO;NO;P;PJM;PX;V
      0ρ⎕← 'MOMENT FIT:  [MUO,NUO]  FOR  [MU,NU]' ,[0.5]36ρ'-',⎕←
      0ρCPTIME
      M←2 FREQSAMM FX
      →(M[3]≠M[2])/LBL1
      →LBL2,P←2ρ10000000000,0ρ⎕← 'MEAN = SECOND MOMENT ABOUT MEAN
      : NO MOMENT FIT' ,⎕←''
     LBL1:→(M[3]>M[2])/LBL3
      'MEAN > SECOND MOMENT ABOUT MEAN: NO MOMENT OR M.L. FIT' ,⎕
      ←''
      →
     LBL3:P←MO,NO←M[2]÷MO←M[2]×M[2]÷M[3]-M[2]
      MJP← 2 2 ρNO,MO,(NO×1+NO×1+2×MO),MO×1+2×NO×1+MO
      →(0=(MJP[1;1]×MJP[2;2])-MJP[1;2]*2)/LBL6
      V←M[1]COVMO PMOΔFC MO×NO*ι4
      →LBL2, 'MUO NUO ' PEST2 P,PJM+.×V+.×⍉PJM←⌹MJP
     LBL6: 'MUO = ' ,(4⍕MO), ' AND  NUO = ' ,(4⍕NO), ', BUT NO ST
      ANDARD ERRORS BECAUSE OF A SINGULAR MATRIX'
     LBL2:→(∧/(P>0.001)∧P<200)/LBL4
      'UNACCEPTABLE PARAMETER VALUES: UNIFORM ''FITTED'' PROBABIL
      ITIES ASSIGNED, ARBITRARILY' ,⎕←''
      →LBL5,PX←(1+ρFX)÷1+ρFX
     LBL4:PX←(ρFX)NYLEFT P
      '''2×LN RELATIVE LIKELIHOOD'' = ' ,(2⍕I[1]), ' PERHAPS WITH
       ' ,(⍕(I←2 TWRELLIKE(FX,0),[1.5]PX)[2]), ' DF' ,⎕←''
     LBL5:0ρ⎕← 'CPU TIME: ' ,⍕CPTIME,⎕←''
      R←( ¯1 ¯1 0 ,ιρFX),( ¯1 ¯1 ,FX,0),[1.5]P,PX
      ∇

      ∇ R←CPTIME;T
      R← 60 60 1000 ⊤(T←⎕AI[1+⎕IO])-TIMER
      TIMER←T
      ∇

      TIMER
113

      ∇ R←P FREQSAMM FXM;M;N;T;Q;X
      →(2=ρρFXM)/LBL
      FXM←(¯1+ιρ,FXM),[1.5]FXM
     LBL:R←(FXM[;2]+.×(FXM[;1]-M←(T←+/×/FXM)÷N)∘.*1+ιP-1)÷N←+/FXM
      [;2]
      ( 'M1'' = ' ),(⍕M),( '    ,     N    = ' ),(⍕N), ' ,          N×
      M1''  = ' ,⍕T,⎕←''
      (Qρ'M'),( 0 1 ↓⍕(Q,1)ρ1+ιQ),((Q,3)ρ '  =' ),⍕((Q←P-1),1)ρR
      R←N,M,R,⎕←''
      ∇
```

```
      ∇ R←PMO∆FC V;M;J
[1]    J←L1,L2,L3,L4,R←''
[2]    →(4≥ρV←,V)/LBL
[3]    V←4↑V,ρ⎕← 'ONLY FOURTH AND LOWER ORDER CUMULANTS USED'
[4]   LBL:→J[ρV],M←V[1]
[5]   L4:R←R,V[4]+(V[3]×6+4×M)+(3×V[2]*2)+(V[2]×M⊥ 6 18 7 )+M⊥ 1 6
       7 1 0
[6]   L3:R←R,V[3]+(V[2]×3×1+M)+M⊥ 1 3 1 0
[7]   L2:R←R,V[2]+M⊥ 1 1 0
[8]   L1:R←⌽R,M
      ∇

      ∇ R←N COVMO MO;P
[1]    R←(MO[P∘.+P←⍳P]-(P↑MO)∘.×(P←0.5×ρMO)↑MO)÷N
      ∇

      ∇ R←N NYLEFT PARA;PRO;I;A;MU;NU;PX
[1]    PX←PX,(PX←*MU×¯1+*-NU)×PRO←(MU←PARA[1])×NU×*-NU←PARA[2],0ρA
       ←,1+I←0
[2]   LBL:→(N>ρPX←PX,PRO×(÷I+1)×+/PX×⌽A←A,A[I]×NU÷I←I+1)/LBL
[3]    R←PX,1-+/PX
      ∇

      ∇ R←K TWRELLIKE FP
[1]    →(∧/0≠(FP←(×FP[;1])⌿FP)[;2])/LBL
[2]    →ρ⎕← 'AT LEAST ONE EVENT OCCURRED WITH ZERO PROBABILITY ON
       THE PROPOSED HYPOTHESIS'
[3]   LBL:R←(2×+/FP[;1]×(-⍟FP[;2])+⍟FP[;1]÷+/FP[;1]),¯1-K-1↑ρFP
      ∇
```

FIGURE 16 Moment fitting for the Neyman Distribution.

the stopped model, repetition of the experiment with double the quadrat size ought to alter the estimates of μ and ν in a predictable - or at least explicable - way. Unless some stability of estimation exists, and supplementary experimentation may be necessary to make its nature clear, it is not possible to use the estimates for analytical purposes and a merely descriptive fitting of observed frequencies is not scientifically very illuminating.

An illustration of the kind of analytical use which can be made of meaningful parameters may be given with reference to a spraying programme in which the precise mechanism of the spray is unknown. In an over simplified way, does the spray reduce the number of clusters laid? Or does it reduce the numbers of larvae per cluster (by increasing infertility of eggs, or by being repellent to larvae)? Clearly the answers to questions such as these will determine effective modes and timing of spraying.

However, returning to the equations

$$\overset{o}{\mu} = \frac{\bar{X}^2}{M_2 - \bar{X}} \;, \quad \overset{o}{\nu} = \frac{M_2 - \bar{X}}{\bar{X}} \;,$$

since μ and ν are essentially positive no admissible solutions will exist unless $M_2 > \bar{X}$ (and $\bar{X} > 0$), naturally enough since the second moment $\mu\nu(1+\nu)$ strictly exceeds the mean $\mu\nu$.

Considering $\overset{o}{\mu}$, where for brevity the distribution of $\overset{o}{\mu}$ rather than the joint distribution of $(\overset{o}{\mu},\overset{o}{\nu})$ will be referred to, the distribution of real interest is that of $\overset{o}{\mu}$ given that $M_2 > \bar{X} > 0$, i.e., $\overset{o}{\mu} \mid M_2 > \bar{X} > 0$; what use will an experimenter make of a sample from a postulated Neyman distribution in which $M_2 < \bar{X}$? However, it is reasonable to expect that for large n the distributions of $\overset{o}{\mu}$, and of $\overset{o}{\mu}$ given $M_2 > \bar{X} > 0$, will be very similar (see the Appendix on Asymptotic Expansions); in particular the formal expansion for $\mathcal{E}(\overset{o}{\mu})$, $\alpha_0 + \alpha_1/n + \alpha_2/n^2 + \ldots$, will be the same as that for $\mathcal{E}(\overset{o}{\mu} \mid M_2 > \bar{X} > 0)$. Since it is not analytically practicable to investigate directly the distribution of $\overset{o}{\mu} \mid M_2 > \bar{X} > 0$, internal evidence of the appropriateness of the use of early terms of this expansion is afforded by their behaviour. Bowman and Shenton (1967), Shenton and Bowman (1967, 1977), with extremely arduous analysis, have determined some of these terms, and several examples follow:

Terms in

$1/n^0$	$1/n$	$1/n^2$	$1/n^3$	$1/n^4$	$1/n^5$

(These headings index the terms in the expansions shown below with a rule)

a) Taking the case $n = 100$, $(\mu,\nu) = (1,1)$ leads to:

$$\mathcal{E}(\mathring{\mu}) = \overline{1 + 0.1300 + 0.0142 + 0.0061 + 0.000040 + 0.0025}$$

Hopefully, one would assert $\mathcal{E}(\mathring{\mu}) = 1.150$; the usual assertion of $\mathcal{E}(\mathring{\mu}) = 1 + 0(1/n)$ though technically correct is numerically not very useful.

b) Correspondingly for $n = 100$, $(\mu,\nu) = (4,3)$:

$$\mathcal{E}(\mathring{\mu}) = \overline{4 + 0.2011 + 0.0105 + 0.0010 + 0.000086 + 0.000020}$$

c) Small values of the parameters require unexpectedly large values of n for useful expansions - an extreme example is provided by the instance of $n = 1000$, $(\mu,\nu) = (0.1, 0.1)$.

$$\mathcal{E}(\mathring{\mu}) = \overline{0.1 + 0.235 + 1.049 + 7.85 + 78.0 + 1004.9}$$

$$\mathcal{E}(\mathring{\nu}) = \overline{0.1 - 0.011 - 0.0009' + 0.00000}$$

$$\mathcal{V}(\mathring{\mu}) = \overline{0 + 0.0225 + 0.3010}$$

$$\mathcal{V}(\mathring{\nu}) = \overline{0 + 0.0234 - 0.0024}$$

$$\mathcal{C}(\mathring{\mu},\mathring{\nu}) = \overline{0 - 0.0224 - 0.1023 - 0.7735}$$

For $\overset{o}{\nu}$ it is reasonable to assert $\mathcal{E}(\overset{o}{\nu}) = 0.089$ and $\mathcal{V}(\overset{o}{\nu}) = 0.02$ (or 0.021) with some confidence; but it would be very speculative to use the customary formula in $1/n$ for the large sample variance of $\overset{o}{\mu}$ and assert $\mathcal{V}(\overset{o}{\mu}) = 0.02$.

In the example of the European cornborer larvae, tables are not available for $n = 120$ nor for (μ,ν) very near their estimated values of (2.2,1.4). However, taking $n = 100$ (a correction would be easy but not worthwhile), tabulated values (Bowman and Shenton (1967)) for $(\mu,\nu) = (1,1)$ should give results which indicate in general terms how appropriate the first order figures are. The results are: $n = 100$, $(\mu,\nu) = (1,1)$.

	Terms in n^{o}	n^{-1}	n^{-2}		Terms in n^{o}	n^{-1}	n^{-2}
$\mathcal{E}(\overset{o}{\mu}) =$	1.00 +	0.13 +	0.01,	$\mathcal{E}(\overset{o}{\nu}) =$	1.00 -	0.03 -	0.00*
$\mathcal{V}(\overset{o}{\mu}) =$		0.11 +	0.05,	$\mathcal{V}(\overset{o}{\nu}) =$		0.11 -	0.00

(*Bowman and Shenton give figures for $n = 20$)

Using standard errors rather than variances, these are:

$$\mathcal{E}(\overset{o}{\mu}) = 1.14, \qquad \mathcal{E}(\overset{o}{\nu}) = 0.97$$

$$\text{s.e.}(\overset{o}{\mu}) = 0.40, \qquad \text{s.e.}(\overset{o}{\nu}) = 0.33$$

suggesting that the use of a moment estimate on the average

overestimates μ by about 15% ,
underestimates ν by about 3% ,

while the standard error of $\overset{o}{\nu}$ is not grossly affected, that of $\overset{o}{\mu}$ being rather more so.

Since part of the difficulty in the interpretation of these expansions is due to the deleted region $M_2 > \bar{X} > 0$, it may be possible to eliminate this by using a sample size n such that this region is of very small probability. From

$$\mathcal{E}(M_2 - \bar{X}) = \mu\nu^2 - \frac{\mu\nu(1+\nu)}{n}$$

and

$$\mathscr{V}(M_2 - \overline{X}) = \frac{\mu\nu^2}{n} (2\mu(1+\nu)^2 + 2 + 4\nu + \nu^2) + 0(1/n^2) ,$$

n can be determined so that

$$\mathscr{E}(M_2 - \overline{X}) \geq c\sqrt{\mathscr{V}(M_2 - \overline{X})}$$

for a suitable number c of standard deviations. Thus, for $c = 5$ and $(\mu,\nu) = (1,1)$, $n \geq 375$ will ensure that $M_2 - \overline{X}$ will be negative very rarely; Bowman and Shenton remark that such a choice of n is very conservative, leading to useful expansions.

In summary (see Bowman and Shenton for fuller detail), the adequacy of the usual formulae for bias (zero order in $1/n$) and variance or covariance (first order) is suggested by the asymptotic expansions to be very roughly as follows for the realistic sample size of $n = 100$. The value of ν is most critical: the first terms are 'adequate' except for 'small' ν , where 'small' means less than 5 for the bias of $\overset{o}{\mu}$ and $\overset{o}{\nu}$, less than 1 for the variance of $\overset{o}{\mu}$, and less than 1/2 for the variance of $\overset{o}{\nu}$ and covariance of $\overset{o}{\mu},\overset{o}{\nu}$ - 'adequate' means that the next term in the expansion is not more than 5% of the first. (It will be recalled that small ν corresponds to a region of the parameter space where the Neyman is well approximated by a Poisson Distribution.) However, with the increasing availability of time on high speed computers, it is possible to generate large numbers of random samples of specified size from specified distributions and so, calculating estimates for each sample, to build up the sampling distributions of the estimators by this simulation method. There is a separate discussion in the section on comparisons between moment and m.l. estimation procedures; but very briefly it can be said here it appears that, provided the sample size and distribution are not such that negative estimates (which are rejected) appear frequently, cases in which the "correction" terms are small can be interpreted in the natural way, but that large "correction" terms are to be taken as general warning signs as to the inadequacy of the usual "large sample" formulae. For the numerical case $n = 100$, $(\mu,\nu) = (1,1)$ already examined, 1000 random samples gave:

observed mean: for $\overset{o}{\mu}$ of 1.18 for $\overset{o}{\nu}$ of 0.97

observed s.d.: for $\overset{o}{\mu}$ of 0.69 for $\overset{o}{\nu}$ of 0.31

and these are not inconsistent with the asymptotic results quoted.

A quick reference also may be made to the previously mentioned asymptotically orthogonal parameterization

$$\alpha = \mu\nu \ , \ \lambda = \mu \, e^{-\nu} \ .$$

The moment estimators of α and λ are

$$\overset{o}{\alpha} = \overset{o}{\mu}\overset{o}{\nu} \quad \text{and} \quad \overset{o}{\lambda} = \overset{o}{\mu} \, e^{-\overset{o}{\nu}} \ ,$$

and the usual asymptotic result

$$\underset{\sim}{\mathcal{V}}\begin{pmatrix} \overset{o}{\alpha} \\ \overset{o}{\lambda} \end{pmatrix} = \underset{\sim}{J}\,\underset{\sim}{\mathcal{V}}\begin{pmatrix} \overset{o}{\mu} \\ \overset{o}{\nu} \end{pmatrix}\underset{\sim}{J}^{t} \quad \text{with} \quad \underset{\sim}{J} = \begin{pmatrix} \dfrac{\partial\alpha}{\partial\mu} & \dfrac{\partial\alpha}{\partial\nu} \\ \dfrac{\partial\lambda}{\partial\mu} & \dfrac{\partial\lambda}{\partial\nu} \end{pmatrix},$$

gives (to the lowest order in the sample size $n = 100$):

α: $\mathcal{E}(\overset{o}{\alpha}) = 1$

s.e. $(\overset{o}{\alpha}) = 0.14$

corr. $(\overset{o}{\alpha},\overset{o}{\lambda}) = 0.00$

λ: $\mathcal{E}(\overset{o}{\lambda}) = 0.37$

s.e. $(\overset{o}{\lambda}) = 0.24$

A typical set of 300 simulated samples gave:

α: ave. $(\overset{o}{\alpha}) = 0.99$

s.d. $(\overset{o}{\alpha}) = 0.15$

obs. corr. $(\overset{o}{\alpha},\overset{o}{\lambda}) = 0.03$

λ: ave. $(\overset{o}{\lambda}) = 0.52$

s.d. $(\overset{o}{\lambda}) = 0.37$

The estimators are indeed almost uncorrelated, and that of α is as good as one would have expected (it is of the distribution mean, with the Central Limit Theorem directly applicable); but the estimator of λ is comparable with that of μ both with respect to bias and standard error. (See also the section on Comparisons between Moment and Maximum Likelihood Estimation Procedures.)

b) Maximum likelihood estimation. For the p.f. p_x and observed frequencies F_x the likelihood L is given by

$$L = p_0^{F_0} p_1^{F_1} \cdots p_x^{F_x} \cdots$$

or, usually more conveniently, via the log-likelihood

$$\ell n\ L = \Sigma\ F_x\ \ell n\ p_x\ .$$

This may be maximized numerically for given F_x via a general purpose optimizing programme on a high speed computer in the variables μ,ν $(p_x = p_x(\mu,\nu))$, and, by fitting a quadratic surface near the maximum, estimates of the covariance matrix of the maximizing values may be obtained. (See Ross (1970) for a detailed account of a Rothamsted system for doing so.)

More usually, the "(maximum) likelihood equations" are obtained by differentiating the log-likelihood and equating to 0:

$$\left.\begin{aligned} \Sigma\ F_x \frac{\partial\ \ell n\ p_x}{\partial\mu} &= 0 \\ \Sigma\ F_x \frac{\partial\ \ell n\ p_x}{\partial\nu} &= 0 \end{aligned}\right\} \quad \text{at} \quad (\mu,\nu) = (\hat{\mu},\hat{\nu})$$

Since these equations cannot be solved explicitly, iterative procedures are needed; but in any case convenient expressions for the derivatives are most readily obtained not directly but via the p.g.f. (Shenton (1949)). From the p.g.f.

$$\begin{aligned} \frac{\partial\ g(z)}{\partial\mu} &= \Sigma\ \frac{\partial\ p_x}{\partial\mu}\ z^x \\ &= g(z)\ (-1 + e^{-\nu+\nu z}) \\ &= -g(z) + \frac{1}{\mu\nu}\ g'(z)\ , \end{aligned}$$

whence equating coefficients of z^x ,

$$\frac{\partial\ p_x}{\partial \mu} = -\ p_x + \frac{x+1}{\mu\nu}\ p_{x+1}\ .$$

Introducing

$$\Pi_x = (x+1)\ \frac{p_{x+1}}{p_x}\ ,$$

this can be written as

$$\frac{\partial\ \ln\ p_x}{\partial \mu} = -\ 1 + \frac{1}{\mu\nu}\ \Pi_x\ ;$$

similarly,

$$\frac{\partial\ \ln\ p_x}{\partial \nu} = \frac{x}{\nu} - \frac{1}{\nu}\ \Pi_x\ .$$

A general two dimensional Newton-Raphson iterative procedure may now be used for the likelihood equations if a computer is available, taking (usually) as starter values those given by the moment equations. For these, (μ_1,ν_1) is computed for successive $x = 0,1,2,\ldots$ until p_x is sufficiently small, where this means that enough terms are taken so that expectation expressions involving p_x are small (see below). One then requires the matrix (evaluated at (μ_1,ν_1))

$$\underset{\sim}{\mathscr{I}}\binom{\mu}{\nu} = n \times \begin{pmatrix} \mathscr{E}\left[\left(\dfrac{\partial\ \ln\ p_X}{\partial \mu}\right)^2\right] & \mathscr{E}\left(\dfrac{\partial\ \ln\ p_X}{\partial \mu} \cdot \dfrac{\partial\ \ln\ p_X}{\partial \nu}\right) \\ \mathscr{E}\left(\dfrac{\partial\ \ln\ p_X}{\partial \mu} \cdot \dfrac{\partial\ \ln\ p_X}{\partial \nu}\right), & \mathscr{E}\left[\left(\dfrac{\partial\ \ln\ p_X}{\partial \nu}\right)^2\right] \end{pmatrix}$$

where, e.g.,

$$\mathscr{E}\left(\frac{\partial\ \ln\ p_X}{\partial \mu}\right)^2 = \Sigma\ p_x\left(-1 + \frac{\Pi_x}{\mu\nu}\right)^2\ ,$$

whence the necessity for accurate evaluation of p_x . From this, second iterates are given by

$$\begin{pmatrix}\mu_2\\ \nu_2\end{pmatrix} = \begin{pmatrix}\mu_1\\ \nu_1\end{pmatrix} + \left[\underset{\sim}{\mathscr{I}}\begin{pmatrix}\mu_1\\ \nu_1\end{pmatrix}\right]^{-1} \times \begin{pmatrix}\dfrac{\partial\ \ell n\ L}{\partial\mu}\\[2ex] \dfrac{\partial\ \ell n\ L}{\partial\nu}\end{pmatrix}_{\mu_1,\nu_1} ,$$

with repetition until the increments are small enough; and the covariance matrix of the ml estimators $\hat{\mu},\hat{\nu}$ is then given to $O(1/n)$ by

$$\underset{\sim}{\mathscr{V}}\begin{pmatrix}\hat{\mu}\\ \hat{\nu}\end{pmatrix} = \left[\underset{\sim}{\mathscr{I}}\begin{pmatrix}\hat{\mu}\\ \hat{\nu}\end{pmatrix}\right]^{-1} ,$$

with $\mathscr{E}\begin{pmatrix}\hat{\mu}\\ \hat{\nu}\end{pmatrix} = \begin{pmatrix}\mu\\ \nu\end{pmatrix} + O(1/n)$.

However, a one dimensional iterative system is usually faster and less sensitive to poor starter values, and this - almost quite essential for hand calculation - can be obtained as follows (Shenton (1949)). The likelihood equations are, for $S_i = \Sigma\, x^i\, F_x$,

$$\left.\begin{aligned} -\,S_0 + \frac{1}{\hat{\mu}\hat{\nu}}\,\Sigma\, F_x\,\hat{\Pi}_x &= 0\\ \frac{S_1}{\hat{\nu}} + \frac{1}{\hat{\nu}}\,\Sigma\, F_x\,\hat{\Pi}_x &= 0 \end{aligned}\right\}\qquad \hat{\Pi}_x = \Pi_x(\hat{\mu},\hat{\nu})$$

or, for $\overline{X} = S_1/S_0$,

$$\left.\begin{aligned} \hat{\mu}\hat{\nu} &= \overline{X}\\ \Sigma\, F_x\,\hat{\Pi}_x &= S_1 \end{aligned}\right\} .$$

Eliminating $\hat{\mu}$ from the second m.ℓ. equation by use of the first (which is also a moment equation, it is seen), $\hat{\mu} = \overline{X}/\hat{\nu}$, leads to the (single) m.ℓ. equation

$$\Sigma\, F_x\,\Pi_x(\hat{\nu}) - S_1 = 0 .$$

Consider then the function

$$H(\nu) = \Sigma F_x \Pi_x(\nu) - S_1$$

(where the m.l. equation is $H(\hat{\nu}) = 0$). Then if ν_1 is a first approximation to the root of

$$H(\nu) = 0 ,$$

a second approximation ν_2 is given by the Newton-Raphson method as

$$\nu_2 = \nu_1 \quad \frac{H(\nu_1)}{H'(\nu_1)} ;$$

the process may of course be iterated.

To obtain a suitable expression for $H'(\nu)$, notice (Shenton (1949)) that

$$H'(\nu) = \Sigma F_x \frac{d\Pi_x}{d\nu} = \Sigma F_x \Pi_x \frac{d \ln \Pi_x}{d\nu} .$$

But $\dfrac{d \ln \Pi_x}{d\nu} = \dfrac{d \ln p_{x+1}}{\partial\mu} - \dfrac{d \ln p_x}{d\nu}$

and $\dfrac{d \ln p_x}{d\nu} = \dfrac{\partial \ln p_x}{\partial\mu} \cdot \dfrac{d\mu}{d\nu} + \dfrac{\partial \ln p_x}{\partial\nu}$, with $\mu\nu = \overline{X}$,

$$= \left(-1 + \frac{\Pi_x}{\mu\nu}\right) \cdot \left(-\frac{\mu}{\nu}\right) + \frac{x}{\nu} - \frac{\Pi_x}{\nu}$$

$$= \frac{\mu}{\nu} + \frac{x}{\nu} - \frac{\nu+1}{\nu^2} \Pi_x ,$$

whence

$$\Pi_x \frac{d \ln \Pi_x}{d\nu} = \frac{1}{\nu} \Pi_x - \frac{\nu+1}{\nu^2} \Pi_x (\Pi_{x+1} - \Pi_x) .$$

Writing $\chi_x = \Pi_x(\Pi_{x+1} - \Pi_x)$,

then finally

$$H'(\nu) = \frac{1}{\nu} \Sigma F_x \Pi_x - \frac{\nu+1}{\nu^2} \Sigma F_x \chi_x .$$

The calculations for the application of this method are by no means trivial if done by hand, and can be rather troublesome with respect to accuracy: $\Pi_x = (x+1)p_{x+1}/p_x$ is difficult to determine accurately for large x, since each probability is then small, and χ_x requires the differences of these not very reliable ratios.

For hand computation in appropriate ranges, it is possible to rewrite the probabilities as follows (Douglas (1955)):

$$p_x = p_0 \frac{\nu^x}{x!} \mu'_{x:\lambda} ,$$

where $\mu'_{x:\lambda}$ is the xth moment of a Poisson(λ) variate, $\lambda = \mu e^{-\nu}$, whence

$$p_{x+1} = \nu \frac{p_x}{x+1} p^*_x , \quad p^*_x = \frac{\mu'_{x+1:\lambda}}{\mu'_{x:\lambda}} = \frac{1}{\nu} \Pi_x ,$$

and tables of p^*_x are available; so are tables of

$$q_x = p^*_x(p^*_{x+1} - p^*_x) ,$$

in each case for $x = 0(1)19$, $\lambda = 0(0.001)0.03(0.01)0.3(0.1)3$. But the m.ℓ. equations in this notation are

$$\left.\begin{aligned} \hat{\mu}\hat{\nu} &= \bar{X} \\ \hat{\nu} \Sigma F_x \hat{p}^*_x &= S_1 \end{aligned}\right\}$$

with

$$H(\nu) \equiv \nu \Sigma F_x p^*_x - S_1 \qquad \text{(written } F(\nu) \text{ in Douglas (1955))}$$

and

$$H'(\nu) = \Sigma F_x p^*_x - (1+\nu) \Sigma F_x q_x .$$

Hence the iterative procedure can be applied relatively rapidly and easily by use of the tables if these cover the range required.

As has been remarked in the Appendix on Iterative Procedures, convergence from moment (or other) starter values is speculative; but it is usually better for the one variable case. If the moment estimates do not lead to a m.l. solution, other empirical, convenient, starter values can be taken -

e.g. $(\mu_1,\nu_1) = ((\text{mean})^{1/2}, (\text{mean})^{1/2})$

or $(\mu_1,\nu_1) = (1, \text{mean})$, and the obvious interchange.

(Should the second moment be less than the first, the likelihood will not have a maximum in the region of interest in which both μ and ν are positive, and no further work is necessary.) As a last resort, the likelihood can of course be calculated over a grid of μ,ν values and the maximum point so determined.

"Explicit" asymptotic formulae can be given for the variances and covariances of $\hat{\mu}$ and $\hat{\nu}$, in terms of

$$\phi = \mathscr{E}(\Pi_X^2) = \Sigma\, p_x\, \Pi_x^2 .$$

As in Shenton (1949), noting that $\mathscr{E}(\Pi_X) = \mathscr{E}(X)$ and $\mathscr{E}(X\Pi_x) = \mathscr{E}\{X(X-1)\}$, it follows that

$$\mathscr{E}\left(\frac{\partial\, \ln p_X}{\partial \mu}\right)^2 = \frac{\phi}{\mu^2\nu^2} - 1,$$

$$\mathscr{E}\left(\frac{\partial\, \ln p_X}{\partial \mu} \cdot \frac{\partial\, \ln p_X}{\partial \nu}\right) = -\frac{\phi}{\mu\nu^2} + 1 + \mu ,$$

$$\mathscr{E}\left(\frac{\partial\, \ln p_X}{\partial \nu}\right)^2 = \frac{\phi}{\nu^2} + \frac{\mu}{\nu} - \mu(1+\mu) ;$$

the determinant of the information matrix $\underset{\sim}{\mathscr{I}}$ is thus

$$\frac{n^2}{\mu\nu^3}\{(1+\nu)\phi - \mu\nu^2(\mu+\mu\nu+\nu)\}$$

(the reciprocal of the generalized variance of $(\hat{\mu},\hat{\nu})^t$), asymptotically and its inverse, the first order covariance matrix, is

$$\frac{1}{n} \cdot \frac{1}{(1+\nu)\phi-\mu\nu^2(\mu+\mu\nu+\nu)} \begin{pmatrix} \mu\nu\{\phi+\mu\nu-\mu\nu^2(1+\mu)\}, & -\nu(\mu\nu^2(1+\mu)-\phi \\ -\nu(\mu\nu^2(1+\mu)-\phi) & , \frac{\nu}{\mu}(\phi-\mu^2\nu^2) \end{pmatrix}.$$

Hence

$$\mathscr{V}(\hat{\mu}) = \frac{1}{n} \cdot \frac{\mu\nu\{\phi+\mu\nu-\mu\nu^2(1+\mu)\}}{(1+\nu)\phi-\mu\nu^2(\mu+\mu\nu+\nu)} + 0(1/n^2) = \frac{\beta_1}{n} + \frac{\beta_2}{n^2} + 0(1/n^3) ,$$

say,

$$\mathscr{C}(\hat{\mu},\hat{\nu}) = \frac{-1}{n} \cdot \frac{\nu\ \mu\nu^2(1+\mu)-\phi)}{(1+\nu)\phi-\mu\nu^2(\mu+\mu\nu+\nu)} + 0(1/n^2) ,$$

$$\mathscr{V}(\hat{\nu}) = \frac{1}{n} \cdot \frac{\nu(\phi-\mu^2\nu^2)}{\mu\{(1+\nu)\phi-\mu\nu^2(\mu+\mu\nu+\nu)\}} + 0(1/n^2) = \frac{\delta_1}{n} + \frac{\delta_2}{n^2} + 0(1/n^3),$$

say. These variances and covariances are tabulated and shown graphically by Bowman and Shenton (1967), Shenton and Bowman (1967, 1977)), and some of the graphs follow a little below. The first order generalized variance has been used in efficiency calculations by Katti and Gurland (1962) and Bowman and Shenton (1967). (Katti and Gurland compare moment estimators and minimum chi-squared estimators with m.l. estimators to higher order in $1/n$, giving detailed tabulations.) It may also be noticed that since

$$\phi = \mathscr{E}(\Pi_X^2) = \nu^2\ \mathscr{E}(p_X^2) = \nu^2\ \Sigma\ p_x^*\ p_x ,$$

use of the tables for p_x (at $\hat{\mu},\hat{\nu}$) enables fairly direct calculation of ϕ for use in these expressions.

Application of these results to the numerical example of Neyman (1939) on European cornborers needs the value of ϕ for $\hat{\mu},\hat{\nu}$. This is easy to calculate on a high speed computer at the same time as the m.l. fit is being computed. and the procedure is illustrated with an APL printout for the same data (European cornborers) as shown before (Table 5). The iteration was continued until the moduli of the changes in the estimates was less than 0.00001, and the total computing time on an Amdahl 470V/6 computer was

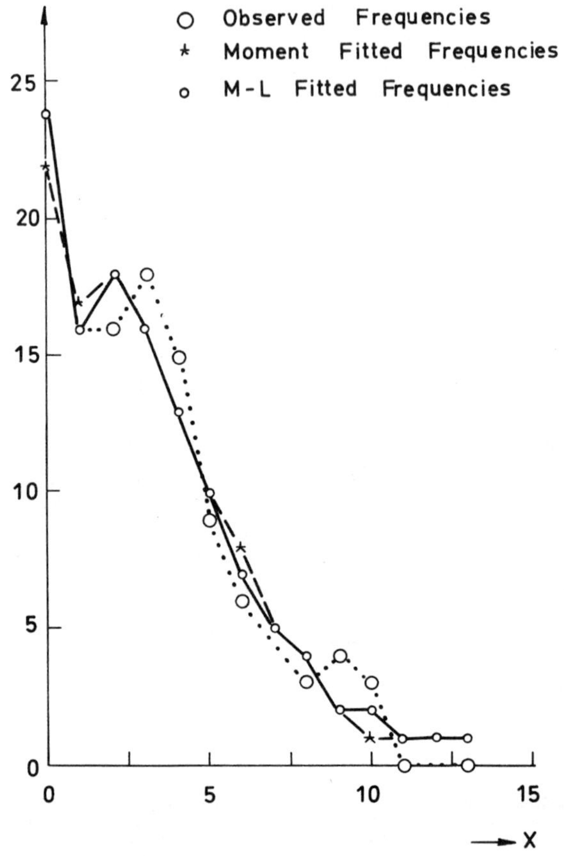

FIGURE 17 European Cornborers: Neyman (1939).

about 0.2 seconds. The maximum likelihood estimates of MU and NU are typed as MUL and NUL, and their estimated standard errors and correlation are not very different in this case from the moment estimates - certainly not in the light of the standard errors. The agreement between the observed and fitted frequencies, also shown graphically, is excellent by both methods. (The m.l. estimates of the Indices of Cluster Frequency and Cluster Size are of course 2.06 clusters per sampling unit and 1.54 larvae per cluster (per sampling unit) respectively.) Ignoring the correlation, "large sample" marginal 95% m.l. confidence intervals are

for μ: (1.37, 2.75)

for ν: (1.03, 2.04)

and it will be noticed that these are shorter than those based on the moment estimates. (But see the remarks in the next section; and an example of Haldane (1942) at the foot of his p. 180.)

However, these results are of low order in $1/n$; and there is no automatic reason to expect for sample sizes ordinarily encountered that higher order terms may not be important. A single illustration from Bowman and Shenton (1967), Shenton and Bowman (1967, 1977) relating to bias is given. Considering $\hat{\mu}$, write

$$\mathscr{E}(\hat{\mu}) = \mu + \frac{\alpha_1}{n} + \frac{\alpha_2}{n^2} + \dots .$$

Then the following diagrams give contours of α_1 and α_2 . It will be noticed that α_1 is large for small ν and that there is a slight irregularity in the region of high modality, but no really unexpected behaviour appears. However, for α_2 the situation is much more complicated, with rapid changes in the value of α_2 in the high modality region (where it is both very large positive and negative and zero). So, for example, for a sample size of 20 (small, but not absurdly so) with $(\mu,\nu) = (4,6)$ the common assertion that for the m.l. estimator $\hat{\mu}$ from such samples

$$\mathscr{E}(\hat{\mu}) = 4 + O(1/n) ,$$

though probably technically valid, is not useful; from the graphs

$$\mathscr{E}(\hat{\mu}) = 4 + \frac{13}{20} + \frac{100}{400} + O(1/n^3) ,$$

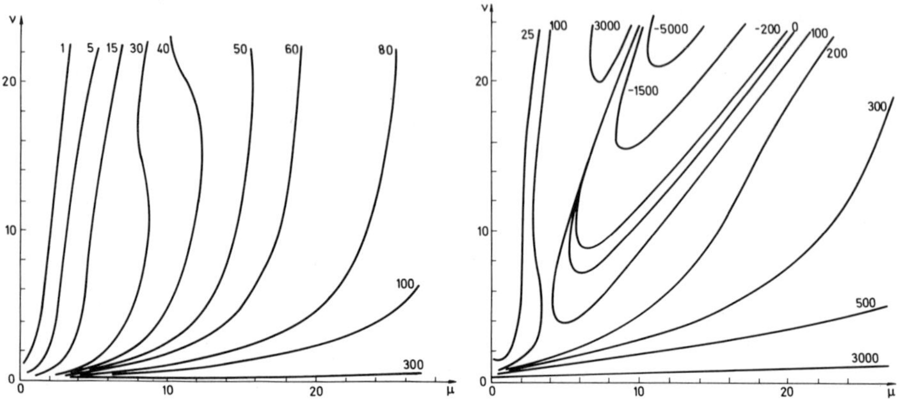

FIGURE 18 Contours of α_1 (*left*) and α_2 (*right*) in the expansion $\mathcal{E}(\hat{\mu}) = \mu + \alpha_1/n + \alpha_2/n^2 + \cdots$. (Bowman and Shenton, 1967)

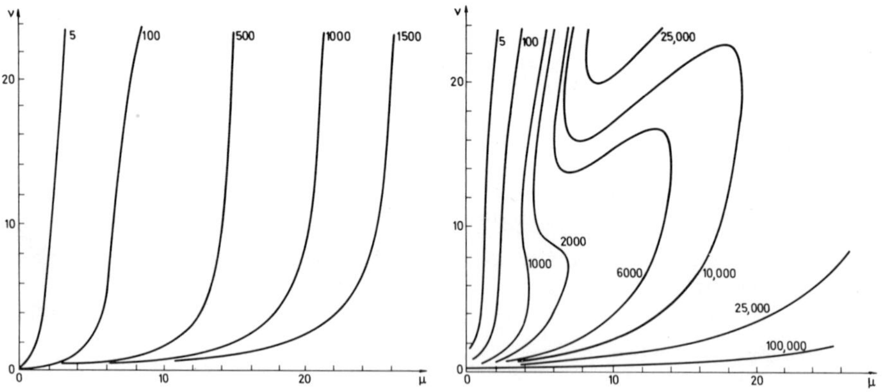

FIGURE 19 Contours of β_1 (*left*) and β_2 (*right*) in the expansion $\mathcal{V}(\hat{\mu}) = \beta_1/n + \beta_2/n^2 + \cdots$. (Bowman and Shenton, 1967)

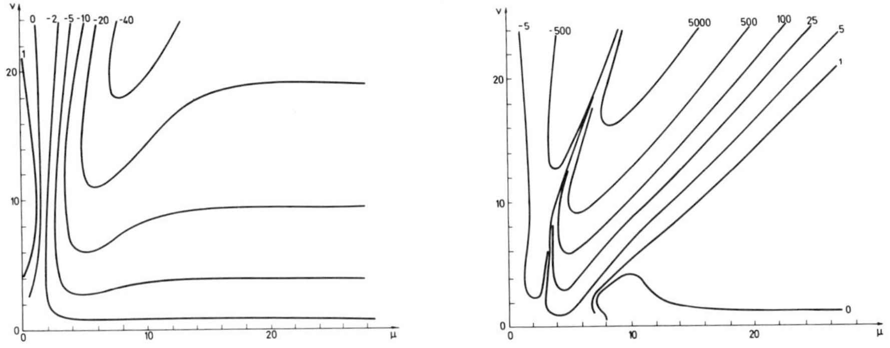

FIGURE 20 Contours of γ_1 (*left*) and γ_2 (*right*) in the expansion $\mathscr{E}(\hat{\nu}) = \nu + \gamma_1/n + \gamma_2/n^2 + \cdots$. (Bowman and Shenton, 1967)

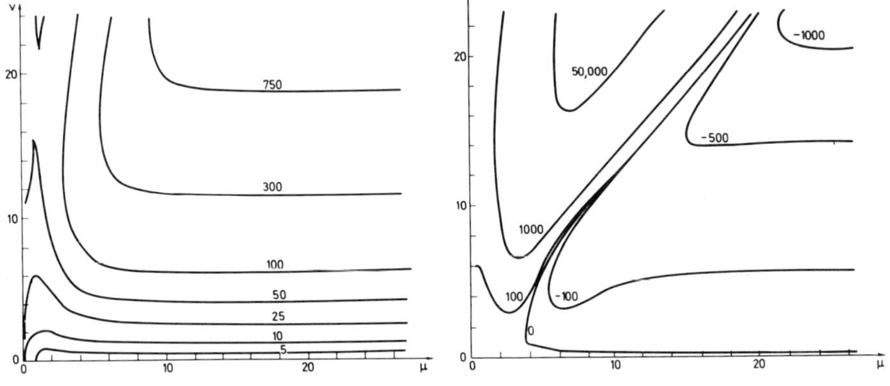

FIGURE 21 Contours of δ_1 (*left*) and δ_2 (*right*) in the expansion $\mathscr{V}(\hat{\nu}) = \delta_1/n + \delta_2/n^2 + \cdots$. (Bowman and Shenton, 1967)

throwing doubt both on the zero order term 4 and the first order result 4.6. But for a sample size of 100,

$$\mathcal{E}(\hat{\mu}) = 4 + \frac{13}{100} + \frac{100}{10000} + O(1/n^3)$$

gives support to the first order result

$$\mathcal{E}(\hat{\mu}) = 4.1 + O(1/n^2) \;;$$

for a sample size of 1000,

$$\mathcal{E}(\hat{\mu}) = 4 + O(1/n)$$

is likely to be useful, since the first order contribution is seen to be 0.01. Of course remarks corresponding to these can be made for the variances; by inversion, as with the moment estimators, for given parameter values a sample size can be determined so that approximations of specified order may be used, though in this case the complicated structure of the estimators does not permit any explicit recognition of regions of the parameter space which are likely to produce problems.

A rough summary for the behaviour of the usual formulae for means and variances of m.l. estimates is rather like (though not of course identical with) that for moment estimators except that, as well as poor behaviour for small ν, the region of high modality leads to uncertain expansions. It is interesting that the more complex m.l. equations appear to reflect the complexity of the distribution in a way that the simpler moment equations do not.

Taking the same numerical example as before, for samples of 100 from a Neyman (1,1) distribution, the Bowman and Shenton tables give the following:

	Terms in n^0	n^{-1}	n^{-2}		Terms in n^0	n^{-1}	n^{-2}
$\mathcal{E}(\hat{\mu}) =$	1.000	+ 0.057	+ 0.009	$\mathcal{E}(\hat{\nu}) =$	1.000	− 0.008	− 0.000
$\mathcal{V}(\hat{\mu}) =$		0.059	+ 0.021	$\mathcal{V}(\hat{\nu}) =$		0.059	+ 0.002
		$(= 0.28^2)$				$(=0.25^2)$	

These suggest an upwards bias of 6 or 7% in $\hat{\mu}$, and a downward of bias of 1% in $\hat{\nu}$, with more "error" in the use of the first terms for the standard error of $\hat{\mu}$ than for $\hat{\nu}$; again, these results are consistent with those from the sampling experiment referred to under moment estimation. It should be observed that the use of the standard error for $\hat{\mu}$ from the term in n^{-1} would lead to a fictitiously high impression of accuracy of estimation.

In order to appreciate the peculiarities of these contours, it may be helpful to look at a simpler case. Consider the estimation of the variance $\mu\nu(1+\nu)$ by the sample second moment about the mean M_2. Standard theory (e.g. Kendall and Stuart I (1963)) gives

$$\mathscr{E}(M_2) = \kappa_2 - \frac{1}{n}\kappa_2 ,$$

$$\mathscr{V}(M_2) = \frac{\kappa_4 + 2\kappa_2^2}{n} - \frac{2(\kappa_4 + \kappa_2^2)}{n^2} + \frac{\kappa_4}{n^3} ,$$

where κ_r is the rth distribution cumulant, and in the present case

$$\kappa_r = \mu \Sigma \mathfrak{S}_r^{(s)} \nu^s , \quad r = 1,2,\ldots .$$

Hence

$$\mathscr{E}(M_2) = \mu\nu(1+\nu)(1 - \tfrac{1}{n}) = \theta_0 + \theta_1/n + \theta_2/n^2 + \ldots$$

$$\text{where } \theta_2 = \theta_3 = \ldots = 0 ,$$

$$\theta_0 = \mu\nu(1+\nu) = -\theta_1 ;$$

$$\mathscr{V}(M_2) = \phi_1/n + \phi_2/n^2 + \phi_3/n^3 + \ldots$$

$$\text{where } \phi_4 = \phi_5 = \ldots = 0$$

$$\text{and for } \Phi = 1 + 7\nu + 6\nu^2 + \nu^3 ,$$

$$\phi_1 = \mu\nu(\Phi + 2\mu\nu(1+\nu)^2) ,$$

$$\phi_2 = -2\mu\nu(\Phi + \mu\nu(1+\nu)^2) ,$$

$$\phi_3 = \Phi .$$

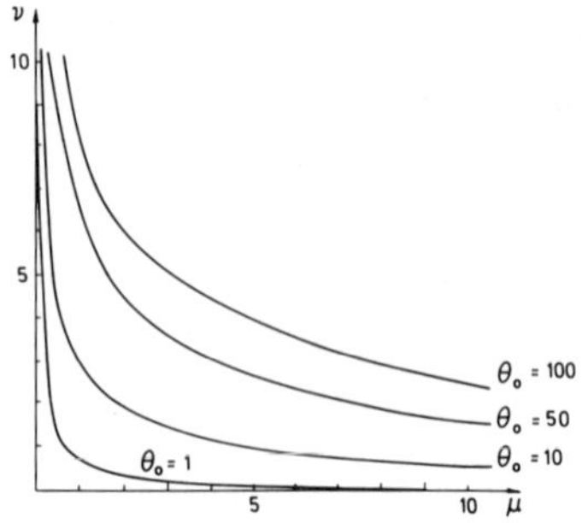

FIGURE 22 Contours of θ_0 in $\mathscr{E}(M_2) = \theta_0 + \theta_1/n + \cdots$ for the Neyman(μ,ν) distribution. (These are also contours of θ_1, $\theta_1 = -\theta_0$).

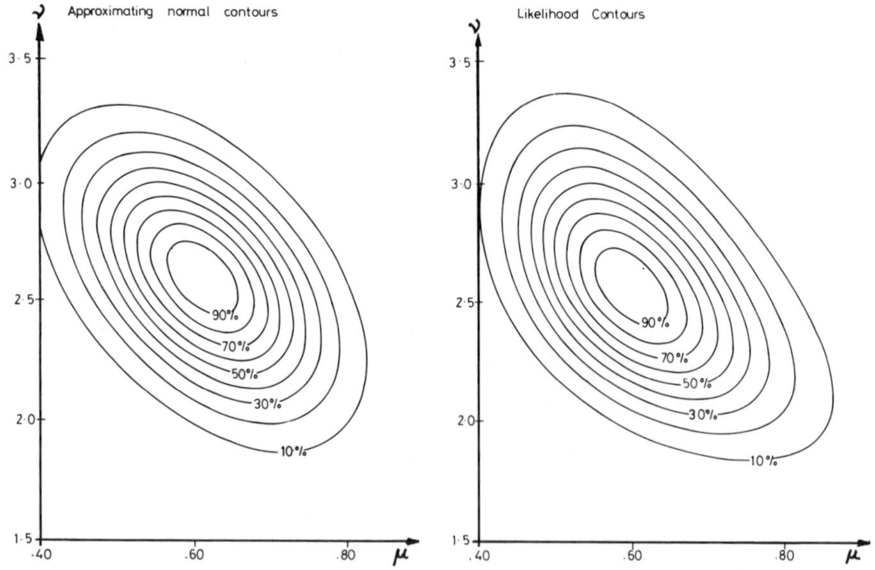

FIGURE 23 Approximating normal contours (*left*) and likelihood contours (*right*) for the Neyman(μ,ν) distribution fit to *Armeria maritima* data. (Thomas, 1949)

Graphs showing the contours of θ_0 and θ_1 (and ϕ_1, ϕ_2 and ϕ_3) are easily constructed - for $\mathcal{E}(M_2)$ they are as in Figure 22. It is to be noticed that they are "smooth" with no loops, and the coefficients large only where they would be expected to be.

As set out in the Appendix on Estimation, likelihood ratio contours can also be used both to convey an idea of the accuracy of estimation and to indicate with faith might be placed in the use of an approximating bivariate normal distribution for that of $(\hat{\mu},\hat{\nu})$. For data from Thomas (1949), of Armeria maritima, the estimates are:

$$\hat{\mu} = 0.581 \quad \text{with} \quad \text{s.e.} = 0.087$$
$$\text{correlation} = -0.357$$
$$\hat{\nu} = 1.717 \quad \text{with} \quad \text{s.e.} = 0.259$$

The contours of the elliptical confidence regions for the bivariate normal approximation are exhibited in Figure 23. As well, in Figure 23, the contours are shown of the ratio of the likelihood at each point to that at the maximizing point, in each case in terms of fractions of the appropriate maximum. (It may also be recalled that minus twice the natural logarithm of the likelihood to the maximum likelihood is asymptotically distributed as chi-squared, in this case with 2 degrees of freedom.) The two sets of contours are different, but not strikingly so; simulation of the sampling distribution of $(\hat{\mu},\hat{\nu})$ leads at any rate quite often to "banana shaped" contours with the convex side towards the origin.

c) Comparison of moment and M.L. Estimators. The most commonly used comparison between estimators is that provided by the ratio of (first order asymptotic terms of) their sampling variances in the case of single parameters, and of their corresponding generalized variances for several parameters. However, whether point or interval estimates are of interest, comparisons with respect to bias also may be of relevance; and various other aspects are examined below.

<u>Bias</u>

The following examples, with theoretical results mostly from Bowman and Shenton (1967), illustrate the kind of behavior present. In each case the sample size is $n = 100$.

(i) $(\mu,\nu) = (1,5)$ (The distribution has 1½ modes.)

	Term which includes n^0	n^{-1}	n^{-2}		Term which includes n^0	n^{-1}	n^{-2}
$\mathcal{E}(\mathring{\mu})$ =	1	+ 0.044	+ 0.001,	$\mathcal{E}(\mathring{\nu})$ =	5	− 0.058	+ 0.000
$\mathcal{E}(\hat{\mu})$ =	1	+ 0.005	+ 0.000,	$\mathcal{E}(\hat{\nu})$ =	5	+ 0.002	+ 0.001

The terms of zero order appear to provide fairly adequate representation; the first order result

$$\mathcal{E}(\mathring{\nu}) = 4.94 + 0(1/n^2) ,$$

appears to be a satisfactory numerical approximation when $n = 100$, e.g., the "correction" of −0.06 (about 1%) being in any case small compared with the standard error.

A small sampling experiment of 100 samples each with $n = 100$, confirmed the reasonable nature of these results: The observed averages of the estimates of μ and ν were:

$$\text{average } (\mathring{\mu}) = 1.04, \qquad \text{average } (\mathring{\nu}) = 4.94$$

$$\text{average } (\hat{\mu}) = 1.01, \qquad \text{average } (\hat{\nu}) = 5.01$$

(ii) $(\mu,\nu) = (10,20)$ (3½ modes)

	n^0	n^{-1}	n^{-2}		n^0	n^{-1}	n^{-2}
$\mathcal{E}(\mathring{\mu})$ =	10	+ 0.376	+ 0.010,	$\mathcal{E}(\mathring{\nu})$ =	20	+ 0.044	+ 0.000
$\mathcal{E}(\hat{\mu})$ =	10	+ 0.381	+ 0.388,	$\mathcal{E}(\hat{\nu})$ =	20	= 0.81	+ 1.21

The greater "instability" of the m.l. estimator is illustrated here.

Efficiency

The first order asymptotic efficiency of the moment with respect to the m.l. estimators ("the" efficiency) is given by

$$\frac{\varepsilon_2}{\phi_2} ,$$

where the determinants of the covariance matrices have the expansions

$$\begin{vmatrix} \mathcal{V}(\hat{\mu}), & \mathcal{C}(\hat{\mu},\hat{\nu}) \\ \mathcal{C}(\hat{\mu},\hat{\nu}), & \mathcal{V}(\hat{\nu}) \end{vmatrix} = \frac{\varepsilon_2}{n^2} + \frac{\varepsilon_3}{n^3} + \ldots$$

$$\begin{vmatrix} \mathcal{V}(\mathring{\mu}), & \mathcal{C}(\mathring{\mu},\mathring{\nu}) \\ \mathcal{C}(\mathring{\mu},\mathring{\nu}), & \mathcal{V}(\mathring{\nu}) \end{vmatrix} = \frac{\phi_2}{n^2} + \frac{\phi_3}{n^3} + \ldots .$$

Contours of this ratio are shown in the Figure 24 (Katti and Gurland (1962), Bowman and Shenton (1967)); and from it "the" efficiency of the moment fit in the case of the Neyman (1939) cornborer data can be seen to be about 50% (for μ about 2 and ν about 1.3). However, any useful interpretation of this depends on the adequacy of the representation of the covariance determinants by ε_2 and ϕ_2 ; for example, including the next terms ε_3 and ϕ_3 gives a second order asymptotic efficiency

$$\frac{\varepsilon_2 + \varepsilon_3/n}{\phi_2 + \phi_3/n} = \frac{\varepsilon_2}{\phi_2} \cdot \frac{1 + (\varepsilon_3/\varepsilon_2)/n}{1 + (\phi_3/\phi_2)/n} .$$

For $n = 100$ and $(\mu,\nu) = (1,1)$ (for which the Bowman and Shenton tables are available) ε_2/ϕ_2 gives a first order efficiency of 52% while that of the second order is

$$52 \cdot \frac{1 + 0.44}{1 + 1.21}\%$$
$$= 34\%$$

throwing considerable suspicion on the validity of the 52%. (In fact, for the Neyman European cornborer data there is little difference - compared with their standard errors - between the moment and m.l. estimates.) Moreover, an examination of the tabulation shows that, perversely, when the first order efficiency is

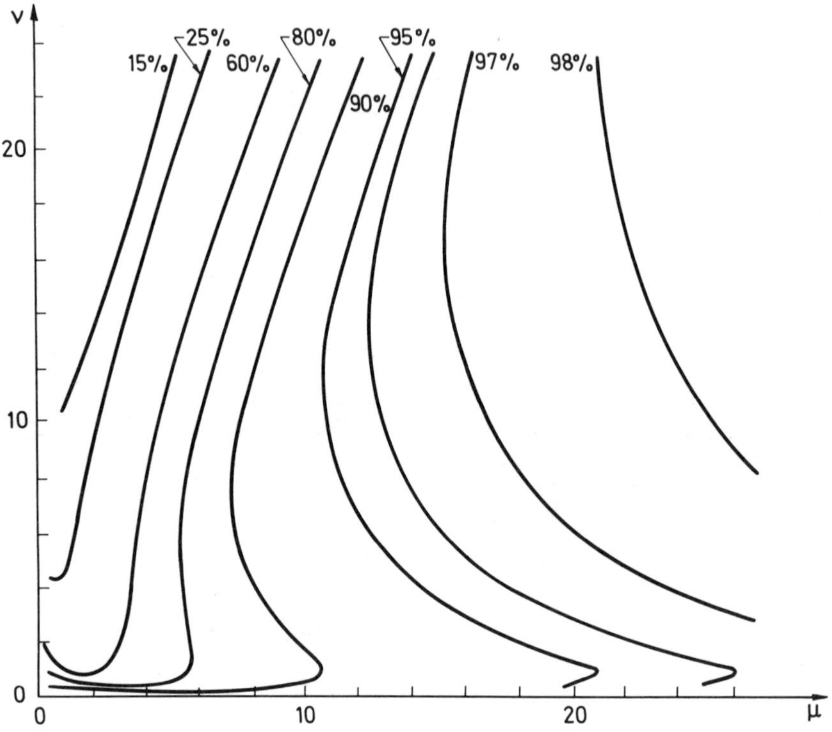

FIGURE 24 First order asymptotic efficiency of moments with resp. to maximum likelihood. (Bowman and Shenton, 1967; Katti and Gurland, 1962)

high the second order terms are often large, and conversely: i.e. a high first order efficiency often corresponds to circumstances where the first order terms in the variances and covariance may represent very poorly the actual variances and covariance.

A small sampling experiment of 300 samples with $n = 100$ for $(\mu,\nu) = (1,1)$, for which "the" efficiency is 52%, reduced to 34% with the second order terms, gave a ratio of determinants of the use of the estimated covariance matrices as 26%. (The covariance matrices were estimated by direct use of the second order sample moments of the observed joint sampling distribution of the estimates, in this and similar cases.) This certainly does not support an automatic appeal to "the" efficiency of a method of estimation, and particularly not in the common application when the parameter values used for the comparison are themselves estimated; there is also strong dependence on the correlations when these are large.

In such an application, the European cornborer data (Neyman (1939)) already quoted, an idea of the implications may be gained by looking at the tabulation for $n = 100$ and $(\mu,\nu) = (1,1)$, these being values not far from those in the experiment. Then the expansions are:

	Terms in n^0	n^{-1}	n^{-2}		Terms in n^0	n^{-1}	n^{-2}
$\mathcal{E}(\overset{o}{\mu}) =$	1	+ 0.130	+ 0.014,	$\mathcal{E}(\overset{o}{\nu}) =$	1	− 0.030	− 0.000,
$\mathcal{E}(\hat{\mu}) =$	1	+ 0.057	+ 0.009,	$\mathcal{E}(\hat{\nu}) =$	1	− 0.007	− 0.000,
$\mathcal{V}(\overset{o}{\mu}) =$		0.110	+ 0.047,	$\mathcal{V}(\overset{o}{\nu}) =$		0.110	− 0.005,
$\mathcal{V}(\hat{\mu}) =$		0.059	+ 0.021,	$\mathcal{V}(\hat{\nu}) =$		0.059	+ 0.002.

Based on leading terms, confidence intervals for ν are clearly likely to be satisfactory (though if the moment estimate were $\overset{o}{\nu} = 1$ perhaps the interval would be better centered at 0.97 rather than at 1 - since the estimated (first order) standard error is 0.3, the difference in unimportant); those for μ are much less satisfactory, the second terms being substantial percentages of the first for both bias and variance. A sample size of 120 is thus not likely to be adequate for reasonable estimation of μ (and, indeed, if one wishes to be at all precise in its estimation, not for ν).

It may also be remarked that although m.l. is (correctly enough) regarded as a method of estimating parameters, it actually proceeds by maximizing a certain function of the probabilities (themselves functions of the parameters) in a way which lays a good deal of emphasis on the agreement of fitted and observed frequencies. (Readers may recall the asymptotic equivalence of the classical minimum chi-squared and m.l. procedures.) A fairly considerable experience of fitting a variety of contagious distributions by a variety of methods leads to the empirical observation that, compared with moments (for example), while m.l. usually leads to a no worse fit as measured by chi-squared (always no worse as measured by the likelihood, of course), the difference is greatly exaggerated in importance when looked at through the frequencies rather than through the estimates of the parameters. In simulations, with known values of the parameters in the parent distribution, it is common to find that while the fit to a random sample as observed is better for m.l. estimates than for the moment estimates, the m.l. estimates are further from the correct parameter values than are the moment estimates; in the majority of cases the differences between the two methods are much smaller than appears to be commonly supposed. (These remarks are not meant to imply that the sampling properties of m.l. estimators are not preferable to those of moment estimators: m.l. estimators are usually to be preferred on this score. But the advantage is smaller than is generally thought to be the case, and certainly less than referees and editors of papers in many journals appear to think; since the computational effort for a m.l. solution often vastly exceeds that for other methods, the point is worth making.)

Finally, and still more qualitatively, attention may again be drawn to the facts, that although corresponding terms in the asymptotic expansions for moments of moment estimators are often greater than for those of m.l. estimators this is not invariably the case, and that there is a region (of high modality) of instability for the m.l. expansions which does not exist for moment expansions. But even more important in this context is the lack of justification for supposing that solely the first terms of asymptotic expansions of the sampling moments of an estimator can be used to recommend or otherwise a particular method of estimation: much more attention is necessary to the appropriateness of the expansions in relation to sample size.

Asymptotically Uncorrelated Estimators

Regarding the estimation question more abstractly, one may seek a re-parameterization of the distribution to simplify the analysis and interpretation of questions of the kind raised above. A particular nuisance is the high degree of correlation in the joint distribution of $\overset{o}{\mu}$ and $\overset{o}{\nu}$ (and of $\hat{\mu}$ and $\hat{\nu}$); it is natural to seek to reduce or eliminate this. As already mentioned, the choice of

$$\alpha = \mu\nu \ , \ \lambda = \mu e^{-\nu}$$

leads to asymptotically uncorrelated estimators:

$$\overset{o}{\alpha} = \overset{o}{\mu}\overset{o}{\nu} \qquad\qquad \hat{\alpha} = \hat{\mu}\hat{\nu}$$

and

$$\overset{o}{\lambda} = \overset{o}{\mu} e^{-\overset{o}{\nu}} \qquad\qquad \hat{\lambda} = \hat{\mu} e^{-\hat{\nu}}$$

are of this character.

In other areas it sometimes happens that an adjustment for one deficiency helps another - as variance stabilization may partly normalize. It might thus have been hoped that the large standard errors characteristic even for large sample sizes for μ and ν might have been less troublesome with respect to α and λ. (Note, however, that if the experimenter really does need to estimate μ and ν, dealing with α and λ can help only in the mathematical or numerical analysis: inevitably, in transforming back to μ and ν both the high variabilities and correlation will be re-introduced.) It is clear that "the" efficiency of moment estimation remains unchanged, since writing

$$\underset{\sim}{0} = (\alpha,\lambda)^t \quad , \quad \underset{\sim}{N} = (\mu,\nu)^t$$

(orthogonal) (natural)

then $$\underset{\sim}{\mathcal{V}}(\overset{o}{\underset{\sim}{0}}) = \frac{\partial \underset{\sim}{0}}{\partial \underset{\sim}{N}} \underset{\sim}{\mathcal{V}}(\overset{o}{\underset{\sim}{N}}) \left(\frac{\partial \underset{\sim}{0}}{\partial \underset{\sim}{N}}\right)^t$$

and $$\underset{\sim}{\mathcal{V}}(\hat{\underset{\sim}{0}}) = \frac{\partial \underset{\sim}{0}}{\partial \underset{\sim}{N}} \underset{\sim}{\mathcal{V}}(\hat{\underset{\sim}{N}}) \left(\frac{\partial \underset{\sim}{0}}{\partial \underset{\sim}{N}}\right)^t ,$$

to the first terms of the expansions, whence the ratios of the generalized variances are the same. Nevertheless, the interpretation of the generalized variances of $\overset{o}{\underset{\sim}{0}}$ and $\hat{\underset{\sim}{0}}$ is much more direct, since only the diagonal elements - the variances - contribute to them, and so to this extent such a comparison is more meaningful.

To illustrate the behavior, a sampling experiment consisting of two sets of 300 samples of $n = 100$ observations from a parent with $(\mu,\nu) = (1,1)$ (or $\alpha = 1$, $\lambda = 0.3678\ldots$) will be quoted. The theoretical (first terms of the asymptotic expressions) and observed results were:

	Moment		m.l.	
	Theoretical	Observed	Theoretical	Observed
α mean	1	1.00	1	1.00
α s.d.	0.14	0.14	0.14	0.14
λ mean	0.37	0.50	0.37	0.44
λ s.d.	0.24	0.37	0.17	0.24
corr. (α,λ)	0	0.01	0	0.01

Since α is in fact the mean (and the Central Limit Theorem applies very powerfully for $n = 100$), the results for α are excellent. The correlations between the estimators have also been reduced very satisfactorily; but both the means and standard errors of the moment and m.l. estimators of λ are as misleading as for μ and ν in the previous parameterization. Taking the ratios of the covariance determinants, "the" efficiency of $(\mathring{\alpha},\mathring{\lambda})$, with respect to $(\hat{\alpha},\hat{\lambda})$, is 51.7% (as is $(\mathring{\mu},\mathring{\nu})$ with respect to $(\hat{\mu},\hat{\nu})$); for the observed covariance determinants the ratio is 42.6% (compared with 25.7% for the ratio for the observed covariance in terms of (μ,ν). It is perhaps plausible that efficiency calculations make more sense for uncorrelated, or nearly uncorrelated, estimators.

d) Minimum Chi-squared estimation (generalized least squares). A general account of this procedure is in the Appendix on Estimation; a description is given here for this particular case. Writing $\underset{\sim}{\lambda} = (\mu,\nu)^t$ for the indexing parameter vector of the Neyman(μ,ν) distribution, one seeks the minimum with respect to $\underset{\sim}{\lambda}$ of the asymptotically chi-squared distributed quadratic form

$$(\underset{\sim}{T} - \underset{\sim}{\tau})^t \; \underset{\sim}{i} \; (\underset{\sim}{T} - \underset{\sim}{\tau}) ,$$

where $\underset{\sim}{T}$ is a vector of asymptotically normally distributed estimators $T_1,\ldots,T_m$ with asymptotic mean vector $\underset{\sim}{\tau} = (\tau_1,\ldots,\tau_m)^t = \underset{\sim}{\tau}(\underset{\sim}{\lambda})$ and asymptotic covariance matrix $\mathcal{V}(\underset{\sim}{T})$, $\underset{\sim}{i}$ being an estimate of $\mathcal{V}(\underset{\sim}{T})^{-1}$ based on a consistent estimator, the asymptotic properties being in powers of $1/n$ where n is the sample size. The conditions for a minimum so obtained will usually lead to non-linear equations, and thus a choice of $\underset{\sim}{T}$, at this stage rather arbitrary, may be made to obtain linear, and hence readily solved, equations.

In order to do this, re-parameterization of the original distribution is convenient: in place of $\underset{\sim}{\lambda} = (\mu,\nu)^t$ consider the specification (in general)

$$\underset{\sim}{\theta} = (\theta_1,\theta_2)^t$$

where

$$\underset{\sim}{\lambda} = \underset{\sim}{\lambda}(\underset{\sim}{\theta})$$

gives the functional relation between $\underset{\sim}{\lambda}$ and $\underset{\sim}{\theta}$, and $\underset{\sim}{\theta}$ is chosen so that the relation between $\underset{\sim}{\tau} \sim \mathcal{E}(\underset{\sim}{T})$ and $\underset{\sim}{\theta}$ is linear:

$$\underset{\sim}{\tau} = \underset{\sim}{w}\,\underset{\sim}{\theta}\ , \text{ where } \underset{\sim}{w} \text{ is constant.}$$

As a first choice, because of their simplicity (Hinz and Gurland (1967)), took $\underset{\sim}{\tau}$ to be the ratios of factorial cumulants, written as $\underset{\sim}{\eta}$ for this particular choice:

$$\eta_0 = \kappa_{(1)}\ ,\ \eta_r = \kappa_{(r+1)}/\kappa_{(r)}$$

$$\text{for } r = 1,2,\ldots,m-1\ ,$$

with $\underset{\sim}{T}$, written as $\underset{\sim}{H}$ here, for the corresponding sample estimators. Then

$$\eta_0 = \mu\nu \quad \text{and} \quad \eta_r = \nu \quad \text{for} \quad r = 1,2,\ldots,$$

and hence the choice of the reparameterization

$$\underset{\sim}{\theta} = (\theta_1,\theta_2)^t \quad \text{with} \quad \theta_1 = \mu\nu,\ \theta_2 = \nu\ ,$$

for which $\underset{\sim}{\eta} = \underset{\sim}{w}\,\underset{\sim}{\theta}$ is certainly satisfied by

$$\underset{\sim}{\eta} \equiv \begin{pmatrix} \mu\nu \\ \nu \\ \cdots \\ \nu \end{pmatrix} = \begin{pmatrix} 1 & 0 \\ 0 & 1 \\ & \cdots \\ 0 & 1 \end{pmatrix} \begin{pmatrix} \theta_1 \\ \theta_2 \end{pmatrix} \equiv \underset{\sim}{w}\,\underset{\sim}{\theta}$$

Then the linear normal equations (ℓmc[A]) (see the Appendix for equations labelled thus) for the minimizing $\underset{\sim}{\theta}$, written $\underset{\sim}{\theta}*$, are

$$\underset{\sim}{\theta}* = (\underset{\sim}{w}^t \underset{\sim}{i}\, \underset{\sim}{w})^{-1}\, \underset{\sim}{w}^t\, \underset{\sim}{i}\, \underset{\sim}{H}$$

where $\underset{\sim}{i}$ is an estimate of $\underset{\sim}{\mathcal{V}}(\underset{\sim}{H})^{-1}$; and hence $\underset{\sim}{\lambda}*$ is obtained from $\underset{\sim}{\theta}*$ as

$$\underset{\sim}{\lambda}^* = (\mu^*, \nu^*)^t = (\theta_1^*/\theta_2^*\ ,\ \theta_2^*)^t\ .$$

From (ℓmc[C]) it follows that

$$\underset{\sim}{\mathcal{V}}(\underset{\sim}{\lambda}^*) = \begin{pmatrix} 1/\nu\ , & -\mu/\nu \\ 0 & 1 \end{pmatrix} \underset{\sim}{\mathcal{V}}(\underset{\sim}{\theta}^*) \begin{pmatrix} 1/\nu\ , & 0 \\ -\mu/\nu\ , & 1 \end{pmatrix},$$

where (ℓmc[B])

$$\underset{\sim}{\mathcal{V}}(\underset{\sim}{\theta}^*) = (\underset{\sim}{w}^t \underset{\sim}{\mathcal{V}}(\underset{\sim}{H})^{-1}\, \underset{\sim}{w})^{-1}$$

and finally

$$\underset{\sim}{\mathcal{V}}(\underset{\sim}{H}) = \frac{\partial \underset{\sim}{\eta}}{\partial \underset{\sim}{\kappa}_{(\ )}} \cdot \frac{\partial \underset{\sim}{\kappa}_{(\ )}}{\partial \underset{\sim}{\mu}'} \cdot \underset{\sim}{\mathcal{V}}(\underset{\sim}{M}') \left(\frac{\partial \underset{\sim}{\eta}}{\partial \underset{\sim}{\kappa}_{(\ )}} \cdot \frac{\partial \underset{\sim}{\kappa}_{(\ )}}{\partial \underset{\sim}{\mu}'} \right)^t ,$$

$\underset{\sim}{M}'$ and $\underset{\sim}{\mu}'$ being the m-dimensional vectors of sample and distribution moments about the origin, respectively. In applications it is of course necessary to use estimates of all of these parametric expressions.

As a practical rule, the choice of $m = 3$, i.e. of $\underset{\sim}{H} = (H_0, H_1, H_2)^t$, is often likely to be satisfactory (unless, perhaps, the zero class proportion is large, for which see the next section), and so the formulae are written out explicitly for this case. With the reparameterization $\underset{\sim}{\theta} = (\theta_1,\theta_2)^t$ where $\theta_1 = \mu\nu$, $\theta_2 = \nu$, and $\underset{\sim}{\eta} = (\kappa_{(1)}, \kappa_{(2)}/\kappa_{(1)}, \kappa_{(3)}/\kappa_{(2)})^t$, $\underset{\sim}{\eta} = \underset{\sim}{w}\,\underset{\sim}{\theta}$ is satisfied by choosing

$$w = \begin{pmatrix} 1 & 0 & 0 \\ 0 & 1 & 1 \end{pmatrix}^t .$$

Since

$$H_0 = K_{(1)} = M_1' = M ,$$

$$H_1 = K_{(2)}/K_{(1)} , \text{ where } K_{(2)} = M_2' - M^2 - M ,$$

$$H_2 = K_{(3)}/K_{(2)} , \text{ where } K_{(3)} = M_3' + M_2'(-3M-3) + 2M^2 + 3M^2 + 2M ,$$

and $\underset{\sim}{\eta} = (\mu\nu,\nu,\nu)^t = (\theta_1,\theta_2,\theta_2)^t$,

$$\underset{\sim}{\mathcal{V}}(\underset{\sim}{H}) = \frac{\partial\underset{\sim}{\eta}}{\partial\underset{\sim}{\kappa}_{(\,)}} \cdot \frac{\partial\underset{\sim}{\kappa}_{(\,)}}{\partial\underset{\sim}{\mu}}\, \underset{\sim}{\mathcal{V}}(\underset{\sim}{M}')\left(\frac{\partial\underset{\sim}{\eta}}{\partial\underset{\sim}{\kappa}_{(\,)}} \cdot \frac{\partial\underset{\sim}{\kappa}_{(\,)}}{\partial\underset{\sim}{\mu}}\right)^t , = \underset{\sim}{J}\,\underset{\sim}{\mathcal{V}}(\underset{\sim}{M}')\underset{\sim}{J}^t , \text{ say,}$$

$$\frac{\partial\underset{\sim}{\eta}}{\partial\underset{\sim}{\kappa}_{(\,)}} = \begin{pmatrix} 1 & , & 0 & , & 0 \\ -\kappa_{(2)}/\kappa_{(1)}^2 & , & 1/\kappa_{(1)} & , & 0 \\ 0 & , & -\kappa_{(3)}/\kappa_{(2)}^2 & , & 1/\kappa_{(2)} \end{pmatrix}$$

$$\frac{\partial\underset{\sim}{\kappa}_{(\,)}}{\partial\underset{\sim}{\mu}} = \begin{pmatrix} 1 & , & 0 & , & 0 \\ -1-2\mu_1' & , & 1 & , & 0 \\ -3\mu_2' + 6\mu_1'^2 + 6\mu_1' + 2 & , & -3\mu_1' - 3 & , & 1 \end{pmatrix},$$

and $$\underset{\sim}{\mathcal{V}}(\underset{\sim}{M}') = \frac{1}{n}\begin{pmatrix} \mu_2' - \mu_1'^2 & , & \mu_3' - \mu_2'\mu_1' & , & \mu_4' - \mu_3'\mu_1' \\ \mu_3' - \mu_2'\mu_1' & , & \mu_4' - \mu_2'^2 & , & \mu_5' - \mu_3'\mu_2' \\ \mu_4' - \mu_3'\mu_1' & , & \mu_5' - \mu_4'\mu_1' & , & \mu_6' - \mu_3'^2 \end{pmatrix} .$$

Hence an estimate of θ^* is obtained from

$$\underset{\sim}{\theta}^* = (\underset{\sim}{w}^t \, \underset{\sim}{\mathcal{V}}(\underset{\sim}{H})^{-1} \, \underset{\sim}{w})^{-1} \, \underset{\sim}{w}^t \underset{\sim}{\mathcal{V}}(\underset{\sim}{H})^{-1} \, \underset{\sim}{H}$$

by replacing the elements of $\underset{\sim}{H}$ and $\underset{\sim}{\mathcal{V}}(\underset{\sim}{H})$ by their sample estimates and so an estimate of λ^* from

$$\underset{\sim}{\lambda}^* = (\theta_1^*/\theta_2^*, \theta_2^*)^t \ , = (\mu^*, \nu^*)^t \ .$$

Finally, the estimated covariance of $\underset{\sim}{\lambda}^*$ is found by using these estimates in

$$\underset{\sim}{\mathcal{V}}(\underset{\sim}{\lambda}^*) = \begin{pmatrix} 1/\nu \ , & -\mu/\nu \\ 0 \ , & 1 \end{pmatrix} (\underset{\sim}{w}^t \, \underset{\sim}{\mathcal{V}}(\underset{\sim}{H})^{-1} \, \underset{\sim}{w})^{-1} \begin{pmatrix} 1/\nu \ , & 0 \\ -\mu/\nu \ , & 0 \end{pmatrix} .$$

Applied to the European Cornborer data (Neyman (1939)) already quoted this procedure yields the following Table 6. Without repetition of all the previous cautions, it will again be noticed that the estimated standard errors are rather large, and the correlation typically numerically large; no analysis is given, e.g., with respect to bias.

However, if the zero class is large, commonsense suggests this should be specifically included, and indeed Hinz and Gurland (1967) found that the first order asymptotic efficiency of (H_0, H_1, A) , where A is a function (given explicitly below) of the zero class proportion P_0 , exceeded that for (H_0, H_1, H_2) and even that for (H_0, H_1, H_2, H_3) ; that of (H_0, H_1, H_2, A) was even better, and is shown in the short Table 7 following.

Hence both commonsense and asymptotic efficiency calculations suggest the use of low order sample cumulants and the zero class proportion P_0 . To preserve linearity, the sample function

$$A = P_0 H_1 \exp\left[\frac{H_0}{H_1} \{1 - \exp(-H_1)\}\right]$$

is adjoined to $\underset{\sim}{H}$, and the vector of statistics written

$$\underset{\sim}{T} = (H_0, H_1, \ldots, H_{m-1} \ \vdots \ A)^t = (\underset{\sim}{H} \ \vdots \ A)^t \ ;$$

the corresponding distributional expression is

TABLE 6

EUROPEAN CORNBORERS

(Neyman, 1939)

	Observed	Fit on (H_0,H_1,H_2)
0	24	24.3
1	16	17.2
2	16	18.6
3	18	16.3
4	15	13.1
5	9	9.9
6	6	7.1
7	5	4.9
8	3	3.3
9	4	2.1
10	3	1.3
11	0	.8
12	1	.5
13+	0	.6
	120	120.0

ℓn likelihood = -269.8531

Chi-squared = 5.03 with 12 classes

Hinz and Gurland chi-squared = 1.88, 1 d.f.

$\mu^* = 2.08$ with s.e. = 0.41
correlation = -0.92
$\nu^* = 1.46$ with s.e. = 0.27

TABLE 7

FIRST ORDER ASYMPTOTIC PERCENTAGE EFFICIENCY OF (H_0,H_1,H_2,A)

(Hinz and Gurland, 1967)

ν \ μ	0.1	0.3	0.5	1.0	1.5	2.0
5.0	99	99	97	92	91	-
2.0	100	100	100	99	99	99
1.5	100	100	100	100	99	100
0.1-1.0	100	100	100	100	100	100

$$\underset{\sim}{\tau} = (\underset{\sim}{\eta} \vdots \alpha)^t$$

where, for the Neyman(μ,ν) distribution, $\alpha = \nu$. The linear equation $\underset{\sim}{\tau} = \underset{\sim}{w}\underset{\sim}{\theta}$ is satisfied by taking

$$\underset{\sim}{w} = \begin{pmatrix} 1 & 0 & \dots & 0 & \vdots & 0 \\ 0 & 1 & \dots & 1 & \vdots & 1 \end{pmatrix}^t ,$$

and, for $\underset{\sim}{i}$ an estimate of $\mathcal{V}(\underset{\sim}{T})^{-1}$, the previous equations (with the new meanings for $\underset{\sim}{T}$, $\underset{\sim}{w}$ and $\underset{\sim}{i}$) still apply:

[ℓmc(A)] $\quad \underset{\sim}{\theta}^* = (\underset{\sim}{w}^t \underset{\sim}{i}\, \underset{\sim}{w})^{-1}\, \underset{\sim}{w}^t \underset{\sim}{i}\, \underset{\sim}{T}$ with $\underset{\sim}{T}$ in place of $\underset{\sim}{H}$,

$$\underset{\sim}{\lambda}^* = (\theta_1^*/\theta_2^*, \theta_2^*)^t ,$$

and [ℓmc(B)] and [ℓmc(C)] . Then, by the partitioning indicated above, the results for $\mathcal{V}(\underset{\sim}{T})$ can be expressed in terms of those known for $\underset{\sim}{H}$: they are given in the Appendix as equations [ℓmc(E)] , [(F)] and [(G)] .

An explicit setting out of the formulae for the practically recommended choice of $\underset{\sim}{T} = (H_0, H_1, A)^t$ is given, corresponding to the earlier choice of $\underset{\sim}{H} = (H_0, H_1, H_2)^t$. They are

$$H_0 = K_{(1)} = M_1' = M ,$$

$$H_1 = K_{(2)}/K_{(1)} \text{ , where } K_{(2)} = M_2' - M^2 - M ,$$

$$A = P_0\, H_1 \exp\left[\frac{H_0}{H_1}\ \{1 - \exp(-H_1)\}\right] ,$$

with the reparameterization $\underset{\sim}{\theta} = (\theta_1, \theta_2)^t$ where $\theta_1 = \mu\nu$, $\theta_2 = \nu$,

and $\underset{\sim}{\tau} = (\kappa_{(1)}, \kappa_{(2)}/\kappa_{(1)}, \alpha)^t = (\eta_0, \eta_1, \alpha)^t$

with $\alpha = p_0 \eta_1 \exp\left[\frac{\eta_0}{\eta_1}\{1 - \exp(-\eta_1)\}\right], = \nu$

so $\underset{\sim}{\tau} = \underset{\sim}{w}\ \underset{\sim}{\theta}$ is (still) satisfied by choosing

$$\underset{\sim}{w} = \begin{pmatrix} 1 & 0 & 0 \\ 0 & 1 & 1 \end{pmatrix}^t ,$$

$$\underset{\sim}{\tau} = (\mu\nu, \nu, \nu)^t = (\theta_1, \theta_2, \theta_2)^t .$$

Then $\underset{\sim}{\mathcal{V}}(\underset{\sim}{T}) = \dfrac{\partial \underset{\sim}{\tau}}{\partial \underset{\sim}{\sigma}} \underset{\sim}{\mathcal{V}}(\underset{\sim}{S}) \left(\dfrac{\partial \underset{\sim}{\tau}}{\partial \underset{\sim}{\sigma}}\right)^t$, where

$$\frac{\partial \underset{\sim}{\tau}}{\partial \underset{\sim}{\sigma}} = \left(\begin{array}{cc:c} 1 & 0 & 1 \\ 0 & 1 & 0 \\ \hdashline \dfrac{\partial \alpha}{\partial \eta_0} & \dfrac{\partial \alpha}{\partial \eta_1} & \dfrac{\partial \alpha}{\partial p_0} \end{array}\right) ,$$

$$(\underset{\sim}{S}) = \begin{pmatrix} \underset{\sim}{}(\underset{\sim}{\not{\mu}})^t & \dfrac{-p_0}{n} \underset{\sim}{\not{J}} \underset{\sim}{\not{\mu}}' \\ \dfrac{-p_0}{n} (\underset{\sim}{\not{J}} \underset{\sim}{\not{\mu}})^t & \dfrac{p_0(1-p_0)}{n} \end{pmatrix} , \quad \underset{\sim}{\not{\mu}}' = \begin{pmatrix} \mu_1' \\ \mu_2' \end{pmatrix} ,$$

$$\underset{\sim}{\mathcal{V}}(\underset{\sim}{\not{\mu}}) = \underset{\sim}{\not{J}} \underset{\sim}{\mathcal{V}}(\underset{\sim}{\not{M}}') \underset{\sim}{\not{J}}^t ,$$

$$\underset{\sim}{\not{J}} = \begin{pmatrix} 1 & , & 0 \\ -\kappa_{(2)}/\kappa_{(1)}^2 & , & 1/\kappa_{(1)} \end{pmatrix} \begin{pmatrix} 1 & , & 0 \\ -1-2\mu_1' & , & 1 \end{pmatrix} , \quad \underset{\sim}{\not{M}}' = \begin{pmatrix} M \\ M_2' \end{pmatrix} ,$$

and $$\underset{\sim}{\mathcal{V}}(\underset{\sim}{\not{M}}) = \frac{1}{n} \begin{pmatrix} \mu_2' - \mu_1'^2 & , & \mu_3' - \mu_2'\mu_1' \\ \mu_3' - \mu_2'\mu_1' & , & \mu_4' - \mu_2'^2 \end{pmatrix} .$$

(It will be noticed that $\underset{\sim}{\not{J}}$, and $\underset{\sim}{\not{M}}'$ are upper sub-matrices and vectors of $\underset{\sim}{J}$ and $\underset{\sim}{M}'$ in the previous section.) As before, estimates of $\underset{\sim}{\theta}^*$ follow from

$$\underset{\sim}{\theta}^* = (\underset{\sim}{w}^t \underset{\sim}{\mathcal{V}}(\underset{\sim}{T})^{-1} \underset{\sim}{w})^{-1} \underset{\sim}{w}^t \underset{\sim}{\mathcal{V}}(\underset{\sim}{T})^{-1} \underset{\sim}{T}$$

by replacing the elements of $\underset{\sim}{T}$ and $\underset{\sim}{\mathcal{V}}(\underset{\sim}{T})$ by their sample estimates whence estimates of

$$\underset{\sim}{\lambda}^* = (\theta_1^*/\theta_2^*, \theta_2^*)^t \ , = (\mu^*, \nu^*)^t \ ,$$

and $$\underset{\sim}{\mathcal{V}}(\underset{\sim}{\lambda}^*) = \begin{pmatrix} 1/\nu \ , & -\mu/\nu \\ 0 \ , & 1 \end{pmatrix} (\underset{\sim}{w}^t \ \underset{\sim}{\mathcal{V}}(\underset{\sim}{T})^{-1} \underset{\sim}{w})^{-1} \begin{pmatrix} 1/\nu & , & 0 \\ -\mu/\nu & , & 1 \end{pmatrix} .$$

(If the vector of statistics were taken to be $\underset{\sim}{T} = (H_0, H_1, H_2 \ \vdots \ A)^t$, the above results are unchanged except for the obvious

$$\underset{\sim}{\tau} = (\eta_0, \eta_1, \eta_2 \ \vdots \ \alpha)^t \ , \ \underset{\sim}{w} = \begin{pmatrix} 1 & 0 & 0 & \vdots & 0 \\ 0 & 1 & 1 & \vdots & 1 \end{pmatrix}^t ,$$

$$\frac{\partial \underset{\sim}{\tau}}{\partial \underset{\sim}{\sigma}} = \left[\begin{array}{ccc|c} 1 & 0 & 0 & 0 \\ 0 & 1 & 0 & 0 \\ 0 & 0 & 1 & 0 \\ \hline \frac{\partial \alpha}{\partial \eta_0} & \frac{\partial \alpha}{\partial \eta_1} & 0 & \frac{\partial \alpha}{\partial p_0} \end{array}\right] ,$$

with $\underset{\sim}{\not J}$, $\underset{\sim}{\not M}$ being replaced by $\underset{\sim}{J}$, $\underset{\sim}{M}$.)

For comparison with the previous fits, the European Cornborer data (Neyman (1939)) serve to illustrate in a case for which the zero class frequency is not relatively large. The fitted frequencies are not reproduced (they are really very similar), but the estimates obtained are as follows.

$$\mu^* = 2.09 \quad \text{with} \quad \text{s.e.} = 0.50$$
$$\text{correlation} = -0.95$$
$$\nu^* = 1.50 \quad \text{with} \quad \text{s.e.} = 0.24$$

ℓn-likelihood = -269.7101

chi-squared = 4.17 with 12 classes

Hinz and Gurland chi-squared = 0.30 with 1 d.f. (see the Appendix).

An illustration of the fitting procedures is given by data on the distribution of retail food stores in Ljubljana, Yugoslavia, taken from Rogers (1969), Table 1.2. Here 144 quadrats were taken over a map of the city and the numbers of stores per quadrat counted, leading to the observed frequency distribution. There is reason to expect some degree of clustering with retail stores, and one purpose in carrying out the fitting is to investigate this, and to compare it, e.g., with other types of stores and across other cities. Thus, interpreting ν as the number of stores per cluster, these data suggest there are about 2 per cluster; corresponding data for non-food stores (not quoted here) suggest there are about 4 per cluster, a rather higher figure consistent with the notion that food being a more or less daily purchase food stores are more widely dispersed and hence less clustered. (See Table 8.)

The estimates of the parameters are accompanied by their estimated standard errors, estimated correlation (which is high, as usual) and the fourth root (for dimensional comparability) of their estimated generalized variance. The chi-squared goodness of fit criterion is calculated so that no cell has a fitted frequency of less than unity; the HG chi-squared (Hinz and Gurland (1970)), appropriate to Minimum Chi-Squared fitting, is described in the Appendix on Goodness of Fit Testing. It is not surprising to find that, with more than half the observations in the zero class, the fit based on (H_0, H_1, A) is excellent.

All the cautions already expressed concerning the interpretations of standard errors, "asymptotic zero bias" and the like, apply of course no less to these methods of estimation than they do to moments and maximum-likelihood. Many of these can be summarized by asking "How large is 'large n'?" when a result is announced as valid for 'large n'.

The first publication of results along the above lines (Katti and Gurland (1962b)) did not seek linear minimum chi-squared equations. Not requiring linear normal equations simplifies the preliminary once-for-all analysis, but substantially increases the arithmetic necessary in each specific application actually to determine estimates with no guarantee that these have any advantages over those from linear equations, and hence no details of the procedures are included apart from a result in the Exercises.

e) Miscellaneous methods of estimation. Since the Neyman distribution is a two-parameter distribution sometimes used in circumstances in which the zero class is large, two other intuitive estimating procedures are to equate:

TABLE 8

Food Stores in Ljubljana
Neyman Distribution — Rogers (1969)

Numbers of stores x	Observed frequency	Moment fit	M.L. fit	Minimum Chi-squared H_0,H_1,H_2 fit	Minimum Chi-squared H_0,H_1,A fit
0	83	97.6	81.5	74.6	85.1
1	18	4.4	13.4	21.6	13.3
2	13	7.6	15.3	19.0	14.9
3	9	8.9	12.5	12.6	11.8
4	7	8.0	8.4	7.4	7.8
5	7	6.0	5.3	4.2	4.7
6	2	4.1	3.2	2.3	2.8
7	1	2.6	1.9	1.2	1.6
8	2	1.7	1.1	.6	.9
10	1				
19	1	3.1	1.4	.5	1.1
20+	0				
	144				
estimate of μ		0.402	0.646	0.855	0.601
and s.e.		0.084	0.092	0.36	0.13
estimate of ν		3.42	2.13	1.466	2.08
and s.e.		0.60	0.26	0.68	0.33
correlation		-0.710	-0.558	-0.943	-0.754
(generalized variance)$^{1/4}$		0.189	0.141	0.284	0.167
ℓn (likelihood)		-235.81	-223.56	-227.55	-223.84
Chi-squared		52.0	5.5	13.0	4.9
and d.f.		8	7	6	6
HG chi-squared		-	-	3.95	1.36
and d.f.				1	1

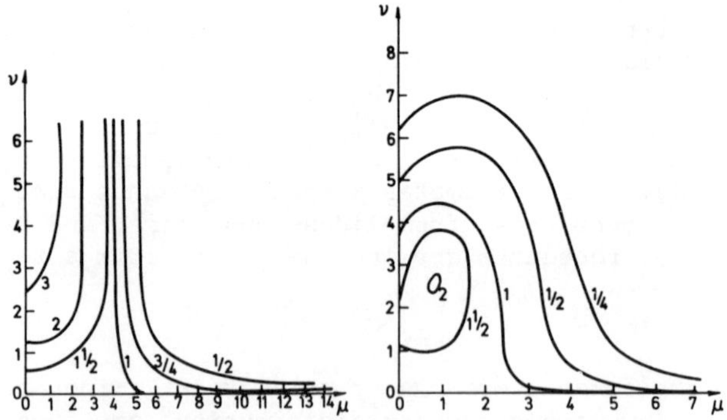

FIGURE 25 Asymptotic first order efficiency contours (as fractions) with respect to moments for (*left*) first moment and zero-class frequency and (*right*) first moment and ratio of zero- and first-class frequencies. (Katti and Gurland, 1962)

a) sample and distribution means (as with moments and maximum likelihood) and the proportion P_0 and probability p_0 in the zero class;

b) sample and distribution means, and the ratios which use both zero and first classes: P_1/P_0 and p_1/p_0 .

These lead to the equations:

$$\text{a)} \quad \frac{1-e^{-\tilde{\nu}}}{\tilde{\nu}} = \frac{-\ln P_0}{\overline{X}} \quad \text{and} \quad \tilde{\mu} = \overline{X}/\tilde{\nu} \;;$$

$$\text{b)} \quad \tilde{\nu} = \ln \frac{\overline{X}\, P_0}{P_1} \quad \text{and} \quad \tilde{\mu} = \overline{X}/\tilde{\nu} \;.$$

The same methods as before can be applied to investigate the asymptotic properties of these estimators, and since they are "pencil and paper" rather than "computer" methods it is perhaps more reasonable to compare them with moment than maximum likelihood estimators. This has been done by Katti and Gurland (1962), and the contours of the ratios of the first order asymptotic generalized variances are shown. In terms of these there is not much to choose between the two in general, though a) is to be preferred (on this criterion) to moments if $\mu < 5$, while b) is to be preferred to moments if both $\mu < 3$ and $\nu < 4$.

It may also be noted that these have sometimes been described as "maximum likelihood" methods (with only $\overline{X}$ and P_0 , or P_0 and P_1 available); they are of course also special cases of minimum chi-squared methods.

On a literal interpretation of the model, ν , the mean cluster size (or Index of Cluster Size) can be estimated directly without reference to μ , as was done by Pielou (1957) with method a).

However, in view of such methods' limited applicability, no further discussion will be included here; although the Exercises include some results on standard errors.

f) Estimation for a common μ *or* ν. Suppose random samples of larvae have been taken on different fields $i = 1,2,\ldots,f$, giving quadrat counts in the zero class, first class,... as shown:

Field 1 : $F_{01}, F_{11}, \ldots, F_{x1}, \ldots, \; \Sigma F_{x1} = n_1$,

....

Field f : $F_{0f}, F_{1f}, \ldots, F_{xf}, \ldots, \; \Sigma F_{xf} = n_f$.

Suppose further that a Neyman clustering model is reasonable, with the additional assumption that the mean number of individuals per cluster is a species characteristic with a (common) Poisson(ν) distribution, while the mean number of clusters per quadrat is a habitat characteristic differing from field to field with Poisson(μ_i) distributions, $i = 1,2,\ldots,f$. (In different circumstances a common μ and differing $\nu_1,\ldots,\nu_f$ may be more natural assumptions.)

How can these parameters be estimated?

Common ν , variable $\mu = \mu_1,\ldots,\mu_f$

a) Moment estimation

The most obvious method is to write, as before, for the ith field

$$\overset{o}{\mu}_i \overset{o}{\nu}_i = \overline{X}_i ,$$

$$\overset{o}{\mu}_i \overset{o}{\nu}_i (1 + \overset{o}{\nu}_i) = M_{2:i} ,$$

whence $$\overset{o}{\mu}_i = \frac{\overline{X}_i^2}{K_{(2):i}} = \frac{\overline{X}_i^2}{M_{2:i} - \overline{X}_i} , \quad i = 1,2,\ldots,f ,$$

and $$\overset{o}{\nu}_i = \frac{K_{(2):i}}{\overline{X}_i} ,$$

from which the common ν may be estimated by

$$\bar{\overset{o}{\nu}}_i = \frac{1}{f} \Sigma \overset{o}{\nu}_j .$$

The f+1 parameters are thus estimated by $\overset{o}{\mu}_1, \overset{o}{\mu}_2, \ldots, \overset{o}{\mu}_f$ and $\bar{\overset{o}{\nu}}$. However this procedure pays no attention to the usual situation: the various $\overset{o}{\nu}_i$ will have different variabilities. It would of course be possible to use estimated "first order" weights obtained from reciprocals of

$$\mathcal{V}(\overset{o}{\nu}_i) = \frac{1}{n_i} \frac{2+\nu+2\mu_i(1+\nu)^2}{\mu_i} \quad \text{to} \quad O(1/n_i) ,$$

but nothing is known about such a procedure except that it would be at best approximate. A less "estimation dependent" method would be to use weights to the numbers n_i of individuals observed in the ith field:

$$\frac{1}{n} \Sigma n_j \overset{o}{\nu}_j , \text{ where } n = \Sigma n_j .$$

Confidence intervals of the usual approximate kind would follow - e.g.

$$\mathcal{V}\left(\frac{1}{n} \Sigma n_j \overset{o}{\nu}_j\right) = \frac{1}{n^2} \Sigma n_j^2 \mathcal{V}(\overset{o}{\nu}_j) .$$

b) Maximum likelihood estimation

Methods analogous to those just described for moments can be used: having the usual m.l. estimators $\hat{\mu}_i$ and $\hat{\nu}_i$ for each field, the combined estimator of ν could be taken as

$$\frac{1}{f} \Sigma \hat{\nu}_j ;$$

or $$\frac{\Sigma \hat{\nu}_j / \mathcal{V}(\hat{\nu}_j)}{\Sigma 1/\mathcal{V}(\hat{\nu}_j)} ,$$

where $$\mathcal{V}(\hat{\nu}_i) = \frac{1}{n_i} \cdot \frac{\nu(\phi_i - \mu_i^2 \nu^2)}{\mu_i\{(1+\nu)\phi_i - \mu_i \nu^2(\mu_i + \mu_i \nu + \nu)\}}$$

is estimated, perhaps iteratively for ν ;

or $\frac{1}{n} \Sigma n_j \hat{\nu}_j$. The difficulties are also analogous.

A "full" maximum likelihood solution is obtained from the log-likelihood

$$\ln L = \sum_x F_{x1} \ln p_x(\mu_1,\nu) + \sum_x F_{x2} \ln p_x(\mu_2,\nu) + \ldots$$

$$= \sum_j \sum_x F_{xj} \ln p_x(\mu_j,\nu) ,$$

where the previous m.l. results give

$$\frac{\partial \ln L}{\partial \mu_i} = \sum_x F_{xi} \{-1 + \frac{1}{\mu_i \nu} \Pi_x(\mu_i,\nu)\}$$

$$= -n_i + \frac{1}{\nu\mu_i} \sum_x F_{xi} \Pi_x(\mu_i,\nu),$$

$$i = 1,2,\ldots,f , \quad n_i = \sum_x F_{xi} ,$$

$$\Pi_x = (x+1)\, p_{x+1}/p_x ,$$

and

$$\frac{\partial \ln L}{\partial \nu} = \Sigma \Sigma F_{xj} \{\frac{x}{\nu} - \frac{1}{\nu} \Pi_x(\mu_j,\nu)\}$$

$$= \frac{1}{\nu} \sum_j n_j \bar{X}_j - \frac{1}{\nu} \Sigma \Sigma F_{xj} \Pi_x(\mu_j,\nu) .$$

Equating these derivatives to zero and rearranging leads to

$$\hat{\mu}_i \hat{\nu} = \frac{1}{n_i} \Sigma F_{xi} \Pi_x(\hat{\mu}_i,\hat{\nu}) ,$$

$$i = 1,2,\ldots,f$$

and

$$\hat{\nu} = \frac{T}{\Sigma n_j \hat{\mu}_j} \quad \text{where } T = \Sigma n_j \bar{X}_j .$$

It is interesting to notice that if the mean

$$\bar{\hat{\mu}} = \frac{1}{n} \Sigma n_j \hat{\mu}_j$$

is introduced, with weighting by numbers of observations only, these equations can be written

$$\bar{\hat{\mu}} \hat{\nu} = \bar{X} \quad (= T/n)$$

and $\sum_x F_{xi} \Pi_x(\hat{\mu}_i,\hat{\nu}) = \hat{\nu}\, n_i \hat{\mu}_i$, $i = 1,2,\ldots,f$,

with some resemblance to the single field case.

The practical solution of these equations can be achieved iteratively by using first approximations

$$\mu_{11},\mu_{21},\ldots,\mu_{f1} \text{ (for example } \overset{o}{\mu}_1,\overset{o}{\mu}_2,\ldots,\overset{o}{\mu}_f)$$

and obtaining the first approximation

$$\nu_1 = \frac{T}{\Sigma\, n_j \mu_{j1}} \quad ;$$

second approximations follow from

$$\mu_{i2} = \frac{1}{\nu_2 n_i} \Sigma\, F_{xi}\, \Pi_x(\mu_{i1},\nu_1) \quad , \quad i = 1,2,\ldots,f$$

and $$\nu_2 = \frac{T}{\Sigma\, n_j \mu_{j2}} \quad ,$$

with iteration until changes are sufficiently small.

The covariance matrix can be found as the inverse of the matrix with elements like

$$n\mathcal{E}\left(\frac{\partial\, \ln p_X}{\partial \mu_i}\;\frac{\partial\, \ln p_X}{\partial \mu_j}\right) \quad ;$$

this can be done at the same time as the calculation of fitted probabilities for the distributions.

c) <u>Minimum chi-squared estimation</u>

Methods like those described for moment estimation can be used; the variance which might be used for weighting would be obtained from the covariance matrices for the individual fields.

Common μ , variable $\nu = \nu_1, \nu_2, \ldots, \nu_f$

The same procedures may be applied.

a) Moment estimation

$$\overset{o}{\nu}_i = \frac{K_{(2):i}}{\bar{X}_i} \;,\; i = 1,2,\ldots,f,$$

$$\bar{\overset{o}{\mu}} = \frac{1}{f} \Sigma \overset{o}{\mu}_j \quad \text{with} \quad \overset{o}{\mu}_i = \frac{\bar{X}_i^2}{K_{(2):i}}$$

$$\left(\text{or} \quad \frac{\Sigma\, \overset{o}{\mu}_j / \mathcal{V}(\overset{o}{\mu}_j)}{\Sigma\, 1/\mathcal{V}(\overset{o}{\mu}_j)} \right.$$

where $\mathcal{V}(\overset{o}{\mu}_i) = \frac{1}{n_i} \cdot \frac{2+\nu_i + 2\mu\,(1+\nu_i)^2}{\mu}$;

or $\left. \frac{1}{n} \Sigma\, n_j \overset{o}{\mu}_j \right)$.

b) Maximum likelihood estimation

$$n\hat{\mu} = \Sigma \frac{T_j}{\hat{\nu}_j} \;,$$

where $n = \Sigma\, n_j$, $T_i = n_i \bar{X}_i$, $T = \Sigma\, T_j$,

$$T_i = \sum_x F_{xi}\, \Pi_x(\hat{\mu}, \hat{\nu}_i) \;,\; i = 1,2,\ldots,f \;.$$

These $f+1$ equations may be solved iteratively by taking first approximations $\nu_{11}, \ldots, \nu_{f1}$ for $\nu_1, \ldots, \nu_f$ (from a), for example), and obtaining a first approximation μ_1 for μ from

$$\mu_1 = \frac{1}{n} \Sigma \frac{T_j}{\nu_{j1}} \;.$$

Then the second approximations $\nu_{12},\ldots,\nu_{f2}$ are to be obtained from

$$\nu_{i2} = \nu_{i1} - \frac{F_i(\mu_1,\nu_{11},\ldots,\nu_{f1})}{[\partial F_i/\partial \nu_i]_{\mu_1,\nu_{11},\ldots,\nu_{f1}}}$$

where $F_i(\mu,\nu_1,\ldots,\nu_f) = \Sigma\, F_{xi}\, \Pi_x(\mu,\nu_i) - T_i$,

and $$\frac{\partial F_i}{\partial \nu_i} = \frac{1}{\nu_i}\left\{\Sigma\, F_{xi}\, \Pi_x(\mu,\nu_i) - \Sigma\, F_{xi}\, \chi_x(\mu,\nu_i)\right\},$$

$$\chi_x = \Pi_x(\Pi_{x+1} - \Pi_x) ;$$

and the second approximation μ_2 from

$$\mu_2 = \frac{1}{n}\,\Sigma\,\frac{T_j}{\nu_{j2}} .$$

As usual, iteration is carried out until the changes between iterates are sufficiently small.

c) Minimum chi-squared estimation

The procedures are as for a), with minimum chi-squared and not moment formulae.

Comparisons of Neyman Distributions

For comparisons of this kind, of which the preceding may be regarded as an important special case, Hinz and Gurland (1971) have proposed procedures based on minimum chi-squared fittings which lead to tests of hypotheses in a familiar form, even including an analysis of chi-squared ("ANOVA") without transformation of the observations.

If there are f distributions specified by

$$\exp(-\mu_i + \mu_i e^{-\nu_i+\nu_i z}) , \quad i = 1,2,\ldots,f ,$$

and the fittings are based on the statistics

$$\underset{\sim}{T}_i = (K_{(1),i}, H_{1,i}, H_{2,i},\ldots,H_{m_i-1,i}, A_i)^t$$

(these are described in detail under Linear Minimum Chi-squared Estimation:

$K_{(r),i}$ is the rth sample factorial cumulant for the ith distribution

$$H_{r,i} = K_{(r+1,i}/K_{(r),i} ,$$

$$A_i = P_{0,i}H_{1,i} \exp\left\{ \frac{K_{(1),i}}{H_{1,i}} (-\exp(-H_{1,i}) + 1) \right\}),$$

with distribution analogues $\underset{\sim}{\tau}_i$, a typical hypothesis about the distribution means

$$\mu_i \nu_i = \kappa_{(1),i}$$

is $H_0 : \kappa_{(1),i} = \dot{\kappa}_{(1),i}$ say, $i = 1,2,\ldots,f_0 < f$, where $\dot{\kappa}_{(1),i}$ are specified. On H_0 , suppose the minimum of the quadratic form

$$(\underset{\sim}{T}-\underset{\sim}{\tau})^t \underset{\sim}{i} \cdot (\underset{\sim}{T}-\underset{\sim}{\tau}) ,$$

where $\underset{\sim}{i}$ is a consistent estimate of $\underset{\sim}{\mathcal{V}}(\underset{\sim}{T})^{-1}$, is obtained at $\underset{\sim}{\tau} = \underset{\sim}{\dot{t}}^*$; using $\underset{\sim}{\tau}^*$ for the unrestricted minimum chi-squared estimator, then

$$(\underset{\sim}{T}-\underset{\sim}{\dot{t}}^*)^t \underset{\sim}{i} \cdot (\underset{\sim}{T}-\underset{\sim}{\dot{t}}^*) = (\underset{\sim}{T}-\underset{\sim}{\tau}^*)^t \underset{\sim}{i} \cdot (\underset{\sim}{T}-\underset{\sim}{\tau}^*)$$

has on H_0 an asymptotic central chi-squared distribution with f_0 degrees of freedom. (The degrees of freedom are not unexpected, from $f-(f-f_0)$; see Gurland (1948).) Hence a test of H_0 is obtained by rejecting H_0 if the above difference is too large. Equally, since $\eta_{1i} = \nu_i$, a corresponding formulation could be given to test $\nu_i = \dot{\nu}_i$ (specified) for $i = 1,2,\ldots,f_0 < f$. An extension to investigate a linear structure in the distribution means $\kappa_{(1),i}$ is given in Hinz and Gurland (1971), and it is this which can be written in a standard "analysis of variance" table:

the results are not given in detail here because of the emphasis in this work on estimation rather than testing.

4.5 COMPARISONS WITH OTHER DISTRIBUTIONS

a) With the Poisson distribution. The first and simplest distribution ordinarily thought of for counts is the Poisson. Bateman (1950) has investigated the power of the test of a Poisson(μ) hypothesis against a Neyman(μ,ν) alternative when the test statistic used is the Index of Dispersion

$$Z = \frac{\sum_{i=1}^{n} (X_i - \bar{X})^2}{\bar{X}} ;$$

on the Poisson hypothesis Z is approximately chi-squared distributed with $n-1$ degrees of freedom (Hoel (1943)).

By expanding in powers of $1/\mu\nu$, Bateman showed that the first four moments of Z on the Neyman(μ,ν) hypothesis are approximately those of $(1+\nu)Q^2$ where Q^2 is a chi-squared variate with $n-1$ degrees of freedom. Using these approximations the power of the Z-test was calculated, and results for a 5% level test are shown on the accompanying graphs. The power is attractively high: if $\nu = 3$ a 5% level test gives a power of 90% for a sample size of 10; for $\nu = 1$ a sample size of between 30 and 40 achieves the same power.

As Bateman mentions, it is also possible to use the above distributional result for Z $(=(1+\nu)Q^2$ on the Neyman hypothesis) to test hypotheses about or construct confidence intervals for ν when μ is supposed known.

Some further results along similar lines are given by Kathirgamatamby (1953), for Neyman, negative binomial and Thomas alternative hypotheses.

b) With the negative binomial distribution. The negative binomial (Pascal) distribution and the Neyman distribution are often thought of as "competitors", and hence comparisons between them are of particular interest.

One such comparison is of the probabilities in the zero class when the mean and variance are equal - apart from its intrinsic interest, this is particularly relevant to fitted distributions

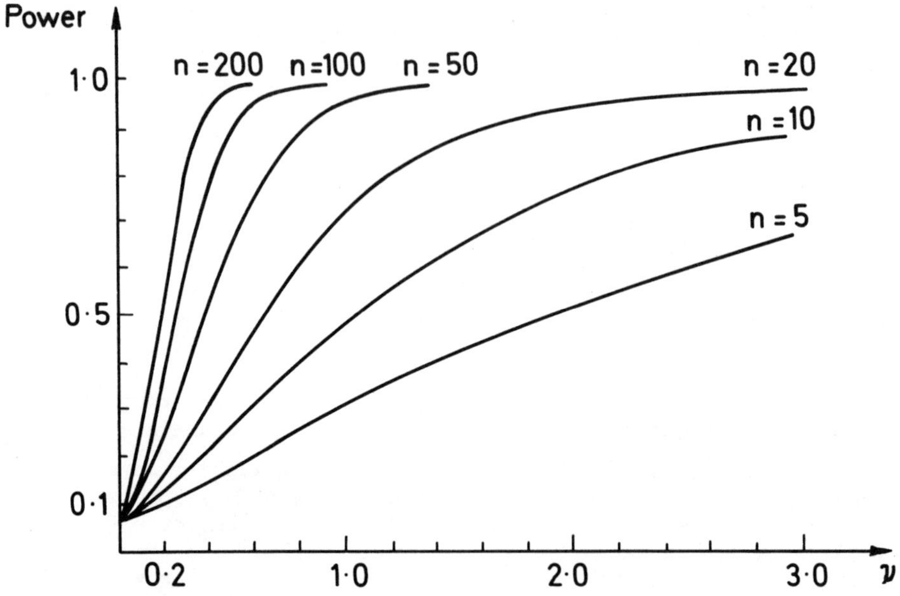

FIGURE 26 Power of a 5% Index of Dispersion test of H: Poisson(μ) against H': Neyman(μ,ν). (Bateman, 1950)

when moment methods are used. Writing

$$e^{-\mu+\mu e^{-\nu+\nu z}} = \Sigma\, n_x z^x \quad , \text{ with } n_0 = e^{-\mu+\mu e^{-\nu}} \quad ,$$

$$(1+\nu-\nu z)^{-\mu} = \Sigma\, p_x z^x \quad , \text{ with } p_0 = (1+\nu)^{-\mu}$$

(in the usual notation, $\mu = \kappa$, $\nu = \rho$ for the Pascal case), using Taylor expansions in powers of μ and ν gives

$$n_0 = 1 - \mu\nu + \frac{1}{2}\mu\nu^2 + \frac{1}{2}\mu^2\nu^2 - \frac{1}{6}\mu\nu^3 + 5^{th} \text{ degree terms} ,$$

and

$$p_0 = 1 - \mu\nu + \frac{1}{2}\mu\nu^2 + \frac{1}{2}\mu^2\nu^2 - \frac{1}{3}\mu\nu^3 + 5^{th} \text{ degree terms} .$$

(A Poisson distribution with the same mean has a corresponding expansion

$$1 - \mu\nu + \frac{1}{2}\mu^2\nu^2 + 6^{th} \text{ degree terms} .)$$

It is not surprising, even from this single example, that the two distributions are often competitors; similar expansions can of course be obtained for other probabilities, and another result for the zero class probabilities is given in the exercises.

Another comparison concerns functions of moments, such as coefficients of skewness and kurtosis. Taking factorial moments for convenience the first 4 are as follows,

	Neyman(μ,ν)	Pascal(μ,ν)
$\mu'_{(1)}$	$\mu\nu$	$\mu\nu$
$\mu'_{(2)}$	$\mu(\mu + 1)\nu^2$	$\mu(\mu + 1)\nu^2$
$\mu'_{(3)}$	$\mu(\mu^2 + 3\mu + 1)\nu^3$	$\mu(\mu^2 + 3\mu + 2)\nu^3$
$\mu'_{(4)}$	$\mu(\mu^3 + 6\mu^2 + 7\mu + 1)\nu^4$	$\mu(\mu^3 + 6\mu^2 + 11\mu + 6)\nu^4$

Obviously these are very similar, and especially so for large μ . In fact

$$\text{Neyman:} \qquad \mu'_{(r)} = \nu^r \, \Sigma \, \mathfrak{S}_r^{(t)} \, \mu^t$$

$$\text{Pascal:} \qquad \mu'_{(r)} = \nu^r \, \Sigma |s_r^{(t)}| \mu^t$$

and so the coefficients of μ^r and μ^{r-1}, but no others, are identical.

However, factorial cumulants would be preferred if one were concerned to display differences rather than similarities. For here the results are:

	Neyman(μ,ν)	Pascal(μ,ν)
$\kappa_{(1)}$	$\mu\nu$	$\mu\nu$
$\kappa_{(2)}$	$\mu\nu^2$	$\mu\nu^2$
$\kappa_{(3)}$	$\mu\nu^3$	$2\mu\nu^3$
$\kappa_{(4)}$	$\mu\nu^4$	$6\mu\nu^4$

and this has suppressed the similar terms, leaving those which differ. (See the comparisons of "skewness" and "kurtosis" for a number of distributions in the section concerned with Moments and Cumulants; for the Pascal case,

$$\kappa_{(r)} = (r-1)!\mu\nu^r \text{ .)}$$

It may also be noted that whereas the Pascal distribution has 1 mode (or a half mode at the origin), the Neyman distribution may have any number.

An application of the results relating to power given in the Appendix on Chi-Squared Goodness of Fit Testing has been made by Rogers (1969). Suppose the hypothesis tested H_0 is that the observations are Neyman(μ_0,ν_0) against the alternative hypothesis H_1 that they are negative binomial(μ_0,ν_0), where the test is to be carried out at the 5% level and $n = 400$ observations are to be taken. If the grouping of class frequencies is such that no expected frequency is less than 5 (which is rather high), leading to a certain number of degrees of freedom (necessary to enter the non-central chi-squared distribution tables), what is the power of the chi-squared goodness of fit test? Rogers finds, e.g.,

for $\mu_0 = 1$, $\nu_0 = 1.5$	then the power is	10% approximately
$\nu_0 = 2.0$		50%
$\nu_0 = 2.5$		100%

His graph is reproduced below.

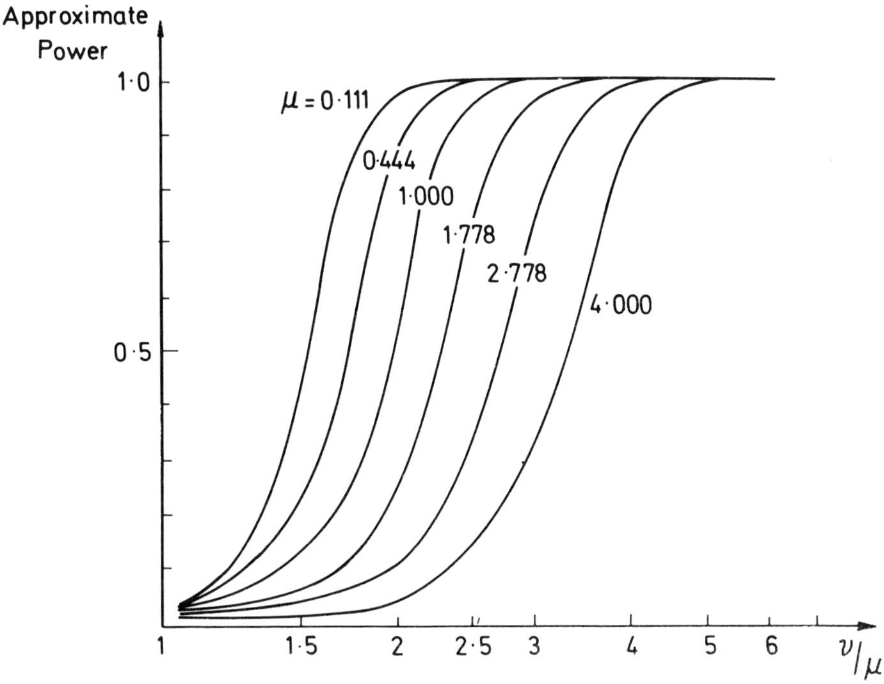

FIGURE 27 Approximate power of 5% level chi-squared goodness of fit test of Neyman(μ,ν) hypothesis against negative binomial(μ,ν) hypothesis (sample size n=400; no expected frequency < 5).
(Rogers, 1969)

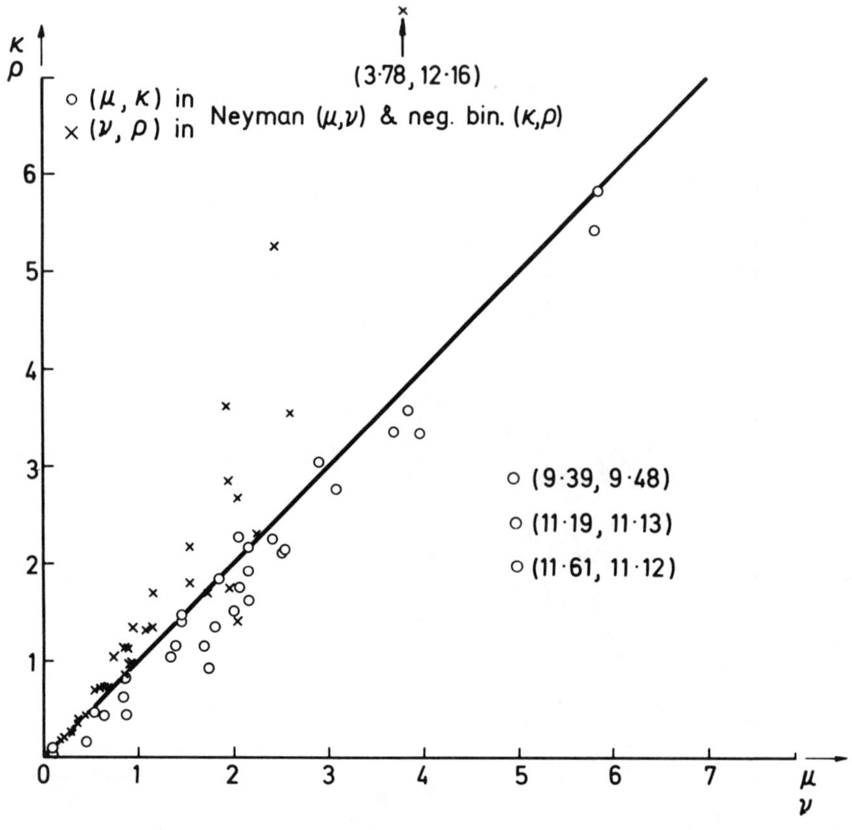

FIGURE 28 Thirty five empirical distributions.
(Katti and Rao, 1965)

At a practical level, Katti and Rao (1965) selected 35 published distributions of empirically observed data of counts of insects, bacteria, cells, plants and animals, and fitted to these by maximum likelihood a number of different distributions, including the Neyman and negative binomial. The fits were often excellent, and very often were comparable across these two distributions (as measured by a chi-squared goodness of fit criterion). More surprising than this was the very high degree of correspondence between values of the Neyman μ and negative binomial κ, and the Neyman ν and negative binomial ρ. These are shown on the accompanying Figure 28, where their concentration on the line $\kappa = \mu$, or $\rho = \nu$, is generally high even for those for which the fits were not necessarily very satisfactory. (There is also a tendency for ρ to exceed ν, or, equivalently, for μ to exceed slightly κ.) Whatever the interpretations which may be given to these parameters, it is comforting to observe this degree of robustness under model variation.

c) Limiting results. At various places, approximations and limiting results are mentioned, such as

$$\text{Neyman}(\mu,\nu) \to \text{Poisson}(\mu\nu) \quad \text{as} \quad \nu \to 0,$$

and

$$\text{Poisson-binomial}(\mu,\nu,\pi) \to \text{Neyman}(\mu,\nu\pi)$$
$$\text{as} \quad \nu \to \infty,\ \pi \to 0 \quad \text{with} \quad \nu\pi \quad \text{finite.}$$

Although these have an obvious relation with comparisons of such distributions, they are not discussed separately in this section.

4.6 THE TRUNCATED NEYMAN DISTRIBUTION

In some circumstances the frequency in the zero class is either not observable in principle or at least is not observed. An example is given by counts of the number of eggs in egg clusters where a cluster can be identified only by the presence of eggs.

Such a zero-truncated variate Y, taking values $1,2,3,\ldots$, has probabilities related to those of the original variate X, taking values $0,1,2,\ldots$, by

$$\Pr(Y=y) = \frac{\Pr(x=y)}{1-\Pr(X=0)}, \quad y=1,2,3,\ldots .$$

In the case of the Neyman(μ,ν) distribution, the p.g.f. of the (zero-) truncated variate is

$$\frac{\exp(-\mu+\lambda e^{\nu z}) - \exp(-\mu+\lambda)}{1-\exp(-\mu+\lambda)} , \quad \lambda = \mu e^{-\nu} .$$

Expressions for recurrence relations, moments etc. follow directly from these:

e.g. $$\mathscr{E}(Y^r) = \frac{1}{1-\Pr(X=0)} \mathscr{E}(X^r) ,$$

but central moments and cumulants take no simple shape because of the factor $1-\Pr(X=0)$.

Moment and maximum likelihood estimators can be found by arguments corresponding to those for the truncated case. Thus for moments, writing $\overset{o}{p}_0 = \exp\{-\overset{o}{\mu} + \overset{o}{\mu} \exp(-\overset{o}{\nu})\}$ (corresponding to a moment estimator of the zero class) and

$$S'_r = \sum_{y>0} y^r F_y ,$$

the moment estimators $(\overset{o}{\mu},\overset{o}{\nu})$ are given by the solutions of

$$\left.\begin{aligned} \frac{\overset{o}{\mu}\overset{o}{\nu}}{1-\overset{o}{p}_0} &= \frac{S'_1}{S'_0} \\ \overset{o}{\nu}(1+\overset{o}{\mu}) &= \frac{S'_2}{S'_1} - 1 \end{aligned}\right\}$$

However, positive solutions of these equations need not exist, nor can explicit solutions be obtained. The results for a direct iteration are given in Exercise 16 (with the notation there defined): for a starter value $\underset{\sim}{\lambda}_1 = (\mu_1,\nu_1)^t$, then

$$\underset{\sim}{\lambda}_2 = \underset{\sim}{\lambda}_1 - \left(\frac{\partial \underset{\sim}{m}'}{\partial \underset{\sim}{\lambda}}\right)^{-1}\Bigg|_{\underset{\sim}{\lambda}=\underset{\sim}{\lambda}_1} (\underset{\sim}{m}'(\underset{\sim}{\lambda}_1) - M' ,$$

with repetition. Maximum likelihood estimators are similarly the solutions (Douglas (1955)) of

$$\left.\begin{aligned} \frac{\hat{\mu}\hat{\nu}}{1-\hat{p}_0} &= \frac{S'_1}{S'_0} \\ 1-\hat{p}_0\, e^{-\hat{\nu}} &= \frac{1}{S'_1}\sum_{y>0} F_y \hat{\Pi}_y \end{aligned}\right\}$$

where $\Pi_y = (y+1)\dfrac{p_{y+1}}{p_y}$ as before. Using a Newton-Raphson iterative procedure on

$$H(\nu) = 1 - p_0 e^{-\nu} - \frac{1}{S'_1}\Sigma\, F_y\, \Pi_y$$

requires the derivative

$$H'(\nu) = e^{-\nu}\; p_0(1+\mu e^{-\nu} - (1-e^{-\nu})\frac{\mu}{\nu}\cdot\frac{1-\mu\nu e^{-\nu}p'_0}{1-\mu(1-e^{-\nu})p'_0})$$

$$-\frac{1}{\nu\, S'_1}\Sigma\, F_y\, \Pi_y$$

$$+\frac{1}{\nu\, S'_1}(1+\frac{1}{\nu}\cdot\frac{1-\mu\nu e^{-\nu}p'_0}{1-\mu(1-e^{-\nu})p'_0})\;\Sigma\, F_y \chi_y$$

in which $\chi_y = \Pi_y(\Pi_{y+1} - \Pi_y)$ and $p'_0 = p_0/(1-p_0)$. Thus given first approximations (μ_1, ν_1) the second approximation for ν is

$$\nu_2 = \nu_1 - \frac{H(\nu_1)}{H'(\nu_1)}$$

and for μ

$$\mu_2 = \frac{S'_1/S'_0}{\nu_2}(1-\exp\{-\mu_1(1-e^{-\nu_2})\}) ,$$

with of course repetition of the process until successive iterates differ by sufficiently little. The results for a direct iteration on both parameters are given in Exercise 17.

If hand calculation is employed, the tabulation already referred to of ratios p_x^* of Poisson(λ) moments about the origin can reduce the labour if it covers the appropriate range of values. In the notation already introduced the maximum likelihood equations are

$$\left.\begin{aligned} \frac{\hat{\mu}\hat{\nu}}{1-\hat{p}_0} &= \frac{S_1'}{S_0'} \\ \frac{1-\hat{p}_0 e^{-\hat{\nu}}}{\hat{\nu}} &= \frac{\Sigma F_y p_y^*}{S_1'} \end{aligned}\right\}$$

Iteration is accomplished from

$$G(\nu) = 1 - p_0 e^{-\nu} - \frac{\nu}{S_1'} \Sigma F_y p_y^* ,$$

$$G'(\nu) = e^{-\nu} \; p_0(1+\mu e^{-\nu} - (1 - e^{-\nu}) \frac{\mu}{\nu} \cdot \frac{1-\mu\nu e^{-\nu} p_0'}{1-\mu(1-e^{-\nu})p_0'}$$

$$- \frac{1}{S_1'} \Sigma F_y p_y^*$$

$$+ \frac{1}{S_1'} (\nu + \frac{1-\mu\nu e^{-\nu} p_0'}{1-\mu(1-e^{-\nu})p_0'}) \Sigma F_y q_y .$$

The question as to the choice of starting values for any of the iterative procedures is more troublesome: moment estimates, too, in this case are to be found iteratively. Perhaps the best procedure is to insert a guessed value for the zero frequency and use the moment equations for the non-truncated distribution to obtain starter values (μ_1, ν_1). These are used in the first moment (first maximum likelihood) equation for the truncated case:

$$\frac{\mathring{\mu}\mathring{\nu}}{1-\mathring{p}_0} = \frac{S_1'}{S_0'} ,$$

to obtain a second approximation μ_2 from

$$\mu_2 = \mu_1 - \frac{f(\mu_1)}{f'(\mu_1)}$$

where $f(\mu) = 1 - \mu \dfrac{\nu_1}{S_1'/S_0'} - \exp\{-\mu(1 - e^{-\nu_1})\}$,

$$f'(\mu) = -\frac{\nu_1}{S_1'/S_0'} + (1 - e^{-\nu_1}) \exp\{-\mu(1 - e^{-\nu_1})\} .$$

Leaf counts of Leucopogon Virgatus (unpublished data due to D. W. Goodall) may be used to illustrate the numerical procedure. The observed frequency distribution is as follows.

y	1	2	3	4	5	6	7	8	9	10	12	14	15+
F_y	70	41	33	29	11	7	3	4	1	1	1	1	0

There are no positive solutions to the moment equations, so analogues of the moment equations

$$\tilde{\nu} = \frac{S_2'}{S_1'} - \frac{S_1'}{S_0'} - 1 , \quad \tilde{\mu} = \frac{S_1'}{S_0'} \cdot \frac{1}{\tilde{\nu}}$$

were used, the values then being adjusted to satisfy the first moment (m.1.) equation, these being $(\mu_1,\nu_1) = (0.293,2.17)$. Substitution of these in the maximum likelihood iterative procedure (using p_x^* and q_x) led to third approximations $(\hat{\mu},\hat{\nu}) = (1.49,1.21)$ and hence fitted frequencies

$$\phi_y' = S_0' \, p_y'$$

as follows.

F_y	70	41	33	29	11	7	11
ϕ_y'	58.6	51.2	36.1	23.2	14.0	8.1	10.8

The value of chi-squared in a goodness of fit test is 6.8 with 4 degrees of freedom - an adequate fit in terms of overall frequencies, though not informative about the estimators of μ and ν .

Expressions for the covariance matrices corresponding to the various methods of estimation follow by application of the methods already set out: they are, however, rather complex (see, e.g., Exercise 16).

4.7 THE NEYMAN FAMILY OF CONTAGIOUS DISTRIBUTIONS

When the Neyman Type A - strictly, the 2 parameter Type A - distribution was introduced (Neyman (1939)) it was as one member of a family, though it has turned out to be much the most popular member. The generalization which comprises the family is the multiparameter Types A, B and C (etc.), in each case obtained through a rather specific biological model, partially described below.

Neyman considered individuals which occur in clusters, with counts being made of individuals in randomly selected quadrats: the distribution of these counts is sought. In the present notation and terminology, and largely deprived of its geometric formulation, the derivation was as follows. For a specific quadrat the simplest classification of clusters is into those which are represented on the quadrat, of proportion π say, and those which are not, of proportion $1-\pi$. A represented cluster is characterized by its number $N = n$ (with distribution $F_N(n)$) of individuals and by the probability $P = p$ (distribution $F_P(p)$) that a single individual (from the cluster) appears on the quadrat. (In fact Neyman supposed P to be a function of a position vector, of the centre of the cluster; but in the applications direct assumptions about P were made.) Then the p.g.f. of the number of individuals from this represented cluster is the binomial

$$(q+pz)^n \ , \ q = 1-p \ ,$$

and so for a randomly selected represented cluster the p.g.f. is the "double" mixture

$$g(z) = \int_0^1 \int_0^\infty (q+pz)^n \, dF_N(n) \, dF_P(p) \ .$$

(It might be more convenient to write the integral over n explicitly as a sum.) Hence the p.g.f. for the number of individuals from a randomly selected cluster on a randomly selected quadrat is the "triple" mixture

$$1-\pi+\pi f(z) \ ,$$

and for m such clusters

$$\{1-\pi+\pi f(z)\}^m = [1-\pi\{1-f(z)\}]^m .$$

Considering $m \to \infty$, corresponding to a "large" area from which clusters are drawn, π must tend to 0 or the number of individuals will simply increase indefinitely. So, writing $m\pi = \mu$, say (non-zero finite), the p.g.f. of this total number of individuals is

$$[1-\frac{\mu}{m}\{1-f(z)\}]^m$$
$$\to \exp\{-\mu+\mu f(z)\} = h(z) \text{ , say ,}$$

which has the form of a Poisson-stopped p.g.f., though this is not its derivation above.

Neyman then chose

$$dF_P(p_0) = 1 \quad \text{(so the individuals of all represented clusters are equivalent) ,}$$

and $$dF_N(n) = e^{-\theta}\theta^n/n! \text{ , } n = 0,1,2,\ldots \quad \text{(for cluster sizes) ,}$$

giving $$f(z) = \exp(-\theta p_0 + \theta p_0 z) ,$$

and for $\nu = \theta p_0$,

$$h(z) = \exp\{-\mu+\mu \exp(-\nu+\nu z)\} ,$$

the p.g.f. for a 2 parameter Neyman Type A distribution. (The variations for other cases are set out after an alternative formulation, in terms of stopped variates, has been given below.)

It may be observed that the "simpler" choice $dF_N(\nu) = 1$, rather than the Poisson assumption above, would have led to

$$f(z) = (1-p_0+p_0 z)^{\nu} = (q_0+p_0 z)^{\nu}$$

and hence to

$$h(z) = \exp\{-\mu+\mu(q_0+p_0 z)^{\nu}\} ,$$

a Poisson-binomial. In fact Neyman suggested, though he did not set out the details of, substituting

$$dF_N(n) = \binom{\nu}{n} \beta^n (1-\beta)^{\nu-n} , \quad 0 < \beta < 1 ,$$

corresponding to an upper limit ν for cluster size, leading to

$$f(z) = (1 - p_0\beta + p_0\beta z)^{\nu}$$

and

$$h(z) = \exp\{-\mu + \mu(1 - p_0\beta + p_0\beta z)^{\nu}\} ;$$

he therefore came very close to an explicit exhibition of "another family of contagious distributions" (p. 41), the Poisson-binomial.

Feller (1943) gave a somewhat different account of Neyman's work using at the third stage (involving the appearance of π) not a general mixing but a stopping procedure. Using exactly the preceeding notation and description up to the appearance of $f(z)$, all clusters are considered as represented on the quadrat, the number $M = m$ of clusters having the distribution $dF_M(m)$. Then in this case the p.g.f. of the number of individuals taken over all clusters is (the $F_M(m)$-stopped)

$$\int_0^{\infty} \{f(z)\}^m \, dF_M(m) .$$

If M is a Poisson(μ) variate the p.g.f. is thus

$$\exp\{-\mu + \mu\, f(z)\} ,$$

a "genuine" Poisson-stopped p.g.f.

It is convenient to list here some of the results which follow on making specific assumptions for the various distributions.

1. $dF_M(m) = e^{-\mu} \mu^m/m!$, $m = 0,1,\ldots,$

   $dF_P(p_0) = 1$,

   $dF_N(n) = e^{-\theta} \theta^n/m!$, $n = 0,1,\ldots$

   These lead to the p.g.f.

   $$\exp\{-\mu + \mu \exp(-\nu+\nu z)\} \quad \text{for} \quad \nu = \theta p_0 ,$$

   which is of course a 2 parameter Neyman Type A p.g.f., obtained as a stopped mixture.

2. As in 1. except $dF_P(p_1) = dF_P(p_2) = \frac{1}{2}$.
The p.g.f. is

$$\exp[-\mu + \frac{1}{2}\,\mu\{\exp(-\nu_1 + \nu_2 z) + \exp(-\nu_2 + \nu_2 z)\}] \;,\; \nu_i = \theta p_i \;:$$

a 3 parameter Neyman Type A distribution. The generalization, with positive normed weights w_i corresponding to $dF_P(p_i) = w_i$, $i = 1,2,\ldots,s$, leads to the p.g.f.

$$\exp(-\mu + \mu\,\Sigma\, w_i\, e^{-\nu_i + \nu_i z}) \quad \text{with} \quad \nu_i = \theta p_i \;,$$

a multi-parameter Neyman Type A p.g.f.

3. As in 1. except $dF_P(p) = 1$, $0 < p < 1$, a rectangular distribution. This leads to

$$\exp\{-\mu + \mu\,\frac{\exp(-\nu + \nu z) - 1}{-\nu + \nu z}\} \;,\; \nu = \theta \;,$$

a Neyman Type B p.g.f.

4. As in 1. except $dF_P(p) = 2(1-p)$, $0 < p < 1$, leading to

$$\exp\left\{-\mu + \mu\,\frac{\exp(-\nu + \nu z) - 1 - (-\nu + \nu z)}{\frac{1}{2}(-\nu + \nu z)^2}\right\} ,\; \nu = \theta \;,$$

a Neyman Type C p.g.f. The generalization where $dF_P(p)$ is a suitable polynomial of degree $\beta-1$ can also be obtained from 2. as a limit; taking

$$dF_P(p) = \beta(1-p)^{\beta-1} \;,\; 0 < p < 1 \;,$$

gives the p.g.f.

$$\exp\left\{\mu\,\frac{\beta!}{\Delta^\beta}\,(e^\Delta - 1 - \Delta - \ldots - \frac{\Delta^\beta}{\beta!})\right\}$$

where $\Delta = -\nu + \nu z$, $\nu = \theta$.

5. Returning to 1. except for

$$dF_N(n) = \binom{\nu}{n}\,\rho^n(1-\rho)^{\nu-n} \qquad \text{(binomial)}$$

leads to the p.g.f.

$$\exp\{-\mu+\mu(1-\rho_0\rho+\rho_0\rho z)^{\nu}\} ,$$

which is that of a Poisson-binomial variate.

Specific derivations, not embedded in the general structure of the family, may help to indicate circumstances in which various distributions can be applied.

For example, the 3 parameter Type A distribution (with equal weights) of the form

$$\exp[-\mu+\frac{1}{2}\mu\{\exp(-\nu_1+\nu_1 z)+\exp(-\nu_2+\nu_2 z)\}]$$

can be generated most simply in the following two ways.

i) Stopping (or clustering)

Suppose insects of two kinds lay eggs in clusters, the numbers of clusters per quadrat laid by each kind being Poisson($\frac{1}{2}\mu$) distributed, and the numbers of eggs per cluster Poisson(ν_i) distributed where $i=1,2$ correspond to the kinds of insect. Then for the ith kind of insect the numbers of eggs per quadrat will have the stopped p.g.f.

$$\exp\{-\frac{1}{2}\mu\exp(-\nu_i+\nu_i z)\} , \quad i=1,2,$$

whence for independence of the two kinds the p.g.f. of the total number of eggs per quadrat will be the product of these two p.g.f.'s, that of the 3 parameter Type A.

ii) Mixing

In i) the "cluster laying capacity" $\frac{1}{2}\mu$ was supposed constant both across and within kinds. Further, although the mean number of eggs laid per cluster depended on the kind of insect, within a kind there was homogeneity of insects: all had the same mean. Putting aside any suppositions about eggs being laid in clusters and homogeneity of insects within a kind, suppose instead that for the ith kind of insect the insects under observation lay mean numbers of eggs which are one of

$$0,\ \nu_i,\ 2\nu_i,\ 3\nu_i,\ldots,u\nu_i,\ldots$$

where the proportion of insects with "mean egg-laying capacity" $u\nu_i$ is given by the Poisson

$$\Pr(U=u) = e^{-\frac{1}{2}\mu} \frac{(\frac{1}{2}\mu)^u}{u!} \quad , \; u = 0,1,2,\ldots,$$

and hence has p.g.f. $\exp(-\frac{1}{2}\mu + \frac{1}{2}\mu z) = g(z)$, say. The variation of numbers of eggs laid about the mean $u\nu_i$ is also supposed Poisson, and so has p.g.f.

$$\exp(-u\nu_i + u\nu_i z) \quad , \; i = 1,2, \;;$$

whence for an insect of the ith kind selected randomly (from the distribution over "mean egg-laying capacity") the p.g.f. of the numbers of eggs laid is (the mixture)

$$\sum_u e^{-u\nu_i + u\nu_i z} e^{-\frac{1}{2}\mu} \frac{(\frac{1}{2}\mu)^u}{u!}$$

$$= \exp\{-\tfrac{1}{2}\mu + \tfrac{1}{2}\mu \exp(-\nu_i + \nu_i z)\}$$

$$= g(e^{-\nu_i + \nu_i z}) \quad , \; \text{say.}$$

Hence if one of each kind of insect is selected randomly the p.g.f. of the total numbers of eggs is

$$g(e^{-\nu_1 + \nu_1 z}) \cdot g(e^{-\nu_2 + \nu_2 z}) \; ,$$

the above 3 parameter Type A p.g.f.

A recurrence relation for this three parameter case is given in the Exercises.

In a similar way, the p.g.f. of the Type B can be obtained by supposing a Poisson(νP) distribution is selected by uniform weights for P over the unit interval - the resulting mixture has p.g.f.

$$\int_0^1 e^{-\nu p + \nu p z}\, dp = \frac{e^{\nu(z-1)} - 1}{\nu(z-1)} \; .$$

If this is the p.g.f. of the identically independently distributed $X_1, X_2, \ldots,$ then the p.g.f. of $X_1 + X_2 + \ldots + X_N$ where N is Poisson(μ) is that of the Type B variate.

Recurrence relations for the probabilities for Types B and C are given in the Exercises.

A rather comprehensive account, not since developed to any extent, of interpretation and fitting practice (by moments) of the Neyman family specified by the p.g.f.

$$\exp\left\{-\mu+\mu\,\frac{e^{\nu(z-1)}-1-\nu(z-1)-\ldots-\dfrac{\nu^{\beta-1}(z-1)^{\beta-1}}{(\beta-1)!}}{\dfrac{\nu^{\beta}(z-1)^{\beta}}{\beta!}}\right\}$$

has been given by Beall and Rescia (1953). This three parameter distribution (in which $\beta = 0,1,2$ corresponds to Types A, B, C) is there fitted by trial over successive values of β, using moment estimators for μ and ν, finally choosing a value for β for which the fit (measured by a chi-squared goodness of fit test) is best or at least good enough. Fortunately, other than the case $\beta \to \infty$, only small values of β appear to be of interest. (Writing the coefficient of μ in the above p.g.f. as

$$\Gamma(\beta+1)\sum_{i=0}^{\infty}\frac{\nu^{i}(z-1)^{i}}{\Gamma(\beta+1+i)} = f(z)\ ,\ \text{say,}$$

there is no necessity for β to be integral - no applications appear to have been made of this generalization, however,)

The moment estimators for any fixed β can be obtained from

$$\kappa_{(r)} = \mu\nu^{r}\cdot\frac{r!}{(\beta+1)^{\{r\}}}\ ,$$

so that $\mathscr{E}(X) = \dfrac{\mu\nu}{\beta+1}$, and $\mathscr{V}(X) = \dfrac{\mu\nu}{\beta+1}\left\{1+\dfrac{2\nu}{\beta+2}\right\}$;

hence the moment estimators are

$$\overset{o}{\mu} = \frac{\beta+1}{\dfrac{S_2}{S_1}-\dfrac{S_1}{S_0}}\ ,$$

$$\overset{o}{\nu} = \frac{1}{2}(\beta+2)\left(\frac{S_2}{S_1}-\frac{S_1}{S_0}-1\right)$$

where as usual $S_i = \Sigma\, x^{i}F_x$, F_x being the observed frequency in the xth class.

(When considering the family for fitting purposes it is of some interest to notice that, given a constant mean and variance, the probability of the zero class decreases as β increases. In the case $\beta \to \infty$ with finiteness of $\nu/(\beta+1)$ and μ, the Polya-Aeppli distribution is obtained as a limiting distribution with p.g.f.

$$\exp(-\mu + \frac{1}{1+\rho-\rho z})$$

where $\frac{\nu}{\beta+1} \to \rho$ and β (and ν) $\to \infty$.)

For the general family, the calculation of probabilities can be carried out recursively from the analysis appropriate to the general Poisson stopped distribution

$$\exp\{-\mu+\mu f(z)\} = \Sigma\, p_x z^x ,$$

where $f(z) = \Sigma\, f_x z^x$ is defined above. Initially

$p_0 = \exp(-\mu+\mu f_0)$ where

$$f_0 = \frac{\beta!}{(-\nu)\beta} \left\{ e^{-\nu} \sum_{r=0}^{\beta-1} \frac{(-\nu)^r}{r!} \right\} ;$$

$$f_1 = (\beta+\nu)\, f_0 - \nu ,$$

and

$$(i+1) f_{i+1} = (\nu+\beta+i) f_i - \nu f_{i-1} , \quad i \geq 1 ,$$

whence

$$(x+1) p_{x+1} = \mu \sum_{i=0}^{x} (i+1) f_{i+1} p_{x-i} ,$$

the general recurrence relation for Poisson-stopped distributions. (A method involving the use of forward differences is also described by Beall and Rescia (1953).)

In Gurland's (1958) generalization, where $f(z)$ above is the confluent hypergeometric function $M(\alpha,\alpha+\beta,\nu(z-1))$ defined by

$$M(p,q,x) = 1 + \frac{p}{q^x} + \frac{p(p+1)}{q(q+1)} \cdot \frac{x^2}{2!} + \ldots + \frac{p^{\{r\}}}{q^{\{r\}}} \cdot \frac{x^r}{r!} + \ldots) ,$$

similar procedures can be applied, most simply based on the factorial cumulants

$$\kappa_{(r)} = \mu\nu^r \frac{\alpha^{\{r\}}}{(\alpha+\beta)^{\{r\}}}$$

From these, moment estimators can be obtained, and again probabilities from

$$(x+1)p_{x+1} = \mu \sum_{i=0}^{x} (i+1)\, f_{i+1}\, p_{x-i}$$

where $(i+1)f_{i+1} = (\nu+\alpha+\beta+i-1)\, f_i - \nu \frac{\alpha+i-1}{i} f_{i-1}\ , \ i \geq 1\ ,$

the initial values f_0 and f_1 being obtained from tables of the confluent hypergeometric function or of the incomplete gamma function. But this four parameter distribution has not been extensively applied - Gurland suggests that the first two moments be used in conjunction with P_0 and P_1 if all are to be estimated, though little information is available about the usefulness of the procedure. (Beall and Rescia report a lack of success in the use of three moments to estimate μ, ν and β in their generalization.)

Arguments which are extensions of Neyman's original rather geometric model are given in Thompson (1954), including a report of results due to J. H. Darwin - they tend to rest on somewhat specific assumptions regarding the nature of the spread of the individuals and perhaps therefore do not seem to have been developed further. Somewhat similar extensions have also been given by Katti and Sly (1965), and Warren (1971), where more of the difficulties of "cluster-satellite" models are discussed; see also Skellam (1958).

4.8 EXERCISES

1. Use the recurrence relation

$$h_{x+1} = \frac{\mu}{x+1} \sum_{p=0}^{x} (p+1)\, h_{x-p}\, f_{p+1}\ ,$$

obtained for a Poisson(μ)-stopped distribution with sum distribution $f_0, f_1, \ldots,$ to obtain the recurrence relation for the Neyman distribution.

2. From the expression

$$\kappa_r = \sum_{n=1}^{r} \mathscr{S}_r^{(n)} \nu^n$$

for the rth cumulant of the Neyman distribution, obtain the recurrence relations for κ_r :

$$\kappa_{r+1} = \nu(\kappa_r + \frac{\partial\kappa_r}{\partial\nu}) ,$$

$$\kappa_{r+1} = \nu(\Sigma\binom{r}{s} \kappa_r + \mu) .$$

3. For $e^{-\mu+\mu e^{-\nu+\nu z}} = \Sigma\, n_x z^x$

and $(1+\nu-\nu z)^{-\mu} = \Sigma\, p_x z^x$,

by expanding in powers of ν show that

$$0 < n_0 - \{1-\mu\nu + \frac{1}{2}\mu(\mu+1)\nu^2 - \frac{1}{6}\mu(1+3\mu+\mu^2)\nu^3\} < \frac{1}{24}\mu(1+7\mu+6\mu^2+\mu^3)\nu^4,$$

$$0 < p_0 - \{1-\mu\nu + \frac{1}{2}\mu(\mu+1)\nu^2 - \frac{1}{6}\mu(2+3\mu+\mu^2)\nu^3 < \frac{1}{24}\mu^{\{4\}}\nu^4 .$$

4. For the Neyman(μ,ν) distribution with p.g.f. $g(z)$ show that $\{\ell n\, g(z)\}' = \mu\,\nu e^{-\nu+\nu z}$, the differentiation being wtih respect to z , and hence that

$$\frac{\partial\, \ell n\, g(z)}{\partial\mu} = -1 + \frac{\{\ell n\, g(z)\}'}{\mu\nu} ,$$

$$\frac{\partial\, \ell n\, g(z)}{\partial\nu} = \frac{-1+z}{\nu} \{\ell n\, g(z)\}' .$$

In the general case, with

$$g(z) = \Sigma\, z^x\, p_x$$

and $\ell n\ g(z) = \ell n\ p_0 + \ell n(1 + \rho_1 z + \ldots + \rho_x z^x + \ldots)$

where $\rho_x = p_x/p_0$

$$= \ell n\ p_0 + \sum_{r=1}^{\infty} \tau_r \frac{z^r}{r!} ,$$

show that $\dfrac{\tau_r}{r!} = \rho_r - \dfrac{1}{r!} \displaystyle\sum_{s=1}^{r-1} \frac{(r-1)!}{(r-1-s)!} \rho_s \tau_{r-s}$.

5. Write M_1', M_2' for the usual sample moments and $\overset{o}{\mu}, \overset{o}{\nu}$ for the moment estimators for μ, ν in Neyman(μ,ν). Show that the covariance matrix $\underset{\sim}{\mathcal{V}}(\overset{o}{\mu}, \overset{o}{\nu})$ for $\overset{o}{\mu}, \overset{o}{\nu}$ is given to first order in $1/n$, n being the sample size, by

$$\underset{\sim}{\mathcal{V}}(\overset{o}{\mu}, \overset{o}{\nu}) = \left(\frac{\partial \underset{\sim}{\mu}'}{\partial \underset{\sim}{\lambda}}\right)^{-1} \underset{\sim}{\mathcal{V}}(\underset{\sim}{M}') \left[\left(\frac{\partial \underset{\sim}{\mu}'}{\partial \underset{\sim}{\lambda}}\right)^{-1}\right]^t$$

where $\underset{\sim}{\mu}' = \begin{pmatrix} \mu_1' \\ \mu_2' \end{pmatrix} = \begin{pmatrix} \mu\nu \\ \mu\nu(1+\nu) \end{pmatrix}$ is the vector of distribution moments,

$\underset{\sim}{\lambda} = \begin{pmatrix} \mu \\ \nu \end{pmatrix}$ is the vector of parameters,

$\dfrac{\partial \underset{\sim}{\mu}'}{\partial \underset{\sim}{\lambda}} = \begin{pmatrix} \dfrac{\partial \mu_1'}{\partial \mu} & \dfrac{\partial \mu_1'}{\partial \nu} \\ \dfrac{\partial \mu_2'}{\partial \mu} & \dfrac{\partial \mu_2'}{\partial \nu} \end{pmatrix}$ is the Jacobian shown ,

$\underset{\sim}{\mathcal{V}}(\underset{\sim}{M}') = \begin{pmatrix} \mathcal{V}(M_1') \ , & \mathcal{C}(M_1' \ , \ M_2') \\ \mathcal{C}(M_1' \ , \ M_2') \ , & \mathcal{V}(M_2') \end{pmatrix}$ is the covariance matrix of the sample moments, and t denotes the transpose. (In applications the Jacobian and covariance matrices have to be replaced by estimates from the sample; it can be shown that this does not alter the order of the asymptotic result.)

6. For the observed frequency distribution

x	0	1	2	3+
F_x	4	1	2	0

and the assumption of a Neyman(μ,ν) distribution show that the moment estimates are $\overset{o}{\mu} = 8.33\ldots$ and $\overset{o}{\nu} = 0.0857\ldots$.

Use these in Shenton's Newton-Raphson procedure as starter values for maximum likelihood estimation: observe that the process fails to converge.

Write down explicitly the likelihood of the above sample, and show by direct differentiation that the likelihood equation for $\hat{\nu}$ is

$$1 + 5\hat{\lambda} + \frac{2(1+3\hat{\lambda}+\hat{\lambda}^2)}{1+\hat{\lambda}} - \frac{5}{\hat{\nu}} = 0$$

$$\text{where} \quad \hat{\lambda} = \frac{\bar{x}e^{-\hat{\nu}}}{\hat{\nu}}, = \frac{5e^{-\hat{\nu}}}{7\hat{\nu}},$$

for which the root is

$$\hat{\nu} = 0.519\ldots$$

and so $\hat{\mu} = 1.375\ldots$

(The estimated error structures, using the corresponding estimates, are:

	s.e.	correlation
$\overset{o}{\mu}$	59. ...	-0.99...
$\overset{o}{\nu}$	0.61...	
$\hat{\mu}$	2.0 ...	-0.93...)
$\hat{\nu}$	0.80...	

7. By using a table of random numbers and a table of the cumulative Poisson distribution, generate a sample of say $n = 50$ from a Neyman distribution. For example, taking

$\mu = 1.2$, $\nu = 4.0$, select an observation N from a Poisson(1.2) distribution, and then select N observations $X_1,\ldots,X_N$ from a Poisson(4.0) distribution, recording these and their sum X (which is thus Neyman(1.2,4.0)): 4 typical observations are:

N_j	$X_{1j} +\ldots+ X_{N_j j}$	=	X_j
3	7+2+5	=	14
0	0	=	0
1	0	=	0
2	3+2	=	5

Use these as data to calculate estimates of the parameters μ and ν:

a) from the frequency distribution of the X_j (the usual data situation) by any convenient method, such as moments;

b) of μ from the mean of the N_j, and of ν from the mean of the individual X_{ij} values actually observed $(\Sigma\Sigma X_{ij}/\Sigma N_j)$.

What of the standard errors of these estimators (see also Exercise 15)? (For the example above with $n = 50$, the frequency distribution will probably have a large half mode at $X = 0$ and a mode near $X = 3$ or 4. For one such simulation, $\overset{o}{\mu} = 1.28$, $\overset{o}{\nu} = 3.80$ and the estimates of b) 1.24, 3.85. The s.e. of the moment estimator of μ is 0.34, and of the b) estimator 0.15. It is more effective to use a typewriter computer terminal, with slow typeout, to go through the above complete sequence.)

8. For the parameterization of the Neyman distribution in which the mean is σ and the variance $\sigma(1+\tau)$, so the p.g.f. is

$$\exp\{-\frac{\sigma}{\tau}(1+e^{-\tau+\tau z})\},$$

obtain the following estimation results.

Moment Estimators

$$\overset{\circ}{\sigma} = \bar{X} , \quad \overset{\circ}{\tau} = \frac{M_2 - \bar{X}}{\bar{X}} ,$$

$$\mathscr{V}(\overset{\circ}{\sigma}) = \frac{\sigma(1+\tau)}{n} ,$$

$$\mathscr{V}(\overset{\circ}{\tau}) = \frac{1}{n} \{2(1+\tau)^2 + \frac{\tau(2+\tau)}{\sigma}\} + 0(1/n^2) ,$$

$$\mathscr{C}(\overset{\circ}{\sigma},\overset{\circ}{\tau}) = \frac{\tau}{n} + 0(1/n^2) .$$

Mean and Zero Class Estimators

$$\tilde{\sigma} = \bar{X} , \quad \frac{1-e^{-\tilde{\tau}}}{\tilde{\tau}} = \frac{-\ell n\ P_0}{\bar{X}} ,$$

$$\mathscr{V}(\tilde{\sigma}) = \mathscr{V}(\overset{\circ}{\sigma}) ,$$

$$\mathscr{V}(\tilde{\tau}) = \frac{1}{n}\left[\frac{\tau^2}{\sigma(1+\tau)} + \frac{\tau^4}{\sigma^2\{(1+\tau)e^{-\tau}-1\}^2} \times \left\{e^{\sigma(1-e^{-\tau})/\tau} - 1 - \frac{\sigma}{1+\tau}\right\}\right] + 0(1/n)^2) ,$$

$$\mathscr{C}(\tilde{\sigma},\tilde{\tau}) = \frac{\tau}{n} + 0(1/n^2) .$$

(Evans(1953)).

9. Show that the recurrence relation for the probabilities p_x in the 3 parameter Neyman Type A distribution is

$$p_{x+1} = \frac{\mu\nu_1 e^{-\nu_1}}{2(x+1)} \sum_{m=0}^{x} \frac{\nu_1^m}{m!} p_{x-m} + \frac{\mu\nu_2 e^{-\nu_2}}{2(x+1)} \sum_{m=0}^{x} \frac{\nu_2^m}{m!} p_{x-m} .$$

(Neyman (1939)).

10. Obtain the recurrence relations listed for the probabilities in the Neyman Type B and Type C distributions:

Type B

$$p_{x+1} = \frac{\mu}{\nu(x+1)} \sum_{r=0}^{x} (r+1) \left(1-e^{-\nu} \sum_{s=0}^{r+1} \frac{\nu^s}{s!}\right) p_{x-r} \,.$$

Type C

$$p_{x+1} = \frac{2\mu e^{-\nu}}{\nu^2(x+1)} \sum_{r=0}^{x} (r+1) \left\{\nu\left(e^{\nu} - \sum_{s=0}^{r} \frac{\nu^s}{s!}\right) - (r+2)\right.$$

$$\left.\times\left(e^{\nu} - \sum_{s=0}^{r+1} \frac{\nu^s}{s!}\right)\right\} p_{x-r} \,.$$

(Beall (1940)).

11. From the factorial cumulant g.f. $-\mu+\mu e^{\nu t}$ for the Neyman Type A distribution, obtain the Type B by replacing ν by x, then μ by μ/ν and integrating from 0 to ν with respect to x; similarly the Type C by replacing ν by x, then μ by $2\mu(\nu-x)/\nu^2$ and integrating from 0 to ν. Interpret these as mixtures of Neyman Type A distributions.

(Anscombe (1950)).

12. Show that a Poisson(μ)-stopped {mixed Poisson($u\nu$) with uniform weight for u on $(0,\alpha)$} variate has a 2 parameter Neyman Type B distribution, with p.g.f.

$$\exp\left\{-\mu+\mu \frac{e^{-\alpha\nu+\alpha\nu z}-1}{-\alpha\nu+\alpha\nu}\right\} .$$

Show that the 2 parameter Type B distribution can also be obtained as a

{Poisson(μ)-stopped unit uniform} - stopped Poisson(ν) variate .

13. For the Poisson(λ)-stopped Poisson(μ)-stopped Poisson(ν) distribution with p.g.f.

$$\exp[-\lambda+\lambda \exp\{-\mu+\mu \exp(-\nu+\nu z)\}]$$

show that the mean and variance are

$\kappa_1 = \lambda\mu\nu$ and $\kappa_2 = \lambda\mu\nu(1+\nu+\mu\nu)$ respectively.

Obtain the rth factorial cumulant as

$$\kappa_{(r)} = \lambda\, \nu^r \sum \mathcal{S}_n^{(n)}\, \mu^n$$

and the rth power cumulant as

$$\kappa_r = \lambda \sum_{n=1}^{r} \mathcal{S}_r^{(m)}\, \nu^m \sum_{n=1}^{m} \mathcal{S}_m^{(n)}\, \mu^n ,$$

$\mathcal{S}_m^{(n)}$ being a Stirling Number of the Second Kind.

Introducing $\beta = \mu e^{-\nu}$ and $\alpha = \lambda e^{-\mu+\beta}$, obtain for the probability function p_x the recurrence relation

$$p_{x+1} = \frac{\lambda}{x+1} \sum_{r=0}^{x} (x-r+1)\, p_r\, p^*_{x-r+1}$$

where p^*_s is the Neyman(μ,ν) probability given by

$$p^*_s = e^{-\mu+\beta} \frac{\nu^s}{s!} \sum_{t=1}^{s} \mathcal{S}_s^{(t)}\, \beta^t .$$

(Douglas (1970)).

14. On the (μ,ν) diagram exhibiting the occurrence of modes for the Neyman distribution, show that the contour separating the $\frac{1}{2}$ and $1\frac{1}{2}$ mode regions meets the contour of the 1 mode region at the point $(e,1)$.

(Shenton).

15. Use the results from the exercise (Chapter 3) relating to estimation of the means of Y and N in

$$X_i = Y_{i1} + Y_{i2} + \ldots + Y_{iN_i} \qquad i = 1,2,\ldots,n,$$

given observations on Y_{ij} and N_i, to show that for N_i:Poisson(μ) and Y_{ij}:Poisson(ν) (so X_i:Neyman(μ,ν)), then

$$\overline{N} = \Sigma N_i/n \quad \text{is an estimator of} \quad \mathscr{E}(N) = \mu$$

$$\text{with} \quad \mathscr{V}(\overline{N}) = \mu/n \ ,$$

$$\tilde{Y} = \Sigma X_i/\Sigma N_i \quad \text{is an estimator of} \quad \mathscr{E}(Y) = \nu$$

$$\text{with} \quad \mathscr{V}(\tilde{Y}) = \nu/n\mu + 0(1/n^2) \ .$$

Compare these sampling variances with those of the moment estimators $\overset{o}{\mu}$ and $\overset{o}{\nu}$ based on observations on X_i, showing that always

$$\mathscr{V}(\overset{o}{\mu}) > \mathscr{V}(\overline{N}) \quad \text{and} \quad \mathscr{V}(\overset{o}{\nu}) > \mathscr{V}(\tilde{Y}) \ .$$

An insurance company's records for a certain class of accidents show the frequency distribution for the total amounts of claims (in units of $100) for each day of 1970 to be as follows.

x	0	1	2	3	4	5	6	7	8	9	10	11	12
F_x	56	34	41	39	27	34	22	20	16	13	13	10	5

13	14	15	16	17	18	19	20	23	28	29+
6	3	9	2	5	4	2	1	2	1	0

(E.g. A total of $1500 was claimed on each of 9 days.)
Assuming that the total amount of claims X_i on the ith day, $i = 1,2,\ldots,365$, can be represented in the above form, the estimates and estimated standard errors for μ, the mean number of claims per day, and ν, the mean amount of an individual claim, turn out to be

	μ and s.e.	ν and s.e.
Minimum Chi-squared	1.986	2.158
(H_0,H_1,A)	0.18	0.21
Maximum Likelihood	1.987	2.625
	0.15	0.18

Of course the records <u>also</u> show the frequency distributions of the numbers of claims per day and of the amounts of the individual claims ($100). These are as follows.

Claims per day N

n	0	1	2	3	4	5	6	7	8	9+
F_n	47	97	109	62	25	16	4	3	2	0

Individual claim amounts Y

y	0	1	2	3	4	5	6	7	8	9	10
F_y	68	228	170	106	68	39	17	13	9	2	8

11	12	13	15	18	23	24+
4	5	1	2	1	1	0

A cursory inspection shows that it is indeed reasonable to take N to be Poisson distributed, but highly implausible to suppose Y to be Poisson distributed. In any case, the means and variances for these observed distributions can be found by direct calculation, and they turn out to be:

Claims per day

Mean	2.03	(cf.	m.l.	1.09
s.e.	0.078			0.15)
Variance	2.20			

Individual claims

Mean	2.57	(cf.	m.l.	2.62
s.e.	0.090			0.18)
Variance	6.08			

(The expression for $\tilde{\mathcal{V}}(Y)$ above was not used to calculate the s.e. because Y was so much more dispersed than a Poisson variate with mean 2.57. Formally, the use of $\mathcal{V}(\tilde{Y})$ leads to a s.e. of 0.059.)

Since in many circumstances (e.g. larvae counts, specks of impurity broken from larger particles) only observations on X are available it is pleasing to notice the stability of the estimation of the means of N and Y even when the distribution for Y is far from Poisson.

There were 56 days on which the total claim was 0 ; but these days are of two different types, corresponding to 0 claims and to some claim/s but of 0 amount. (cf. larvae sampling: larvae from no egg cluster and no larvae hatched from cluster/s.) The numbers of these can be estimated - e.g. the number of days with 0 claims from

$$365e^{-1.99} = 49.9$$

(for which the observed number was 46) with estimated s.e. $49.9 \times 0.182 = 9.0$.

16. For the moment fit of a zero truncated Neyman(μ,ν) distribution show that the covariance matrix for $\overset{o}{\mu},\overset{o}{\nu}$ is given by the following:

$$\underset{\sim}{\mathcal{V}}\begin{pmatrix}\overset{o}{\mu}\\ \overset{o}{\nu}\end{pmatrix} = \left(\frac{\partial\underset{\sim}{\mu}'}{\partial\underset{\sim}{\lambda}}\right)^{-1} \underset{\sim}{\mathcal{V}}(\underset{\sim}{M}')\left[\left(\frac{\partial\underset{\sim}{\mu}'}{\partial\underset{\sim}{\lambda}}\right)^{-1}\right]^{t}$$

where $'$ denotes moment about the origin,

^{t} denotes transpose,

$\underset{\sim}{M}' = (M_1',M_2')^t$, the sample moment vector,

$\underset{\sim}{\lambda} = (\mu,\nu)^t$

and $\underset{\sim}{\mu}' = (\mu_1',\mu_2')^t = \left(\dfrac{\mu\nu}{1-p_0}, \dfrac{\mu\nu(1+\nu+\mu\nu)}{1-p_0}\right)^t$, the distribution moment vector.

In particular, with $p_0 = \exp(-\mu+\mu e^{-\nu})$,

$$\mathcal{V}(M_1') = \frac{\mu_2'-\mu_1'^2}{n} , \mathcal{C}(M_1',M_2') = \frac{\mu_3'-\mu_2'\mu_1'}{n} , \mathcal{V}(M_2') = \frac{\mu_4'-\mu_2'^2}{n} ,$$

give the elements of $\underset{\sim}{\hat{\mathcal{V}}}(\underset{\sim}{M}')$ on substitution for these moments of the truncated Neyman, and

$$\frac{\partial\mu_1'}{\partial\mu} = \frac{\nu}{1-p_0} - \frac{\mu\nu(1-e^{-\nu})p_0}{(1-p_0)^2},$$

$$\frac{\partial\mu_1'}{\partial\nu} = \frac{\nu}{1-p_0} - \frac{\mu^2\nu e^{-\nu}p_0}{(1-p_0)^2},$$

$$\frac{\partial\mu_2'}{\partial\mu} = \frac{\nu+\nu^2+2\mu\nu^2}{1-p_0} - \frac{\mu\nu(1+\nu+\mu\nu)(1-e^{-\nu})p_0}{(1-p_0)^2},$$

$$\frac{\partial\mu_2'}{\partial\nu} = \frac{\mu+2\mu\nu+2\mu^2\nu}{1-p_0} - \frac{\mu^2\nu(1+\nu+\mu\nu)e^{-\nu}p_0}{(1-p_0)^2}$$

give the elements of

$$\frac{\partial\underset{\sim}{\mu}}{\partial\underset{\sim}{\lambda}} = \begin{pmatrix} \partial\mu_1'/\partial\mu, & \partial\mu_1'/\partial\nu \\ \partial\mu_2'/\partial\mu, & \partial\mu_2'/\partial\nu \end{pmatrix}.$$

17. If a direct iterative procedure on both estimators (not involving the preliminary elimination of one) is used for m.ℓ. estimation in the zero-truncated Neyman(μ,ν) distribution, with variate $Y = 1,2,\ldots$ the familiar iteration formula for the method of scoring,

$$\underset{\sim}{\lambda}_{r+1} = \underset{\sim}{\lambda}_r + \underset{\sim}{\mathscr{V}}(\underset{\sim}{\lambda}_r)^{-1} \cdot \frac{\partial\, \ell n\, L}{\partial\underset{\sim}{\lambda}}, \quad r = 1,2,\ldots,$$

applies. In this formula,

$\underset{\sim}{\lambda} = (\mu,\nu)^t$, using ^{t} for transpose,

$$\underset{\sim}{\mathscr{I}}(\underset{\sim}{\lambda}) = \begin{bmatrix} n\,\mathscr{E}\left(\frac{\partial\, \ell n\, p_Y'}{\partial\mu}\right)^2, & n\,\mathscr{E}\left(\frac{\partial\, \ell n\, p_Y'}{\partial\mu}\cdot\frac{\partial\, \ell n\, p_Y'}{\partial\nu}\right) \\ n\,\mathscr{E}\left(\frac{\partial\, \ell n\, p_Y'}{\partial\mu}\cdot\frac{\partial\, \ell n\, p_Y'}{\partial\nu}\right), & n\,\mathscr{E}\left(\frac{\partial\, \ell n\, p_Y'}{\partial\nu}\right)^2 \end{bmatrix}$$

and (for any zero truncated distribution)

$$\mathscr{E}\left(\frac{\partial\ \ell n\ p'_Y}{\partial\theta_1} \cdot \frac{\partial\ \ell n\ p'_Y}{\partial\theta_2}\right) = \frac{1}{q_0}\left(\iota_{\theta_1\theta_2} - \frac{p_0}{q_0} \cdot \frac{\partial\ \ell n\ p_0}{\partial\theta_1} \cdot \frac{\partial\ \ell n\ p_0}{\partial\theta_2}\right)$$

where $\iota_{\theta_1\theta_2}$ is the appropriate element of the information matrix for a single observation for the corresponding untruncated distribution, and $\theta_1 = \theta_2$ is permitted. Hence in the present instance, with $\phi = \mathscr{E}(\Pi_X^2)$ for the untruncated distribution;

$$\iota_{\mu\mu} = \frac{\phi}{\mu^2\nu^2} - 1\ , \quad \iota_{\mu\nu} = \frac{-\phi}{\mu\nu^2} + 1 + \mu, \quad \iota_{\nu\nu}\ \frac{\phi}{\nu^2} + \frac{\mu}{\nu} - \mu(1+\mu)\ ,$$

and

$$\frac{\partial\ \ell n\ p_0}{\partial\mu} = -1 + \frac{\Pi_0}{\mu\nu}\ , \quad \frac{\partial\ \ell n\ p_0}{\partial\nu} = -\frac{\Pi_0}{\nu}\ ,$$

$$\Pi_x = (x+1)\ p_{x+1}/p_x = (x+1)\ p'_{x+1}/p'_x\ ,$$

p_x referring to the Neyman(μ,ν) distribution and

$$p'_y = p_y/(1-p_0) = p_y/q_0$$

with $$\frac{\partial\ \ell n\ L}{\partial\underset{\sim}{\lambda}} = \begin{pmatrix} \dfrac{n'}{q_0}\left(\dfrac{p_1}{\mu\nu} - 1\right) + \dfrac{1}{\mu\nu}\,\Sigma\, F_y\, \Pi_y \\[2ex] \dfrac{n'}{\nu}\left(\bar{Y} - \dfrac{p_1}{q_0}\right) - \dfrac{1}{\nu}\,\Sigma\, F_y\, \Pi_y \end{pmatrix}, \quad n' = \Sigma\, F_y.$$

The estimated covariance matrix of the m.l. estimator $\hat{\underset{\sim}{\lambda}}$ is then $\underset{\sim}{\mathscr{I}}(\hat{\underset{\sim}{\lambda}})^{-1}$.

18\. For the "mean and zero-class" estimators $\tilde{\mu},\tilde{\nu}$, show that in the notation of the text,

$$\mathscr{C}(P_x, P_y) = \begin{cases} \dfrac{p_x\,(1-p_x)}{n}, & x = y, \\[2ex] \dfrac{-p_x\,p_y}{n}, & x \neq y \end{cases} \qquad *$$

by appeal to the properties of the multinomial distribution; and hence $\mathscr{V}(\tilde{\mu})$, $\mathscr{V}(\tilde{\nu})$ and $\mathscr{C}(\tilde{\mu},\tilde{\nu})$ can be found by solving

$$\nu^2\,\mathscr{V}(\tilde{\mu}) + 2\mu\nu\,\mathscr{C}(\tilde{\mu},\tilde{\nu}) + \mu^2\,\mathscr{V}(\tilde{\nu}) = \Sigma\Sigma\,\frac{\partial f}{\partial p_x}\,\frac{\partial f}{\partial p_y}\,\mathscr{C}(P_x,P_y)$$

$$\nu\,h(\nu)\,\mathscr{C}(\tilde{\mu},\tilde{\nu}) + \mu h'(\nu)\,\mathscr{V}(\tilde{\nu}) = \Sigma\Sigma\,\frac{\partial f}{\partial p_x}\,\frac{\partial g}{\partial p_y}\,\mathscr{C}(P_x,P_y)$$

$$\{h'(\nu)\}^2\,\mathscr{V}(\tilde{\nu}) = \Sigma\Sigma\,\frac{\partial g}{\partial p_x}\,\frac{\partial g}{\partial p_y}\,\mathscr{C}(P_x,P_y),$$

where $\quad \tilde{\mu}\,\tilde{\nu} = \bar{X} = \Sigma\, x\, P_x = f(P_0, P_1, \ldots, P_x \ldots),$

$$h(\tilde{\nu}) = \frac{\tilde{\nu}}{1-e^{-\tilde{\nu}}} = \frac{-1}{\ln P_0}\,\Sigma\, x\, P_x = g(P_0, P_1, \ldots, P_x, \ldots).$$

(*N.B. $\mathscr{C}(P_x,P_x) = \mathscr{V}(P_x)$)

19. Non-linear Generalized Least Squares, or non-linear Minimum Chi-squared.

Suppose for a sample of n from the Neyman(μ,ν) distribution the estimator

$$\underset{\sim}{T} = (K_{(1)}, K_{(2)}, \ln P_0)^t$$

is chosen, where $P_0 = F_0/n$ is the proportion in the zero class. Hence its asymptotic expectation is

$$\underset{\sim}{\tau} = (\mu\nu,\ \mu\nu^2,\ -\mu + \mu e^{-\nu})^t,$$

and for $\underset{\sim}{\lambda} = (\mu,\nu)^t$

$$\frac{\partial \underset{\sim}{\tau}}{\partial \underset{\sim}{\lambda}} = \begin{pmatrix} \nu & \nu^2 & -1+e^{-\nu} \\ \mu & 2\mu\nu & -\mu e^{-\nu} \end{pmatrix}^t ,$$

so the minimum chi-squared equation obtained by minimizing the quadratic form

$$(\underset{\sim}{T} - \underset{\sim}{\tau})^t \cdot \underset{\sim}{i} \cdot (\underset{\sim}{T} - \underset{\sim}{\tau}) ,$$

$\underset{\sim}{i}$ being an estimate from a consistent estimator of $\underset{\sim}{}(\underset{\sim}{T})$, is

$$\left(\frac{\partial \underset{\sim}{\tau}}{\partial \underset{\sim}{\lambda}}\right)^t_{\underset{\sim}{\lambda}^*} \underset{\sim}{i} \cdot \underset{\sim}{T} = \left(\frac{\partial \underset{\sim}{\tau}}{\partial \underset{\sim}{\lambda}}\right)^t_{\underset{\sim}{\lambda}^*} \underset{\sim}{} \cdot \begin{pmatrix} \mu^* \nu^* \\ \mu^* \nu^{*2} \\ -\mu^* + \mu^* e^{-\nu^*} \end{pmatrix} .$$

By partitioning $\partial \underset{\sim}{\tau} / \partial \underset{\sim}{\lambda}$ it is possible to write this in a somewhat simpler form for a numerical solution; and

$$\underset{\sim}{\mathscr{V}}(\underset{\sim}{\lambda}^*) = \left(\frac{\partial \underset{\sim}{\tau}}{\partial \underset{\sim}{\lambda}} \mathscr{V}(\underset{\sim}{\tau})^{-1} \left(\frac{\partial \underset{\sim}{\tau}}{\partial \underset{\sim}{\lambda}}\right)^t \right)^{-1} \quad \text{to first order,}$$

where on returning to the sample moments M_1' and M_2', for

$$\underset{\sim}{S} = (M_1', M_2', \ell n\ P_0)^t$$

and $\underset{\sim}{\sigma}$ the corresponding distribution expression,

$$\underset{\sim}{\mathscr{V}}(\underset{\sim}{T}) = \frac{\partial \underset{\sim}{\tau}}{\partial \underset{\sim}{\sigma}} \underset{\sim}{\mathscr{V}}(\underset{\sim}{S}) \left(\frac{\partial \underset{\sim}{\tau}}{\partial \underset{\sim}{\sigma}}\right)^t \quad \text{to first order ,}$$

with $\mathscr{C}(M_r', \ell n\ P_0) = -\dfrac{\mu_r'}{n} + 0(1/n^2)$,

$$\mathscr{V}(\ell n\ P_0) = \frac{q_0}{np_0} + 0(1/n)^2 , \quad q_0 = 1-p_0 ,$$

$$\frac{\partial \underset{\sim}{\tau}}{\partial \underset{\sim}{\sigma}} = \begin{pmatrix} 1 & 0 & 0 \\ -1-2\mu_1' & 1 & 0 \\ 0 & 0 & 1 \end{pmatrix} .$$

(Katti and Gurland (1962b))

Chapter 5

OTHER DISTRIBUTIONS

5.1 THE POISSON-BINOMIAL DISTRIBUTION

As is so commonly the case, this distribution has two immediate interpretations, both of which are useful. As a stopped distribution it can be obtained from the structure:

Poisson(μ)-stopped summed binomial(ν,π), with p.g.f.
$\exp\{-\mu+\mu(\chi+\pi z)^{\nu}\}$,

and presumably this form accounts for the common name, which will be used here in the form Poisson-binomial(μ,ν,π). As a mixed distribution it can be obtained as:

mixed-binomial($k\nu,\pi$) on k by Poisson(μ),

with p.g.f. $$\sum_{k=0}^{\infty} (\chi+\pi z)^{k\nu} e^{-\mu} \frac{\mu^k}{k!}$$

$$= \exp\{-\mu+\mu(\chi+\pi z)^{\nu}\} .$$

Its introduction is due to Polya (1950) or Skellam (1952) (see some further remarks in the section on the Neyman Family of Contagious Distributions); examples of its use, further mentioned later, were given by McGuire, Brindley and Bancroft (1957 and 1958).

It is a three parameter distribution, ν necessarily being a positive integer, and indeed in most cases of application a rather small integer, although in terms of a physical interpretation as a simple Poisson-stopped distribution this is rather restrictive, since ν is then the (fixed) number in a cluster from which a binomially distributed number of individuals is observed. For $\nu = 1$, the p.g.f. reduces to $\exp(-\mu\pi + \mu\pi z)$, a Poisson($\mu\pi$) p.g.f., and this reduction will generally be excluded from what follows. Since it is a Poisson-stopped distribution, it has all the properties of that class. (It should not be confused with Poisson's generalization of the binomial distribution, with p.g.f. $(\chi_1 + \pi_1 z)(\chi_2 + \pi_2 z)\ldots(\chi_\nu + \pi_\nu z)$.) The distribution thus has a reproductive property: if $X_1, X_2, \ldots, X_m$ are independently Poisson-stopped binomial(μ_i, ν, π) variates, then their convolution is Poisson-stopped binomial($\Sigma\mu_i, \nu, \pi$).

Somewhat comparably with the Poisson-Pascal and Polya-Aeppli distributions, the special case with $\nu = 2$ has been separately studied by Kemp and Kemp (1965), by whom it was called the Hermite distribution because its probability function can be expressed in terms of Hermite Polynomials. Only passing references to this special case will be made in the following sections; the original paper is self-contained in its exposition and may be referred to for a rather complete account, including of fitting procedures. Further details are also to be found in Patil (1971), Patil, Shenton and Bowman (1974), and Shenton and Bowman (1977).

Fundamental Properties

The Probability Function

i) From the interpretation as a mixed-binomial by Poisson variate, the p.g.f. is

$$\sum_{k=0}^{\infty} (\chi + \pi z)^{k\nu} e^{-\mu} \frac{\mu^k}{k!} .$$

Writing $\lambda = \mu\chi^\nu$ and $\rho = \pi/\chi$, the p.g.f. is

$$e^{-\mu} \sum_{k=0}^{\infty} \frac{\lambda^k}{k!} (1 + \rho z)^{\nu k}$$

$$= e^{-\mu} \sum_{k} \frac{\lambda^k}{k!} \sum_{x=0}^{\infty} \binom{\nu k}{x} \rho^x z^x ,$$

where in fact the upper terminal of summation for x is νk , but since the binomial coefficient $\binom{p}{q} = 0$ for $q > p$ the sum can be shown as unrestricted. Hence, interchanging the order of summation gives

$$e^{-\mu} \sum_{x=0}^{\infty} \left(\sum_{k=0}^{\infty} \binom{\nu k}{x} \frac{\lambda^k}{k!} \right) \rho^x z^x$$

(where now k runs from $k \geq x/\nu$) and the coefficient of z^x is

$$p_x = e^{-\mu} \left(\frac{\pi}{\chi}\right)^x \sum_{k \geq x/\nu}^{\infty} \binom{\nu k}{x} \frac{(\mu\chi^\nu)^k}{k!} .$$

The result is obviously not at all convenient for direct use.

ii) Recurrence Relation. Differentiating

$$g(z) = \exp\{-\mu + \mu(\chi + \pi z)^\nu\}$$

gives

$$g'(z) = \mu\nu\pi\, g(z)\, (\chi + \pi z)^{\nu-1} ,$$

i.e. $$\Sigma(x+1)p_{x+1}z^x = \mu\nu\pi(\Sigma\, p_r z^r)(\Sigma\binom{\nu-1}{s} \chi^{\nu-1-s} \pi^s z^s) .$$

Equating coefficients of z^x $(r+s=x)$ results in

$$p_{x+1} = \frac{\mu\nu\pi}{x+1} \sum_{s=0}^{x} \binom{\nu-1}{s} \chi^{\nu-1-s} \pi^s p_{x-s} , \quad x = 0,1,2,\ldots,$$

together with $p_0 = \exp(-\mu + \mu\chi^\nu)$. (See also Figure 29.)
For small values of ν a simpler relation can be found - see e.g. section iv below.

iii) Probabilities in terms of Poisson moments and Stirling Numbers. Analogously to the Neyman distribution, Poisson binomial probabilities can be expressed in terms of functions closely related to the factorial moments of a Poisson$(\mu\chi^\nu)$ distribution (Shumway and Gurland (1960b). Rewriting the explicit expression for p_x , with the binomial coefficient $\binom{\nu k}{x}$ expressed in terms of factorial powers,

```
      ∇ R←X PBI PARA;RA;CH;I;A;MU;NU;PI;SUM;PRO
[1]     →(((NU-2)=|⌊¯2+NU←PARA[2])∧(PI←PARA[3])<1)/LBL1
[2]     →ρ0ρ⎕← 'AT LEAST ONE OF  NU  AND  PI  IS NOT ADMISSIBLE.
        START AGAIN.'
[3]   LBL1:R←R,(PRO←MU×NU×(RA←PI÷CH)×CH*NU)×R←*(MU←PARA[1])×((CH←
        1-PI)*NU)-1,0ρA←,1+I←0
[4]   LBL2:→((ρR←R,PRO×(÷I+1)×+/(A←A,A[I]×(NU-I)×RA÷I←I+1)×⌽R)≤⌈/
        X)/LBL2
[5]     R←R[1+X]
      ∇
```

```
      ∇ R←TOL PB PARA;RA;CH;I;A;MU;NU;PI;SUM:PRO
[1]     →(((NU-2)=|⌊¯2+NU←PARA[2])∧(PI←PARA[3])<1)/LBL1
[2]     →ρ0ρ⎕← 'AT LEAST ONE OF  NU  AND  PI  IS NOT ADMISSIBLE.
        START AGAIN.'
[3]   LBL1:R←R,(PRO←MU×NU×(RA←PI÷CH)×CH*NU)×R←*(MU←PARA[1])×((CH←
        1-PI)*NU)-1,0ρA←,1+I←0
[4]   LBL2:→((SUM←+/R←R,PRO×(÷I+1)×+/(A←A,A[I]×(NU-I)×RA÷I←I+1)×⌽
        R)≤1-10*-TOL)/LBL2
[5]     R←R,1-SUM
      ∇
```

FIGURE 29 APL functions for calculating Poisson-binomial(PARA←μ,ν,π) probabilities.

$$p_x = e^{-\mu} \frac{\rho^x}{x!} \sum_k (\nu k)^{(x)} \frac{\lambda^k}{k!} , \quad \lambda = \mu\chi^\nu ,$$

$$(\nu k)^{(x)} = \nu k(\nu k-1)\ldots(\nu k-x+1) ,$$

$$= e^{\lambda-\mu} \frac{\rho^x}{x!} \sum_k (\nu k)^{(x)} e^{-\lambda} \frac{\lambda^k}{k!}$$

$$= e^{\lambda-\mu} \frac{\rho^x}{x!} m_{(x)} , \text{ say} ,$$

where $m_{(x)} = \mathscr{E}[(\nu K)^{(x)}]$ is the expectation with respect to a Poisson(λ) distribution of the factorial power $(\nu K)^{(x)}$.

Introducing $p_{(x)} = \dfrac{m_{(x+1)}}{m_{(x)}}$,

then

$$p_{x+1} = \frac{\pi}{\chi} \cdot \frac{p_{(x)}}{x+1} p_x ,$$

and Shumway and Gurland (1960b) tabulate $p_{(x)}$, which they write $p_{[x]}$, for $\nu = 2$, $x = 0(1)9$ and $\mu\chi^\nu = 0.1(0.02)1.1$ to 5 decimal places.

It is also possible to express $m_{(x)}$ in terms of Stirling Numbers. For

$$m_{(i)} = \mathscr{E}\{(K\nu)^{(i)}\} = \mathscr{E}(\sum_r S_i^{(r)} K^r \nu^r)$$

from the definition of $S_i^{(r)}$

$$= \sum_r S_i^{(r)} \nu^r \mathscr{E}(K^r)$$

$$= \sum_r S_i^{(r)} \nu^r \sum_j \mathcal{S}_r^{(j)} \lambda^j$$

from the result for a Poisson distribution,

$$= \sum_j a_{ij} \lambda^j \quad \text{for} \quad a_{ij} = \sum_r S_i^{(r)} \mathcal{S}_r^{(j)} \nu^r .$$

Shumway and Gurland (1960a) have tabulated the matrix (a_{ij}), which they write $A = (A_j^i)$, for $\nu = 2,3,4$ and $i = 1(1)10$ (on pages 103 and 105, wrongly said in their paper to be pages 26, 27, and 28). The limited ranges of tabulation should be noted.

For some purposes it is convenient to write these results in matrix form - e.g. for computers which readily accept this - by introducing the following notation.

$\underset{\sim}{\lambda} = (\lambda,\lambda^2,\ldots,\lambda^m)^t$, p_m being the highest probability sought,

$\underset{\sim}{\nu} = \text{diag.}(\nu,\nu^2,\ldots,\nu^m)$,

$\underset{\sim}{S}$ and $\underset{\sim}{\$}$ the m×m matrices of Stirling Numbers of the First and Second Kinds,

$\underset{\sim}{p} = (p_1,p_2,\ldots,p_m)^t$,

$b_x = p_0 \frac{\rho^x}{x!}$ and $\underset{\sim}{b} = \text{diag.}(b_1,b_2,\ldots,b_m)$.

Then the column vector of probabilities $\underset{\sim}{p}$ is given by

$$\underset{\sim}{p} = \underset{\sim}{b}(\underset{\sim}{S}\,\underset{\sim}{\nu}\,\underset{\sim}{\$})\underset{\sim}{\lambda}$$

where $\underset{\sim}{S}\,\underset{\sim}{\nu}\,\underset{\sim}{\$} = (a_{ij})$ of the previous paragraph. But it should be noticed that the maximum of a_{xj} increases, and b_x decreases, very fast with x, and so the method is only practicable for fairly small x, though non-recursive once the Tables are available.

As well as the common single mode or half-mode appearance, the shapes the p.f. can take for the Poisson-binomial distribution are rather unusual: for values of π near 1 a saw tooth profile at approximate multiples of ν is very striking. To illustrate this behavior a selection of graphs is given: these show "curious" plots; but it should not be inferred that the unimodal appearance is uncommon for the Poisson-binomial because of this selection. (The graph with the logarithmic ordinate scale is included to show that the multi-modality may persist into the tail.)

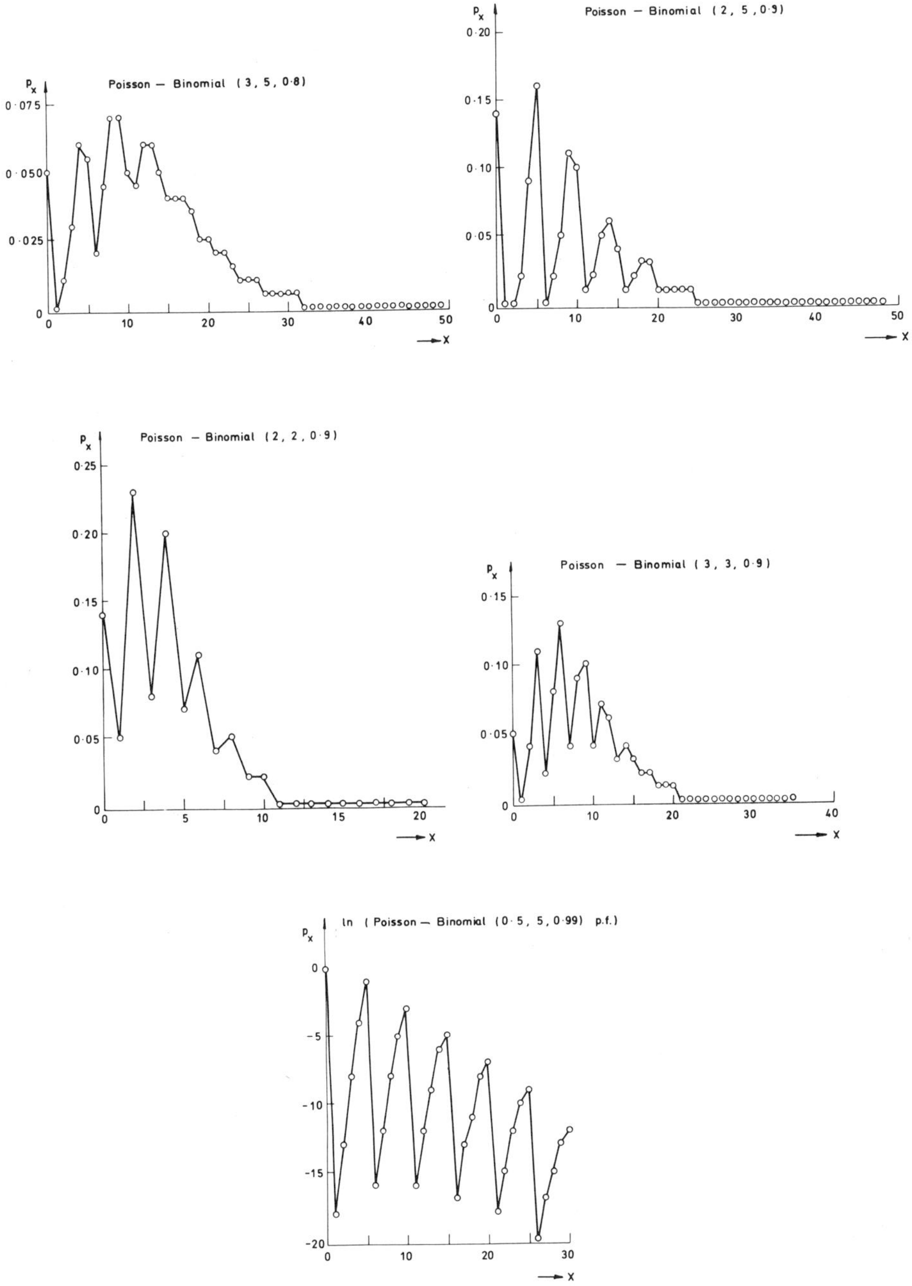

FIGURE 30 Graphs of some Poisson-binomial probability functions.

iv) Some results are given for the Hermite distribution, the case for which $\nu = 2$, from Kemp and Kemp (1965). The 'modified' Hermite polynomials $H_x^*(u)$ are defined as the coefficients in the generating function

$$\exp(uz + \frac{1}{2} z^2) = \sum_{x=0}^{\infty} H_x^*(u) \frac{z^x}{x!},$$

so that

$$H_0^* = 1,\ H_1^* = u,\ H_2^* = u^2 + 1,\ H_3^* = u^2 + 3u, \dots .$$

Re-writing the p.g.f. in this form leads to

$$\exp(-\mu+\mu\chi^2) \cdot \exp[\surd (2\mu)\chi\{\surd (2\mu)\pi z\} + \frac{1}{2} \{\surd (2\mu)\pi z\}^2]$$

$$= \exp(-\mu+\mu\chi^2) \Sigma H_x (\surd (2\mu)\chi) \frac{\{\surd (2\mu)\pi\}^x z^x}{x!},$$

whence $\quad p_0 = \exp(-\mu+\mu\chi^2)$,

$$p_x = p_0 H_x(\surd (2\mu)\chi) \frac{\{\surd (2\mu)\pi\}^x}{x!}, \quad x = 1,2,\dots$$

A rather simple recurrence relation

$$p_{x+1} = \frac{2\mu}{x+1} (\chi p_x + \pi p_{x-1})$$

with $p_1 = 2\mu\pi\chi p_0$, also follows, together with other less simple recurrence relations and explicit representations, for example in terms of confluent hypergeometric functions.

Moments and Cumulants

The factorial cumulants are particularly simple: replacing z by $1+t$ in the logarithm of the p.g.f., the factorial cumulant g.f. is

$$-\mu + \mu\{\chi + (1 + t)\} = -\mu + \mu(1 + \pi t)^\nu$$

$$= \sum_{r=1}^{\nu} \nu^{(r)} \pi^r \frac{t^r}{r!},$$

whence

$$\kappa_{(r)} = \mu\nu^{(r)}\pi^r ;$$

i.e.

$$\kappa_{(1)} = \mu\nu\pi$$

$$\kappa_{(2)} = \mu\nu(\nu - 1)\pi^2$$

$$\kappa_{(3)} = \mu\nu(\nu - 1)(\nu - 2)\pi^3$$

$$\kappa_{(4)} = \mu\nu(\nu - 1)(\nu - 2)(\nu - 3)\pi^4 \text{ , etc.}$$

From these the first four cumulants can be found by the standard transformations; alternatively they can be found by using the recurrence relation

$$\kappa_{r+1} = \pi(1-\pi)\frac{\partial\kappa_r}{\partial\pi} + \nu\pi\kappa_r \text{ , } \kappa_1 = \mu\nu\pi \text{ ,}$$

(see the Appendix on Moments and Cumulants; or directly):

$$\kappa_1 = \mu\nu\pi$$

$$\kappa_2 = \mu\nu\pi(\chi+\nu\pi) = \mu\nu\pi \quad \text{with} \quad \tau = \chi + \nu\pi \text{ ,}$$

$$\kappa_3 = \mu\nu\pi(\tau^2 + \chi\tau - \chi) \text{ ,}$$

$$\kappa_4 = \mu\nu\pi(\tau^3 + 3\chi\tau^2 + 2\chi(\chi - 2)\tau - 2\chi^2 + \chi) \text{ ,}$$

and moments are readily found from these.

The Index of Cluster Size is $\iota_{CS} = \nu\pi-\pi$, and the Index of Cluster Frequency $\iota_{CF} = \mu \frac{\nu}{\nu-1}$: these are 'close to' the means of the summed and stopping variates $\nu\pi$ and μ respectively only if ν is 'large'.

For the Hermite distribution, Kemp and Kemp (1961) show that

$$\kappa_r = 2\mu\pi(\chi + 2)^{r-1}\pi) \text{ ,}$$

$$\mu'_{(r)} = [\surd(2\mu)\chi]^r \, H_r^*[\surd(2\mu)] \text{ ,}$$

and

$$\mu'_{(r+1)} = 2\mu\pi[\mu'_{(r)} + \pi\mu'_{(r-1)}] \text{ .}$$

Tables and Computing Formulae

These have been mentioned in discussion of the probability function: they are of functions from which the distribution may be obtained, and not direct tabulations of the probability function.

Computing formulae are exhibited in Figure 29.

Limits and Approximations

It is simplest to examine the factorial cumulant g.f.

$$-\mu + \mu(1 + \pi t)^{\nu} .$$

i) $\pi \to 0$.

Supposing μ, ν are fixed (so the mean and variance tend to zero), the g.f.

$$= -\mu + \mu(1 + \pi t + 0(\pi^2))$$

$$= \mu\nu\pi t + 0(\pi^2) ,$$

corresponding to a Poisson($\mu\nu\pi$) approximation. More interestingly, suppose $\nu \to \infty$ in such a way that $\nu\pi$ is finite. Then the g.f. is

$$-\mu + \mu(1 + \nu\pi t \frac{1}{\nu})^{\nu} \to -\mu + \mu \exp(\nu\pi t) ,$$

corresponding to a Neyman($\mu, \nu\pi$) distribution. (This follows directly by recalling that in these circumstances a binomial tends to a Poisson variate.) The approximation is rather good, even for quite small ν, as was soon found in applications. If $\nu = 1$, the distribution is Poisson($\mu\pi$), without approximation.

ii) $\pi \to 1$.

The factorial cumulant g.f. then tends to $-\mu + \mu(1 + t)^{\nu}$, and this is the g.f. of a "scaled Poisson(μ) variate", with Poisson(μ) probabilities at $0, \nu, 2\nu, 3\nu, \ldots$ instead of $0, 1, 2, 3, \ldots$. The saw tooth graphs shown earlier are examples of cases in which π is close to 1.

A direct application of the steepest descent formula leads to

$$p_x \sim \frac{\exp\left\{-\mu + \frac{x(\chi+\pi z_x)}{\nu\pi z_x}\right\}}{\surd(2\pi) z_x^{\,x}} \cdot \sqrt{\frac{\chi+\pi z_x}{x(\chi+\nu\pi z_x)}}$$

where z_x is a root of

$$\mu\nu\pi z_x(\chi + \pi z_x)^{\nu-1} = x ;$$

further reduction is needed before this likely to be useful, however.

Modality

There will obviously be a (half) mode at the origin if

$$(\text{mean}) < \frac{1}{(1-\pi)^{\nu-1}} .$$

Further, since the distribution tends to the Neyman for large ν, any number of modes may occur, though no detailed knowledge of the occurrence is available.

Clearly multimodes, if they occur for large ν, will be roughly at $\nu, 2\nu, 3\nu, \ldots$ because of the Neyman limiting result; equally when $\pi \to 1$ the same result follows.

Estimation

In addition to the complexity which follows simply because three parameters may need to be estimated, the problem is made more difficult from a theoretical point of view by the restriction that ν should be a positive integer. In practice this is almost helpful, for what has been done is either to choose rather arbitrarily a value of ν or else to choose a sequence of values like $\nu = 2,3,4$ and to fit μ and π for each value, choosing the value of ν corresponding to the best fit (on whatever criteria may be adopted). Little or nothing is known about estimation procedures based on the simultaneous estimation of all three parameters - however, it should be noted that not all ν values are possible for distributions with certain specifications: for example, if a Poisson-binomial distribution has a mean of 16 and a variance of 144 then its ν value cannot be less than 9. While procedures are not available for the estimation of ν (together with at least large sample variabilities) any physically or biologically based interpretations of the parameters could be meaningful only exceptionally.

There are no two dimensional sufficient estimators.

Moment Estimation

For a given value of ν (= 2,3,4,...) the moment equations for μ and π are given by

$$\left.\begin{aligned} \bar{X} &= \overset{\circ}{\mu}\nu\overset{\circ}{\pi} \\ M_2 &= \overset{\circ}{\mu}\nu\overset{\circ}{\pi}(1+(\nu-1)\overset{\circ}{\pi}) \end{aligned}\right\} ,$$

leading to the estimators

$$\overset{\circ}{\mu} = \frac{\nu-1}{\nu} \cdot \frac{\bar{X}^2}{M_2-\bar{X}} = \frac{\nu-1}{\nu} \cdot \frac{K_{(1)}^2}{K_{(2)}} ,$$

$$\overset{\circ}{\pi} = \frac{\nu}{\nu-1} \cdot \frac{M_2-\bar{X}}{\bar{X}^2} = \frac{\nu}{\nu-1} \cdot \frac{K_{(2)}}{K_{(1)}} .$$

(The similarity of these to the moment equations for certain other distributions is due to the expression for the distribution variance as

(distribution mean) (1 + constant × parameter).

Note also the requirement $1 < M_2/\bar{X} \le \nu$.)

Sprott (1958) gives the first order generalized variance of these estimators:

$$\frac{1}{n^2} \cdot \frac{(1+7\delta\pi+6\delta^{(2)}\pi^2+\delta^{(3)}\pi^3)(1+\delta\pi)+2\mu\nu\pi(1+\delta\pi)^3-(1+3\delta\pi+\delta^{(2)}\pi^2)^2}{\delta^2\nu^2\pi^2} ,$$

where $\delta = \nu-1$. Exactly the same procedures for other cases yield the first order expansions for the sampling variances, with the same cautions.

A short table of the first order relative efficiency of moment with respect to maximum likelihood estimation is given by Sprott (1958) for $\nu = 2$; see also Patel (1971).

TABLE 9

First Order % Efficiency of Moment with Respect to Maximum Likelihood Fitting, $\nu = 2$ (Hermite)

π \ μ	0.1	0.5	0.8	1.0	2.0
0.3	73	53	51	53	64
0.2	84	68	67	68	75
0.1	93	89	84	84	87

(Sprott (1958))

McGuire et al. (1957) suggested that since

$$\frac{\kappa_{(1)}\kappa_{(3)}}{\kappa^2_{(2)}} = \frac{\nu-2}{\nu-1}\ , = \rho\ , \text{ say },$$

this may be inverted: $\nu = \frac{\rho-2}{\rho-1}$; and so sample factorial cumulants (or k-statistics) could be used to estimate ν , presumably with a rounding rule, to obtain an integer. Other experience with similar moment functions does not suggest much reliance is to be placed on such an estimator; and in fact they used as an estimate of ν the smallest integer such that the inequality following the moment equations is satisfied. The usual formulae for sampling variances are of course not strictly applicable when this procedure is used.

An example is provided by McGuire et al. (1957,8) Distribution 6, for which the moment estimates are obtained with $\nu = 2$, since $M_2/\overline{X} = 0.512/0.410 = 1.2 < 2$. They are $\overset{o}{\mu} = 0.83$, $\overset{o}{\pi} = 0.25$, with estimated standard errors of 0.16, 0.047 respectively - the details are shown in Table 10, where MUO is used for the moment estimate of μ , etc., and a sequence of fits of these data is given later, under Maximum Likelihood Estimation.

A literal interpretation as a Poisson(μ)-stopped binomial (ν,π) distribution is hardly biologically significant: larvae are assumed to occur in clusters of $\nu = 2$, with the mean number present per plant from a single cluster estimated as $\nu\overset{o}{\pi} = 0.49$, and the mean number of clusters per plant estimated as $\overset{o}{\mu} = 0.83$: 1296 plants were randomly selected. In terms of Indices of Cluster Size and Frequency, estimating from moments leads to an ι_{CS} of 0.25 and an ι_{CF} of 1.66 ; since $\nu = 2$ is as small as it can be, these are as different as is possible from the means $\nu\overset{o}{\pi}$ and $\overset{o}{\mu}$.

An alternative simple interpretation as a mixed binomial is that the area on which sampling is carried out can be thought of as made up of subareas on which the 'actual' numbers are $0,\nu,2\nu,\ldots$ (with $\nu = 2$) and that the proportion of the subarea corresponding to $k\nu$ is given by $e^{-0.83}(0.83)^k/k!$; for a particular subarea, k , the number of larvae observed have a binomial($k\nu$,0.25) distribution. The experiment leading to the frequency distribution consists of 1296 independent repetitions of choice of a k , with Poisson probabilities, followed by a choice from the binomial distribution so obtained. This is hardly likely to correspond to the way the observations were obtained.

TABLE 10

MOMENT FIT OF POISSON-BINOMIAL(μ,2,π)

EUROPEAN CORNBORE

(McGuire *et al.*, 1957, 8)

```
      PBMO

ENTER  NU
⎕:
      2
ENTER FREQUENCIES
⎕:
      907 275 88 23 3

N    = 1296      N×M1' = 532
M1'  = 0.4104938272
M2   = 0.5120503734

NU   = 2
MUO  = 0.8296 WITH S.E. = 0.1607
  CORRELATION = ¯0.9685          [GEN. VAR.]*1÷4 = 0.0436
PIO  = 0.2474 WITH S.E. = 0.0475

'2×LN RELATIVE LIKELIHOOD' =  2.62, PERHAPS WITH 2 DF

CPU TIME: 0 0 31

                 OBSERVED          FITTED

      0          907               904.4
      1          275               279.4
      2           88                89.1
      3           23                18.6
      4            3                 3.7
      5            0                  .7

                 1296              1296.0

LN LIKELIHOOD = ¯1098.960165

CHISQUARED = 1.57 WITH 5 CLASSES
```

Maximum Likelihood Estimation

Since the m.ℓ. equations are

$$\left.\begin{aligned} \Sigma\, F_x \frac{\partial\, \ell n\, p_x}{\partial \mu} &= 0 \\ \Sigma\, F_x \frac{\partial\, \ell n\, p_x}{\partial \pi} &= 0 \end{aligned}\right\} \quad \text{for} \quad \mu = \hat{\mu}\ ,\ \pi = \hat{\pi}$$

ν being supposed fixed, expressions for these derivatives are needed (Sprott (1958)). From the p.g.f.

$$g(z) = \exp\{-\mu + \mu(\chi + \pi z)^{\nu}\}\ ,\ \chi = 1 - \pi\ ,$$

$$\frac{\partial g}{\partial \mu} = g(z)\{-1 + (\chi + \pi z)^{\nu}\} \qquad (= \Sigma\, \frac{\partial p_s}{\partial \mu}\, z^x)$$

$$= -g(z) + \frac{\chi + \pi z}{\mu\, \nu\, \pi}\, g'(z)\ .$$

Hence expanding the p.g.f., noting

$$g'(z) = \Sigma(x+1)p_{x+1}\, z^x\ ,$$

and equating coefficients of z^x, gives

$$\frac{\partial p_x}{\partial \mu} = -p_x + \frac{\chi}{\mu\nu\pi}\,(x+1)\; p_{x+1} + \frac{xp_x}{\mu\nu}$$

i.e. $$\frac{\partial\, \ell n\, p_x}{\partial \mu} = -1 + \frac{x}{\mu\nu} + \frac{\chi}{\mu\nu\pi}\, \Pi_x$$

$$\text{for} \quad \Pi_x = (x+1)\, \frac{p_{x+1}}{p_x}\ .$$

Similarly,

$$\frac{\partial g}{\partial \pi} = \frac{z}{\pi}\, g'(z) - \frac{g'(z)}{\pi}$$

and

$$\frac{\partial p_x}{\partial \pi} = \frac{1}{\pi}\,(xp_x - (x+1)\; p_{x+1})\ ,$$

i.e.

$$\frac{\partial\, \ell n\, p_x}{\partial \pi} = \frac{1}{\pi}\,(x - \Pi_x)\ .$$

Substitution in the m.l. equations thus leads to

$$\left.\begin{aligned} \hat{\mu}\nu\hat{\pi} &= \frac{\Sigma\, x\, F_x}{\Sigma\, F_x} = \overline{X} \\ \text{and} \qquad \Sigma\, F_x\, \hat{\Pi}_x &= \Sigma\, x\, F_x = S_1 \end{aligned}\right\}$$

where in the previous notation $S_r = \Sigma x^r F_x$, equations which are strikingly like those for the Neyman distribution. (The second equation in both cases, is obtained because in the p.g.f.

$$\exp\{-\mu + \mu f(z)\}$$

where $f(z)$ is a function of a parameter α, the p.g.f. $f(z)$ has the form $f^*\{(1-z)v(\alpha)\}$ where $v(\alpha)$ is an arbitrary function of α - see Maximum Likelihood Estimation in Stopped Distributions.)

It will be recalled that if an observed distribution has a mean and second moment such that, e.g., if $(M_2 - \overline{X})/\overline{X} > 1$ there is no moment fit possible for a Poisson-bionomial distribution with $\nu = 2$. This restriction does not hold for a maximum likelihood fit, although the equations are too complex to be able to make a corresponding general statement if indeed it needs to be made.

An iterative solution of these equations is necessary, and the procedure can be arranged to be very like that used previously for the Neyman distribution. Eliminating μ with $\mu = \overline{X}/\nu\pi$ in order to use the Newton-Raphson method on a single variable, write

$$H(\pi) \equiv \Sigma\, F_x\, \Pi_x - S_1$$

where $H(\hat{\pi}) = 0$ defines the m.l. solution.

Then $$\frac{dH(\pi)}{d\pi} = \Sigma\, F_x\, \Pi_x\, \frac{d\,\ell n\, \Pi_x}{d\pi}$$

and $$\frac{d\,\ell n\, \Pi_x}{d\pi} = \frac{d\,\ell n\, p_{x+1}}{d\pi} - \frac{d\,\ell n\, p_x}{d\pi}.$$

From $$\frac{d\,\ell n\, p_x}{d\pi} = \frac{\partial\,\ell n\, p_x}{\partial\mu}\cdot\frac{d\mu}{d\pi} + \frac{\partial\,\ell n\, p_x}{\partial\pi}$$

$$= \left(-1 + \frac{x}{\mu\nu} + \frac{\chi}{\mu\nu\pi}\,\Pi_x\right)\left(-\frac{\mu}{\nu}\right) + \frac{x}{\pi} - \frac{\Pi_x}{\pi} \;,$$

$$= \frac{\mu}{\pi} + \frac{\nu-1}{\mu\pi} - \frac{\nu\pi+\chi}{\nu\pi^2}\,\Pi_x \;,$$

$$\Pi_x\,\frac{d\,\ell n\,\Pi_x}{d\pi} = \frac{\nu-1}{\nu\pi}\,\Pi_x - \frac{(\nu-1)\pi+1}{\nu\pi^2}\,\chi_x \;,$$

where $\chi_x = \Pi_x(\Pi_{x+1} - \Pi_x)$,

and so

$$H'(\pi) = \frac{\nu-1}{\nu\pi}\,\Sigma\, F_x\Pi_x - \frac{(\nu-1)\pi+1}{\nu\pi^2}\,\Sigma\, F_x\,\chi_x \;.$$

As usual, if a first approximation (perhaps from $\overset{o}{\pi}$) to $\hat{\pi}$ is π_i with $i = 1$, then a second approximation is

$$\pi_{i+1} = \pi_i - \frac{H(\pi_i)}{H'(\pi_i)} \;,\quad i = 1 \;;$$

the process will usually be iterated, for $i = 2,3,\ldots,$ and at every stage

$$\mu_i = \frac{\bar{X}}{\nu\,\pi_i} \;.$$

For hand computational purposes in certain ranges, the ratios $p_{(x)}$ of factorial moment-like functions already defined are useful. Writing

$$p_{(x)} = \frac{\chi}{\pi}\,\Pi_x \;,$$

the m.l. equations can be written as

$$\left.\begin{aligned} \hat{\mu}\hat{\nu}\hat{\pi} &= \overline{X} \\ \frac{\hat{\pi}}{1-\hat{\pi}} \Sigma F_x \hat{p}_{(x)} &= S_1 \end{aligned}\right\} ,$$

together with

$$H(\pi) = \frac{\pi}{1-\pi} \Sigma F_x p_{(x)} - S_1$$

and $$H'(\pi) = \frac{\nu-1}{\nu\chi} \Sigma F_x p_{(x)} - \frac{\nu\pi+\chi}{\nu\chi^2} \Sigma F_x q_{(x)}$$

where $q_{(x)} = p_{(x)} (p_{(x+1)} - p_{(x)})$.

Shumway and Gurland (1960b), who introduced this treatment, tabulate $p_{(x)}$ and $q_{(x)}$ (which they write $p_{[x]}$ and $q_{[x]}$) for $\nu = 2$, $\mu\chi^{\nu} = 0.1(0.02)1.1$ for $x = 0(1)9$ for $p_{(x)}$ and $x = 0(1)8$ for $q_{(x)}$, to 5 decimal places.

An analysis like that for previous cases gives the first order asymptotic covariance matrix of the maximum likelihood estimators (Sprott (1958)).

For $\psi = -1 + \frac{1}{\mu^2\nu^2\pi^2} \mathscr{E}(\Pi_X^2)$, $\Pi_x = (x+1) \frac{p_{x+1}}{p_x}$,

and $\delta = \nu-1$,

$$\mathscr{V}(\hat{\mu}) = \frac{1}{n} \cdot \frac{\mu^2\nu^2\psi-\mu\nu(\delta-1/\pi)}{\mu\nu(\delta+1/\pi)\psi-\delta^2} + O(1/n^2) ,$$

$$\mathscr{V}(\hat{\pi}) = \frac{1}{n} \cdot \frac{\chi^2\psi+(\nu\pi+\delta\pi\chi)/\mu\nu}{\mu\nu(\delta+1/\pi)\psi-\delta^2} + O(1/n^2) ,$$

$$\mathscr{C}(\hat{\mu},\hat{\nu}) = \frac{-1}{n} \cdot \frac{-\mu\nu\chi\psi+\nu\chi+\pi}{\mu\nu(\delta+1/\pi)\psi-\delta^2} + O(1/n^2) .$$

Hence the first order generalized variance is

$$\frac{1}{n^2} \cdot \frac{1}{\mu\nu(\delta+1/\pi)\psi-\delta^2} .$$

It should be noticed that, since

$$\mathscr{E}(\Pi_x^2) = \frac{\pi^2}{\chi^2} \mathscr{E}(p_{(x)}^2)$$

$$= \frac{\pi^2}{\chi^2} \Sigma p_{(x)}^2 p_x ,$$

it is not difficult to calculate numerically ψ for $\hat{\mu},\hat{\pi}$ from the tables of $p_{(x)}$ or directly on a computer and hence to estimate these leading terms.

Some comments on the solution procedures and their results can be found in Martin and Katti (1965) - e.g., in fitting 35 sets of published data, the Newton-Raphson procedure using moment estimates as starter values failed to converge in about half the cases.

The fitting procedure may be illustrated by repeated use of the previous data (McGuire et al. (1957-1958)) on European Corn-borers. Taking $\nu = 2$, the approximate 95% confidence intervals for μ and $\tilde{\omega}$ (with many cautions being necessary, and ignoring the correlation) based on the m.l. estimates are:

for μ : (0.54, 1.06)

for $\tilde{\omega}$: (0.17, 0.34)

However, as can be surmised from Table 11, it is indeed true that the likelihood is maximized for the choice $\nu = 2$, and thus one should presumably write $\hat{\nu} = 2$ if this was the method used to arrive at the value of ν ; then, for example, if the degrees of freedom of chi-squared were being taken seriously (if approximately) these should be dropped by unity. (Note the usual maximum likelihood theory does not carry over to the case of a discontinuously varying parameter.)

Returning for a moment to the Hermite case, $\nu = 2$, Kemp and Kemp (1965) give m.l. methods specific to this case together with numerical examples; see also Patel (1971).

TABLE 11

Fitted Frequencies

	Observed Frequencies	$\nu = 2$ Moment	$\nu = 2$ M.-L.	$\nu = 3$ Moment	$\nu = 3$ M.-L.	$\nu = 4$ Moment	$\nu = 4$ M.-L.
0	907	904.4	906.0	902.6	907.2	902.0	907.1
1	275	279.4	276.7	284.5	277.2	286.0	278.0
2	88	89.1	89.9	85.0	86.4	83.9	85.3
3	23	18.8	18.8	19.3	20.1	19.3	20.3
4	3	3.7	3.8	3.9	4.2	3.9	4.3
5+	0	0.7	0.7	0.8	0.9	0.9	1.0
Estimate of μ		0.830	0.802	1.106	0.997	1.244	1.105
s.e.		.16	.13	.21	.18	.24	.20
Estimate of $\tilde{\omega}$		0.247	0.256	0.124	0.137	0.082	0.0928
s.e.		.047	.043	.024	.025	.016	.017
correl.		-.97	-.96	-.97	-.96	-.97	-.96
ℓn likelihood		-1098.96	-1098.94	-1099.15	-1098.99	-1099.29	-1099.11
Chi-squared		1.6	1.5	1.8	1.3	2.0	1.9
# classes		5	5	5	5	5	5

European Cornborers - McGuire et al. (1957)

An asymptotic marginal confidence interval for ν might be written down by use of likelihood ratio results. For the parameter space $\Omega = \{(\mu,\nu,\pi);\ \mu > 0\ ,\ \text{integral}\ \nu \geq 2,\ 0 < \pi < 1\}$, and the reduced parameter space Ω_0 in which ν is specified as ν_0,

$$R = \frac{\max_{\Omega_0} L(\underset{\sim}{F};\ \mu,\nu,\pi)}{\max_{\Omega} L(\underset{\sim}{F};\ \mu,\nu,\pi)}$$

is such that $-2\ \ell n\ R$ is asymptotically chi-squared distributed with $3-2 = 1$ df ,

$$L = \Pi\{(p_x(\mu,\nu,\pi)\}^{F_x}$$

being the likelihood.

For the above data, the following is obtained:

ν	2	3	4	5	10	20	100	∞ (Neyman)
$-2\ \ell n\ R$	0	0.11	0.33	0.48	0.75	0.87	0.97	0.99

Since all of these are less than 1, the expectation of $-2\ \ell n\ R$, all values of ν are included in a confidence interval at any ordinary level, and the data are uniformative for estimating ν, and unpalatable but perhaps useful result.

Minimum Chi-Squared Estimation (Generalized Least Squares)

As usual, the choice of sample statistics is arbitrary, and hence a variety of different estimators can be generated - e.g. Katti and Gurland (1962a) concluded that those based on the first two sample factorial cumulants and the logarithm of the zero class frequency are generally adequate. However, the equations obtained there are non-linear and hence often tedious to solve, so that a linearization procedure is worth looking for. Using the methods (Hinz and Gurland (1967)) set out in the Appendix on Estimation, consider the use of ratios of factorial cumulants:

$$\eta_0 = \kappa_{(1)} = \mu\nu\pi\ ,\ \eta_r = \kappa_{(r+1)}/\kappa_{(r)} = (\nu-r)\pi\ .$$

A suitable parameterization is then

$$\underset{\sim}{\theta} = (\theta_1,\theta_2)^t = (\mu\pi,\pi)^t\ ,$$

so when estimators θ_1^*, θ_2^* are found, $\mu^* = \theta_1^*/\theta_2^*$, $\pi^* = \theta_2^*$ provide the estimators for μ and π.

One choice of a set of statistics uses only these ratios of factorial cumulants: writing H_0 and H_r for the ratios of the sample factorial cumulants (corresponding to η_r)

$$\underset{\sim}{H} = (H_0, H_1, H_2)^t$$

(more components can be used), and

$$\underset{\sim}{\eta} = (\eta_0, \eta_1, \eta_2)^t = (\mu\nu\pi, (\nu-1)\pi, (\nu-2)\pi)^t ,$$

and the linear equation $\underset{\sim}{\eta} = \underset{\sim}{w}\underset{\sim}{\theta}$ is satisfied by taking

$$\underset{\sim}{w} = \begin{pmatrix} \nu & 0 & 0 \\ 0 & \nu-1 & \nu-2 \end{pmatrix}^t .$$

Then because of this choice, the minimum chi-squared estimator is

$$\underset{\sim}{\theta}^* = (\underset{\sim}{w}^t \underset{\sim}{i}\, \underset{\sim}{w})^{-1}\, \underset{\sim}{w}^t \underset{\sim}{i}\, \underset{\sim}{H}$$

where $\underset{\sim}{i}$ is a consistent estimate of $\underset{\sim}{\mathcal{V}}(\underset{\sim}{H})^{-1}$, and

$$\underset{\sim}{\mathcal{V}}(\underset{\sim}{\theta}^*) = (\underset{\sim}{w}^t\, \underset{\sim}{\mathcal{V}}(\underset{\sim}{H})^{-1}\, \underset{\sim}{w})^{-1} ,$$

while for $\underset{\sim}{\lambda}^* = (\mu^*, \pi^*)^t$

$$\underset{\sim}{\mathcal{V}}(\underset{\sim}{\lambda}^*) = \frac{\partial \underset{\sim}{\lambda}}{\partial \underset{\sim}{\theta}}\, \underset{\sim}{\mathcal{V}}(\underset{\sim}{\theta}^*) \left(\frac{\partial \underset{\sim}{\lambda}}{\partial \underset{\sim}{\theta}}\right)^t$$

to the first order where

$$\frac{\partial \underset{\sim}{\lambda}}{\partial \underset{\sim}{\theta}} = \begin{pmatrix} 1/\theta_2, & -\theta_1/\theta_2^2 \\ 0 \;, & 1 \end{pmatrix} .$$

As usual (equation [ℓmc(D)] from the Appendix),

$$\underset{\sim}{\mathcal{V}}(\underset{\sim}{H}) = \frac{\partial \underset{\sim}{\eta}}{\partial \underset{\sim}{\kappa}_{(\,)}} \cdot \frac{\partial \underset{\sim}{\kappa}_{(\,)}}{\partial \underset{\sim}{\mu}}\, \underset{\sim}{\mathcal{V}}(\underset{\sim}{M}') \left(\frac{\partial \underset{\sim}{\eta}}{\partial \underset{\sim}{\kappa}_{(\,)}} \cdot \frac{\partial \underset{\sim}{\kappa}_{(\,)}}{\partial \underset{\sim}{\mu}} \right)^t .$$

$$= \underset{\sim}{J}\, \underset{\sim}{\mathcal{V}}(\underset{\sim}{M}')\, \underset{\sim}{J}^t \text{, say,}$$

where $\underset{\sim}{M}'$ is the vector of sample moments about the origin, and $\mathcal{V}(\underset{\sim}{M}')$ and these Jacobians are given in the Appendix on Estimation. Explicitly stated formulae are effectively given in the section on Linear Minimum Chi-Squared Estimation for the Neyman Distribution, with rather slight modifications. They are:

$\mathcal{V}(\underset{\sim}{H}) = \underset{\sim}{J}\,\mathcal{V}(\underset{\sim}{M}')\,\underset{\sim}{J}^t$ without change (except of course that μ_2' etc. differ for the two distributions),

$$\underset{\sim}{w} = \begin{pmatrix} \nu & 0 & 0 \\ 0 & \nu-1 & \nu-2 \end{pmatrix}^t ,$$

$$\underset{\sim}{\theta}^* = (\underset{\sim}{w}^t\,\mathcal{V}(\underset{\sim}{H})^{-1}\underset{\sim}{w})^{-1}\,\underset{\sim}{w}^t\,\mathcal{V}(\underset{\sim}{H})^{-1}\,\underset{\sim}{H} \quad \text{with } \underset{\sim}{i} \text{ for } \mathcal{V}(\underset{\sim}{H})^{-1},$$

without change,

$$\underset{\sim}{\lambda}^* = (\theta_1^*/\theta_2^*, \theta_2^*)^t \quad \text{without change}, = (\mu^*, \pi^*)^t ,$$

$$\mathcal{V}(\underset{\sim}{\lambda}^*) = \begin{pmatrix} 1/\pi, & -\mu/\pi \\ 0, & 1 \end{pmatrix} (\underset{\sim}{w}^t\,\mathcal{V}(\underset{\sim}{H})^{-1}\underset{\sim}{w})^{-1} \begin{pmatrix} 1/\pi, & 0 \\ -\mu/\pi, & 1 \end{pmatrix} .$$

A second choice, particularly appropriate when the zero class proportion P_0 is large, explicitly uses this proportion (in a rather involved way, to preserve linearity). Writing

$$A = P_0 H_1 \exp\left[-\frac{\nu-1}{\nu} \cdot \frac{H_0}{H_1} \left\{ \left(\frac{\nu-1-H_1}{\nu-1} \right)^{\nu} - 1 \right\} \right]$$

with a distributional analogue α (in which P_0 is replaced by p_0 and H_r by η_r), which reduces to $(\nu-1)\pi$, the set of statistics

$$\underset{\sim}{T} = (H_0, H_1, H_2, A)^t$$

(or $(H_0, H_1, A)^t$, etc.) is taken with $\underset{\sim}{\tau} = (\eta_0, \eta_1, \eta_2, \alpha)^t$. Requiring $\underset{\sim}{\tau} = \underset{\sim}{w}\underset{\sim}{\theta}$, for unchanged $\underset{\sim}{\theta}$ and $\underset{\sim}{\lambda}$, leads to

$$\underset{\sim}{w} = \begin{pmatrix} \nu & 0 & 0 & 0 \\ 0 & \nu-1 & \nu-2 & \nu-1 \end{pmatrix}^t ;$$

the normal equation [ℓmc(A)]

$$\underset{\sim}{\theta}^* = (\underset{\sim}{w}^t \underset{\sim}{i}\, \underset{\sim}{w})^{-1}\, \underset{\sim}{w}^t \underset{\sim}{i}\, \underset{\sim}{T}\ ,\quad \underset{\sim}{i} = \text{estimate of } \underset{\sim}{\mathcal{V}}(\underset{\sim}{T})^{-1}\ ,$$

is unchanged in form, but with these new meanings of $\underset{\sim}{w}$, $\underset{\sim}{i}$ and $\underset{\sim}{T}$; hence a further transformation is needed to obtain $\underset{\sim}{\mathcal{V}}(\underset{\sim}{T})$. The results can be conveniently expressed in terms of these relating to $\underset{\sim}{H}$; the formulae are given in the Appendix as [ℓmc(E)] , [(F)] and [(G)] , where $\underset{\sim}{J}$ is defined in [ℓmc(D)] , and [ℓmc(B)] and [(C)] still apply since $\underset{\sim}{\lambda}$ and $\underset{\sim}{\theta}$ are unchanged.

Again reference can be made to the Linear Minimum Chi-Squared Estimation section of the Neyman Distribution. The formulae correspond very closely to those given except that here they will be based on (H_0,H_1,H_2,A) and not (H_0,H_1,A) , so instead of $\underset{\sim}{\not H}$, $\underset{\sim}{\not J}$ and $\underset{\sim}{\not M}'$ of that section they will contain $\underset{\sim}{H}$, $\underset{\sim}{J}$ and $\underset{\sim}{M}'$ of the previous section. The formulae are:

$$\underset{\sim}{T} = (H_0,H_1,H_2 \,\vdots\, A)^t = (\underset{\sim}{H} \,\vdots\, A)^t,\quad \underset{\sim}{S} = (\underset{\sim}{H} \,\vdots\, P_0)^t$$

$$A = P_0 H_1 \exp\left[-\frac{\nu-1}{\nu}\cdot\frac{H_0}{H_1}\left\{\left(\frac{\nu-1-H_1}{\nu-1}\right)^{\nu} - 1\right\}\right] ,$$

$$\underset{\sim}{\theta} = (\theta_1,\theta_2)^t \text{ unchanged, } = (\mu\pi,\pi)^t ,$$

$$\underset{\sim}{\tau} = (\eta_0,\eta_1,\eta_2 \,\vdots\, \alpha)^t ,\quad \underset{\sim}{\sigma} = (\eta \,\vdots\, p_0)^t ,$$

$$\alpha = \text{analogue of } A, = (\nu-1)\pi ,$$

$$\underset{\sim}{w} = \begin{pmatrix} \nu & 0 & 0 & 0 \\ 0 & \nu-1 & \nu-2 & \nu-1 \end{pmatrix}^t ,$$

$$\underset{\sim}{\mathcal{V}}(\underset{\sim}{T}) = \frac{\partial \underset{\sim}{\tau}}{\partial \underset{\sim}{\sigma}}\ \underset{\sim}{\mathcal{V}}(\underset{\sim}{S}) \left(\frac{\partial \underset{\sim}{\tau}}{\partial \underset{\sim}{\sigma}}\right)^t ,$$

$$\frac{\partial \underset{\sim}{\tau}}{\partial \underset{\sim}{\sigma}} = \left(\begin{array}{ccc|c} 1 & 0 & 0 & 0 \\ 0 & 1 & 0 & 0 \\ 0 & 0 & 1 & 0 \\ \hline \dfrac{\partial\alpha}{\partial\eta_0} & \dfrac{\partial\alpha}{\partial\eta_1} & 0 & \dfrac{\partial\alpha}{\partial p_0} \end{array}\right) ,$$

$$\underset{\sim}{\mathcal{V}}(\underset{\sim}{S}) = \left(\begin{array}{c:c} \underset{\sim}{\mathcal{V}}(\underset{\sim}{H}) & -\dfrac{p_0}{n}\,\underset{\sim}{J}\,\underset{\sim}{M} \\ \hdashline -\dfrac{p_0}{n}\,(\underset{\sim}{J}\,\underset{\sim}{M})^t & \dfrac{p_0(1-p_0)}{n} \end{array}\right)$$

$\underset{\sim}{\mathcal{V}}(\underset{\sim}{H})$, $\underset{\sim}{J}$ and $\underset{\sim}{M}'$ from the previous section without change (except μ_2' etc. differ for the two distributions),

$$\underset{\sim}{\theta}^* = (\underset{\sim}{w}^t\,\underset{\sim}{\mathcal{V}}(\underset{\sim}{T})^{-1}\,\underset{\sim}{w})^{-1}\,\underset{\sim}{w}^t\,\underset{\sim}{\mathcal{V}}(\underset{\sim}{T})^{-1}\,\underset{\sim}{T} \quad \text{without change,}$$

$$\underset{\sim}{\lambda}^* = (\theta_1^*/\theta_2^*, \theta_2^*)^t \quad \text{without change,} = (\mu^*, \pi^*)^t \ ,$$

$$\underset{\sim}{\mathcal{V}}(\underset{\sim}{\lambda}^*) = \begin{pmatrix} 1/\pi, & -\mu/\pi \\ 0 & 1 \end{pmatrix} (\underset{\sim}{w}^t\,\underset{\sim}{\mathcal{V}}(\underset{\sim}{T})^{-1}\,\underset{\sim}{w})^{-1} \begin{pmatrix} 1/\pi\ , & 0 \\ -\mu/\pi\ , & 1 \end{pmatrix} .$$

Miscellaneous Methods of Estimation

As for other distributions, these can be considered as special cases of the foregoing methods; they will here be treated as pencil-and-paper procedures.

a) Mean and zero class frequency. Equating these sample functions and their distributional equivalents leads to

$$\left.\begin{aligned} \overline{X} &= \tilde{\mu}\nu\tilde{\pi} \\ P_0 &= \exp(-\tilde{\mu} + \tilde{\mu}\,\tilde{\chi}^{\nu}) \end{aligned}\right\} \ ;$$

eliminating $\tilde{\mu}$ gives the implicit equation for $\tilde{\pi}$:

$$\frac{1-(1-\tilde{\pi})^{\nu}}{\tilde{\pi}} = \frac{-\nu \,\ln P_0}{\overline{X}} \ ;$$

whence

$$\tilde{\mu} = \overline{X}/\nu\tilde{\pi} \ .$$

Application of the usual Taylor formulae to Sprott's (1958) results leads to the first order variances and covariances:

$$\mathscr{V}(\tilde{\pi}) = \frac{1}{n} \cdot \frac{\pi}{\mu^2 \nu p_0} \frac{p_0\{\mu(1-\chi^\nu)(1+\delta\pi)-2\mu\nu\pi(1-\chi^\nu)-\nu\pi\}+\nu\pi}{(1-\chi^\nu-\nu\pi\chi^{\nu-1})^2} ,$$

$$\mathscr{V}(\tilde{\mu}) = \frac{1}{n}\left\{\frac{\mu(1+\delta\pi)}{\nu\pi}-\frac{2\mu}{\nu\pi}\cdot\frac{(1+\delta\pi)(1-\chi^\nu)-\nu\pi}{1-\chi^\nu-\nu\pi\chi^{\nu-1}}+\frac{\mu^2}{\pi^2}\, n\, \mathscr{V}(\tilde{\pi})\right\},$$

$$\mathscr{C}(\tilde{\mu},\tilde{\pi}) = \frac{1}{n}\left\{\frac{1}{\nu}\cdot\frac{(1+\delta\pi)(1-\chi^\nu)-\nu\pi}{1-\chi^\nu-\nu\pi\chi^{\nu-1}}-\frac{\mu}{\pi}\, n\, \mathscr{V}(\tilde{\pi})\right\} ,$$

where $\delta = \nu-1$;

from these, first order efficiencies can be calculated. With respect to maximum likelihood the following efficiency Table is given by Sprott for the case $\nu = 2$ (see Table 12).

b) Mean, and ratio of first and zero class frequencies. The equations obtained by equating these sample and distributional expressions are

$$\left.\begin{aligned} \bar{X} &= \tilde{\mu}\,\nu\,\tilde{\pi} \\ \frac{P_1}{P_0} &= \tilde{\mu}\,\nu\,\tilde{\pi}\,\chi^{\nu-1} \end{aligned}\right\}$$

whence explicitly

$$\tilde{\pi} = 1 - \left(\frac{P_1}{\bar{X}\,P_0}\right)^{\frac{1}{\nu-1}}$$

and

$$\tilde{\mu} = \bar{X}/\nu\tilde{\pi} \ .$$

(These $\tilde{\mu}$ and $\tilde{\nu}$ are different from those in a).)

An application of the usual Taylor expansion will lead to asymptotic expressions for the covariance matrix and efficiency, as before.

TABLE 12

First Order % Efficiency of Mean and Zero Class Proportion with Respect to Maximum Likelihood Fitting $\nu = 2$ (Hermite)

π \ μ	0.1	0.5	0.8	1.0	1.5	2.0
0.4	86	70	70	72	74	85
0.3	94	88	90	92	-	-
0.2	96	94	96	97	-	-
0.1	99	98	99	99	-	-

(Sprott (1958))

Truncated-at-Zero Distribution

As usual, the p.g.f. is

$$\frac{g(z) - g(0)}{1 - g(0)} ,$$

where $g(z)$ is that of the untruncated distribution. Much of the previous analysis applies, with a greater degree of algebraic complexity since

$$g(0) = \exp(-\mu+\mu\chi^{\nu}) = p_0 .$$

Shumway and Gurland (1960) exhibit explicitly the maximum likelihood fit, the equations for μ and π being

$$\left.\begin{aligned} \frac{\hat{\mu}\,\nu\,\hat{\pi}}{1-\hat{p}_0} &= \frac{S_1'}{S_0'} \\ \frac{1}{\hat{\mu}\,\nu\,\hat{\pi}}\,\Sigma'\,F_x\hat{p}_{(x)} - \frac{\hat{p}_0}{1-\hat{p}_0}\,(1-\hat{\chi}^{\nu-1})S_0' - S_0' &= 0 \end{aligned}\right\}$$

where ' denotes summing with $x = 0$ excluded, and $p_{(x)} = \chi\Pi_x/\pi$.

Of course no explicit solution can be obtained; writing $H(\pi)$ for the left hand side of the second equation, with $\hat{\mu}$ eliminated (in principle with the first and then $\hat{\pi}$ replaced by π , the Newton-Raphson procedure can be applied to $H(\delta) = 0$: for an approximation π_1 the next is

$$\pi_2 = \pi_1 - H(\pi_1)/H'(\pi_1).$$

But an expression for $H'(\pi)$ will be rather involved and so a finite difference approximation may be worth using: given any two approximations π_1 and π_2 , and writing $\Delta\pi_2 = \pi_2 - \pi_1$, $\Delta H(\pi_2) = H(\pi_2) - H(\pi_1)$ the next approximation is

$$\pi_3 = \pi_2 - \frac{H(\pi_2)}{\Delta H(\pi_2)/\Delta\pi_2} .$$

The process may be iterated; but Shumway and Gurland suggest a quadratic interpolate:

$$\pi_4 = \pi_3 - \frac{(\pi_3 - \pi_2)^2}{\pi_3 - 2\pi_2 + \pi_1} = \pi_3 - \frac{(\Delta\pi_3)^2}{\Delta^2\pi_3} .$$

Alternatively, the usual two variable Newton-Raphson iteration may be used. Writing the m.l. equations as

$$\left.\begin{aligned} H(\hat{\mu},\hat{\pi}) &\equiv \hat{\mu}\ \nu\ \hat{\pi} - (1-\hat{p}_0)\ S_1'/S_0' = 0 \\ K(\hat{\mu},\hat{\pi}) &\equiv (1-\hat{p}_0)\ \Sigma'\ F_x \hat{p}_{(x)} - S_0'\hat{\mu}\ \nu\ \hat{\pi}\ \{\hat{p}_0(1-\hat{\chi}^{\nu-1}) - 1\} = 1 \end{aligned}\right\},$$

for starter values μ_1, π_1 the second set is given by

$$\begin{pmatrix} \mu_2 \\ \pi_2 \end{pmatrix} = \begin{pmatrix} \mu_1 \\ \pi_1 \end{pmatrix} - \left.\begin{pmatrix} \frac{\partial H}{\partial \mu} & \frac{\partial H}{\partial \pi} \\ \frac{\partial K}{\partial \mu} & \frac{\partial K}{\partial \pi} \end{pmatrix}^{-1}\right|_{\mu_1,\pi_1} \times \begin{pmatrix} H(\mu_1,\pi_1) \\ K(\mu_1,\pi_1) \end{pmatrix} ,$$

with iteration as necessary.

In both cases the rest of the analysis is along standard lines.

5.1a EXERCISE

1. Refer to the Exercise 12 in Chapter II dealing with multivariate generating functions for a derivation of the Hermite distribution from a bivariate Poisson distribution.

   In the notation of that Exercise, by recalling that $U+V$ is Poisson($\lambda+\mu$) and $2W$ a 'doublet' Poisson variate, deduce that the Hermite is a convolution of independent Poisson and doublet Poisson variates. An example of a situation for which this model might apply is as follows: In copying digits, the relative frequency of single digit errors may be supposed Poisson, while the relative frequency of interchanging a pair of adjacent digits (a doublet error) may also be supposed Poisson, no other errors being supposed plausible. Then the total number of errors will have a Hermite distribution. (Kemp and Kemp (1965)).

5.2 THE POISSON-PASCAL DISTRIBUTION

The simplest structure for this distribution is perhaps as a Poisson(μ)-stopped Pascal(κ,ρ) distribution, with p.g.f. therefore

$$\exp\{-\mu+\mu(\gamma-\rho z)^{-\kappa}\} ,$$

whence its most common name. As a mixture it is a mixed-Pascal($u\kappa,\rho$) on u by Poisson(μ) distribution, whence by Gurland's equivalence theorem, or directly, the same p.g.f. follows.

It is also sometimes called a generalized Polya-Aeppli distribution (Skellam (1952)), since Polya (1930) refers to the special case (of which more below) with $\kappa = 1$ as having been studied by his student Aeppli in 1924.

An alternative form takes the negative binomial as $z^{\kappa}(\gamma-\rho z)^{-\kappa}$, in which the variate values are $\kappa,\kappa+1,\kappa+2$, in place of $0,1,2,\ldots$. κ is then a positive integer and this can be interpreted as a "waiting" distribution: e.g., if a coin with probability $1/\gamma$ of falling heads at a single toss is tossed until κ heads are obtained, then the number X of tosses one has to wait to achieve this κ has the p.g.f. $\mathcal{E}(z^X) = z^{\kappa}(\gamma-\rho z)^{-\kappa}$. A different parameterization of these two forms writes $\pi = 1/\gamma$ for the probability of a single "success": the negative binomial p.g.f.'s then became

$$(\gamma-\rho z)^{-\kappa} = \pi^{\kappa}(1-\chi z)^{-\kappa}$$

and

$$z^{\kappa}(\gamma-\rho z)^{-\kappa} = \pi^{\kappa} z^{\kappa}(1-\chi z)^{-\kappa} ,$$

where $\chi = 1-\pi$. However, in the experiments here contemplated there is no such underlying "waiting" distribution and indeed κ is not generally an integer; the first stated p.g.f. is therefore taken as the standard form for the Poisson-Pascal distribution.

The distribution is obviously reproductive on μ: if $X_1,X_2,\ldots$ are independent Poisson-Pascal(μ_i,κ,ρ) variates, ΣX_i is a Poisson-Pascal($\Sigma\mu_i,\kappa,\rho$) variate.

Multimodality to any degree may occur with the distribution, but little is known in a general way regarding the variation of these modes with the parameter values, although there will be a half mode at $x = 0$ if $\mu\kappa\rho < (1+\rho)^{\kappa+1}$.

The Probability Function

i) As a mixed distribution, the p.g.f. has the form

$$\sum_{u=0}^{\infty} (\gamma-\rho z)^{-\kappa u} e^{-\mu} \frac{\mu^u}{u!} ;$$

introducing $\lambda = \mu\gamma^{-\kappa}$, as with the Poisson-binomial, reduces this to

$$e^{-\mu} \sum_u \frac{\lambda^u}{u!} (1-\alpha z)^{-\kappa u} \quad \text{for} \quad \alpha = \rho/\gamma$$

$$= e^{-\mu} \sum_u \frac{\lambda^u}{u!} \sum_{x=0}^{\infty} \binom{-\kappa u}{r} (-\alpha)^x z^x$$

$$= e^{-\mu} \sum \frac{\lambda^u}{u!} \sum \binom{\kappa u+x-1}{x} \alpha^x z^x$$

$$= e^{-\mu} \sum_x \{\sum_u \binom{\kappa u+x-1}{x} \frac{\lambda^u}{u!}\} \alpha^x z^x .$$

Hence the coefficient of z^x is

$$p_x = e^{-\mu} (\frac{\rho}{\gamma})^x \sum_{u \geq 1/\kappa} \binom{\kappa u+x-1}{x} \frac{(\mu\gamma^{-\kappa})^u}{u!} ;$$

but this is essentially impracticable for direct use.

ii) Recurrence Relation. Since the distribution is Poisson-stopped, it has a simple recurrence relation. Writing

$$g(z) = \exp\{-\mu+\mu(\gamma-\rho z)^{-\kappa}\} ,$$

$$g'(z) = \kappa\mu\rho\, g(z)(\gamma-\rho z)^{-\kappa-1}$$

or

$$\sum (x+1) p_{x+1} z^x = \kappa\mu\rho\gamma^{-\kappa-1} \sum p_r z^r \sum \binom{\kappa+s}{s} (\frac{\rho}{\gamma})^s z^s ,$$

i.e.

$$(x+1) p_{x+1} = \kappa\mu\rho\gamma^{-\kappa-1} \sum_{s=0}^{x} p_{x-s} \binom{\kappa+s}{s} (\frac{\rho}{\gamma})^s ,$$

with of course $p_0 = \exp(-\mu+\mu\gamma^{-\kappa}) = e^{\lambda-\mu}$, $\lambda = \mu\gamma^{-\kappa}$. This relation is implemented in APL as shown in Figure 31 where, e.g., 2 7 8 PPI 1 2 3 returns the probabilities for Poisson-Pascal(1,2,3) at x = 2,7,8 and 3 PP 1 2 3 returns the probabilities until the sum from the left is at least 0.999.

For small integral κ , another useful reucrrence relation involving differences may be obtained - see the Polya-Aeppli case for an example.

iii) Probabilities in Terms of Poisson Moments and Stirling Numbers. The explicit expression for p_x can be put (Shumway and Gurland (1960b)) in terms of ascending factorial powers, since

$$\binom{q+x-1}{x} = \frac{(q+x-1)(q+x-2)\ldots(q-1)q}{x!} = \frac{(q+x-1)^{(x)}}{x!}$$

can also be written as

$$(-1)^x \frac{q(q+1)\ldots(\overline{q-x-2})(\overline{q-x-1})}{x!} = (-1)^x \frac{q^{\{x\}}}{x!} .$$

So

$$p_x = e^{-\mu} \left(\frac{\rho}{\gamma}\right)^x \frac{1}{x!} \sum \frac{(\kappa u)^{\{x\}}}{u!} \ell^u , \quad \ell = \mu\gamma^{-\kappa} ,$$

$$= e^{\ell-\mu} \left(\frac{\rho}{\gamma}\right)^x \frac{1}{x!} m_{\{x\}} ,$$

where $m_{\{x\}} = \mathscr{E}[(\kappa U)^{\{x\}}]$ is the expectation with respect to a Poisson(ℓ) distribution of the ascending factorial power $(\kappa U)^{\{x\}}$.

Introducing $p_{\{x\}} = m_{\{x+1\}}/m_{\{x\}}$, a simple recurrence relation follows:

$$p_{x+1} = \frac{p_x}{x+1} \frac{\rho}{\gamma} p_{\{x\}} ,$$

but since there are no Tables of $p_{\{x\}}$ available (unlike the corresponding cases for the Neyman and Poisson-binomial distributions) this is not practically useful even for small x.

If one introduces "ascending Stirling Numbers of the First Kind" $S_m^{\{r\}}$ by

```
      ∇ R←X PPI PARA;RA;GA;PRO;I;A;MU;KA;RH;SUM
[1]     R←R,(PRO←KA×MU×(RA←RH÷GA)×GA*-KA)×R←*(MU←PARA[1])×((GA←1+
        RH←PARA[3])*-KA←PARA[2])-1,0ρA←,1+I←0
[2]   LBL:→((ρR←R,PRO×(÷I+1)×+/(A←A,A[I]×(KA+I)×RA÷I←I+1)×⌽R)≤⌈/X
        )/LBL
[3]     R←R[1+X]
      ∇

      ∇ R←TOL PP PARA;RA;GA;PRO;I;A;MU;KA;RH;SUM
[1]     R←R,(PRO←KA×MU×(RA←RH÷GA)×GA*-KA)×R←*(MU←PARA[1])×((GA←1+
        RH←PARA[3])*-KA←PARA[2])-1,0ρA←,1+I←0
[2]   LBL:→((SUM←+/R←R,PRO×(÷I+1)×+/(A←A,A[I]×(KA+I)×RA÷I←I+1)×⌽R
        )≤1-10*-TOL)/LBL
[3]     R←R,1-SUM
      ∇
```

FIGURE 31 APL functions for calculating Poisson-Pascal(PARA←μ,κ,ρ) probabilities.

$$x^{\{m\}} = \Sigma\, S_m^{\{r\}}\, x^r ,$$

a parallelism with the Poisson-binomial case is obtained, and a convenient matrix representation follows. However, since

$$\begin{aligned} x^{\{m\}} &= x(x+1)\ldots(x+m-1) \\ &= (x+m-1)(x+m-2)\ldots(x+1)x = (x+m-1)^{(m)} \\ &= (-1)^m(-x)(-x-1)\ldots(-x-m+2)(-x-m+1) \\ &= (-1)^m(-x)^{(m)} \\ &= (-1)^m\, \Sigma\, S_m^{(r)}(-x)^r \\ &= \Sigma\, (-1)^{m+r}\, S_m^{(r)}\, x^r \\ &= \Sigma\, |S_m^{(r)}|\, x^r , \end{aligned}$$

thus

$$S_m^{\{r\}} = |S_m^{(r)}| ,$$

and familiar properties of $S_m^{(r)}$ alone are needed. Defining

$\underset{\sim}{K} = \text{diag.}(\kappa,\kappa^2,\ldots,\kappa^m)$, $\underset{\sim}{p} = (p_1,p_2,\ldots,p_m)^t$ with $p_0 = e^{\lambda-\mu}$,

where p_m is the highest probability sought,

$$b = \text{diag.}(b_1,b_2,\ldots,b_m) \quad \text{with} \quad b_r = p_0(\tfrac{\rho}{\gamma})^r \frac{1}{r!} ,$$

$$\underset{\sim}{S}^{\{\ \}}\underset{\sim}{K}\, \underset{\sim}{\$} = \underset{\sim}{\alpha}^* ,$$

$$\underset{\sim}{\Lambda} = (\lambda,\lambda^2,\ldots,\lambda^m)^t ,$$

then

$$\underset{\sim}{p} = \underset{\sim}{b}\, \underset{\sim}{\alpha}^*\, \underset{\sim}{\Lambda}.$$

In order to make this usable Shumway and Gurland (1960b) tabulate $\underset{\sim}{\alpha}^*$ (which they write A^*) in an extended form, as follows. Since $\underset{\sim}{\alpha}^*$ is lower triangular,

$$\underset{\sim}{\alpha}^* = \begin{bmatrix} \alpha_{11}^* & 0 \; \ldots & 0 \\ \alpha_{21}^* & \alpha_{22}^* \ldots & 0 \\ \ldots & \ldots & \ldots \\ \alpha_{m1}^* & \alpha_{m2}^* \ldots & \alpha_{mm}^* \end{bmatrix} ,$$

and what is tabulated in turn for $r = 1,2,\ldots,10$ is

$$\alpha_{r1}^*, \alpha_{r2}^*, \ldots, \alpha_{rr}^*$$

in the form

$$\begin{bmatrix} \alpha_{r1}^* \\ \alpha_{r2}^* \\ \ldots \\ \alpha_{rr}^* \end{bmatrix} = \begin{bmatrix} \alpha_{11} & \alpha_{12} \ldots \alpha_{1r} \\ 0 & \alpha_{22} \ldots \alpha_{2r} \\ & \ldots\ldots \\ 0 & 0 \; \ldots \alpha_{rr} \end{bmatrix} \begin{bmatrix} \kappa \\ \kappa^2 \\ \ldots \\ \kappa^r \end{bmatrix} ;$$

the tabular entries are the elements α_{ij} .

Moments and Cumulants

Replacing z by $1+t$ in the logarithm of the p.g.f. gives the factorial cumulant g.f.:

$$-\mu+\mu(1-\rho t)^{-\kappa} = \mu \sum_{r=1}^{\infty} \binom{\kappa+r-1}{r} \rho^r t^r$$

$$= \mu \sum \kappa^{\{r\}} \rho^r \frac{t^r}{r!} ,$$

and so

$$\kappa_{(r)} = \mu \kappa^{\{r\}} \rho^r .$$

I.e.,

$$\kappa_{(1)} = \mu\kappa\rho ,$$

$$\kappa_{(2)} = \mu\kappa(\kappa+1)\rho^2,$$

$$\kappa_{(3)} = \mu\kappa(\kappa+1)(\kappa+2)\rho^3 ,$$

$$\kappa_{(4)} = \mu\kappa(\kappa+1)(\kappa+2)(\kappa+3)\rho^4 .$$

From these the first four cumulants can be found by the standard transformations; alternatively they can be found by using the recurrence relation

$$\kappa_{r+1} = \rho(1+\rho)\frac{\partial\kappa_r}{\partial\rho} + \kappa\rho\,\kappa_r\ , \quad \kappa_1 = \mu\kappa\rho\ ,$$

(see the Appendix on Moments and Cumulants; or directly):

$$\kappa_1 = \mu\kappa\rho,$$

$$\kappa_2 = \mu\kappa\rho(\gamma+\kappa\rho)\ ,\ = \mu\kappa\rho\tau \quad \text{with} \quad \tau = \gamma+\kappa\rho$$

$$\kappa_3 = \mu\kappa\rho(\tau^2+\gamma\tau-\gamma)\ ,$$

$$\kappa_4 = \mu\kappa\rho(\tau^3+3\gamma\tau^2+2\gamma(\gamma-2)\tau-2\gamma^2+\gamma)\ .$$

The Index of Cluster Size is $\iota_{CS} = \kappa\rho+\rho$, and the Index of Cluster Frequency $\iota_{CF} = \mu\kappa/(\kappa+1)$; these are close to the means of the summed and stopping variates $\kappa\rho$ and μ respectively if κ is 'large'.

Limits and Approximations

It may be noticed immediately that limiting forms of the distribution include the Poisson, negative binomial and Neyman distributions, each being suggested by other known results (Katti and Gurland (1961)). Among these are the following.

i) $\rho \to 0$, $\kappa \to \infty$ so that $\kappa\rho = \lambda$, say, is non-zero finite. Then

$$(1+\rho-\rho z)^{-\kappa} = \left(1+\frac{\lambda(1-z)}{\kappa}\right)^{-\kappa}$$

$$\to e^{-\lambda+\lambda z} \quad \text{as} \quad \kappa \to \infty\ ;$$

hence

$$\exp\{-\mu+\mu(\gamma-\rho z)^{-\kappa} \to \exp(-\mu+\mu e^{-\lambda+\lambda z})\ ,$$

the p.g.f. of a Neyman(μ,λ) distribution.

ii) $\kappa \to 0$, $\mu \to \infty$ so that $\kappa\mu = \lambda$ is non-zero finite. Then

$$\exp\left\{-\frac{\lambda}{\kappa}+\frac{\lambda}{\kappa}\frac{1}{(\gamma-\rho z)^{\kappa}}\right\} = \left\{\exp\frac{m^{-\kappa}-1}{\kappa}\right\}^{\lambda}$$

for $m = \gamma-\rho z$

$$\rightarrow \{\exp(-\ln m)\}^{\lambda} \quad \text{as} \quad \kappa \rightarrow 0$$

$$= (\gamma-\rho z)^{-\lambda} ,$$

the p.g.f. of a negative binomial(λ,ρ) distribution.

iii) $\rho \rightarrow 0$, $\mu \rightarrow \infty$ so that $\mu\rho=\lambda$ is non-zero finite. Then

$$\exp\left[-\frac{\lambda}{\rho}+\frac{\lambda}{\rho}\left\{1+\rho(1-z)\right\}^{-\kappa}\right]$$

$$= \exp\left\{\lambda\,\frac{(1+\rho(1-z))^{-\kappa}-1}{\rho}\right\}$$

$$\rightarrow \exp(-\kappa\lambda+\kappa\lambda z) \quad \text{as} \quad \rho \rightarrow 0 ,$$

the p.g.f. of a Poisson$(\kappa\lambda)$ distribution.

A direct application of the steepest descent formula leads to

$$p_x \sim \frac{\exp\left\{-\mu+\frac{x(\gamma-\rho z_x)}{\kappa\rho z_x}\right\}}{\sqrt{(2\pi)\,z_x^{x}}}\sqrt{\frac{\gamma-\rho z_x}{x(\gamma+\kappa\rho z_x)}} ,$$

where z_x is a root of

$$\kappa\mu\rho z_x = x(\gamma-\rho z_x)^{\kappa+1} ;$$

further reduction is needed before this is likely to be useful, however.

Estimation

As the distribution has three parameters, all of which will usually have to estimated in applications, there will be a good deal of numerical work involved in its use. No three dimensional sufficient estimators exist.

Moment Estimation

Using the factorial cumulants because of their relative simplicity, with $K_{(1)}$, $K_{(2)}$ and $K_{(3)}$ the sample factorial cumulants, the moment estimators $\overset{o}{\mu}$, κ, $\overset{o}{\rho}$ are the solutions of

$$\overset{o}{\mu}\,\overset{o}{\kappa}\,\overset{o}{\rho} = K_{(1)} ,$$
$$\overset{o}{\mu}\,\overset{o}{\kappa}(\overset{o}{\kappa}+1)\,\overset{o}{\rho}^{2} = K_{(2)} ,$$
$$\overset{o}{\mu}\,\overset{o}{\kappa}(\overset{o}{\kappa}+1)(\overset{o}{\kappa}+2)\,\overset{o}{\rho}^{3} = K_{(3)} ;$$

as usual,

$$K_{(1)} = M_1', = \bar{X},$$
$$K_{(2)} = M_2' - \bar{X}^2 - \bar{X},$$
$$K_{(3)} = M_3' - 3(\bar{X}+1)\,M_2' + 2\bar{X}^3 + 3\bar{X}^2 + 2\bar{X} .$$

Hence by division and re-arrangement

$$\left.\begin{aligned} \overset{o}{\kappa} &= \frac{2K_{(2)}^2 - K_{(1)}\,K_{(3)}}{K_{(1)}\,K_{(3)} - K_{(2)}^2} \\ \overset{o}{\rho} &= \frac{K_{(2)}}{K_{(1)}} \cdot \frac{1}{\overset{o}{\kappa}+1} \\ \overset{o}{\mu} &= \frac{K_{(1)}}{\overset{o}{\kappa}\,\overset{o}{\rho}} \end{aligned}\right\} .$$

There are obvious restrictions imposed, such as $\tfrac{1}{2} K_{(1)}\,K_{(3)} < K_{(2)}^2 < K_{(1)}\,K_{(3)}$, for a meaningful solution.

An example of the fit is given under Maximum Likelihood Estimation; and there are further details in the Exercises.

Straightforward Taylor expansions will lead to the first order (in $1/n$) terms for the variances and covariances of these estimators. It is perhaps hardly worth doing this algebraically, since the formulae obviously will be complex - instead, direct computer evaluation can be carried out in the course of the fitting operations. Thus, using the standard formulae, with the notation

$$\underset{\sim}{\lambda} = \begin{pmatrix} \lambda_1 = \mu \\ \lambda_2 = \kappa \\ \lambda_3 = \rho \end{pmatrix}, \quad \overset{\circ}{\underset{\sim}{\lambda}} = \begin{pmatrix} \overset{\circ}{\mu} \\ \overset{\circ}{\kappa} \\ \overset{\circ}{\rho} \end{pmatrix},$$

$$\mathcal{V}(\overset{\circ}{\underset{\sim}{\lambda}}) = \frac{\partial \underset{\sim}{\lambda}}{\partial \underset{\sim}{\kappa}_{(\,)}} \cdot \frac{\partial \underset{\sim}{\kappa}_{(\,)}}{\partial \underset{\sim}{\mu}'} \; \underset{\sim}{\mathcal{V}}(\underset{\sim}{M}') \left(\frac{\partial \underset{\sim}{\lambda}}{\partial \underset{\sim}{\kappa}_{(\,)}} \cdot \frac{\partial \underset{\sim}{\kappa}_{(\,)}}{\partial \underset{\sim}{\mu}'} \right)^t \text{ to } 0(1/n)$$

where $\underset{\sim}{\kappa}_{(\,)} = (\kappa_{(1)}, \kappa_{(2)}, \kappa_{(3)})^t$,

$\underset{\sim}{\mu}' = (\mu_1', \mu_2', \mu_3')^t$,

and $\underset{\sim}{\mathcal{V}}(\underset{\sim}{M}')$ is the covariance matrix of $(M_1', M_2', M_3')^t$, therefore involving 6th order moments. These estimated variances may then be used to construct approximate confidence intervals, etc. (see the Exercises).

The first order asymptotic efficiency of the moment compared with the maximum likelihood fit has been investigated by Katti and Gurland (1961). A tabulation is unwieldy because of the three parameters, but a selection of values is as follows in Table 13.

The usual cautions about a literal interpretation of these figures apply.

An illustration of the fitting is given from the data of Shenton (1949). These refer to insurance claims:

x	0	1	2	3	4	5	6	7	8	9	10	11	12	13	14	15	16+
F_x	187	185	200	164	107	68	49	39	21	12	11	2	5	2	3	1	0

The first three moments turn out to be

$$M_1' = 2.8059$$

$$M_2 = 6.4045$$

$$M_3 = 21.8442$$

TABLE 13

First Order Asymptotic % Efficiency of Moments with Respect to Maximum Likelihood (* doubtful accuracy)

		$\mu = 0.1$	$\mu = 1.0$	$\mu = 5.0$
$\rho = 2.0,\ \kappa =$	2.0	13	-	-
	1.0	12	-	-
	0.1	13	-	-
$\rho = 1.0\ \ \kappa =$	2.0	18	5	-
	1.0	22	15	-
	0.1	26	27	29*
$\rho = 0.1,\ \kappa =$	2.0	76	63	21*
	1.0	81	75	58*
	0.1	84	81	94*

(Katti and Gurland (1961))

and the moment estimates:

$$\overset{\circ}{\mu} = 10.2$$

$$\overset{\circ}{\kappa} = 0.273$$

$$\overset{\circ}{\rho} = 1.01$$

But since the estimated standard errors are

$$\text{s.e. } \overset{\circ}{\mu} : 13.6$$

$$\text{s.e. } \overset{\circ}{\kappa} : 0.47$$

$$\text{s.e. } \overset{\circ}{\rho} : 0.40$$

these are effectively meaningless; the estimated correlations are

$$\text{corr. } \overset{\circ}{\mu}, \overset{\circ}{\kappa} : -0.9988$$

$$\text{corr. } \overset{\circ}{\kappa}, \overset{\circ}{\rho} : -0.97$$

$$\text{corr. } \overset{\circ}{\rho}, \overset{\circ}{\mu} : 0.96 ,$$

still further complicating any interpretation. A tabulation of the fitted frequencies follows in the section on maximum likelihood fitting; it is perhaps disconcerting that the fit is quite good. (See that section for further comments.)

Maximum Likelihood Estimation

The device which in a number of cases with Poisson-stopped distributions leads to relatively simple maximum likelihood equations is the expression of the derivative of the logarithm of the probability function with respect to a parameter in terms of the probability function. (Shenton has exploited these recursive relations in a number of different ways and especially in computerized mathematical work.) However this can only be done neatly for the parameters μ and ρ of the Poisson-Pascal(μ,κ,ρ) distribution.

The full set of maximum likelihood equations is 3 equations like

$$\Sigma\, F_x \frac{\partial\, \ell n\ p_x}{\partial \mu} = 0 \quad \text{at} \quad \hat{\kappa},\ \hat{\mu},\ \hat{\rho}\ ;$$

by differentiation with respect to μ and ρ simple recursive expressions are found.

From $\Sigma\, p_x z^x = \exp\{-\mu+\mu(1+\rho-\rho z)^{-\kappa}\}$,

$$g'(z) = \kappa\mu\rho\; g(z)(1+\rho-\rho z)^{-\kappa-1} ,$$

and

$$\frac{\partial g(z)}{\partial\mu} = \Sigma\, \frac{\partial p_x}{\partial\mu}\, z^x = g(z)\{-1+(\gamma-\rho z)^{-\kappa}\}$$

$$= -g(z) + \frac{1}{\kappa\mu\rho}\,(\gamma-\rho z)\; g'(z)$$

$$= -\Sigma\, p_x z^x + \frac{\gamma}{\kappa\mu\rho}\,\Sigma(x+1)p_{x+1}z^x - \frac{1}{\kappa\mu}\,\Sigma\, x\, p_x z^x ,$$

whence

$$\frac{\partial p_x}{\partial\mu} = -\,p_x\left(1+\frac{x}{\kappa\mu}\right) + \frac{\gamma}{\kappa\mu\rho}(x+1)\;p_{x+1}$$

or

$$\frac{\partial\, \ell n\, p_x}{\partial\mu} = -\,\frac{x+\kappa\mu}{\kappa\mu} + \frac{\gamma}{\kappa\mu\rho}\,\Pi_x$$

$$\text{for}\quad \Pi_x = (x+1)p_{x+1}/p_x .$$

Similarly,

$$\frac{\partial\, \ell n\, p_x}{\partial\rho} = \frac{x}{\rho} - \frac{\Pi_x}{\rho} .$$

Multiplying by F_x , summing and equating to zero leads with a little re-arrangement to the pair of equations

$$\left.\begin{aligned} \hat{\kappa}\,\hat{\mu}\,\hat{\rho} &= \overline{X} \qquad (1)\\ \Sigma\, F_x\,\hat{\Pi}_x &= n\,\overline{X} \qquad (2)\end{aligned}\right\} ,$$

a very familiar set.

There is rather more trouble with κ. Direct differentiation leads to

$$\frac{\partial g(z)}{\partial\kappa} = -\,\mu\;\ell n(\gamma-\rho z)\;g(z)(\gamma-\rho z)^{-\kappa} .$$

As in Shumway and Gurland (1960a), write

$$\ell n(\gamma-\rho z) = \ell n\ \gamma - \sum_{x=1}^{\infty} \frac{\alpha^x z^x}{x}\ , \quad \alpha = \rho/\gamma\ ,$$

and

$$(\gamma-\rho z)^{-\kappa} = \sum_{x=0}^{\infty} n_x z^x$$

for the negative binomial probabilities, for which there is the recurrence relation

$$n_x = \frac{1}{\alpha\kappa}\left\{(x+1)n_{x+1} - \alpha x\ n_x\right\} .$$

Thus

$$\frac{\partial g(z)}{\partial\kappa} = -\mu\ \ell n\ \gamma\ \Sigma\ p_r z^r\ \Sigma\ n_s z^s + \mu\ \Sigma\ \frac{\alpha^t}{t}\ z^t\ \Sigma\ p_r z^r\ \Sigma\ n_s z^s\ ,$$

and, equating the coefficients of z^x ,

$$\frac{\partial p_x}{\partial\kappa} = -\mu\ \ell n\ \gamma\ \sum_{r=0}^{x} p_r\ n_{x-r} + \mu \sum_{t=0}^{x-1} \frac{\alpha^{x-t}}{x-t} \sum_{r=0}^{t} p_r\ n_{t-r}\ .$$

But, quite generally, for the Poisson-stopped distribution, if

$$g(z) = \Sigma\ p_x\ z^x = \exp\{-\mu+\mu\ n(z)\}$$

$$\text{where}\quad n(z) = \Sigma\ n_x\ z^x\ ,$$

then $p_x = \frac{\mu}{x} \sum_{r=0}^{x-1} (x-r)n_{x-r}\ p_r$

$$\left(\text{and}\quad p_{x+1} = \frac{\mu}{x+1} \sum_{r=0}^{x} (x-r+1)\ n_{x-r+1}\ p_r\right) .$$

Hence

$$\Sigma\, p_r\, n_{x-r} = \frac{1}{\alpha\kappa}\,\Sigma\, p_r(x-r+1)n_{x-r+1} - \frac{1}{\kappa}\,\Sigma\, p_r(x-r)n_{x-r}$$

$$= \frac{1}{\alpha\kappa\mu}\left\{(x+1)\,p_{x+1} - \alpha x p_x\right\} ;$$

also

$$\sum_{t=0}^{x-1} \frac{\alpha^{x-t}}{x-t} \sum_{r=0}^{t} p_r\, n_{t-r}$$

$$= \frac{1}{\kappa\mu}\,\Sigma\, \frac{\alpha^{x-t-1}}{x-t}\left\{(t+1)\,p_{t+1} - \alpha t p_t\right\}$$

$$= \frac{1}{\kappa\mu}\,\Sigma\left\{\frac{\alpha^{x-(t+1)}}{x+1-(t+1)}\,(t+1)\,p_{t+1} - \frac{\alpha^{x-t}}{x-t}\, t\, p_t\right\}$$

$$= \frac{1}{\kappa\mu}\left\{x p_x + \sum_{t=0}^{x-2} \frac{\alpha^{x-s}}{x-s+1}\, s p_s - \sum_{t=1}^{x-1} \frac{\alpha^{x-t}}{x-t}\, t p_t\right\} \quad \text{for} \quad s=t+1$$

$$= \frac{1}{\kappa\mu}\left\{x p_x - \sum_{t=1}^{x-1} \frac{\alpha^{x-t}\, t p_t}{(x-t)(x-t+1)}\right\} .$$

So

$$\frac{\partial\, \ell n\, p_x}{\partial\kappa} = \frac{1}{\kappa}\left\{x(1+\ell n\,\gamma) - \frac{\ell n\,\gamma}{\alpha}\,\Pi_x - \frac{1}{p_x}\,\Sigma\, b_{xt}\, p_t\right\}$$

$$\text{where} \quad b_{xt} = \begin{cases} \dfrac{t}{(x-t)(x-t+1)}\left(\dfrac{\rho}{1+\rho}\right)^{x-t} , & x \geq t , \\ 0 \quad \text{for} \quad x < t , \end{cases}$$

and multiplying by F_x , summing, equating to zero and re-arranging (with the use of the earlier equations)

$$n\,\hat{\kappa}\,\hat{\mu}\,\hat{\rho}\left(1 - \frac{\ell n(1+\hat{\rho})}{\hat{\rho}}\right) = \sum_{x=2}^{\infty} \frac{F_x}{\hat{p}_x} \sum_{t=1}^{x-1} \hat{b}_{xt}\,\hat{p}_t . \qquad (3)$$

Since this can be written in the form

$$\hat{\kappa} = f(\hat{\kappa},\hat{\mu},\hat{\rho})$$

by division through by

$$n\,\hat{\mu}\,\hat{\rho}\left(1 - \frac{\ell n\,\hat{\gamma}}{\hat{\rho}}\right),$$

iteration can be used fairly readily to solve the three maximum likelihood equations starting from the moment estimators. As above it is convenient first to regard κ as fixed. From the first two maximum likelihood equations, writing

$$H(\rho) = \Sigma\, F_x\, \Pi_x - n\,\overline{X}$$

where μ is regarded as eliminated by use of $\mu = \overline{X}/\kappa\rho$, the Newton-Raphson procedure can be applied. For

$$H'(\rho) = \Sigma\, F_x \frac{d\,\Pi_x}{d\rho} = \Sigma\, F_x\, \Pi_x \frac{d\,\ell n\,\Pi_x}{d\rho}$$

and

$$\frac{d\,\ell n\,\Pi_x}{d\rho} = \frac{d\,\ell n\,p_{x+1}}{d\rho} - \frac{d\,\ell n\,p_x}{d\rho}.$$

But

$$\frac{d\,\ell n\,p_x}{d\rho} = \frac{d\,\ell n\,p_x}{\partial\rho} + \frac{\partial\,\ell n\,p_x}{\partial\mu}\cdot\frac{d\mu}{d\rho},$$

and since

$$\frac{d\mu}{d\rho} = -\frac{\overline{X}}{\kappa\rho^2} = -\frac{\mu}{\rho},$$

$H'(\rho)$ follows readily. In detail,

$$\frac{d\,\ell n\,p_x}{d\rho} = \frac{1}{\rho}(x-\Pi_x) + \left\{\left(\frac{x}{\kappa\mu}+1\right) - \frac{\gamma}{\mu\kappa\rho}\,\Pi_x\right\}\frac{\mu}{\rho}$$

$$= \frac{\mu}{\rho} + \frac{x}{\rho}\left(1+\frac{1}{\kappa}\right) - \frac{1}{\rho}\left(1+\frac{\gamma}{\kappa\rho}\;\Pi_x\right)$$

whence

$$\frac{d\,\ell n\,\Pi_x}{d\rho} = \frac{1}{\rho}\,(1+\frac{1}{\kappa}) - \frac{1}{\rho}\,(1+\frac{\gamma}{\kappa\rho})(\Pi_{x+1} - \Pi_x)$$

and so

$$H'(\rho) = \frac{\kappa+1}{\kappa\rho}\,\Sigma\,F_x\,\Pi_x - \frac{\gamma+\kappa\rho}{\kappa\rho^2}\,\Sigma\,F_x\,\chi_x$$

where $\chi_x = \Pi_x(\Pi_{x+1} - \Pi_x)$.

Hence, given first approximations μ_1, κ_1, ρ_1 (e.g., from $\overset{o}{\mu}$, $\overset{o}{\kappa}$, $\overset{o}{\rho}$), a second approximation ρ_2 is obtained from

$$\rho_2 = \rho_1 - H(\rho_1/H'(\rho_1)$$

(using also κ_1 and μ_1), whence

$$\mu_2 = \bar{X}/\kappa_1\rho_2$$

and

$$\kappa_2 = f(\kappa_1,\mu_2,\rho_2)\ ;$$

the complete sequence of calculations can be iterated until changes in the approximations are as small as necessary. It is of course not obvious that convergence will occur, nor whether the suggested starting values will be good enough to achieve a reasonable rate of convergence in most or even many cases.

Shumway and Gurland (1960) note that the equation (3) for $\hat{\kappa}$ can be put in matrix form:

$$\hat{\kappa} = \frac{1}{n\hat{\mu}\ \hat{\rho}-\ell n\ \hat{\gamma})}\begin{bmatrix}\frac{F_2}{\hat{\rho}_2}, & \frac{F_3}{\hat{\rho}_3}, & \frac{F_4}{\hat{\rho}_4}, & \cdots\end{bmatrix}\begin{bmatrix}\hat{b}_{21} & 0 & 0 & 0\cdots \\ \hat{b}_{31} & \hat{b}_{32} & 0 & 0\cdots \\ \hat{b}_{41} & \hat{b}_{42} & \hat{b}_{43} & 0 \\ & \cdots\cdot & & \end{bmatrix}\begin{bmatrix}p_1 \\ \hat{p}_2 \\ \hat{p}_3 \\ \vdots\end{bmatrix}\ ;$$

and they give a short table of (b_{ij}) as a function of ρ/γ, for $i = 2(1)9$ and $j = 1(1)7$. (There are misprints in their formula (5), in which they write B^i_j for b_{ij}.)

The first order covariance matrix

$$\begin{pmatrix} \mathscr{V}(\hat{\kappa}), & \mathscr{C}(\hat{\kappa},\hat{\mu}), & \mathscr{C}(\hat{\kappa},\hat{\rho}) \\ \mathscr{C}(\hat{\kappa},\hat{\mu}), & \mathscr{V}(\hat{\mu}), & \mathscr{C}(\hat{\mu},\hat{\rho}) \\ \mathscr{C}(\hat{\kappa},\hat{\rho}), & \mathscr{C}(\hat{\mu},\hat{\rho}), & \mathscr{V}(\hat{\rho}) \end{pmatrix}$$

is obtained as usual by inverting the information matrix with i,j^{th} element

$$-\mathscr{E}\left(\frac{\partial^2 \ \ell n\ L}{\partial\lambda_i \partial\lambda_j}\right) = n\mathscr{E}\left(\frac{\partial\ \ell n\ p_X}{\partial\lambda_i} \cdot \frac{\partial\ \ell n\ p_X}{\partial\lambda_j}\right)$$

where $\lambda_1 = \mu$, $\lambda_2 = \kappa$ and $\lambda_3 = \rho$, these derivatives having been dealt with already. In fact, since what is required is the estimated covariance matrix, this is obtained by estimating the (i,j^{th}) element of the information matrix with

$$n\ \Sigma\ \hat{p}_x \frac{\partial\ \ell n\ p_x}{\partial\lambda_i} \frac{\partial\ \ell n\ p_x}{\partial\lambda_j} ,$$

the derivatives being evaluated at $\lambda_i = \hat{\lambda}_i$: some of the numerical difficulties in doing so are described in Katti and Gurland (1961).

The fitting procedure is illustrated on the insurance claim data used for a moment fit. The m.l. estimates and the estimated error structure are:

$\hat{\mu}$ = 6.01 with s.e. 4.1

$\hat{\kappa}$ = 0.59 with s.e. 0.68

$\hat{\rho}$ = 0.79 with s.e. 0.38

corr. $(\hat{\mu},\hat{\kappa})$ −0.99

corr. $(\hat{\kappa},\hat{\rho})$ −0.99

corr. $(\hat{\rho},\hat{\mu})$ 0.96

Although these look rather different from the moment estimates, they are not so in the light of the standard errors; moreover, in spite of the large absolute differences (e.g. $\overset{\circ}{\mu} = 10.2$ and $\hat{\mu} = 6.0$), the fitted frequencies are very similar, as shown in Table 14. At least partly, this is attributable to the high correlations: a large value for one estimate may be quite closely "corrected" for by a large value of another if their correlation is positive (or small if negative). For a given set of frequencies this makes the estimation problem inherently all but impossible to solve: there will be widely differing sets of parameter values which will closely reproduce these frequencies. Partly because of this, iteration procedures need particular care: they may terminate not because a maximum has been attained but because the surface is very nearly flat - e.g. Gebski (1975), where the nominated maximum is at most on a ridge.

Consequently, any precise interpretation of the estimates in terms of a Poisson(μ)-stopped summed-Pascal(κ,ρ) distribution is certainly unrealistic; but to illustrate on the m.l. estimates, if the data are interpreted as arising in this manner, the numbers of claims on a policy coming in clusters, the Poisson distributed number of clusters has a mean μ of about 6, and the negative-binomially distributed number of claims per cluster has a mean ($\kappa\rho$) of about 0.5 (with individual estimates of κ and ρ of about 0.6 and 0.8). The Indices of Cluster Frequency and Size are estimated as follows:

	ι_{CF}	ι_{CS}
Moments or Moment-analogue	2.19	1.28
M.ℓ.	2.23	1.26

where the moment-analogue estimates have the estimated standard errors 0.19 and 0.12 respectively (estimated correlation -0.95).

Minimum Chi-squared Estimation

Of the variety of estimators which may be used, two will be described in detail (Hinz and Gurland (1967)), one based on the ratio of factorial cumulants

TABLE 14

Poisson-Pascal Fittings
Moment [1]
Max. Like. [2]

	Observed	Fitted [1]	Fitted [2]
0	187	180.3	183.9
1	185	208.4	204.0
2	200	187.0	184.8
3	164	148.6	148.7
4	107	109.8	110.9
5	68	77.1	78.3
6	49	52.2	53.1
7	39	34.4	34.9
8	21	22.1	22.3
9	12	14.0	14.0
10	11	8.7	8.6
11	2	5.3	5.2
12	5	3.2	3.1
13	2	1.9	1.8
14	3	1.2	1.1
15	1	.7	.6
16	0	1.0	.8
	1056	1056.0	1056.0

Ln-Likelihood [1] = -2265.33074
Chi-squared [1] = 14.51 with 12 df

Ln-Likelihood [2] = -2265.216781
Chi-squared [2] = 14.83 with 12 df

Insurance Claims - Shenton (1949)

$$\underset{\sim}{\eta} = \begin{bmatrix} \eta_0 = \kappa_{(1)} \\ \eta_1 = \kappa_{(2)}/\kappa_{(1)} \\ \eta_2 = \kappa_{(3)}/\kappa_{(2)} \\ \eta_3 = \kappa_{(4)}/\kappa_{(3)} \end{bmatrix} = \begin{bmatrix} \kappa\mu\rho \\ (\kappa+1)\rho \\ (\kappa+2)\rho \\ (\kappa+3)\rho \end{bmatrix} ,$$

and the other based on these with the inclusion of another row:

$$\alpha = \left(\frac{2\eta_1-\eta_2}{\eta_0} \; \ell n \; p_0 + 1 \right) \left(\eta_2-\eta_1+1 \right)^{\frac{\eta_1}{\eta_2-\eta_1} - 1} , = \rho \; .$$

Both of these lead to Linear Minimum Chi-squared Estimation - see the Appendix.

For the first choice, the linear relation

$$\underset{\sim}{\eta} = \underset{\sim}{w} \, \underset{\sim}{\theta}$$

is satisfied by choosing the reparameterization

$$\underset{\sim}{\theta} = \begin{bmatrix} \theta_1 = \kappa\mu\rho \\ \theta_2 = \kappa\rho \\ \theta_3 = \rho \end{bmatrix} ,$$

and

$$\underset{\sim}{w} = \begin{bmatrix} 1 & 0 & 0 \\ 0 & 1 & 1 \\ 0 & 1 & 2 \\ 0 & 1 & 3 \end{bmatrix} .$$

Writing $\underset{\sim}{i}$ for an estimate of $\underset{\sim}{\mathcal{V}}(\underset{\sim}{H})^{-1}$, where $\underset{\sim}{H} = (H_0, H_1, H_2, H_3)^t$ is the sample analogue in terms of sample factorial cumulants, the solution of the minimum chi-squared (normal) equation for $\underset{\sim}{\theta}$ is

$$\underset{\sim}{\theta}^* = (\underset{\sim}{w}^t \, \underset{\sim}{i} \, \underset{\sim}{w})^{-1} \, \underset{\sim}{w}^t \, \underset{\sim}{i} \, \underset{\sim}{H}$$

with

$$\underset{\sim}{\mathcal{V}}(\overset{*}{\underset{\sim}{\theta}}) = (\underset{\sim}{w}^t \, \underset{\sim}{\mathcal{V}}(\underset{\sim}{H})^{-1} \, \underset{\sim}{w})^{-1} ;$$

Hence the solution for $\underset{\sim}{\lambda}$, $\underset{\sim}{\lambda}^*$, is

$$\underset{\sim}{\lambda}^* = \begin{pmatrix} \lambda_1^* = \mu^* \\ \lambda_2^* = \kappa^* \\ \lambda_3 = \rho^* \end{pmatrix} = \begin{pmatrix} \theta_1^*/\theta_2^* \\ \theta_2^*/\theta_3^* \\ \theta_3^* \end{pmatrix} ,$$

and its covariance matrix is to first order

$$\underset{\sim}{\mathcal{V}}(\underset{\sim}{\lambda}^*) = \frac{\partial \underset{\sim}{\lambda}}{\partial \underset{\sim}{\theta}} \, \underset{\sim}{\mathcal{V}}(\underset{\sim}{\theta}^*) \left(\frac{\partial \underset{\sim}{\lambda}}{\partial \underset{\sim}{\theta}}\right)^t .$$

As set out in the Appendix,

$$\frac{\partial \underset{\sim}{\lambda}}{\partial \underset{\sim}{\theta}} = \begin{pmatrix} 1/\theta_2 & -\theta_1/\theta_2^2 & 0 \\ 0 & 1/\theta_3 & -\theta_2/\theta_3^2 \\ 0 & 0 & 1 \end{pmatrix} , \quad \text{and}$$

$$\underset{\sim}{\mathcal{V}}(\underset{\sim}{H}) = \frac{\partial \underset{\sim}{\eta}}{\partial \underset{\sim}{\kappa}_{(\,)}} \cdot \frac{\partial \underset{\sim}{\kappa}_{(\,)}}{\partial \underset{\sim}{\mu}} \, \underset{\sim}{\mathcal{V}}(\underset{\sim}{M}') \left(\frac{\partial \underset{\sim}{\eta}}{\partial \underset{\sim}{\kappa}_{(\,)}} \cdot \frac{\partial \underset{\sim}{\kappa}_{(\,)}}{\partial \underset{\sim}{\mu}}\right)^t$$

with $\quad \underset{\sim}{\mathcal{V}}(\underset{\sim}{M}') = \frac{1}{n}(\mu'_{i+j} - \mu'_i \, \mu'_j)$, $i,j = 1,2,3,4;$

consequently, and unfortunately, 8th order moments are involved. It may also be noted that if the final row for each $\underset{\sim}{\eta}$ and $\underset{\sim}{H}$ is deleted, the same formulae apply except that $\underset{\sim}{w}$ loses its last row and $\underset{\sim}{\mathcal{V}}(\underset{\sim}{M}')$ its last row and column.

For the second choice, which is particularly appropriate when the zero class is large, writing

$$\underset{\sim}{\tau} = (\eta_0, \eta_1, \eta_2 \,\vdots\, \alpha)^t , \quad \underset{\sim}{T} = (H_0, H_1, H_2 \,\vdots\, A)^t ,$$

$\underset{\sim}{\theta}$ is unchanged, and $\underset{\sim}{\tau} = \underset{\sim}{w}\,\underset{\sim}{\theta}$ is achieved by choosing

$$\underset{\sim}{w} = \begin{bmatrix} 1 & 0 & 0 \\ 0 & 1 & 1 \\ 0 & 1 & 2 \\ 0 & 0 & 1 \end{bmatrix} .$$

$\underset{\sim}{i}$ is now taken to be an estimate of $\mathcal{V}(\underset{\sim}{T})^{-1}$; $\underset{\sim}{\theta}^*$ and $\mathcal{V}(\underset{\sim}{\theta}^*)$ change only by having $\underset{\sim}{T}$ in place of $\underset{\sim}{H}$; while $\underset{\sim}{\lambda}^*$, $\mathcal{V}(\underset{\sim}{\lambda}^*)$ and $\partial\underset{\sim}{\lambda}/\partial\underset{\sim}{\theta}$ are unchanged. $\mathcal{V}(\underset{\sim}{T})$ is most systematically handled by using the partitioning shown in [ℓmc(E)], [F] and [G] of the Appendix on Linear Minimum Chi-squared Estimation, the notation corresponding exactly. Hinz and Gurland (1967) in fact remark that the first order efficiency of the estimator $\underset{\sim}{T} = (H_0,H_1,H_2,A)^t$ is very high, and thus there may be little advantage in using as well H_3.

An example is quoted from Hinz and Gurland (1967), where $\underset{\sim}{T} = (H_0,H_1,H_2,A)^t$ was used (Table 15).

Yet another method uses the 'probability ratio cumulants' mentioned in the Appendix. However, this appears (Hinz and Gurland (1967)) to have no advantage over the methods already mentioned at least in terms of first order efficiencies, unreliable though this may be for sample sizes ordinarily encountered (and perhaps especially for the case of three parameters).

Miscellaneous Methods of Estimation

It is also possible to use various ad hoc methods, sometimes designed to reduce the formidable amount of calculation involved in the more sophisticated methods, and sometimes to exploit particular features such as the knowledge that large zero and first classes are to be expected. Katti and Gurland (1961) include a treatment of this latter; but because of the increasing availability of automatic computing facilities and the rather considerable arithmetical demands made by the Poisson-Pascal distribution in any circumstances no details will be given here.

TABLE 15

Minimum Chi-squared Fitting of Poisson-Pascal
Colorado Potato Beetles

(Beall (1940), Hinz and Gurland (1967))

Number of beetles per quadrat	Observed frequency	Poisson-Pascal fit
0	33	34.4
1	12	10.7
2	5	6.5
3	6	4.4
4	5	3.2
5	0	2.4
6	2	1.8
7	2	1.4
8	2	1.1
9	0	0.9
10	1	0.7
17	1	2.5 (17, 21, 22+ combined)
21	1	
22+	0	

$$\mu^* = 4.738$$

$$\kappa^* = 0.093$$

$$\rho^* = 4.715$$

5.2a EXERCISES

1. For the moment estimators in the Poisson-Pascal(μ,κ,ρ) show that the explicit expressions are:

$$\left.\begin{aligned} \overset{o}{\mu} &= A^2 B/P \\ \overset{o}{\kappa} &= P/Q \\ \overset{o}{\rho} &= Q/AB \end{aligned}\right\}$$

where $A = K_{(1)}$, $B = K_{(2)}$, $P = 2K_{(2)}^2 - K_{(1)} K_{(3)}$ and $Q = K_{(1)} K_{(3)} - K_{(2)}^2$.

Hence derive the Jacobian matrix $\partial \underset{\sim}{\lambda}/\partial \underset{\sim}{\kappa}_{(\,)}$ - in the notation of the text - expressed in terms of the above random variables and not the corresponding parameters, as:

$$\begin{bmatrix} \dfrac{AB(4B^2 - AC)}{P^2} & \dfrac{-A^2(2B^2 + AC)}{P^2} & \dfrac{A^3B}{P^2} \\ \\ \dfrac{-B^2C}{Q^2} & \dfrac{2\ ABC}{Q^2} & \dfrac{-AB^2}{Q^2} \\ \\ \dfrac{B}{A^2} & \dfrac{-Q}{AB^2} & \dfrac{1}{B} \end{bmatrix}$$

and $\partial \underset{\sim}{\kappa}_{(\,)}/\partial \underset{\sim}{\mu}'$ expressed in terms of distribution moments (for which estimates must be substituted in applications to data) as:

$$\begin{bmatrix} 1 & 0 & 0 \\ -(2\mu_1' + 1), & 1 & 0 \\ -3\mu_2' + 6\mu_1'^2 + 6\mu_1' + 2 & -(3\mu_1' + 1) & 1 \end{bmatrix}.$$

These may then be used to estimate asymptotically the covariance matrix $\underset{\sim}{\mathcal{V}}(\overset{o}{\underset{\sim}{\lambda}})$, = (v_{ij}), say, where, e.g., to first order,

$$\text{s.e.}(\overset{o}{\mu}) = \sqrt{v_{11}} ,$$
$$\text{corr.}(\overset{o}{\mu},\overset{o}{\kappa}) = v_{12}/\sqrt{(v_{11}\, v_{22})} .$$

2. To reduce the three variable system of maximum likelihood equations to a system to which single variable Newton-Raphson iteration can be applied (hopefully, with an increase in stability), suppose the three likelihood equations are:

$$\left.\begin{aligned} f(\hat{\kappa},\hat{\mu},\hat{\rho}) &\equiv \hat{\kappa}\,\hat{\mu}\,\hat{\rho} - \bar{X} = 0 \\ g(\hat{\kappa},\hat{\mu},\hat{\rho}) &\equiv \Sigma\, F_x\, \hat{\Pi}_x - n\bar{X} = 0 \\ h(\hat{\kappa},\hat{\mu},\hat{\rho}) &\equiv n\,\hat{\kappa}\,\hat{\mu}\,\hat{\rho}\left(1 - \frac{\ell n\,\hat{\gamma}}{\hat{\rho}}\right) - \Sigma\,\hat{b}_{xt}\,\hat{p}_t = 0 \end{aligned}\right\} .$$

Then the first equation can be used to eliminate $\hat{\kappa}$ explicitly from the second two functions, with

$$\hat{\kappa} = \bar{X}/\hat{\mu}\,\hat{\rho} \qquad (\text{say } \hat{\kappa} = F(\hat{\mu},\hat{\rho}))$$

leading to:

$$\left.\begin{aligned} g(\bar{X}/\hat{\mu}\,\hat{\rho},\hat{\mu},\hat{\rho}) &\equiv G(\hat{\mu},\hat{\rho}) = 0 \\ h(\bar{X}/\hat{\mu}\,\hat{\rho},\hat{\mu},\hat{\rho}) &\equiv H(\hat{\mu},\hat{\rho}) = 0 \end{aligned}\right\} .$$

Implicit "elimination" of $\hat{\mu}$ by use of the former of these equations enables the third function to be regarded as a function of ρ alone, say $N(\rho)$, and hence single variable Newton-Raphson can be applied:

$$\rho_{r+1} = \rho_r - \frac{N(\rho_r)}{N'(\rho_r)} ,$$

where

$$N'(\rho) = h_\rho + h_\kappa\, F_\rho - \frac{(h_\mu + h_\kappa F_\mu)(g_\rho + g_\kappa F_\rho)}{g_\mu + g_\kappa F_\mu}$$

(in which $h_\rho = \partial h/\partial\rho$, etc.).

5.3 THE POLYA-AEPPLI DISTRIBUTION

Taking $\kappa = 1$ in the Poisson-Pascal(μ,κ,ρ) distribution leads to a special case which, probably because it is then a two parameter distribution, has been separately studied - this was more important before computer facilities became generally available. The distribution, called the Polya-Aeppli because Polya (1930) refers to its appearance in the 1924 thesis of his student Aeppli, can be generated as follows:

i) Poisson(μ)-stopped Pascal($1,\rho$) or Poisson(μ)-stopped geometric($\frac{\rho}{1+\rho}$) , and

ii) Pascal(u,ρ) mixed on u by Poisson(μ), leading to the p.g.f.

$$\exp\left(-\mu + \frac{\mu}{\gamma-\rho z}\right), \quad \mu \text{ and } \rho > 0 \ , \ \gamma = 1+\rho .$$

Sometimes this is taken more conveniently in the form

$$\exp\left(- \frac{\theta}{1-\pi} + \frac{\theta}{1-\pi z}\right), \quad 0 < \pi < 1 \ , \ \theta > 0 \ ,$$

where $\pi = \rho/\gamma$, $\theta = \mu/\gamma$; $\mu = \theta/\chi$, $\rho = \pi/\chi$. As will appear, it is it is often desirable to change from one specification to the other, since the moment functions are naturally expressed in terms of μ and ρ but the probability functions in terms of θ and π . This does not help in applications, for which direct interpretations of the parameters are usually very desirable.

It also has appeared in the literature (Beall and Rescia (1953)) a limit of the Neyman Type A, B, C,... family: for index β where $\beta = 0,1,2,...$ correspond to the Types A, B, C,... the general family has p.g.f. (see Chapter III)

$$\exp\left\{-\mu + \mu\beta! \sum_{r=0}^{\infty} \frac{\nu^r (z-1)^r}{(\beta+r)!}\right\} .$$

If then $\beta \to \infty$ in such a way that $\nu/\beta \to \rho$ (non-zero finite), the second term in the exponent function can be reduced by Stirling's Formula:

$$\mu \; \Sigma \; \nu^r (z-1)^r \; \frac{\beta!}{(\beta+r)!}$$

$$\sim \mu \; \Sigma (\frac{\nu}{\beta})^r \; (z-1)^r \; \frac{e^r}{(1+\frac{r}{\beta})^{r-\frac{1}{2}}}$$

$$\sim \mu \; \frac{1}{1-\rho(z-1)} = \frac{\mu}{\gamma-\rho z} \quad ,$$

and the stated p.g.f. is obtained.

Galliher et al. (1959) give some details about the distribution, as a particular Poisson-stopped distribution (called by them a stuttering Poisson distribution), some of which are mentioned below; two other important papers are Anscombe (1950) and Evans (1953).

Of course all the general Poisson-Pascal theory is applicable; but there are some specializations of intrinsic interest.

The Probability Function

An explicit expression for the p.f. p_x from the p.g.f.

$$\sum_{x=0}^{\infty} p_x z^x = \exp\left(-\frac{\theta}{\chi} + \frac{\theta}{1-\pi z}\right)$$

can be given by expanding in powers of z (Evans (1953)). It is convenient first to take out the factor

$$p_0 = \exp\left(-\frac{\theta\pi}{\chi}\right) = \exp(-\theta\rho) \quad ,$$

hence leading to the p.g.f. in the form

$$p_0 \; \exp\left(\frac{\theta\pi z}{1-\pi z}\right) \quad .$$

Using the exponential expansion for the second factor and dropping p_0 for the moment gives

$$\sum_{m=0}^{\infty} \frac{(\theta\pi z)^m}{m!} (1-\pi z)^{-m}$$

$$= \sum_m \frac{(\theta\pi)}{m!} z^m \sum_{n=0}^{\infty} \binom{m+n-1}{n} (\pi z)^n$$

$$= \sum_m \sum_n \binom{m+n-1}{m-1} \frac{\theta^m \pi^{m+n}}{m!} z^{m+n} ,$$

whence interchanging the order of summation (see the Appendix on Series Manipulation)

$$= \sum_{x=0}^{\infty} \pi^x z^x \sum_{s=1}^{x} \binom{x-1}{x-1} \frac{\theta^s}{s!} , \text{ since } \binom{r}{-1} = 0 .$$

Hence $p_x = p_0 \pi^x \sum_{s=1}^{x} \binom{x-1}{s-1} \frac{\theta^s}{s!}$, $x = 1,2,\ldots$

The first few expressions are:

$$p_0 = \exp(-\theta\pi/\chi) ,$$

$$p_1 = p_0 \theta\pi ,$$

$$p_2 = p_0 \pi^2 \theta(1+\tfrac{1}{2}\theta) ,$$

$$p_3 = p_0 \pi^3 \theta(1+\theta+\frac{1}{6}\theta^2) ,$$

$$p_4 = p_0 \pi^4 \theta(1+\frac{3}{2}\theta+\frac{1}{2}\theta^2+\frac{1}{24}\theta^3) .$$

It is also possible (Evans (1953)) to express the p.f. explicitly in terms of a confluent hypergeometric function:

$$p_x = p_0 \pi^x \theta e^{-\theta} M(x+1,2,\theta) ,$$

and also (Galliher et al. (1959)) in terms of Laguerre polynomials.

Recurrence Relations

Corresponding to two different ways of expanding the p.g.f. two different recurrence relations may be obtained. In either case, for

$$g(z) = \exp\left(-\frac{\theta}{\chi} + \frac{\theta}{1-\pi z}\right) ,$$

then

$$g'(z) = g(z) \frac{\theta\pi}{(1-\pi z)^2} .$$

Multiplying through by $(1-\pi z)^2$,

$$(1-2\pi z + \pi^2 z^2) \ \Sigma(r+1)p_{r+1}z^r = \theta\pi \ \Sigma \ p_x z^x ,$$

whence picking out the coefficients of z^x ,

$$(x+1)p_{x+1} - 2\pi \ x \ p_x + \pi^2(x-1)p_{x-1} = \theta \ \pi \ p_x .$$

Hence the first recurrence relation:

$$p_{x+1} = \frac{\pi}{x+1} \{(\theta + 2x)p_x - \pi(x-1)p_{x-1}\} , \ x = 1,2,\ldots,$$

together with

$$p_0 = \exp(-\theta\pi/\chi) , \ p_1 = \theta\pi p_0 .$$

Alternatively, recalling the binomial expansion

$$(1-y)^{-2} = 1+2y+3y^2+4y^3+\ldots,$$

$$\Sigma(x+1)p_{x+1}z^x = \theta\pi \ \Sigma \ !_r z^r \ \Sigma(s+1)\pi^s z^s .$$

Equating coefficients of z^x , where $r+s=x$,

$$(x+1)p_{x+1} = \theta\pi \sum_{r=0}^{x} (x-r+1) \ \pi^{x-r} \ p_r , \ x = 0,1,\ldots,$$

again with $p_0 = \exp(-\theta\pi/\chi)$.

(Of course both these relations can also be obtained by differentiating $g(z)$ a total of $x+1$ times, and then writing $z = 0$.) Although the second relation has a number of terms which increases directly with x, all these terms are positive, whereas the former involves differencing between two rather small and similar expressions.

As a matter of curiosity rather than of immediate practical application, the introduction of ascending factorial moments of a Poisson(θ) distribution,

$$\mu'_{\{x\}:\theta} = \mathcal{E}\{Y(Y+1)\ldots(Y+x-1)\} ,$$

leads to a further recurrence relation obtained from

$$p_x = p_0 \frac{\pi^x}{x!} \mu'_{\{x\}:\theta} , \quad x = 1,2,\ldots :$$

for

$$p_{\{x\}} = \mu'_{\{x+1\}:\theta}/\mu'_{\{x\}:\theta} ,$$

then

$$p_{x+1} = p_x \frac{\pi}{x+1} p_{\{x\}} .$$

The distribution may have a mode, as well as a half-mode at the origin: for example for $\mu = 4$ and $\rho = 3$ there is a half mode at $x = 0$, and a mode at $x = 6$, though neither is pronounced:

$$p_0 = 0.049\ 79$$

$$p_1 = \quad 37\ 34$$

$$p_2 = \quad 42\ 01$$

$$p_3 = \quad 45\ 51$$

$$p_4 = \quad 47\ 92$$

$$p_5 = \quad 49\ 33$$

$$p_6 = \quad 49\ 86$$

$$p_7 = 0.049\ 62$$

Approximations

For small π (small ρ) the p.g.f. is nearly $\exp(-\theta\pi + \theta\pi z)$; i.e. the distribution may be approximated to by a Poisson($\theta\pi$)

distribution. Similar results can be obtained for other circumstances, though they appear not to have been used very frequently.

Of a somewhat different character is an asymptotic result using the method of steepest descent. Application of this method (see the Appendix on Asymptotic Expansions, and Douglas (1965)) leads to

$$p_x \sim \frac{e^{-\theta/\chi}}{\pi\sqrt{(2\pi)}} \; \frac{\exp \dfrac{\theta}{1-\pi z_x}}{z_x^{x+1} \sqrt{\left\{\dfrac{2\theta}{(1-\pi z_x)^3} + \dfrac{x}{\pi^2 z_x^2}\right\}}}, \quad x = 1,2,\ldots$$

where z_x is the smaller root of

$$\theta\pi z_x = x(1-\pi z_x)^2 \;;$$

and further expansion of this gives

$$p_x \sim \frac{\theta^{\frac{1}{4}} \exp\left\{-\dfrac{\theta(1+\pi)}{2(1-\pi)}\right\}}{2\sqrt{\pi}} \cdot \frac{e^{2\sqrt{(\theta x)}}\pi^x}{x^{3/4}} .$$

Moments and Cumulants

Using the former of the two expressions for the p.g.f., replacing z by $1+t$ leads to the factorial cumulant g.f.

$$-\mu + \frac{\mu}{1-\rho t} = \sum_{r=1}^{\infty} r!\, \rho^r \frac{t^r}{r!} ,$$

whence

$$\kappa_{(r)} = \mu r!\, \rho^r .$$

Hence the power cumulants follow:

$$\kappa_1 = \mu\rho \ ,$$
$$\kappa_2 = \mu\rho(1+2\rho) \ ,$$
$$\kappa_3 = \mu\rho(1+6\rho+6\rho^2) \ ,$$
$$\kappa_4 = \mu\rho(1+14\rho+36\rho^2+24\rho^3) \ .$$

Evans (1953) gives the general expression

$$\kappa_r = \mu\rho \sum_{s=1}^{r} \rho^{s-1} \sum_{u=1}^{s} (-)^{u+s} \binom{s}{u} u^r \ ,$$

and the recurrence relation

$$\kappa_{r+1} = \rho\kappa_r + \rho(\rho+1) \frac{\partial\kappa_r}{\partial\rho}$$

which can also be deduced from the extended Qvale relation given in the Appendix on Moments and cumulants.

Moment Estimation

Equating the sample and distribution factorial cumulants leads immediately to

$$\overset{o}{\mu} = \frac{2\,K_{(1)}^2}{K_{(2)}} = \frac{2\overline{X}^2}{M_2 - \overline{X}} \ , \quad \overset{o}{\rho} = \frac{K_{(2)}}{2\,K_{(1)}} = \frac{M_2 - \overline{X}}{2\overline{X}} \ ,$$

or, equivalently, to

$$\overset{o}{\theta} = \frac{4\,\overline{X}^3}{M_2^2 - \overline{X}^2} \ , \quad \overset{o}{\pi} = \frac{M_2 - \overline{X}}{M_2 + \overline{X}} \ ,$$

where M_2 is the second sample moment about the mean. Obviously these equations will not always have admissible solutions, but, e.g., since $\mathcal{V}(X) > \mathcal{E}(X)$ for the Polya-Aeppli distribution the estimator $\overset{o}{\theta}$ "usually" will not run into difficulties and $\overset{o}{\pi}$ "usually" will satisfy $0 < \overset{o}{\pi} < 1$.

Asymptotic results for the variances and covariances of these estimators are given by expressions like

$$\mathcal{V}(\overset{\circ}{\theta}) = \left(\frac{\partial\theta}{\partial\mu_1'}\right)\mathcal{V}(\bar{X}) + 2\,\frac{\partial\theta}{\partial\mu_1'}\cdot\frac{\partial\theta}{\partial\mu_2}\,\mathcal{C}(\bar{X},M_2) + \left(\frac{\partial\theta}{\partial\mu_2}\right)^2\mathcal{V}(M_2),$$

where

$$\frac{\partial\theta}{\partial\mu_1'} = \frac{1}{2}\,\frac{1+6\rho+6\rho^2}{\rho^2(1+\rho)^2},$$

$$\frac{\partial\theta}{\partial\mu_2} = -\,\frac{1+2\rho}{2\rho^2(1+\rho)^2},$$

$$\mathcal{V}(\bar{X}) = \frac{1}{n}\cdot\mu\rho(1+2\rho),$$

$$\mathcal{C}(\bar{X},M_2) = \frac{1}{n}\cdot\mu\rho(1+6\rho+6\rho^2) + 0(1/n^2),$$

$$\mathcal{V}(M_2) = \frac{1}{n}\cdot\frac{\rho(3(\mu+4) + 12(\mu+3)\rho + 12(\mu+2)\rho^2)}{4(1+2\rho)} + 0(1/n^2).$$

(See also the Exercises.)

A numerical example is taken from Beall and Rescia (1953), Table III, of European Cornborers. The fit is exhibited in Table 16 the moment estimates being $\overset{\circ}{\mu} = 2.06$ and $\overset{\circ}{\beta} = 0.363$ (or $\overset{\circ}{\theta} = 1.51$ and $\tilde{\omega} = 0.266$), but the standard errors are such that little confidence could be exhibited in the values calculated.

Maximum Likelihood Estimation

Using the specification of the distribution in terms of μ and ρ, the analysis is exactly as for the Poisson-Pascal distribution with $\kappa = 1$, the maximum likelihood equations being

$$\left.\begin{aligned} \hat{\mu}\,\hat{\rho} &= \bar{X} \\ \Sigma\, F_x \hat{\Pi}_x &= n\,\bar{X} \end{aligned}\right\}$$

TABLE 16

MOMENT FIT FOR POLYA-AEPPLI(μ,ρ)

EUROPEAN CORNBORERS
(Beall and Rescia, 1953)

```
      PAFIT 33 12 6 3 1 1

P.G.F.     EXP[-MU+MU÷((1+RH)-RH×Z)]
MEAN   MU×RH ,  VARIANCE = MU×RH×(1+RH×Z)


MOMENT FIT:  [MUO,RHO]  FOR  [MU,RH]
------------------------------------


M1' = 0.75 ,  N    = 56 ,  N×M1'   = 42
M2  = 1.294642857


MUO   =  2.0656, WITH S.E. =  1.1907
      CORRELATION =    ¯.9409,     (GEN.VAR.)*1÷4   =   .2956
RHO   =    .3631, WITH S.E. =   .2167

      IN TERMS OF  TH  = MU÷(1+RH)  AND  PI  = RH÷(1+RH)

THO   =  1.5154, WITH S.E. =  1.0692
      CORRELATION =    ¯.8576,     (GEN.VAR.)*1÷4   =   .4705
PIO   =    .2664, WITH S.E. =   .4026

'2×LN RELATIVE LIKELIHOOD' =  1.22 PERHAPS WITH 3 DF

CPU TIME: 0 0 42

                OBSERVED    FITTED[1]

        0          33          32.3
        1          12          13.0
        2           6           6.1
        3           3           2.7
        4           1           1.1
        5           1            .5
        6           0            .3

                   56          56.0

LN-LIKELIHOOD[1] = ¯66.77825281
CHI-SQUARED[1]   =    .15 WITH 2 DF
```

where as usual F_x is the sample frequency for class x and $\Pi_x = (x+1)p_{x+1}/p_x$. The procedure for applying the Newton-Raphson method is also similar: write

$$H(\rho) = \Sigma F_x \Pi_x - n\bar{X}$$

with μ eliminated by use of $\mu = \bar{X}/\rho$,

$$\text{whence} \quad H'(\rho) = \frac{2}{\rho} \Sigma F_x \Pi_x - \frac{1+2\rho}{\rho^2} \Sigma F_x \chi_x$$

(where $\chi_x = \Pi_x(\Pi_{x+1} - \Pi_x)$) .

So for first approximations μ_1, ρ_1 (e.g. from moments), second approximations are given by

$$\rho_2 = \rho_1 - \frac{H(\rho_1)}{H'(\rho_1)} ,$$

$$\mu_2 = \bar{X}/\rho_2 ,$$

with iteration if necessary.

For the alternative specification in terms of θ and π the estimators are given by

$$\hat{\theta} = \hat{\mu}/\hat{\gamma} , \quad \hat{\pi} = \hat{\rho}/\hat{\gamma} ,$$

and no separate account is necessary.

However, for the asymptotic covariance matrix due regard must be paid to the estimators being used. This matrix is the inverse of the information matrix with $(i,j)^{th}$ element

$$-\mathscr{E}\left(\frac{\partial^2 \ln L}{\partial\lambda_i \, \partial\lambda_j}\right) = n\mathscr{E}\left(\frac{\partial \ln p_x}{\partial\lambda_i} \cdot \frac{\partial \ln p_x}{\partial\lambda_j}\right), \qquad i,j = 1,2,$$

with $\lambda_1 = \mu$, $\lambda_2 = \rho$ or $\lambda_1 = \theta$, $\lambda_2 = \pi$, where this element can be estimated by

$$n \, \Sigma \, \hat{p}_x \frac{\partial \ln p_x}{\partial\lambda_i} \cdot \frac{\partial \ln p_x}{\partial\lambda_j}$$

the derivatives also being evaluated at $\hat{\lambda}_i$, $\hat{\lambda}_j$, and

$$\left.\begin{aligned} \frac{\partial \ln p_x}{\partial \mu} &= -1 - \frac{x}{\mu} + \frac{\gamma \Pi_x}{\mu\rho} \\ \frac{\partial \ln p_x}{\partial \rho} &= -\frac{x}{\rho} + \frac{\Pi_x}{\rho} \end{aligned}\right\} ,$$

or

$$\left.\begin{aligned} \frac{\partial \ln p_x}{\partial \theta} &= -\left(\frac{1}{\chi} + \frac{x}{\theta}\right) + \frac{\Pi_x}{\theta\tilde{\omega}} \\ \frac{\partial \ln p_x}{\partial \tilde{\omega}} &= -\frac{\theta}{\chi^2} + \frac{x}{\tilde{\omega}} \end{aligned}\right\} .$$

(See Katti and Gurland (1961) for numerical details relating to the general Poisson-Pascal case.)

Having determined the asymptotic covariance matrix by inversion for one of these parameterizations, that for the other may be obtained without repetition of the whole process from the familiar first order formula

$$\underset{\sim}{\mathcal{V}}(\hat{\underset{\sim}{\lambda}}) = \frac{\partial \underset{\sim}{\lambda}}{\partial \underset{\sim}{\mu}} \underset{\sim}{\mathcal{V}}(\hat{\underset{\sim}{\mu}}) \left(\frac{\partial \underset{\sim}{\lambda}}{\partial \underset{\sim}{\mu}}\right)^t .$$

The use of the recurrence relation for probabilities in terms of ascending factorial moments of a Poisson(θ) distribution enables an alternative statement of the maximum likelihood equations:

$$\left.\begin{aligned} \frac{\hat{\theta}\hat{\pi}}{\hat{\chi}^2} &= \bar{X} \\ \hat{\omega} \Sigma F_x \hat{p}_{\{x\}} &= n \bar{X} \end{aligned}\right\} .$$

Without a tabulation of $p_{\{x\}}$ this form is not useful for calculation purposes.

Minimum Chi-Squared Estimation

Using the same methods as set out for the Poisson-Pascal case, but with $\kappa = 1$ and a possible consequent reduction of the dimensions of the vectors and matrices, linear estimating equations can be obtained.

Based on the ratio of factorial cumulants, writing

$$\underset{\sim}{\eta} = \begin{pmatrix} \eta_0 = \kappa_{(1)} \\ \eta_1 = \kappa_{(2)}/\kappa_{(1)} \\ \eta_2 = \kappa_{(3)}/\kappa_{(2)} \end{pmatrix} = \begin{pmatrix} \mu\rho \\ 2\rho \\ 3\rho \end{pmatrix},$$

the reparameterization

$$\underset{\sim}{\theta} = \begin{pmatrix} \theta_1 = \mu\rho \\ \theta_2 = \rho \end{pmatrix}$$

with

$$\underset{\sim}{w} = \begin{pmatrix} 1 & 0 \\ 0 & 2 \\ 0 & 3 \end{pmatrix}$$

satisfies $\underset{\sim}{\eta} = \underset{\sim}{w}\,\underset{\sim}{\theta}$.

Exactly as before, with $\underset{\sim}{i}$ an estimate of $\mathcal{V}(\underset{\sim}{H})^{-1}$, $\underset{\sim}{H}'$ being the sample analogue of $\underset{\sim}{\eta}$ in terms of the sample factorial cumulants,

$$\underset{\sim}{\theta}^* = (\underset{\sim}{w}^t\,\underset{\sim}{i}\,\underset{\sim}{w})^{-1}\,\underset{\sim}{w}^t\underset{\sim}{i}\,\underset{\sim}{H}$$

is the minimum chi-squared estimator of $\underset{\sim}{\theta}$, with

$$\underset{\sim}{\lambda}^* = \begin{pmatrix} \lambda_1^* = \mu^* \\ \\ \lambda_2^* = \rho^* \end{pmatrix} = \begin{pmatrix} \theta_1^*/\theta_2^* \\ \\ \theta_2^* \end{pmatrix}$$

and

$$\underset{\sim}{\mathcal{V}}(\underset{\sim}{\lambda}^*) = \frac{\partial \underset{\sim}{\lambda}}{\partial \underset{\sim}{\theta}} \underset{\sim}{\mathcal{V}}(\underset{\sim}{\theta}^*) \left(\frac{\partial \underset{\sim}{\lambda}}{\partial \underset{\sim}{\theta}} \right)^t ,$$

$$\underset{\sim}{\mathcal{V}}(\underset{\sim}{\theta}^*) = (\underset{\sim}{w}^t \underset{\sim}{\mathcal{V}}(\underset{\sim}{H})^{-1} \underset{\sim}{w})^{-1} ,$$

to first order.

If it is desired to include specifically the zero class, introducing

$$\alpha = \left(\frac{2\eta_1 - \eta_2}{\eta_0} \ln p_0 + 1 \right) \left(\eta_2 - \eta_1 + 1 \right)^{\frac{\eta_1}{\eta_2 - \eta_1} - 1}$$

(which reduces to ρ on substitution) and

$$\underset{\sim}{\tau} = (\eta_0, \eta_1, \eta_2 \,\vdots\, \alpha)^t , \quad \underset{\sim}{T} = (H_0, H_1, H_2 \,\vdots\, A)^t$$

with

$$\underset{\sim}{\theta} = \begin{pmatrix} \theta_1 = \mu\rho \\ \theta_2 = \rho \end{pmatrix}$$

leads to $\underset{\sim}{\tau} = \underset{\sim}{w}\, \underset{\sim}{\theta}$ with

$$\underset{\sim}{w} = \begin{pmatrix} 1 & 0 \\ 0 & 2 \\ 0 & 3 \\ 0 & 1 \end{pmatrix} .$$

Then $\underset{\sim}{i}$ is an estimate of $\underset{\sim}{\mathcal{V}}(\underset{\sim}{T})^{-1}$, the formulae of the previous paragraph for $\underset{\sim}{\theta}^*$ and $\underset{\sim}{\mathcal{V}}(\underset{\sim}{\theta}^*)$ alter only by having $\underset{\sim}{T}$ in place of $\underset{\sim}{H}$, while $\underset{\sim}{\lambda}^*$, $\underset{\sim}{\mathcal{V}}(\underset{\sim}{\lambda}^*)$ and $\partial\underset{\sim}{\lambda}/\partial\underset{\sim}{\theta}$ are unaltered. A systematic treatment of $\underset{\sim}{\mathcal{V}}(\underset{\sim}{T})$ follows by use of the partitioning exhibited, using the formulae [ℓmc(E)] , [F] and [G] of the Appendix on Linear Minimum Chi-squared Estimation.

5.3a EXERCISES

1. Carry out some calculations, perhaps illustrated graphically, to show the shapes which can be taken by the Polya-Aeppli probability function. For example, observe the transition from the half-mode at the zero class for $(\mu,\rho) = (3,2)$, to half mode and mode at $x = 3$ for $(3.5,2)$, half mode and mode at $x = 4$ for $(4,2)$, to a single mode at $x = 5$ for $(4.5,2)$; while the sequence $(2,0.5)$, $(2,1)$, $(2,1.5)$,... always has a half mode.

2. Use the standard change of variable theory to show for the Polya-Aeppli(μ,ρ) distribution with p.f. $p_x(\mu,\rho)$ in which the parameters are related by $\mu\rho = \overline{X}$, that

$$\frac{d\,\ell n\, p_x}{d\rho} = -\frac{1+2\rho}{\rho}\Pi_x + \frac{\mu+2x}{\rho},$$

where

$$\Pi_x = (x+1)\; p_{x+1}/p_x \,.$$

Apply this result to obtain the expression for $H'(\rho)$ given under Maximum Likelihood Estimation.

3. If the Polya-Aeppli distribution is specified in the form with its mean σ and variance $\sigma(1+\tau)$, so its p.g.f. is

$$\exp\left\{-\frac{2\sigma}{\tau}\left(1+\frac{2}{2+\tau-\tau z}\right)\right\},$$

obtain the following results.

Cumulants

$$\kappa_1 = \sigma$$

$$\kappa_{r+1} = \left\{1+\tau+\tfrac{1}{2}\tau(\tau+2)\frac{\partial}{\partial\tau}\right\}\kappa_r\,, \qquad r = 1,2,\ldots$$

Moment Estimators

$$\overset{o}{\sigma} = \overline{X}\,, \quad \overset{o}{\tau} = \frac{M_2 - \overline{X}}{\overline{X}}\,,$$

$$\mathcal{V}(\overset{o}{\sigma}) = \frac{\sigma(1+\tau)}{n} ,$$

$$\mathcal{V}(\overset{o}{\tau}) = \frac{1}{n} \cdot \left\{ 2(1+\tau)^2 + \frac{(1+\tau)(2+\tau)}{\sigma} \right\} + 0(1/n^2),$$

$$\mathcal{C}(\overset{o}{\sigma},\overset{o}{\tau}) = \frac{\tau(2+\tau)}{n} + 0(1/n^2) .$$

Mean and Zero Class Estimators

$$\tilde{\sigma} = \bar{X} , \quad \tilde{\tau} = \frac{2\bar{X}}{-\ln P_0} - 2 ,$$

$$\mathcal{V}(\tilde{\sigma}) = \mathcal{V}(\overset{o}{\sigma}) ,$$

$$\mathcal{V}(\tilde{\tau}) = \frac{1}{n}\left(\frac{2+\tau}{2\sigma}\right)^2 \left\{(2+\tau)^2 \left(\exp \frac{2\sigma}{2+\tau} - 1\right) - 4\sigma\right\} + 0(1/n^2),$$

$$\mathcal{C}(\tilde{\sigma},\tilde{\tau}) = \frac{\tau(2+\tau)}{2n} + 0(1/n^2) .$$

(Evans (1953))

4. Using the results of the previous exercise and writing $\underset{\sim}{\lambda} = (\sigma,\tau)^t$, $\underset{\sim}{\mu} = (\mu,\rho)^t$ (where μ and ρ are the standard parameters) use

$$\mathcal{V}(\overset{o}{\underset{\sim}{\mu}}) = \frac{\partial \underset{\sim}{\mu}}{\partial \underset{\sim}{\lambda}} \; \mathcal{V}(\overset{o}{\underset{\sim}{\lambda}}) \left(\frac{\partial \underset{\sim}{\mu}}{\partial \underset{\sim}{\lambda}}\right)^t$$

to first order to deduce the first order covariance matrices for the moment estimator $\overset{o}{\underset{\sim}{\lambda}} = (\overset{o}{\mu},\overset{o}{\rho})^t$ and the mean-zero-class estimator $\tilde{\underset{\sim}{\lambda}} = (\tilde{\mu},\tilde{\rho})^t$.

5.4 THE THOMAS DISTRIBUTION

The Thomas distribution was originally introduced (Thomas (1949)) to describe the numbers of clustered individuals per random quadrat when the parent and first generation offspring constituted the cluster - as for example in the second year of a newly drained lake bed for some types of plant growth (see, e.g., Barnes and Stanbury (1951)). As usual the numbers of offspring per parent, and of parents per quadrat, were supposed Poisson distributed, whence the fortunately disappearing name "double Poisson".

The simplest derivation recognizes that if the numbers of offspring per parent (cluster) are Poisson(λ), so the p.g.f. is $\exp(-\lambda+\lambda z)$, then, including the parent, the numbers of individuals per cluster have the p.g.f. $z.\exp(-\lambda+\lambda z)$, with mean $1+\lambda$. So if the numbers of clusters per quadrat are Poisson(μ) the total numbers of individuals per quadrat will be Poisson(μ)-stopped, and hence will have the p.g.f.

$$\exp\{-\mu+\mu z\ \exp(-\lambda+\lambda z)\}\ :$$

a Poisson(μ)-stopped summed-{Poisson(μ) + 1} variate.

A Poisson mixture interpretation follows by writing the p.g.f. as

$$\sum_{u=0}^{\infty} z^u\, e^{-u\lambda+u\lambda z}\ .\ e^{-\mu}\frac{\mu^u}{u!}\ :$$

it is thus a mixture with Poisson(μ) weights of a translated and rescaled Poisson variate, translated so that its "zero class" is at u instead of 0.

A variate with the above p.g.f. will be said to have a Thomas(μ,λ) distribution.

The Probability Function

By direct expansion of the p.g.f., with $\ell\lambda=\mu e^{-\lambda}$,

$$p_0=e^{-\mu}\ ,$$

and

$$p_x = e^{-\mu}\frac{\lambda^x}{x!}\sum_{s=1}^{x}\binom{x}{s}\ell^s\, s^{x-s}\ , \quad x=1,2,\ldots\ .$$

Writing $c_{x,s} = \binom{x}{s} s^{x-s}$, it follows immediately that

$$c_{x,s} = \frac{xs}{x-s} c_{x-1,s} ,$$

and since $c_{ss} = 1$ these coefficients can be built up recursively very readily (in matrix form, with an upper right set of zeroes and a diagonal of ones); consequently

$$p_x = e^{-\mu} \frac{\lambda^x}{x!} \sum_{s=1}^{x} c_{xs} \ell^s$$

can also be found quite easily. The coefficients up to $x = 15$ are given explicitly by Thomson (1952), and the first few probabilities are:

$$p_1 = e^{-\mu} \lambda\ell,$$

$$p_2 = e^{-\mu} \frac{\lambda^2}{2!} (2\ell + \ell^2) ,$$

$$p_3 = e^{-\mu} \frac{\lambda^3}{3!} (3\ell + 6\ell^2 + \ell^3) ,$$

$$p_4 = e^{-\mu} \frac{\lambda^4}{4!} (4\ell + 24\ell^2 + 12\ell^3 + \ell^4) ,$$

$$p_5 = e^{-\mu} \frac{\lambda^5}{5!} (5\ell + 10.2^3\ell^2 + 10.3^2\ell + 5.4\ell^4 + \ell^5) .$$

The usual procedure of differentiating the p.g.f. yields (Thomas (1951)) a recurrence relation:

$$p_0 = e^{-\mu}$$

$$p_{x+1} = \frac{\mu e^{-\lambda}}{x+1} \sum_{r=0}^{x} \frac{(r+1)\ \lambda^r}{r!} p_{x-r} , \quad x = 0,1,2,\ldots,$$

and this is convenient for numerical work; see the APL functions in Figure 32. (An alternative form in terms of probabilities of the Poisson(λ) distribution, with $\pi_x = e^{-\lambda} \lambda^x/x!$, is

$$p_{x+1} = \frac{\mu}{x+1} \sum_{r=0}^{x} (x-r+1)\ p_r\ \pi_{x-r}\ .)$$

```
      ∇ R←X ThI PARA;LA;B;L:MU;I:SUM
[1]     R←R,LA×(L←MU×(*-LA)÷B←,LA←PARA[2])×R←*-MU←PARA[1],0⍴I←0
[2]   LBL:→((⍴R←R,(+/(B←B,B[I]×(I+1)×LA÷I×I)×⌽R)×L÷1+I←I+1)≤⌈/X)/
        LBL
[3]     R←R[1+X]
      ∇
```

```
      ∇ R←TOL Th PARA;LA;B;L;MU;I;SUM
[1]     R←R,LA×(L←MU×(*-LA)÷B←,LA←PARA[2])×R←*-MU←PARA[1],0⍴I←0
[2]   LBL:→((SUM←+/R←R,(+/(B←B,B[I]×(I+1)×LA÷I×I)×⌽R)×L÷1+I←I+1)≤
        1-10*-TOL)/LBL
[3]     R←R,1-SUM
      ∇
```

FIGURE 32 APL functions for calculating Thomas(PARA ← μ,λ) probabilities.

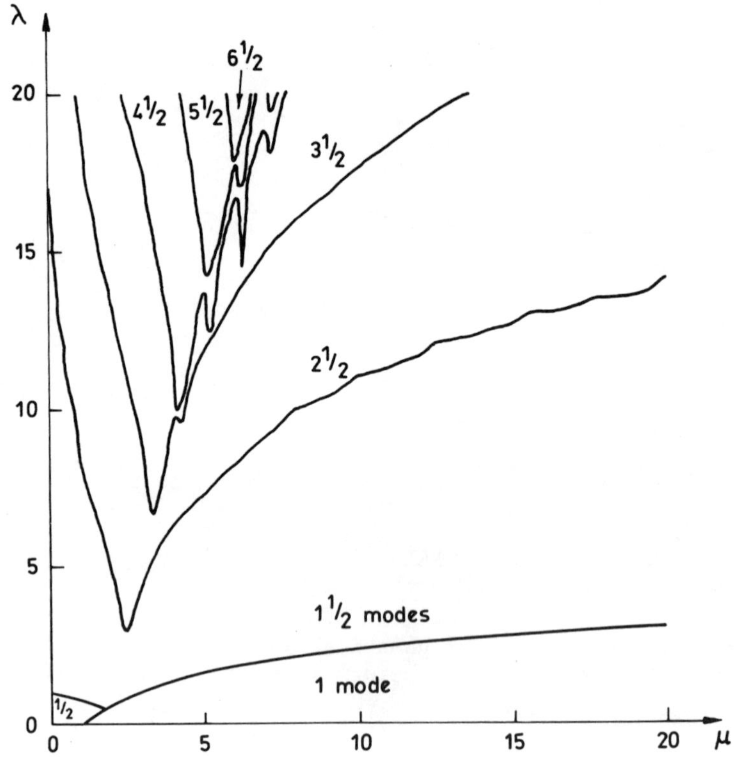

FIGURE 33 Modality contours for Thomas distribution. (Lind, 1974)

In very general terms, the shape of the p.f. has a good deal of resemblance to that of the Neyman distribution, including the occurrence of multiple modes, although the complexity of determining regions of modality for the Thomas case appears even greater than for the Neyman. Numerical investigations by P. Lind (unpublished M.Sc. thesis (1974)) led to a modal diagram like that for the Neyman, until the grid spacing for μ, λ over which the calculations were carried out was reduced from 0.5 to 0.1. The very smooth boundaries of the regions then developed unanticipated "waves" (see Figure 33), which are also suggested for the Neyman case, and the awkwardness implied for estimation, when, for $\lambda = 18.5$, a change of μ from 5 to 8 means the modality changes from $5\frac{1}{2}$ to $6\frac{1}{2}$ to $5\frac{1}{2}$ to $4\frac{1}{2}$ to $3\frac{1}{2}$ to $4\frac{1}{2}$ to $3\frac{1}{2}$, is very apparent.

Moments and Cumulants

The factorial cumulant g.f. is

$$-\mu + \mu(1 + t\, e^{-\lambda+\lambda(1+t)} = \mu\{e^{\lambda t} - 1 + te^{\lambda t}\}$$

$$= \mu \left\{ \sum_{r=1}^{\infty} \frac{\lambda^r t^r}{r!} + \sum_{r=1}^{\infty} \frac{\lambda^{r-1} t^r}{(r-1)!} \right\}$$

$$= \mu\, \Sigma(\lambda + r)\lambda^{r-1} \frac{t^r}{r!} ;$$

i.e.

$$\kappa_{(r)} = \mu(\lambda + r)\lambda^{r-1} .$$

The power cumulants are thus (Thomas (1951))

$$\kappa_1 = \mu(1 + \lambda) ,$$

$$\kappa_2 = \mu(1 + 3\lambda + \lambda^2) ,$$

$$\kappa_3 = \mu(1 + 7\lambda + 6\lambda^2 + \lambda^3) ,$$

and

$$\kappa_4 = \mu(1 + 15\lambda + 25\lambda^2 + 10\lambda^3 + \lambda^4) .$$

It follows that the Indices of Cluster Size and Frequency are given by

$$\iota_{CS} = \lambda \frac{\lambda+2}{\lambda+1}$$

and

$$\iota_{CF} = \frac{\mu}{1-1(\lambda+1)^2} .$$

Johnson and Kotz (1969) point out that $\kappa_2\kappa_4/\kappa_3^2$ is a rational function of λ only, increasing slowly from 1 at $\lambda = 0$ to a maximum of approximately 1.189... near $\lambda = 0.2388...$, and decreasing again to 1 as $\lambda \to \infty$. Similarly, $\kappa_{(2)}\kappa_{(4)}/\kappa_{(3)}^2 = 1 - 1/(\lambda+3)^2$, which increases towards 1 as λ increases, from a smallest value of 8/9 when $\lambda = 0$; cf. the constant value of 1 for the Neyman distribution.

Limits and Approximations

Since there have been few extensive applications of the Thomas distribution there has been little pressure to develop simpler expressions for the probability function.

It is immediately clear that if λ is "sufficiently" small, an approximation to the Thomas(μ,λ) distribution is given by the Poisson(μ) distribution. For the p.g.f. is

$$\begin{aligned}\exp(-\mu+\mu^{-\lambda+\lambda z}) &= \exp\{-\mu+\mu z(1+0(\lambda))\} \\ &= e^{-\mu+\mu z} + 0(\lambda) .\end{aligned}$$

For example, the approximation is quite good for $\lambda = 0.005$ (with $\mu = 1$) , as is illustrated in Table 17. Moreover, thinking of the structure of the Thomas(μ,λ) distribution as a Poisson(μ)-stopped "1 + Poisson(λ)" distribution, it is natural to examine as an approximating distribution the Neyman($\mu,1+\lambda$) , for which the mean is $\mu(1+\lambda)$ as for the Thomas, but the variances are:

Thomas	$\mu(1+3\lambda+\lambda^2)$
Neyman	$\mu(1+3\lambda+\lambda^2) + \mu$.

For the example illustrated in Table 17, the approximation is not at all good. It is also natural to consider as an approximation (not only for small λ) to a Thomas(μ,λ) distribution the Neyman

TABLE 17

	Thomas(1,0.005)	Poisson(1)	Neyman(1,1.005)	Neyman(100.75,0.00998)
0	36788	36788	53049	36788
1	36604	36788	19515	36605
2	18394	18394	13396	18394
3	6223	6131	7333	6223
4	1594	1533	3644	1594
5	330	307	1717	330
6	57	51	774	57
7	9	7	336	9
8	1	1	141	1
9	0	0	57	0

$10^5 \times$ Probabilities in Thomas & Neyman Distributions

distribution with the same mean and variance - i.e., with parameters

$$\left(\frac{\mu(1+\lambda)^2}{\lambda(2+\lambda)}, \frac{\lambda(2+\lambda)}{1+\lambda}\right) .$$

(Equivalently, if the Neyman parameters are (μ,ν), those for the Thomas are

$$\left(\tfrac{1}{2}\mu\nu\{\surd(\nu^2+4) - \nu\}, \tfrac{1}{2}\{\surd(\nu^2+4) + \nu-2\}\right).$$

The probabilities for these are also shown in Table 1, and it is clear in this case that matching the mean and variance gives an excellent approximation.

Similarly, for small μ the p.g.f. is

$$1 - \mu + \mu z e^{-\lambda+\lambda z} + O(\mu^2) ,$$

corresponding to a Poisson-with-zeroes distribution in which the zero class has probability $1-\mu$ and the succeeding classes at $x = 1,2,...$ have probabilities

$$\mu e^{-\lambda} \lambda^{x-1}/(x-1)! .$$

Without exhibiting specific examples, it may be said that the approximation does not appear to be very satisfactory, though the general shape is reproduced reasonably well. A result of Berman (1975) is that for t_x and n_x the corresponding Thomas(μ,λ) and Neyman(μ,λ) probabilities,

$$t_x \to n_{x-1} \quad \text{for} \quad x \geq 1 \quad \text{as} \quad \mu \to 0.$$

The usual limiting result for Poisson stopped distributions applies, giving normality of the standardized variate as the parameter μ increases. Another result of Berman (1975) is that as $\mu \to \infty$

$$t_x \to e^{-\mu} \frac{(\mu e^{-\lambda})^x}{x!} .$$

More generally, and without discussing particular examples in detail, it is remarkable how well the multimodalities of the Thomas and Neyman distributions correspond for λ not small, both with respect to location and magnitude, even when that matching is of the simplest kind:

$$\text{Neyman}(\mu,\nu) \sim \text{Thomas}(\mu,\nu-1) \ .$$

The figures in Table 18 are for the Neyman distributions pictured in the Chapter on that distribution: they give the modal probabilities multiplied by 10^6 and their locations.

Estimation

The algebraic complexity in dealing with the Thomas distribution is quite substantial: few results come out "nicely". As is perhaps inevitable this shows particularly with respect to estimation procedures.

Moment Estimation

Equating the first two sample and distribution factorial cumulants leads to

$$\left.\begin{aligned} K_1 &= \overset{o}{\mu}(1+\overset{o}{\lambda}) \\ K_{(2)} &= \overset{o}{\mu}(2+\overset{o}{\lambda})\overset{o}{\lambda} \end{aligned}\right\}$$

for the moment estimators $\overset{o}{\mu}$ and $\overset{o}{\lambda}$. Introducing $I = K_{(2)}/K_1 = K_{(2)}/\overline{X}$ for the sample moment analogue of the Index of Cluster Size enables the moment estimators to be expressed as

$$\left.\begin{aligned} \overset{o}{\mu} &= \frac{1}{2}\,\overline{X}(\surd(I^2+4) - I) \\ \overset{o}{\lambda} &= \frac{1}{2}\,(\surd(I^2+4) + I - 2) \end{aligned}\right\} .$$

(Recall that $K_{(2)} = M_2' - \overline{X}^2 - \overline{X}$.)

The covariance matrix for $\underset{\sim}{\overset{o}{\theta}} = (\overset{o}{\mu},\overset{o}{\lambda})^t$ is given by the usual asymptotic formula

$$\underset{\sim}{\mathcal{V}}(\underset{\sim}{\overset{o}{\theta}}) = \frac{\partial\underset{\sim}{\theta}}{\partial\underset{\sim}{\mu}'}\ \underset{\sim}{\mathcal{V}}(\underset{\sim}{M}')\left(\frac{\partial\underset{\sim}{\theta}}{\partial\underset{\sim}{\mu}}\right)^t + O(1/n^2),\ \underset{\sim}{\theta} = \binom{\mu}{\lambda} \ , \qquad .$$

TABLE 18

μ,ν	Neyman(μ,ν) $p_x \times 10^6$	x	Thomas(μ,ν-1) $p_x \times 10^6$
1,1	531×10^3	0	368×10^3
		1	368×10^3
1,2	421×10^3	0	368×10^3
	129×10^3	2	160×10^3
1,4	375×10^3	0	368×10^3
	8273	4	90696
2,1	282×10^3	0	
		1	271×10^3
		2	271×10^3
2,12.5	135×10^3	0	135×10^3
	31112	12	32275
	23050	25	23696
4,1	148×10^3	3	195×10^3
		4	195×10^3
5,5	6969	0	6738
	33127	21	
		25	33661
5,17	6738	0	6738
	3288	17	3373
		34	6247
	6147	35	
	9048	53	9106
	10465	70	10530
	10114	82	
		83	10145
12.5,2	46588	23	50820
7,25	912	0	912
	508	25	519
	1284	50	1306
	2550	76	2577
		102	4109
	4086	103	
		128	5418
	5400	129	
		152	6049
	6026	153	
	5898	173	5914
		174	5914

*When $\nu-1=0$, it was taken to be 10^{-15}.

Modal Probabilities $\times 10^6$ in Neyman & Thomas Distributions

where $\underset{\sim}{M}' = \begin{pmatrix} M_1' \\ M_2' \end{pmatrix}$,

$$\underset{\sim}{\mathcal{V}}(\underset{\sim}{M}') = \frac{1}{n}\begin{pmatrix} \mu_2' - \mu_1'^2 \ , & \mu_3' - \mu_1'\mu_2' \\ \mu_3' - \mu_1'\mu_2' \ , & \mu_4' - \mu_2'^2 \end{pmatrix},$$

these moments being obtained from the cumulants given earlier, and

$$\frac{\partial \underset{\sim}{\theta}'}{\partial \underset{\sim}{\mu}} = \begin{pmatrix} \partial\mu_1'/\partial\mu & \partial\mu_1'/\partial\lambda \\ \partial\mu_2'/\partial\mu & \partial\mu_2'/\partial\lambda \end{pmatrix}.$$

This Jacobian matrix is found more readily from its inverse,

$$\frac{\partial \underset{\sim}{\mu}'}{\partial \underset{\sim}{\theta}} = \begin{pmatrix} \partial\mu_1'/\partial\mu & \partial\mu_1'/\partial\lambda \\ \partial\mu_2'/\partial\mu & \partial\mu_2'/\partial\lambda \end{pmatrix},$$

$$= \begin{pmatrix} 1+\lambda & \mu \\ \lambda + (1+2\mu)(1+\lambda)^2 \ , & \mu(1+2(1+\mu)(1+\lambda)) \end{pmatrix},$$

the inversion being carried out numerically in a computer if available.

A numerical example of a moment fitting is given a little further on in Table 19, but it might be remarked that the moment equations are rather unusual for a contagious distribution (including the negative binomial) in that they have no singularities other than when the sample mean is zero - a decidedly uninteresting data set! However, if the second moment about the mean is less than the mean, a negative estimate of λ would be given by a formal use of the equations.

It is also of some importance to investigate the biases of these (and other) estimators. For samples of size $n = 100$, Lind (1974) finds that retaining the term in $1/n$ in the asymptotic expansion leads to a bias of less than 10% (usually much less) for μ, while that for λ becomes large negative (< -10%) for small μ and λ though elsewhere small. These are less than the corresponding results for the Neyman and negative binomial distributions.

A comparison of first order efficiency (with maximum likelihood) is given at the end of the section on maximum likelihood fitting.

Maximum Likelihood Estimation

The likelihood equations written down directly are rather unpleasant. However, by differentiation of the p.g.f. with respect to μ and λ, and by introducing the Poisson probabilities $\pi_x = e^{-\lambda} \lambda^x/x!$, and $c_x = \sum_{r=0}^{x-1} p_r \tilde{\omega}_{x-r-1}$, the following expressions are found:

$$\frac{\partial p_0}{\partial\mu} = -p_0 , \quad \frac{\partial p_x}{\partial\mu} = -p_x + c_x \quad \text{for} \quad x = 1,2,3,\ldots,$$

$$\frac{\partial p_0}{\partial\lambda} = 0 , \quad \frac{\partial p_1}{\partial\lambda} = -p_1 ,$$

$$\frac{\partial p_x}{\partial\lambda} = -\mu\, p_{x-1}\pi_0 + \frac{\mu}{\lambda} \sum_{r=0}^{x-2} (x-r-1-\lambda) p_r \pi_{x-r-1} \quad \text{for} \quad x = 2,3,\ldots .$$

Returning to the p.g.f., differentiation with respect to z yields

$$x\, p_x = \mu \sum_{r=0}^{x-1} (x-r) p_r\, \pi_{x-r-1} , \quad x = 1,2,\ldots,$$

as already noted, and a comparison of this with the above expressions (Russell (1978)) leads to the following rearrangement:

$$x\, p_x = \mu \sum_{r=0}^{x-2} (x-r-1-\lambda) p_r\, \pi_{x-r-1} + \mu(1+\lambda)\, c_x - \mu\lambda p_{x-1}\, \pi_0 ,$$

adding and subtracting the appropriate terms. Hence, dividing by p_x .

$$x = \lambda \frac{\partial\, \ell n\, p_x}{\partial\lambda} + \mu(1+\lambda) + \mu(1+\lambda) \frac{\partial\, \ell n\, p_x}{\partial\mu} , \quad x = 0,1,2,\ldots,$$

whence multiplying by F_x and summing

(recalling

$$\Sigma F_x \frac{\partial \ln p_x}{\partial \mu} = \Sigma F_x \frac{\partial \ln p_x}{\partial \lambda} = 0 \quad \text{at} \quad (\hat{\mu},\hat{\lambda})\Bigg)$$

leads to the equation of the mean:

$$\hat{\mu}(1+\hat{\lambda}) = \bar{X} .$$

This can be used to reduce the likelihood equations to a single (implicit) equation in one variable, leading to faster and surer convergence when iterating.

Choosing the former of the likelihood equations, and inserting the explicit expressions for the derivatives,

$$\sum_{x=1}^{\infty} \frac{F_x}{\hat{p}_x} \hat{c}_x - n = 0.$$

Writing

$$G(\mu,\lambda) = \Sigma \frac{F_x}{p_x} c_x - n$$

and

$$H(\lambda) = G(\bar{X}/(1+\lambda), \lambda) ,$$

$$H'(\lambda) = \frac{\partial G}{\partial \mu} \cdot \frac{d\mu}{d\lambda} + \frac{\partial G}{\partial \lambda} ,$$

where

$$\frac{\partial G}{\partial \mu} = - \Sigma \frac{F_x}{p_x^2} \cdot \frac{\partial p_x}{\partial \mu} c_x + \Sigma \frac{F_x}{p_x} \Sigma \frac{\partial p_r}{\partial \mu} \pi_{x-r-1} ,$$

$$\frac{\partial G}{\partial \lambda} = - \Sigma \frac{F_x}{p_x^2} \cdot \frac{\partial p_x}{\partial \lambda} c_x$$

$$+ \Sigma \frac{F_x}{p_x} \Sigma \left\{ \frac{\partial p_r}{\partial \lambda} \pi_{x-r-1} + p_r \frac{\partial \pi_{x-r-1}}{\partial \lambda} \right\} ,$$

$$\frac{\partial \pi_x}{\partial \lambda} = \frac{x-\lambda}{\lambda} \pi_x , \quad \frac{d\mu}{d\lambda} = - \frac{\bar{X}}{(1+\lambda)^2} .$$

Hence, for starter values (μ_1, λ_1), commonly chosen to be the moment estimates,

$$\lambda_2 = \lambda_1 - \frac{H(\lambda_1)}{H'(\lambda_1)},$$

$$\mu_2 = \frac{\bar{X}}{1+\lambda_2},$$

with repetition until the incremental changes are small enough.

When this is the case, the covariance matrix can be obtained numerically from

$$\underset{\sim}{I}_X = n \times \begin{bmatrix} \left(\frac{\partial \ln p_X}{\partial \mu}\right)^2 & , \frac{\partial \ln p_X}{\partial \mu} \cdot \frac{\partial \ln p_X}{\partial \lambda} \\ \frac{\partial \ln p_X}{\partial \mu} \cdot \frac{\partial \ln p_X}{\partial \lambda} & , \left(\frac{\partial \ln p_X}{\partial \lambda}\right)^2 \end{bmatrix}$$

by recalling that

$$\underset{\sim}{\mathcal{V}}\begin{pmatrix} \hat{\mu} \\ \hat{\lambda} \end{pmatrix} = \left\{\mathcal{E}\left(\underset{\sim}{I}_X\right)\right\}^{-1} \quad \text{asymptotically;}$$

by evaluating p_x and its derivatives at $(\hat{\mu}, \hat{\lambda})$ for sufficiently large x then $\mathcal{E}(\underset{\sim}{I}_X)$ is approximated to by $\Sigma\, p_x\, \underset{\sim}{I}_x$.
(The exercises sketch a two variable iteration sequence.)

Data from Thomson (1952), with counts on golden rod (Solidago rigida), are used to illustrate a fitting - see Table 19.

Table 20 gives the result of first order asymptotic calculations for the moment versus maximum likelihood fitting procedures (Lind (1974)).

Miscellaneous Methods

Minimum chi-squared techniques could be applied without difficulty, but perhaps because of the limited applications made of the distribution no details have been published. (Lind (1974))

TABLE 19

Plants per quadrat x	Observed frequencies	(Moments)	Fitted Frequencies (Max. Like.)	(P_0 and P_1)
0	7229	7230.5	7228.7	7229.0
1	356	351.9	355.2	356.0
2	49	52.2	50.9	50.0
3	5	5.0	4.7	4.6
4	1	0.4	0.4	0.4
5+	0	0.0	0.0	0.0
	7640			
Estimate of μ,		0.0551	0.0553	0.0553
standard error		0.0027	0.0027	0.0027
Estimate of λ,		0.124	0.119	0.116
standard error		0.021	0.019	0.019
Estimated correlation		-0.121	-0.078	-0.074

Distribution of Solidago rigida

Thomas(μ,λ) fitting

(Thomson (1952))

TABLE 20

λ \ μ	0.05	0.1	0.2	0.5	1	2	5	10
10	43	30	21	14	13	18	60	94
5	50	39	29	21	21	32	77	94
2	53	45	36	29	29	41	80	93
1	59	51	43	36	38	52	83	93
0.5	68	62	55	49	53	67	87	94
0.2	81	76	72	71	77	86	94	97
0.1	88	85	83	85	90	94	97	99
0.05	92	91	91	94	96	98	99	100

Percent First Order Asymptotic Efficiency of Method of Moments for Thomas(μ,λ) distribution

Lind (1974)

has worked out the details for a non-linear fit using ratios of factorial cumulants.)

As a hand calculation technique, which also can be used for quick work with observations on merely the zero and first class proportions P_0 and P_1 (together with the total number n of quadrats observed), equating

$$P_0 \quad \text{and} \quad e^{-\mu}$$

$$\frac{P_1}{P_0} \quad \text{and} \quad \mu e^{-\lambda} ,$$

leads (Thomas (1949)) to

$$\tilde{\mu} = - \ln P_0 ,$$

$$\tilde{\lambda} = - \ln \frac{P_1}{P_0 \ln P_0} .$$

The asymptotic variances and covariance are given by

$$\mathscr{V}(\tilde{\mu}) = \frac{1}{n} \cdot \frac{1-e^{-\mu}}{e^{-\mu}} + 0(1/n^2) .$$

$$\mathscr{V}(\tilde{\lambda}) = \frac{1}{n} \cdot \frac{\mu-e^{-\mu-\lambda}+e^{-\lambda}(\mu-1)^2}{\mu^2 e^{-\mu-\lambda}} + 0(1/n^2) ,$$

$$\mathscr{C}(\tilde{\lambda},\tilde{\mu}) = \frac{1}{n} \cdot \frac{1-\mu-e^{-\mu}}{\mu e^{-\mu}} + 0(1/n^2) ,$$

using the usual Taylor expansions; alternatively, as Thomas points out, these can be obtained from maximum likelihood theory based on the availability only of P_0 and P_1. (Note that the expressions for $\mathscr{V}(\tilde{\mu})$ and $\mathscr{C}(\tilde{\lambda},\tilde{\mu})$ are not functions of λ.)

Again, Lind (1974) has investigated the first order in $1/n$ (for $n = 100$) biases for these estimators. That of $\tilde{\mu}$ is a function of μ only, being positive and increasing rapidly for μ greater than about 5. The bias for λ increases numerically as μ and λ increase, though it is large positive for small (say (0,1)) λ and μ; it becomes large negative for small λ and larger (say 2) μ. A short table, Table 21, also due to Lind, is given of the first order efficiency of this method

TABLE 21

λ \ μ	0.05	0.1	0.2	0.5	1	2
5	4	4	4	4	3	2
2	32	33	34	33	27	14
1	60	60	60	56	43	19
0.5	78	78	77	68	49	18
0.2	91	90	86	71	41	13
0.1	95	93	88	78	38	10
0.05	96	94	86	62	33	8

Percent First Order Asymptotic Efficiency of Zero-One Method for the Thomas(μ,λ) Distribution.

Lind (1974)

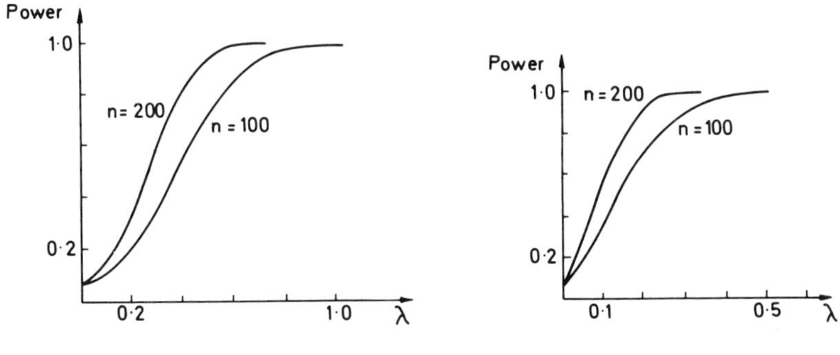

FIGURE 34 Power of two 5% level tests of Poisson(μ) against Thomas(μ,λ), $\mu(1+\lambda) = 2$. (*Left*) power of the test based on $\tilde{\lambda}$; (*right*) power of the Index of Dispersion test. (Thomas, 1951)

(relative to maximum likelihood): naturally it is very low if the probabilities of the first two classes are small.

If the overall mean number of individuals $\mu(1+\lambda)$ is of primary interest, the variance of its estimator is needed. When moment or maximum likelihood estimation is used the estimator is of course $\bar{X}$, with

$$\mathcal{V}(\bar{X}) = \mu(1+3\lambda+\lambda^2)/n;$$

if the 'short' method based on P_0 and P_1 is used the natural estimator is $\tilde{\mu}(1+\tilde{\lambda})$, with

$$\mathcal{V}\{\tilde{\mu}(1+\tilde{\lambda})\} = \frac{1}{n}\left\{\frac{(\mu-\lambda-2)^2}{e^{-\mu}} + \frac{\mu^2}{\mu e^{-\mu-\lambda}} - (\lambda+2)^2\right\} + 0(1/n^2).$$

Thomas (1949) gives a short table of the ratio of the standard errors of these two methods for a range of values of p_0 and p_1, and this table equally can be used to find the numbers of quadrats to be observed in each of the two methods for the same standard error.

A numerical example of the fitting methods is reproduced from data in Thomson (1952) and shown in Table 19. The agreement is obviously excellent, and the 'short' method based on P_0 and P_1 shows up very well, as would be expected from a situation in which more than 99% of the distribution is in these two classes, and these two necessarily reproduced exactly. Thomson also remarks that the "minor clusters of the mathematical models are only related to very local clumping conditions", but that "the major clumps are distinctive and easily differentiated from these local aggregates". There are indeed only very small differences in this example between any of the methods. (Because of the short tail of the distribution, it is hardly realistic to expect much discrimination even between different families of distributions, and in fact the agreement with fitted Neyman frequencies is even "better"; although the tail frequencies fall off too slowly for fitted Poisson frequencies to be regarded as adequate.)

Comparisons with Other Distributions

Thomas (1951) also gives some results for the power of several tests of the hypothesis that a Poisson(μ) distribution applies ("randomness") when the alternative is the Thomas(μ,λ) distributions.

The first test is based on the Index of Dispersion $Z = \Sigma(X_i-\bar{X})^2/\bar{X}$, where X_i is the number of individuals in the ith quadrat, $i = 1,2,\ldots,n$; for the Poisson hypothesis it is known that Z has approximately a chi-squared distribution with $n-1$ degrees of freedom, and for the alternative Thomas hypothesis the distribution of Z was approximated to by a log-normal distribution. The power of 5% level tests turned out not to be heavily dependent on the overall mean (1,2, and 3 were taken) for the sample sizes $n = 100$ and 200. A graphical representation of part of the results is shown in Figure 34.

The second test was based on treating $\tilde{\lambda}$ (defined above) as approximately $N(\tilde{\lambda}, \mathscr{V}(\tilde{\lambda}))$, and so evaluating from normal tables $\Pr(\tilde{\lambda} \geq \ell_\alpha | \lambda)$ where the significance level α defines ℓ_α by the equation

$$\alpha = \Pr\{U \geq \ell_\alpha / \sqrt{\mathscr{V}(\tilde{\lambda} | \lambda = 0)}\} ,$$

U being a standard normal variate. A graphical representation for a 5% level test is shown in Figure 34; however, in this case the power is heavily dependent on the value of $\mu(1+\lambda)$ (as would be expected since p_0 and p_1 decrease as it increases), and Thomas points out that if estimates are used, $\tilde{\mu}(1+\tilde{\lambda})$ and $\tilde{\lambda}$ are of course highly correlated.

A third test, based on the zero class frequency, is also discussed by Thomas, but no details are reproduced here.

5.4 EXERCISES

1. Interpreting the Neyman(μ,ν) variate according to the structure

$$S_N = X_1 + \ldots + X_N$$

where X_i : Poisson(ν) and N : Poisson(μ) , use a conditional expectation argument (for fixed $N=n$) to obtain

$$\mathscr{E}(S_N) \quad \text{as} \quad \mathscr{E}_N\{\mathscr{E}(S_N|N)\}$$

and $\mathscr{E}(S_N^2)$ similarly, whence $\mathscr{V}(S_N)$.

Use the same argument to obtain the mean and variance of a Thomas(μ,λ) variate, recalling that the X_i are Poisson(λ) with the addition of unity.

Recognize that this is the argument which was also used in the general case of a stopped variate.

2. Show that the cumulants of the Thomas(μ,λ) distribution have the recurrence relation

$$\kappa_{r+1} = \lambda(\mu + r\kappa_1 + \binom{r}{2}\kappa_2 + \ldots + r\kappa_{r-1} + \kappa_r) + \kappa_r ,$$

$$r = 1,2,\ldots \quad \text{with} \quad \kappa_1 = \mu(1+\lambda).$$

3. Show that the standard errors of the moment estimators can be obtained from formulae exemplified by the following.

E.g., where $\mu = \frac{1}{2}\,\mu_1'\{\sqrt{(\iota^2+4)} - \iota\}$, $\iota = \frac{\mu_2'}{\mu_1'} - \mu_1' - 1$,

$$\mathscr{V}(\overset{\circ}{\mu}) = \left(\frac{\partial\mu}{\partial\mu_1'} + \frac{\partial\mu}{\partial\iota}\,\frac{\partial\iota}{\partial\mu_1'}\right)^2 \mathscr{V}(\overline{X}) + 2\left(\frac{\partial\mu}{\partial\mu_1'} + \frac{\partial\mu}{\partial\iota}\,\frac{\partial\iota}{\partial\mu_1'}\right)\frac{\partial\mu}{\partial\iota}\,\frac{\partial\iota}{\partial\mu_2'} \times$$

$$\mathscr{C}(\overline{X},M_2') + \left(\frac{\partial\mu}{\partial\iota}\,\frac{\partial\iota}{\partial\mu_2'}\right)^2 \mathscr{V}(M_2') + O(1/n^2) \;;$$

or alternatively, where

$$\mu = \frac{1}{2}\,\mu_1' \left[\sqrt{\left\{\left(\frac{\mu_2'}{\mu_1'} - \mu_1' - 1\right)^2 + 4\right\}} - \left(\frac{\mu_2'}{\mu_1'} - \mu_1' - 1\right)\right] ,$$

$$\mathscr{V}(\overset{\circ}{\mu}) = \left(\frac{\partial\mu}{\partial\mu'}\right)^2 \mathscr{V}(\overline{X}) + 2\,\frac{\partial\mu}{\partial\mu_1'}\,\frac{\partial\mu}{\partial\mu_2'}\,\mathscr{C}(\overline{X},M_2') + \left(\frac{\partial\mu}{\partial\mu_2'}\right)^2 \times$$

$$\mathscr{V}(M_2') + O(1/n^2).$$

4. Using the notation of the text, show that for the maximum likelihood estimators $\hat{\mu},\hat{\lambda}$, with $\underset{\sim}{\theta} = (\mu,\lambda)^t$, the iteration formula can be written as

$$\underset{\sim}{\theta}_{r+1} = \underset{\sim}{\theta}_r + \underset{\sim}{\mathscr{V}}(\hat{\underset{\sim}{\theta}})\bigg|_{\underset{\sim}{\theta}_r} \times \frac{\partial\,\ln L}{\partial\underset{\sim}{\theta}}\bigg|_{\underset{\sim}{\theta}_r}$$

where

$$\underset{\sim}{\mathcal{V}}(\underset{\sim}{\theta})^{-1} = n \times \begin{bmatrix} \mathcal{E}\left(\frac{\partial \ell n\ p_X}{\partial \mu}\right)^2 & , \mathcal{E}\left(\frac{\partial \ell n\ p_X}{\partial \mu} \cdot \frac{\partial \ell n\ p_X}{\partial \lambda}\right) \\ \mathcal{E}\left(\frac{\partial \ell n\ p_X}{\partial \mu} \cdot \frac{\partial \ell n\ p_X}{\partial \lambda}\right), & \mathcal{E}\left(\frac{\partial \ell n\ p_X}{\partial \lambda}\right)^2 \end{bmatrix}$$

asymptotically. Note also that

$$\mathcal{E}\left(\frac{\partial \ell n\ p_X}{\partial \mu}\right)^2 = -\mathcal{E}\left(\frac{\partial^2 \ell n\ p_X}{\partial \mu^2}\right), \quad \text{etc.}$$

and that these second derivatives are

$$\frac{\partial^2 p_0}{\partial \mu^2} = p_0, \quad \frac{\partial^2 p_x}{\partial \mu^2} = -\frac{\partial p_x}{\partial \mu} + \sum_{r=0}^{x-1} \frac{\partial p_r}{\partial \mu} \pi_{x-r}, \quad x = 1, 2, \ldots,$$

$$\frac{\partial^2 p_0}{\partial \lambda \partial \mu} = 0, \quad \frac{\partial^2 p_x}{\partial \lambda \partial \mu} = -\frac{\partial p_x}{\partial \lambda} + \sum_{r=0}^{x-1} \left(\frac{\partial p_r}{\partial \lambda} \pi_{x-r} + p_r \frac{\partial \pi_{x-r}}{\partial \lambda}\right),$$

$$x = 1, 2, \ldots \quad \text{where} \quad \frac{\partial \pi_x}{\partial \lambda} = \frac{x-\lambda}{\lambda} \pi_x,$$

$$\frac{\partial^2 p_0}{\partial \lambda^2} = 0, \quad \frac{\partial^2 p_1}{\partial \lambda^2} = \mu p_0 \pi_0$$

$$\frac{\partial^2 p_x}{\partial \lambda^2} = -\mu \frac{\partial p_{x-1}}{\partial \lambda} \pi_0 + \mu p_{x-1} \pi_0 + \frac{\mu}{\lambda^2} \sum_{r=0}^{x-2} (\lambda - x + r + 1) p_r \pi_{x-r-1}$$

$$- \frac{\mu}{\lambda} \sum_{r=0}^{x-2} \left\{ p_r \pi_{x-r-1} + (\lambda - x + r - 1) \frac{\partial p_r}{\partial \lambda} \pi_{x-r-1} \right.$$

$$\left. + (\lambda - x + r1) p_r \frac{\partial \pi_{x-r-1}}{\partial \lambda} \right\}, \quad x = 2, 3, \ldots$$

5. An alternative "derivation" of the Thomas distribution.

Suppose there are R parents in a quadrat, with p.g.f. $g_R(z)$. For any fixed $R = r$, if the numbers of offspring for each parent are

$$X_1, X_2, \ldots, X_r ,$$

then the number of individuals in the quadrat is

$$(1+X_1) + (1+X_2) + \ldots + (1+X_r) = r + X_1 + \ldots + X_r ,$$

with the p.g.f. of each X_i being $g_X(z)$.

Thus the p.g.f. of the number of individuals in a quadrat, taking account of the distribution of parent numbers, is

$$\mathop{\mathscr{E}}_{R}\left[\mathop{\mathscr{E}}_{\text{all } X_i}(z^{R+X_1+\ldots+X_R}|R)\right]$$

$$= \mathop{\mathscr{E}}_{R}\left[z^R \mathscr{E}(z^{X_1+\ldots+X_R})|R\right]$$

$$= \mathop{\mathscr{E}}_{R}\left[z^R\{g_X(z)\}^R\right]$$

$$= g_R\ [z\ g_X(z)] .$$

Taking R to be Poisson(μ) and X to be Poisson(λ) gives the Thomas(μ,λ) distribution.

5.5 THE LOG-ZERO-POISSON DISTRIBUTION

Most of the work on this distribution has been carried out by S. K. Katti and various collaborators, and what follows is largely a paraphrase in the notation adopted in other sections of this book.

It was originally studied because certain functions, called "skewness" and "kurtosis" in earlier sections, showed substantially greater ranges of permitted variation for this distribution than for other simply derived and manipulated distributions; this apparent wide range of possible applicability was supported when a variety of empirical distributions was actually fitted. However, although a biological interpretation of one parametric specification is given below; the form in which the mathematical analysis of estimation procedures is conveniently put does not lend itself immediately to meaningful application.

Derivation of p.g.f.

Among various descriptions of the distribution, the following sequence is simple. The p.g.f. of a $\log(\alpha)$ distribution is

$$\frac{\ln(1-\alpha z)}{\ln(1-\alpha)} ,$$

and it of course has no zero class. If the log-zero distribution is a $\log(\alpha)$ distribution with a zero class of probability $1-\pi=\chi$ added, its p.g.f. is

$$1-\pi+\pi\,\frac{\ln(1-\alpha z)}{\ln(1-\alpha)}$$

$$= 1+\frac{\pi}{\ln(1-\alpha)}\cdot \ln\left(\frac{1}{1-\alpha}-\frac{\alpha}{1-\alpha}\,z\right) .$$

(The distribution could also be called a binomial$(1,\pi)$-stopped $\log(\alpha)$ distribution, or a mixture of 0 with weight $1-\pi$ and of $\log(\alpha)$ with weight π.) Writing

$$\frac{\alpha}{1-\alpha} = \rho > 0 \quad \text{with} \quad \gamma = 1+\rho = \frac{1}{1-\alpha} ,$$

and

$$-\delta = \frac{\pi}{\ln(1-\alpha)} = -\frac{\pi}{\ln \gamma} \quad (\delta > 0) ,$$

the p.g.f. becomes

$$1 - \delta \ln(\gamma-\rho z) , \qquad \text{(A)}$$

which expands to

$$1 - \delta \ln \gamma + \delta \sum_{x=1}^{\infty} \left(\frac{\rho}{\gamma}\right)^{x} \frac{1}{x}\, z^{x}$$

so that the probability of the zero class is

$$1 - \delta \ln \gamma \geq 0 \quad (\text{whence } 0 \leq \delta \leq 1/\ln \gamma)$$

and of the xth class

$$\frac{\delta}{x}\left(\frac{\rho}{\gamma}\right)^{x} , \quad x = 1,2,\ldots .$$

The distribution with p.g.f. (A) will be called a log-zero(δ,ρ) distribution using δ rather than π to label it.

This distribution may now be used to stop a Poisson distribution, so the

log-zero(δ,ρ)-stopped Poisson(λ) distribution

will have p.g.f.

$$1-\delta\,\ell n(\gamma-\rho e^{-\lambda+\lambda z})\ ,\quad 0\le\delta\le 1/\ell n\,\gamma,\ \rho>0,\ \lambda>0,$$

and the name will usually be abbreviated to <u>log-zero-Poisson</u>(δ,ρ,λ). As will be seen, for some purposes it is convenient to write the probability of the zero class

$$1-\delta\,\ell n(\gamma-\rho e^{-\lambda})=\theta\ ,$$

and introduce θ as a reparameterization: the p.g.f. will then be written as

$$1-\frac{1-\theta}{\ell n(\gamma-\rho e^{-\lambda})}\cdot\ell n(\gamma-\rho e^{-\lambda+\lambda z})\ ,$$

$$0<\theta<1,\ \rho>0,\ \lambda>0.$$

The above derivation can be given an interpretation in terms of the biological insect-larvae illustration previously used. Suppose an insect lays over the counting area (the quadrat) egg clusters from which larvae hatch and are counted. From any one cluster suppose the number of larvae counted has a Poisson(λ) distribution, and that the number of clusters on the quadrat is log(α) distributed (so 'empty' clusters do not occur). Then the number of larvae per quadrat has p.g.f. $\ell n(1-\alpha e^{-\lambda+\lambda z})/\ell n(1-\alpha)=h(z)$, say. If also the probability that the insect selects the quadrat is π, and that it does not do so is $1-\pi$, the overall p.g.f. is $1-\pi+\pi\,h(z)$, that of the log-zero-Poisson.

<u>The Probability Function</u>

Explicit, though not necessarily very useful, formulae can be given for the probabilities.

Writing $g(z)=1-\delta\,\ell n(\gamma-\rho e^{-\lambda+\lambda z})=\Sigma\,p_x z^x$, and introducing $\xi=-\lambda(1-z)$,

$$g(z) = 1-\delta \ln\{1-\rho(e^{\zeta}-1)\}$$

$$= 1+\delta \sum_{s=1}^{\infty} (s-1)!\, \rho^s \frac{(e^{\zeta}-1)^s}{s!}$$

$$= 1+\delta \sum_{s} (s-1)!\, \rho^s \sum_{r=s}^{\infty} \mathfrak{S}_r^{(s)} \frac{\zeta^r}{r!} \quad \text{(Stirling Numbers of the Second Kind)}$$

$$= 1+\delta \sum_{r=1}^{\infty} \frac{\zeta^r}{r!} \sum_{s=1}^{r} \mathfrak{S}_r^{(s)} (s-1)!\, \rho^s$$

$$= 1+\delta \sum_{r} \frac{(-\lambda)^r}{r} \sum_{t=0}^{r} (-1)^t \binom{r}{t} z^t \sum_{s} \mathfrak{S}_r^{(s)} (s-1)!\, \rho^s$$

$$= 1+\delta \left\{ \sum_{r=1}^{\infty} \frac{(-\lambda)^r}{r!} \sum_{s=1}^{r} \mathfrak{S}_r^{(s)} (s-1)!\, \rho^s \right.$$

$$\left. + \sum_{t=1}^{\infty} (-1)^t z^t \sum_{r=t}^{\infty} \frac{(-\lambda)^r}{(r-t)!\,t!} \sum_{s=1}^{r} \mathfrak{S}_r^{(s)} (s-1)!\,\rho^s \right\} .$$

Hence, equating coefficients of t^x gives

$$p_x = (-1)^x \delta \sum_{r=x}^{\infty} \frac{(-\lambda)^r}{(r-x)!\,x!} \sum_{s=1}^{r} \mathfrak{S}_r^{(s)} (s-1)!\, \rho^s , \quad x = 1,2,\ldots$$

with p_0 given by the first two terms, independent of t, above but of course more simply by

$$p_0 = 1-\delta \ln(\gamma-\rho e^{-\lambda}) .$$

More usefully, leaving Poisson probabilities in explicitly,

$$g(z) = 1-\delta \ln \gamma-\delta \ln(1-\frac{\rho}{\gamma} e^{-\lambda+\lambda z})$$

$$= 1-\delta \ln \gamma+\delta \ln \sum_{r=1}^{\infty} \frac{1}{r} \left(\frac{\rho}{\gamma}\right)^r e^{-\lambda r+\lambda r z},$$

whence identifying the coefficient of z^x gives

$$p_x = \delta \sum_{r=1}^{\infty} \frac{1}{r} \left(\frac{\rho}{\gamma}\right)^r \pi_x(r\lambda) \ , \quad x = 1,2,\ldots$$

where

$$\tilde{\omega}_x(r\lambda) = e^{-\lambda r} \frac{(\lambda r)^x}{x!} \ .$$

The usual procedure, differentiating with respect to z and rearranging, leads to a recurrence relation. Thus

$$g'(z)(\gamma-\rho e^{-\lambda+\lambda z}) = \delta\rho\lambda r^{-\lambda+\lambda z} \ ;$$

introducing the Poisson probabilities $\tilde{\omega}_x(\lambda)$ from

$$e^{-\lambda+\lambda z} = \sum_{x=0}^{\infty} \tilde{\omega}_x(\lambda) z^x$$

and equating coefficients of z^x leads to

$$p_0 = 1-\delta \ \ell n(\gamma-\rho e^{-\lambda}) \quad (=\theta) \ ,$$

$$p_1 = \frac{\delta\rho\lambda e^{-\lambda}}{\gamma-\rho e^{-\lambda}} \ ,$$

and

$$p_{x+1} = \frac{\delta\rho\lambda\tilde{\omega}_x(\lambda) + \rho \sum_{r=0}^{x-1} (r+1)p_{r+1} \tilde{\omega}_{x-r}(\lambda)}{(x+1)(\gamma-\rho e^{-\lambda})} \ , \ x = 1,2,\ldots.$$

Bounds can be given for the probabilities (Katti and Rao (1970)):

$$\frac{\delta}{x}\left(\frac{\lambda}{\mu}\right)^x \left\{1 - \frac{\mu}{\sqrt{\{2\pi(x-1)\}}} \exp \frac{-1}{12(x-1)+1}\right\} < p_x$$

$$< \frac{\delta}{x}\left(\frac{\lambda}{\mu}\right)^x \left\{1 + \mu \frac{2}{\pi(x-1)} \exp \frac{-1}{12(x-1)+1}\right\}$$

where $x > 1$ and $\mu = \lambda + \ell n(\gamma/\rho)$.

Tables

Since the distribution has 3 parameters, any direct tabulation would be very bulky; however, because of the parameterization through θ , the probability of the zero class, it is necessary only to tabulate for the distribution truncated at zero and so with p.g.f.

$$\frac{g(z) - g(0)}{1 - g(0)} = 1 - \frac{\ln(\gamma - \rho e^{-\lambda+\lambda z})}{\ln(\gamma - \rho e^{-\lambda})}$$

$$= \Sigma\, n_x z^x \text{ , say,}$$

a function of the two parameters ρ and λ only: the zero truncated log-zero-Poisson distribution.

Hence Tables of this two parameter distribution can be constructed (they are referred to in Katti and Rao (1965)) and from them the probabilities p_x of the non-truncated distribution can be obtained:

$$p_0 = 1 - \ln(\gamma - \rho e^{-\lambda}) \quad (\text{or } \theta)$$

$$p_x = (1 - p_0)n_x \text{ ,} \quad x = 1,2,\ldots.$$

Limits and Approximations

By appealing to the usual continuity theorem (e.g. Feller I (1968)) for p.g.f.s, limiting behaviour can be examined directly through the p.g.f., suggestions for "interesting" behaviour coming from the form of the p.g.f. and of the lower moments.

i) $\lambda \to 0$

Since the mean is $\delta\lambda\rho=\mu > 0$, say, then $\delta\rho$ must increase indefinitely for non-trivial results.

a) $\delta \to \infty$ implies $\rho \to 0$: the result is rather dull, being a two point binomial with

$$p_0 = 1 - \mu,$$

$$p_1 = \mu.$$

b) $\rho \to \infty$ leads to the p.g.f.

$$1 - \delta \ \ell n(1 + \rho\lambda - \rho\lambda z);$$

with finite $\rho\lambda$ this is a log-zero$(\delta, \rho\lambda)$ p.g.f.

ii) λ moderate

a) $\delta \to 0$ with finite non-zero $\delta\rho$ leads to a degenerate variate concentrated at the origin.

b) $\rho \to 0$ with finite non-zero $\delta\rho$ gives

$$1 - \delta \ \ell n \left\{1 - \rho(e^{-\lambda+\lambda z} - 1)\right\}$$

$$= 1 - \delta\rho(e^{-\lambda+\lambda z} - 1) + 0(\rho^2)$$

$$= 1 - \delta\rho + \delta\rho e^{-\lambda+\lambda z} + 0(\rho^2) ,$$

the p.g.f. of a Poisson variate with additional probability at the origin (a "Poisson-zero" variate).

iii) $\lambda \to \infty$.

Since the variate X with the log-zero-Poisson(δ, ρ, λ) distribution has mean $\delta\rho\lambda$ and variance $\delta\rho\lambda(1 + \lambda(\gamma - \delta\rho))$, the probabilities become small and very diffuse for large λ if no restrictions are placed on $\delta\rho$.

a) Consider then the standardized variate

$$Y = \frac{X - \delta\rho\lambda}{\sqrt{\mathscr{V}(X)}} = \frac{\alpha X}{\lambda} - \alpha\delta\rho ,$$

where $\alpha = \dfrac{1}{\sqrt{\{\delta\rho(\gamma - \delta\rho) + \frac{1}{\lambda}\}}} \to \dfrac{1}{\sqrt{\{\delta\rho(\gamma - \delta\rho)\}}} = \beta$ as $\lambda \to \infty$.

Its p.g.f. is

$$z^{-\alpha\delta\rho} \mathscr{E}\left\{(z^{\alpha/\lambda})^X\right\}$$

$$= z^{-\alpha\delta\rho} \left\{1 - \delta \ \ell n(\gamma - \rho e^{-\lambda+\lambda z^{\alpha/\lambda}})\right\}$$

$$= z^{-\alpha\delta\rho} \left[1 - \delta \ \ell n \left\{\gamma - \rho \ \exp\left(\alpha \ \frac{z^{\alpha/\lambda} - 1}{\alpha/\lambda}\right)\right\}\right]$$

$$\to z^{-\beta\delta\rho} \left\{1 - \delta \ \ell n(\gamma \ \rho z^{\beta})\right\} \quad \text{as} \quad \lambda \to \infty. \qquad (*)$$

However, a log-zero(δ,ρ) variate U, with p.g.f. $1-\delta\ \ell n(\gamma-\rho z)$, mean $\delta\rho$ and variance $\delta\rho(\gamma-\delta\rho)$, has for its standardized variate

$$\frac{U-\mathcal{E}(U)}{\sqrt{\mathcal{V}(U)}} = \beta U - \beta\delta\rho$$

the p.g.f. marked (*) above.

In the usual description this would be: a log-zero-Poisson(δ,ρ,λ) distribution is approximately a log-zero(δ,ρ) distribution for large λ; and this is a rather unusual result in that the "large λ" result does not involve λ at all. Contrary to what one might casually think, it therefore follows that no very interesting results are obtained by examining the case $\delta\rho \to 0$ (as $\lambda \to \infty$) in such a way that $\mathcal{E}(X) = \delta\rho\lambda$ remains finite (non zero), for even if this is done $\mathcal{V}(X)$ still increases indefinitely with λ. Such behaviour is usual in distributions with "added zeroes" (or added probability of fixed amount at any fixed point): if the mean of the original distribution goes off to infinity while that of the "added zeroes" distribution remains finite, the variance of the latter must obviously also go to infinity. This also suggests that as an approximation for large λ the result immediately above is unlikely to be very good.

Moments and Cumulants

The factorial moment g.f. is

$$1-\delta\ \ell n(\gamma-\rho e^{\lambda t}),\quad \gamma \doteq 1+\rho,$$

$$= 1-\delta\ \ell n\left\{1-\rho(e^{\lambda t}-1)\right\}$$

$$= 1+\delta \sum_{n=1}^{\infty} (n-1)!\ \rho^n \frac{(e^{\lambda t}-1)^n}{n!}$$

$$= 1+\delta \sum_{n} (n-1)!\ \rho^n \sum_{m=n}^{\infty} \mathcal{S}_m^{(n)} \frac{\lambda^m t^m}{m!}$$

$$= 1+\delta \sum_{m=1}^{\infty} \lambda^m \frac{t^m}{m!} \sum_{n=1}^{m} \mathcal{S}_m^{(n)} (n-1)!\ \rho^n$$

Hence

$$\mu'_{(m)} = \delta\rho\,\lambda^m \sum_{n=1}^{m} \mathcal{S}_m^{(n)}(n-1)!\,\rho^{n-1} , \quad m \geq 1 ,$$

$$= \gamma\delta\rho\,\lambda^m \sum_{n=1}^{m-1} \mathcal{S}_{m-1}^{(n)}\; n!\,\rho^{n-1} , \quad m \geq 2 ;$$

in particular,

$$\mu'_{(1)} = \delta\rho\,\lambda ,$$

$$\mu'_{(2)} = \delta\rho\,\lambda^2(1+\rho) = \gamma\delta\rho\,\lambda^2 ,$$

$$\mu'_{(3)} = \delta\rho\,\lambda^3(1+3\rho+2\rho^2) = \gamma\delta\rho\,\lambda^3(1+2\rho) ,$$

$$\mu'_{(4)} = \delta\rho\,\lambda^4(1+7\rho+12\rho^2+6\rho^3) = \gamma\delta\rho\,\lambda^4(1+6\rho+6\rho^2) .$$

(For further results, see the Exercises.) The variance is therefore $\delta\rho\,\lambda(1+\gamma\lambda-\delta\lambda\rho)$, and the second factorial cumulant $\delta\rho\lambda^2(\gamma-\delta\rho)$; it follows that the distribution is "under" or "over" dispersed (i.e. its variance is less than or exceeds its mean) as $(1+\rho)/\rho = \gamma/\rho$ is less than or exceeds δ. (If $\gamma/\rho = \delta$, although the mean and variance are equal, the distribution is not Poisson.) For the Indices of Cluster Size and Frequency,

$$\iota_{CS} = \lambda(\gamma-\delta\rho)$$

and

$$\iota_{CF} = \frac{\delta\rho}{\gamma-\delta\rho} ;$$

the Index of Crowding is $\iota_C = \gamma\lambda$, and the Index of Patchiness $\iota_P = \gamma\,\delta\rho$.

As has been remarked, if the mean is written as $\alpha\beta$ and $\kappa_{(2)} = \alpha\beta^2$, then the "skewness", $\kappa_{(3)}/\alpha\beta^3$, can range from $-\infty$ to ∞, and the "kurtosis", $\kappa_{(4)}/\alpha\beta^4$, from -3 to ∞ , these being much more extensive than is the case for any other common distribution.

Because the Neyman distribution will often be thought of as an alternative, it is of substantial interest to see if it (or another 2 parameter distribution) is a special case of the log-zero-Poisson distribution - as for example the Poisson is not of the Neyman (though it is a limiting case). That is, for every specified μ,ν in a Neyman(μ,ν) distribution are there values δ,ρ,λ in some log-zero-Poisson(δ,ρ,λ) distribution such that the two distributions are identical?

Since the factorial moments are more tractable than the probabilities, and identity of one set implies identity of the other, it is convenient to examine these moments. For the Neyman

$$\mu'_{(r)} = \nu^r \Sigma \mathcal{S}_r^{(s)} \mu^s ;$$

for the log-zero-Poisson

$$\mu'_{(r)} = \delta\rho\, \lambda^r \Sigma \mathcal{S}_r^{(s)} (s-1)!\, \rho^s .$$

It is apparent that ν and λ, and μ and ρ, would need to correspond for identity for reasons of algebraic degree, and the occurrence of the factor $(s-1)!$ shows that the moments cannot then be the same. (A specific numerical case is exhibited in Katti and Rao (1970).) The same expressions also show that if $\rho \to 0$ and $\delta \to \infty$ so that $\delta\rho=1$, then the log-zero-Poisson(δ,ρ,λ) distribution will be approximated to by a Neyman(ρ,λ) distribution, which in turn is approximated to by a Poisson with added zeroes.

Similar arguments establish that none of the negative binomial, Poisson-binomial or Poisson-Pascal distributions are special cases of the log-zero-Poisson distribution.

Since the log-zero-Poisson distribution has "added zeroes", a comparison with similarly modified versions of familiar distributions is suggested. Katti and Rao consider the common alternative of a Pascal(κ,ρ) distribution, for which the

$$\text{"skewness"} \quad \frac{\kappa_{(3)}}{\kappa\rho^3} = \frac{2\kappa\rho^3}{\kappa\rho^3} = 2 ,$$

and the

$$\text{"kurtosis"} \quad \frac{\kappa_{(4)}}{\kappa\rho^4} = \frac{6\kappa\rho^4}{\kappa\rho^4} = 6 .$$

The Pascal(κ,ρ) mixed with weights $1-\theta$ at 0 and θ elsewhere, the Pascal-with-zeroes, has p.g.f.

$$1-\theta+\theta(\gamma-\rho z)^{-\kappa}$$

with factorial cumulant g.f. $\ln\{1-\theta+\theta(1+\rho t)^{-\kappa}\}$; the mean is thus $\theta\kappa\rho$ and the second factorial cumulant $\theta\kappa\rho(\kappa+1-\theta\kappa\rho)$.

Writing these, respectively, in the forms $\alpha\beta$ and $\alpha\beta^2$, the ranges of possible values for the "skewness" and "kurtosis" are vastly increased:

"skewness" $(-\infty, 2)$

"kurtosis: $(6, \infty)$.

But for the log-zero-Poisson distribution the ranges are

"skewness" $(-\infty, \infty)$

"kurtosis" $(-3, \infty)$,

and of course these include the above.

Estimation

The complexities of estimation for a three parameter distribution are almost automatically much greater than for a two parameter case, and it is very helpful to exploit the parameterization already exhibited in which the probability of the zero class, θ , appears explicitly.

For a method like that of moments, where the procedure is fully defined, no use can be made of this, but the flexibility in minimum chi-squared estimation - sometimes regarded as an embarassment - is able to take full advantage of it.

Moment Estimation

Equating sample and distribution sample factorial moments gives (Katti and Rao (1965))

$$M'_{(1)} = \mathring{\delta}\,\mathring{\rho}\,\mathring{\lambda}$$

$$M'_{(2)} = \mathring{\delta}\,\mathring{\rho}\,\mathring{\lambda}^2\,(1+\mathring{\rho}),$$

and

$$M'_{(3)} = \mathring{\delta}\,\mathring{\rho}\,\mathring{\lambda}^3\,(1+3\mathring{\rho}+\mathring{\rho}^2) ,$$

whence introducing the factorial power sums $S_{(r)} = n\,M'_{(r)}$, n being the sample size,

$$\mathring{\lambda}(1+\mathring{\rho}) = S_{(2)}/S_{(1)}$$

and

$$\mathring{\lambda}\,\frac{1+3\mathring{\rho}+2\mathring{\rho}^2}{1+\mathring{\rho}} = \frac{S_{(3)}}{S_{(2)}} .$$

Hence

$$(1+\mathring{\rho})\,\frac{S_{(3)}}{S_{(1)}} = \mathring{\lambda}(1+\mathring{\rho})(1+2\mathring{\rho}) = (1+2\mathring{\rho})\,\frac{S_{(2)}}{S_{(1)}} ,$$

and thus

$$\left.\begin{aligned} \mathring{\rho} &= \frac{S_{(3)}\,S_{(1)} - S_{(2)}^2}{2S_{(2)}^2 - S_{(3)}\,S_{(1)}} \\ \mathring{\lambda} &= \frac{2S_{(2)}^2 - S_{(3)}\,S_{(1)}}{S_{(2)}\,S_{(1)}} \end{aligned}\right\}$$

$$\mathring{\delta} = \frac{n\,S_{(2)}\,S_{(1)}^2}{S_{(3)}\,S_{(1)} - S_{(2)}^2}$$

are the moment equations. (The efficient calculation by hand of factorial power sums is described in Kendall and Stuart I (1963).)

Clearly, empirically useful solutions require $S_{(2)}^2 < S_{(3)}S_{(1)} < 2\,S_{(2)}^2$ (otherwise the moment estimators may not be positive); also

$$0 < \mathring{\delta} < 1/\ell n(\mathring{\gamma}-\mathring{\rho}e^{-\mathring{\lambda}})$$

must be satisfied.

By analogy with similar cases, there is reason to expect poor behaviour of moment estimators (at least) near points corresponding to zeroes of the denominators. Ignoring this aspect, the usual asymptotic expansions may be carried out, but, since third moments occur in the estimating equations, sixth moments will appear in the asymptotic expressions for the variances of $\mathring{\delta}$, $\mathring{\rho}$ and $\mathring{\lambda}$; they will therefore not be very useful with estimated values.

Maximum Likelihood Estimation

It is immediately effective to take the distribution in the form

$$1-(1-\theta)\ \frac{\ell n(\gamma-\rho e^{-\lambda+\lambda z})}{\ell n(\gamma-\rho e^{-\lambda})}\quad ;$$

truncating the zero class, which has a probability of θ , gives a p.g.f. which is a function of ρ and λ alone. It follows that the likelihood equations

$$\frac{\partial L}{\partial\theta}=\frac{\partial L}{\partial\rho}=\frac{\partial L}{\partial\lambda}=0 \quad \text{at} \quad \hat{\theta},\ \hat{\rho},\ \hat{\lambda}$$

lead to a separate equation for $\hat{\theta}$:

$$\hat{\theta}=P_0\ ,$$

a minimum variance unbiased estimator, and for $\hat{\rho}$ and $\hat{\lambda}$:

$$\left.\begin{aligned} &\frac{(1-\hat{\theta})\ \hat{\rho}\ \hat{\lambda}}{\ell n(\hat{\gamma}-\hat{\rho}e^{-\hat{\lambda}})}=\overline{X} \\ \text{and}\qquad & \\ &\frac{1}{n}\sum_{x=1}^{\infty}F_x\hat{\Pi}_x+\hat{p}_1=\overline{X} \end{aligned}\right\}\ ,$$

where $S_r=\Sigma\, x^rF_x$, $n=S_0$, $\overline{X}=S_1/n$ and $\Pi_x=(x+1)p_{x+1}/p_x$. These equations are derived (Katti and Rao (1965, 1970)) in the usual way, by differentiation of the p.g.f. and equating coefficients.

Since the equations must be solved iteratively, starting values are required, and these may be obtained from moment estimates. (Should these be regarded as doubtful, estimates which may be more stable can be obtained from the use of P_0 and $M'_{(1)}$ and $M'_{(2)}$, omitting the highly variable third moment (as, for example, with very long tailed distributions): more discussion will be found in the following section on Minimum Chi-squared Estimation). Applying the Newton-Raphson procedure to

$$H(\lambda)\equiv\frac{1}{n}\sum_{x=1}^{\infty}F_x\Pi_x+p_1-\overline{X}=0\ ,$$

with derivative

$$H'(\lambda) = \frac{1}{n\lambda}\left\{\Sigma F_x \Pi_x - \Sigma F_x \chi_x + n(p_1 - 2p_2 - \frac{p_1^2}{1-\theta})\right\}$$

where

$$\chi_x = \Pi_x(\Pi_{x+1} - \Pi_x) ,$$

leads to a second approximation λ_2 derived from the first λ_1 by $\lambda_2 = \lambda_1 - H(\lambda_1)/H'(\lambda_1)$, with repetition as usual.

Again the usual procedure may be applied to find the covariance matrix of these m.l. estimators, as given in Rao (1969). The reciprocal of the first order generalized variance is given by Rao and Katti (1970) as

$$\frac{n^3(1-\theta)}{\theta\Lambda}\left\{\frac{(1+\gamma\lambda)\Lambda-\lambda\rho}{\gamma^2\lambda^3\rho\Lambda}\Phi - \frac{(e^\lambda-1)^2 n_1^2}{\lambda^3\rho}(1+2\lambda-\gamma\lambda) - \frac{1}{\Lambda}\right\}$$

where $n_x = p_x/(1-\theta)$, $\Lambda = \ell n(\gamma-\rho e^{-\lambda})$ and

$$\Phi = \sum_{x=1}^{\infty} \Pi_x^2 n_x ,$$

the expectation of Π_x^2 over the zero truncated log-zero-Poisson distribution.

An example of fitting is taken from Rao and Katti (1965), Distribution 5 , on data from Beall (1940).

European Corn-borers

Count of borers	Observed frequencies	Fitted frequencies
0	227	227.00
1	70	69.57
2	21	22.25
3	6	5.03
4+	1	1.15
	325	325.00

The estimates are:

$$\hat{\delta} = 5.558 ,$$

$$\hat{\rho} = 0.123 ,$$

$$\hat{\lambda} = 0.601 ;$$

no estimate of the covariance matrix is given but the log-likelihood is -277.52 (to base e) and the chi-squared goodness of fit criterion 0.38 on 1 degree of freedom.

As a matter of curiosity, fitting a Poisson-zero distribution, also by m.l., with p.g.f.

$$1 - \alpha + \alpha e^{-\beta + \beta z}$$

gave estimates of

$$\hat{\alpha} = 0.606$$

$$\hat{\beta} = 0.662$$

and fitted frequencies of 227.02, 69.10, 22.87, 5.05, 0.96 with the same log-likelihood to this accuracy, and a chi-squared of much the same on 2 degrees of freedom. However, any interpretations of these sets of estimates are not altogether transparent.

<u>Minimum Chi-squared Estimation</u>

A more detailed and extended account of this work is given in Rao and Katti (1970): what follows is a summary in the present notation.

It is convenient to treat the parameter $\theta = p_0$, the probability of the zero class, separately from ρ and λ in the parameter vector

$$\underset{\sim}{\lambda} = (\theta, \rho, \lambda)^t$$

(note the two distinct uses, $\underset{\sim}{\lambda}$ for the vector, and λ for the third component of $\underset{\sim}{\lambda}$). So the statistic

$$\underset{\sim}{T} = (T_0, T_1, \ldots, T_m)^t ,$$

to be used for estimating $\underset{\sim}{\lambda}$, will have throughout $T_0 = P_0$, the zero class sample proportion, while T_i will be supposed to have the form

$$T_i = \Sigma\, a_{ix}\, P_x \ , \ i = 1,2,\dots,m$$

with the a_{ix} specified constants. (E.g., if $a_{ix} = x^i$, then $T_i = M_i'$, the sample moments; if $a_{ix} = x^{(i)}$, then $T_i = M_{(i)}'$.) Write

$$\tau_i = \mathcal{E}(T_i) \quad \text{and} \quad \underset{\sim}{\tau} = (\tau_1,\dots,\tau_m)^t$$

(so $\underset{\sim}{\tau} \neq \mathcal{E}(\underset{\sim}{T})$; $\mathcal{E}(\underset{\sim}{T}) = (\theta,\underset{\sim}{\tau})^t$).

The minimum chi-squared equation is therefore obtained by minimizing the quadratic form

$$\left(\underset{\sim}{T} - \begin{pmatrix}\theta\\ \underset{\sim}{\tau}\end{pmatrix}\right)^t \cdot \underset{\sim}{i} \cdot \left(\underset{\sim}{T} - \begin{pmatrix}\theta\\ \underset{\sim}{\tau}\end{pmatrix}\right)$$

where $\underset{\sim}{i}$ is the numerical value of an at least consistent estimator of $\underset{\sim}{\mathcal{V}}(\underset{\sim}{T})^{-1}$; the equation is thus

$$\left(\frac{\partial(\theta,\underset{\sim}{\tau})}{\partial\,\underset{\sim}{\lambda}}\right)^t \underset{\sim}{i}\left(\underset{\sim}{T} - \begin{pmatrix}\theta\\ \underset{\sim}{\tau}\end{pmatrix}\right) = \underset{\sim}{0} \ , \quad \text{at} \quad \underset{\sim}{\lambda} = \underset{\sim}{\lambda}^* \ ,$$

where for convenience

$$\frac{\partial(\theta,\underset{\sim}{\tau})}{\partial\,\underset{\sim}{\lambda}} \quad \text{is written for} \quad \frac{\partial\begin{pmatrix}\theta\\ \underset{\sim}{\tau}\end{pmatrix}}{\partial\,\underset{\sim}{\lambda}} \ ,$$

the Jacobian matrix.

Because of the special choice of T_0 the covariance matrix $\underset{\sim}{\mathcal{V}}(\underset{\sim}{T})$ can be conveniently partitioned. As before

$$\mathcal{C}(T_0,T_i) = -\frac{\theta\,\tau_i}{n} \ , \ i = 1,\dots,m,$$

$$\mathcal{C}(T_i,T_j) = \frac{d_{ij} - \tau_i\tau_j}{n} \ , \ i,j = 1,\dots,m,$$

where $d_{ij} = \mathscr{E}(a_{iX}\, a_{jX})$. Rao and Katti (1970) denote corresponding functions for the zero truncated distribution by adding an asterisk, so, e.g., they write $p_x = (1-\theta)p_x^*$. To reduce the number of suffixes, and because the asterisk is here used for a minimum chi-squared estimator, the alphabetically preceeding letter will be used, so

$$p_x = (1-\theta)n_x\ ,\quad \underset{\sim}{\tau} = (1-\theta)\underset{\sim}{\sigma}\ ,\quad d_{ij} = (1-\theta)\ c_{ij}$$

$$\left(\text{e.g.}\quad 1 - \frac{\ln(\gamma-\rho e^{-\lambda+\lambda z})}{\ln(\gamma-\rho e^{-\lambda})} = \sum_{x=1}^{\infty} n_x z^x\right) .$$

Then $\mathscr{C}(T_0, T_i) = \dfrac{-\theta(1-\theta)}{n}\sigma_i$,

$$\mathscr{C}(T_i, T_j) = (1-\theta)\ \frac{c_{ij}-\sigma_i\sigma_j}{n} + \theta(1-\theta)\ \frac{\sigma_i\sigma_j}{n}$$

and writing $\underset{\sim}{c} = \left(\dfrac{c_{ij}-\sigma_i\sigma_j}{n}\right)$,

$$\underset{\sim}{\mathscr{V}}(\underset{\sim}{T}) = \left(\begin{array}{c:c} \dfrac{\theta(1-\theta)}{n} & \dfrac{-\theta(1-\theta)}{n}\underset{\sim}{\sigma}^t \\ \hdashline \dfrac{-\theta(1-\theta)}{n}\underset{\sim}{\sigma} & (1-\theta)\underset{\sim}{c} + \dfrac{\theta(1-\theta)}{n}\underset{\sim}{\sigma}\,\underset{\sim}{\sigma}^t \end{array}\right)$$

Hence

$$\underset{\sim}{\mathscr{V}}(\underset{\sim}{T})^{-1} = \left(\begin{array}{c:c} \dfrac{n}{\theta(1-\theta)} + \dfrac{\underset{\sim}{\sigma}^t\underset{\sim}{c}^{-1}\underset{\sim}{\sigma}}{1-\theta} & \dfrac{\underset{\sim}{\sigma}^t\underset{\sim}{c}^{-1}}{1-\theta} \\ \hdashline \dfrac{\underset{\sim}{c}^{-1}\underset{\sim}{\sigma}}{1-\theta} & \dfrac{\underset{\sim}{c}^{-1}}{1-\theta} \end{array}\right)$$

and for $\underset{\sim}{\lambda}^* = (\theta^*, \rho^*, \lambda^*)^t$,

$$\underset{\sim}{\mathscr{V}}(\underset{\sim}{\lambda}^*) = \left(\left(\frac{\partial(\theta, \underset{\sim}{\tau})}{\partial\underset{\sim}{\lambda}}\right)^t \underset{\sim}{\mathscr{V}}(\underset{\sim}{T})^{-1}\ \frac{\partial(\theta, \underset{\sim}{\tau})}{\partial\underset{\sim}{\lambda}}\right)^{-1} ,$$

where the Jacobian can be partitioned also:

$$\frac{\partial(\theta,\underset{\sim}{\tau})}{\partial\underset{\sim}{\lambda}} = \left[\begin{array}{c:c} 1 & \underset{\sim}{0}^t \\ \hdashline -\underset{\sim}{\sigma} & (1-\theta)\dfrac{\partial\underset{\sim}{\sigma}}{\partial(\rho,\lambda)} \end{array}\right]^{-1} .$$

Consequently

$$\underset{\sim}{\mathcal{V}}(\underset{\sim}{\lambda}^*) = \left[\begin{array}{c:c} \dfrac{n}{\theta(1-\theta)} & \underset{\sim}{0}^t \\ \hdashline \underset{\sim}{0} & (1-\theta)\left(\dfrac{\partial\underset{\sim}{\sigma}}{\partial(\rho,\lambda)}\right)^t \underset{\sim}{c}^{-1} \dfrac{\partial\underset{\sim}{\sigma}}{\partial(\rho,\lambda)} \end{array}\right]$$

showing that at least asymptotically θ^* is independent of (ρ^*,λ^*) , very conveniently. (A reminder may be inserted here, that all these results are asymptotic.)

Making comparisons with the corresponding generalized variance for the m. . estimator, Rao and Katti show that the first order asymptotic efficiency of $\underset{\sim}{\lambda}^*$ is a function of ρ and λ alone (and not of θ) ; $(T_0,T_2/T_1,\ldots,T_m/T_{m-1})^t$ in place of $\underset{\sim}{T}$.

Corresponding results follow for observations on the zero-truncated log-zero-Poisson distribution: the "efficiency" of an estimator defined like $\underset{\sim}{\lambda}^*$ above but without T_0 turns out to be the same as that of $\underset{\sim}{\lambda}^*$, and similarly for a ratio estimator.

As a consequence of these results, since $T_0 = P_0$ will always be used as the minimum chi-squared estimator of θ (for the non-truncated case), comparisons of alternative choices of $T_1,T_2,\ldots,T_m$ need only to be made across these estimators of ρ and λ . An exhaustive study by Rao and Katti (1970) based on $P_1,P_2,M,M'_{(2)},M'_{(3)}$, the sample proportions in the first and second classes and the first three factorial moments, considers all combinations of these 2 at a time, 3 at a time, 4 at a time and all 5 together. Comparisons were carried out, between these and with the maximum likelihood estimators, through the ratios of first order generalized variances, with the usual difficulties these raise for interpretation (e.g. finite sample sizes, correlations in generalized variances, and limited regions of tabulation).

Nevertheless, hopefully, such comparisons can be a guide when no others are available. Some very abbreviated tentative conclusions are as follows.

2 statistics

P_1, M is 'best' over a large region.

3 statistics

P_1, P_2, M is 'best' over a large region - but near

$\rho = \frac{1}{2}$ then $M, M'_{(2)}, M'_{(3)}$ is 'better'.

4 statistics

$P_1, P_2, M, M'_{(2)}$ is 'best' over a large region.

5 statistics

Compared with m. . the "efficiency" falls off as ρ, λ increase, but it is generally high. (Rao and Katti also remark that the likelihood is very flat, with consequent difficulty in determining "true" m.l. estimates.)

Since if $T_1, \ldots, T_m$ is a set of statistics and $T_1, \ldots, T_m, T_{m+1}$ a set obtained by adding another component, then the first order generalized variance of the second set does not exceed that of the first (at least if the first set is not sufficient) - given unlimited computing facilities one might therefore always choose the 5 statistics set out above (but why stop at 5?).

5.5a EXERCISE

For the log-zero-Poisson(δ, ρ, λ) distribution

$$g(z) = 1 - \delta \ \ell n(1+\rho-\rho e^{-\lambda+\lambda z})$$

show by differentiation of the factorial moment g.f. with respect to its dummy variable that

$$\mu'_{(r+1)} = \rho \sum_{s=1}^{r} \binom{r}{s} \lambda^s \mu'_{(r-s+1)} + \mu'_{(1)} \lambda^r$$

$$\lambda = 1,2,\ldots;$$

similarly differentiating with respect to ρ ,

$$\mu'_{(r+1)} = \lambda\rho(1+\rho) \frac{\partial\mu'_{(r)}}{\partial\rho} , \; r = 1,2,\ldots .$$

Appendix 1

APPROXIMATIONS FOR MEANS, VARIANCES AND COVARIANCES

Supposing X has a known distribution and g(X) is a known function satisfying mild regularity conditions, then the distribution of g(X) can be found in principle by the usual "change of variable" formulae: the exact result, however, is often analytically too complex to be of use.

Useful asymptotic expansions, and approximations, for the mean and variance and indeed moments of any order can be found by using the early terms of a Taylor expansion: see Cramer (1947) sections 27.7 and 28.4 for some precise statements of regularity conditions, Curtiss (1943), Rao (1965) Chapter 6, and the Appendix on Asymptotic Expansions, for further general remarks.

In the simplest case, writing $\mathscr{E}[X] = \mu$ and proceeding formally,

$$g(x) = g(\mu) + (X-\mu)g'(\mu) + \tfrac{1}{2}(X-\mu)^2 g''(\mu) + \text{higher order terms.}$$

Taking expectations, and neglecting quadratic and higher order terms,

$$\mathscr{E}[g(X)] = g(\mathscr{E}[X]) + \text{higher order terms;}$$

neglecting third and higher order terms,

$$\mathscr{E}[g(X)] = g(\mu) + \tfrac{1}{2}\,\mathscr{V}(X)\; g''(\mu) + \text{higher order terms.}$$

(It may be remarked that if $g(.)$ is a convex function, i.e. if $g(x)$ is never above the chord joining any two points on the graph of the function in the domain considered, so that

$$g(\alpha x_1 + \beta x_2) \le \alpha g(x_1) + \beta g(x_2)\ , \quad \alpha + \beta = 1,\ \alpha,\beta > 0\ ,$$

then $\mathscr{E}[g(X)] \ge g(\mathscr{E}[X])$, by Jensen's Inequality.)

By repetition of the argument on

$$g(X) - g(\mu)\ ,$$

squaring and taking expectations gives when retaining only the terms of lowest order

$$\mathscr{V}[g(X)] = [g'(\mu)]^2\mathscr{V}(X) + \text{higher order terms.}$$

If the next higher order terms are retained, writing

$$\mu_r = \mathscr{E}(X-\mu)^r\ ,\ r = 2,3,4 \quad (\text{and}\ \ \mu_2 = \mathscr{V}(X))$$

for the central moments of X ,

$$\mathscr{V}[g(X)] = [g'(\mu)]^2\mathscr{V}(X) + \tfrac{1}{4}\,[g''(\mu)]^2\,(\mu_4-\mu_2) + g'(\mu)\; g''(\mu)\; \mu_3$$
$$+ \text{higher order terms.}$$

Similarly, for two variates X and Y and the function $g(X,Y)$, proceeding formally and writing $\alpha = \mathscr{E}(X)$, $\beta = \mathscr{E}(Y)$, gives

$$\begin{aligned} g(X,Y) &= g(\alpha,\beta) + (X-\alpha)g'_X(\alpha,\beta) + (Y-\beta)\; g'_Y(\alpha,\beta) \\ &\quad + \tfrac{1}{2}\,(X-\alpha)^2\; g''_{XX}(\alpha,\beta) + (X-\alpha)(Y-\beta)g''_{XY}(\alpha,\beta) \\ &\quad + \tfrac{1}{2}\,(Y-\beta)^2\; g''_{YY}(\alpha,\beta) + \text{higher order terms,} \end{aligned}$$

where $g''_{XY}\ (\alpha,\beta) = \left.\dfrac{\partial^2 g(X,Y)}{\partial X\partial Y}\right|_{X=\alpha,\ Y=\beta}$, etc.

Taking expectations as before gives

$$\mathscr{E}[g(X,Y)] = g(\alpha,\beta) + \tfrac{1}{2}\mathscr{V}(X)g''_{XX}(\alpha,\beta) + \mathscr{C}(X,Y)g''_{XY}(\alpha,\beta)$$

$$+ \tfrac{1}{2}\mathscr{V}(Y)g''_{YY}(\alpha,\beta) + \text{higher order terms.}$$

Introducing the notation

$$\mu_{ij} = \mathscr{E}[(X-\alpha)^i\ (Y-\beta)^j]\ ,$$

$$\mathscr{V}[g(X,Y)] = g'^2_X\mu_{20} + g'^2_Y\mu_{02} + 2g'_X\ g'_Y\mu_{11}$$

$$+ \tfrac{1}{4}\ g''^2_{XX}\ (\mu_{40}-\mu_{20}) + \tfrac{1}{4}\ g''^2_{YY}\ (\mu_{04}-\mu_{02})$$

$$+ g''^2_{XY}\ (\mu_{22}-\mu^2_{11})$$

$$+ g'_X g''_{XX}\mu_{30} + g'_X g''_{YY}\mu_{12} + g''_{XX}g'_Y\mu_{21} + g'_Y g''_{YY}\mu_{03}$$

$$+ 2g'_X g''_{XY}\mu_{21} + 2g'_Y g''_{XY}\mu_{12}$$

$$+ \text{higher order terms}$$

where all the derivatives are evaluated at $X=\alpha$, $Y=\beta$.

As a matter of notational brevity, especially in specific cases for several variables, it is often much more compact to write the Taylor expansion

$$g(X,Y) - g(\alpha,\beta) = (X-\alpha)g'_X(\alpha,\beta) + (Y-\beta)g'_Y(\alpha,\beta) + \ldots$$

as

$$dg = g_X dX + g_Y dY + \tfrac{1}{2}g_{XX}dX^2 + g_{XY}dXdY + \tfrac{1}{2}g_{YY}dY^2 + \ldots$$

whence, e.g.,

$$\mathscr{V}(g) = \mathscr{E}[dg^2] = g^2_X\ \mathscr{V}(X) + g^2_Y\ \mathscr{V}(Y) + 2g_X g_Y\ \mathscr{C}(X,Y) + \text{etc.}$$

Thus a common application is to functions of the observational proportions $P_0, P_1, \ldots, P_c$, where these are F_0/n, $F_1/n, \ldots, F_c/n$ with $\Sigma F_x = n$, the F_x being the observational frequencies: e.g. the sample moment

$$M'_i = \frac{1}{n}\ \Sigma x^i F_x = \Sigma x^i P_x$$

is such a function. For given functions $f(P_0,\ldots,P_c) \equiv f(\underset{\sim}{P})$ and $g(\underset{\sim}{P})$, an approximate expression is sought for $\mathscr{C}\{f(\underset{\sim}{P}),g(\underset{\sim}{P})\}$. As above,

$$df = \Sigma \frac{\partial f}{\partial p_x} dP_x \qquad \left(= \left(\frac{\partial f}{\partial \underset{\sim}{p}}\right)^t d\underset{\sim}{P}\right)$$

$$dg = \Sigma \frac{\partial f}{\partial p_y} dP_y \qquad \left(= \left(\frac{\partial g}{\partial \underset{\sim}{p}}\right)^t d\underset{\sim}{P}\right)$$

and so, all to first order

$$\begin{aligned}\mathscr{C}\{f(\underset{\sim}{P}),\ g(\underset{\sim}{P})\} &= \mathscr{E}(df\ dg)\\ &= \Sigma \frac{\partial f}{\partial p_x}\ \mathscr{E}\{(dP_x)^2\}\\ &+ \underset{x\neq y}{\Sigma\ \Sigma} \frac{\partial f}{\partial p_x}\ \frac{\partial g}{\partial p_y}\ \mathscr{E}(dP_x dP_y)\ .\end{aligned}$$

However, $nP_0, nP_1, \ldots, nP_c$ are multinomially distributed, whence

$$\left.\begin{aligned}\mathscr{V}(P_x) &= \mathscr{E}\{(dP_x)^2\} = \frac{p_x q_x}{n}\ , \quad q_x = 1-p_x\\ \mathscr{C}(P_x,P_y) &= \mathscr{E}(dP_x dP_y) = \frac{-p_x p_y}{n}\ , \quad x \neq y\end{aligned}\right\}$$

and so

$$\mathscr{C}\{f(\underset{\sim}{P}),\ g(\underset{\sim}{P})\} = \frac{1}{n}\,\Sigma \frac{\partial f}{\partial p_x} \frac{\partial g}{\partial p_x} p_x q_x - \frac{1}{n} \underset{x\neq y}{\Sigma\ \Sigma} \frac{\partial f}{\partial p_x} \frac{\partial g}{\partial p_y} p_x p_y$$

to first order.

In particular, writing

$$f(\underset{\sim}{P}) = M_i' \text{ above, and } g(\underset{\sim}{P}) = \ell n\ P_0\ ,$$

an application used in linear minimum chi-squared estimation, $\mathscr{C}(M_i',\ \ln\ P_0) = -\frac{\mu_i'}{n} + 0\left(\frac{1}{n^2}\right)$ and $\mathscr{V}(\ln\ P_0) = \frac{q_0}{np_0} + 0\left(\frac{1}{n^2}\right)$.

For the general case, where the original variate is the n-dimensional vector $\underset{\sim}{X} = (X_1, X_2, \ldots, X_n)^t$ with a vector of expectations $\underset{\sim}{\mu} = \mathcal{E}(\underset{\sim}{X})$, and the transformations are through the m known functions

$$Y_1 = g_1(\underset{\sim}{X}),\ Y_2 = g_2(\underset{\sim}{X}), \ldots, Y_m = g_m(\underset{\sim}{X}),$$

i.e. $\underset{\sim}{Y} = \underset{\sim}{g}(\underset{\sim}{X})$,

it is convenient to use matrix notation. To first order the Taylor expansion is

$$Y_i - g_i(\underset{\sim}{\mu}) = \left(\frac{\partial g_i}{\partial \mu_1}, \frac{\partial g_i}{\partial \mu_2}, \ldots, \frac{\partial g_i}{\partial \mu_n}\right)(\underset{\sim}{X} - \underset{\sim}{\mu}), \quad i = 1,2,\ldots,m,$$

where $\frac{\partial g_i}{\partial \mu_j}$ is used for $\left.\frac{\partial g_i(\underset{\sim}{X})}{\partial X_j}\right|_{\underset{\sim}{X} = \underset{\sim}{\mu}}$.

Thus

$$\underset{\sim}{Y} - \underset{\sim}{g}(\underset{\sim}{\mu}) = \frac{\partial \underset{\sim}{g}}{\partial \underset{\sim}{\mu}}(\underset{\sim}{X} - \underset{\sim}{\mu}) \quad \text{to first order,}$$

introducing the Jacobian matrix

$$\frac{\partial \underset{\sim}{g}}{\partial \underset{\sim}{\mu}} = \begin{pmatrix} \frac{\partial g_1}{\partial \mu_1}, & \frac{\partial g_1}{\partial \mu_2}, & \ldots, & \frac{\partial g_1}{\partial \mu_n} \\ \frac{\partial g_m}{\partial \mu_1}, & \frac{\partial g_m}{\partial \mu_2}, & \ldots, & \frac{\partial g_m}{\partial \mu_n} \end{pmatrix}.$$

Hence

$$(\underset{\sim}{Y} - \underset{\sim}{g}(\underset{\sim}{\mu}))(\underset{\sim}{Y} - \underset{\sim}{g}(\underset{\sim}{\mu}))^t = \frac{\partial \underset{\sim}{g}}{\partial \underset{\sim}{\mu}}(\underset{\sim}{X} - \underset{\sim}{\mu})(\underset{\sim}{X} - \underset{\sim}{\mu})^t \left(\frac{\partial \underset{\sim}{g}}{\partial \underset{\sim}{\mu}}\right)^t,$$

and taking expectations,

$$\underset{\sim}{\mathcal{V}}(\underset{\sim}{Y}) = \frac{\partial \underset{\sim}{g}}{\partial \underset{\sim}{\mu}}\ \underset{\sim}{\mathcal{V}}(\underset{\sim}{X}) \left(\frac{\partial \underset{\sim}{g}}{\partial \underset{\sim}{\mu}}\right)^t,$$

all to first order. The first order relation between generalized variances thus follows directly by "taking determinants".

The above formulae are sometimes called those of the method of statistical differentials, or the propagation of error formulae. One of the most common applications is to replace the distribution of $g(\underset{\sim}{X})$ by that of a normal variate with mean and variance given to some order by the above formulae - this is especially used in the univariate case, and (perhaps obviously) particular caution needs to be exercised if the tails of the distribution are so treated. The sections dealing with Estimation include applications of the technique, especially in finding approximate confidence intervals.

It is also convenient to summarize some much used formulae specifically relating to moments. By standard algebraic manipulation (e.g., the Appendix on Estimation: Moment Estimation; or Kendall and Stuart I (1963)), all the following can be obtained.

$$\mathscr{V}(M_r') = \frac{1}{n}(\mu_{2r}' - \mu_r'^2),$$

$$\mathscr{C}(M_r', M_s') = \frac{1}{n}(\mu_{r+s}' - \mu_r'\mu_s') .$$

These two results are exact - in these and other cases, of course, the existence of the moments is supposed. For moments about the mean

$$\mathscr{V}(M_r) = \frac{1}{n}(\mu_{2r} - \mu_r^2 + r^2\mu_{r-1}^2\mu_2 - 2r\mu_{r+1}\mu_{r-1}) \quad \text{to order } 1/n,$$

$$\mathscr{C}(M_r, M_s) = \frac{1}{n}(\mu_{r+s} - \mu_r\mu_s + rs\mu_{r-1}\mu_{s-1}\mu_2 - r\mu_{r-1}\mu_{s+1} - s\mu_{r+1}\mu_{s-1})$$

$$\text{to order } 1/n,$$

$$\mathscr{C}(\bar{X}, M_r) = \frac{1}{n}(\mu_{r+1} - r\mu_{r-1}\mu_2) \quad \text{to order } 1/n.$$

EXERCISES

1. Obtain the commonly used approximations:

$$\mathscr{E}\left(\frac{X}{Y}\right) \doteq \frac{\mathscr{E}(X)}{\mathscr{E}(Y)}\left[1 + \frac{\mathscr{V}(Y)}{\{\mathscr{E}(Y)\}^2} - \frac{\mathscr{C}(X,Y)}{\mathscr{E}(X)\mathscr{E}(Y)}\right],$$

$$\mathscr{V}\left(\frac{X}{Y}\right) \doteq \frac{\mathscr{V}(X)}{\{\mathscr{E}(Y)\}^2} + \frac{\mathscr{V}(Y)}{\{\mathscr{E}(X)\}^2} - \frac{2\mathscr{C}(X,Y)}{\mathscr{E}(X)\mathscr{E}(Y)} .$$

2. Obtain the result

$$\mathscr{C}\{M_r', g(P_0)\} = -\frac{\mu_r' \, p_0 \, g'(p_0)}{n} + 0\left(\frac{1}{n^2}\right)$$

both by writing out the first order Taylor expansions and by substitution in the formula for $\mathscr{C}\{f(\underset{\sim}{P}), g(\underset{\sim}{P})\}$ derived in the text.

3. Establish the following asymptotic equalities, all correct at least to $O(1/n)$:

$$\mathscr{V}(\bar{X}) = \mu_2/n,$$
$$\mathscr{C}(\bar{X}, M_2) = \mu_3/n,$$
$$\mathscr{V}(M_2) = (\mu_4 - \mu_2^2)/n,$$

and

$$\mathscr{V}(K_{(1)}) = (\kappa_{(2)} + \mu)/n\ , \quad \mu \equiv \kappa_{(1)},$$
$$\mathscr{C}(K_{(1)}, K_{(2)}) = \mathscr{C}(\bar{X}, M_2) - \mathscr{V}(\bar{X}) = (\kappa_{(3)} + 2\kappa_{(2)})/n,$$
$$\mathscr{V}(K_{(2)}) = (\kappa_{(4)} + 4\kappa_{(3)} + 2\kappa_{(2)}^2 + (4\mu + 2)\kappa_{(2)} + 2\mu^2)/n.$$

These are of use for obtaining asymptotic covariance matrices when sample and distribution factorial cumulants take simple forms.

4. Obtain the following exact results (the expectations are of course assumed to exist):

$$\mathscr{E}(XY) = \mathscr{E}(X)\,\mathscr{E}(Y) + \mathscr{C}(X,Y),$$
$$\mathscr{V}(XY) = \mathscr{C}(X^2, Y^2) - \{\mathscr{C}(X,Y)\}^2 - 2\mathscr{C}(X,Y)\,\mathscr{E}(X)\,\mathscr{E}(Y)$$
$$+ \mathscr{V}(X)\mathscr{V}(Y) + \mathscr{V}(X)\{\mathscr{E}(Y)\}^2 + \mathscr{V}(Y)\{\mathscr{E}(X)\}^2\ ,$$

so that for independence or zero correlation

$$\mathscr{E}(XY) = \mathscr{E}(X)\,\mathscr{E}(Y),$$
$$\mathscr{V}(XY) = \mathscr{V}(X)\mathscr{V}(Y) + \mathscr{V}(X)\{\mathscr{E}(Y)\}^2 + \mathscr{V}(Y)\{\mathscr{E}(X)\}^2,$$

and iff $\mathscr{E}(X) = \mathscr{E}(Y) = 0$, then also

$$\mathscr{V}(XY) = \mathscr{V}(X)\mathscr{V}(Y);$$
$$\mathscr{E}\left(\frac{Y}{X}\right) = \mathscr{E}(Y)\mathscr{E}\left(\frac{1}{X}\right) + \mathscr{C}\left(Y, \frac{1}{X}\right)$$
$$= \frac{\mathscr{E}(Y)}{\mathscr{E}(X)} - \frac{\mathscr{C}\left(X, \frac{Y}{X}\right)}{\mathscr{E}(X)}\ .$$

(Frishman (1975) for further results)

Appendix 2

ASYMPTOTIC EXPANSIONS

Most of the asymptotic expansions used in this book are expansions in powers of $1/n$, n being the sample size. For a given function $f(n)$, the formal (Poincare) definition of such an expansion is:

> If $S(n,k) = \alpha_0 + \alpha_1/n + \alpha_2/n^2 + \ldots + \alpha_k/n^k$, and $n^k\{f(n) - S(n,k)\} \to 0$ as $n\to\infty$ for a fixed number k, then $S(n,k)$ is an asymptotic expansion of (at least) order k for $f(n)$.

(Two common notations are:

> $f(n) = S(n,k) + O(1/n^{k+1})$
>
> $f(n) \sim S(n,k)$ or $f(n) \sim \alpha_0 + \alpha_1/n + \ldots$; the loose notation
>
> $f(n) = \alpha_0 + \alpha_1/n + \ldots$ is also not uncommon.)

This definition is quite different from the definition of a convergent expansion of $f(n)$ in powers of $1/n$:

> If $S(n,k)$ is as above, and $f(n) - S(n,k) \to 0$ as $k\to\infty$ for a fixed number n, then $\alpha_0 + \alpha_1/n + \ldots$ is a convergent expansion (for this value) for $f(n)$.

In the former case the number of terms in the expansion is fixed and n increases indefinitely; in the latter n is fixed - though of course it will usually be defined on some domain - and the number of terms increases indefinitely. (It is unfortunate for the present purpose that the most common usage in pure mathematics writes n in place of k and x in place of 1/n; in these comments the standard statistical usage of n for sample size is maintained.)

So, for example, considering the sample second moment about the mean, take

$$f(n) \equiv \mathscr{E}(M_2) = \mathscr{E}\{\frac{1}{n}\Sigma(X_i - \bar{X})^2\}$$

$$= \frac{n-1}{n}\kappa_2 = \kappa_2 - \frac{\kappa_2}{n}$$

$$= \kappa_2 + O(1/n), \quad \text{and}$$

$$f(n) \equiv \mathscr{V}(M_2) = \left(\frac{n-1}{n}\right)^2\left(\frac{\kappa_4}{n} + \frac{2\kappa_2^2}{n-1}\right)$$

$$\sim \frac{1}{n}(\kappa_4 + 2\kappa_2^2) - \frac{2}{n^2}(\kappa_4 + \kappa_2^2) + \frac{\kappa_4}{n^3}$$

$$= \frac{\kappa_4 + 2\kappa_2^2}{n} + O(1/n^2)$$

are asymptotic expansions of frequent occurrence in Statistics. Another much used example is Stirling's Formula for the factorial function, often quoted as

$$\Gamma(n+1) = n! \sim \sqrt{(2\pi)}\, n^{n+\frac{1}{2}} e^{-n}$$

(where the sign ~ is used to mean the ratio of the two expressions tends to unity), or in the present formulation,

$$f(n) \equiv \ln \Gamma(n) - \{\tfrac{1}{2}\ln(2\pi) + (n-\tfrac{1}{2}) \ln n - n\}$$

$$\sim \frac{1}{12n} - \frac{1}{360\, n^3} + \frac{1}{1260\, n^5} - \ldots$$

where this last series, in spite of the initial decrease in the absolute magnitude of successive terms, is divergent (e.g. the coefficient of n^{-59} is about 6×10^{30}).

A review of a variety of statistical applications will be found in Wallace (1958), together with general comments; recent work of particular relevance to the present applications is described in Shenton and Bowman (1977), while a good general reference is to the chapter Asymptotic Theory in Cox and Hinkley (1974).

In spite of the divergence of an expansion, a few early terms may be very useful for numerical work, although increasing the number of terms would eventually lead to nonsensical results. Much of the difference in behavior of convergent power series and asymptotic expansions comes from the following:

> A function has at most one convergent power series expansion; a convergent power series corresponds to only one function.
>
> A function has at most one asymptotic power series expansion; an asymptotic power series expansion corresponds to an infinite class of functions.

For example, $1+\frac{1}{n}+\frac{1}{n^2}+\ldots$ is the asymptotic expansion of arbitrary order for

$$f(n) = \frac{n}{n-1} ;$$

it is also the convergent power series expansion of $f(n)$ for $|n| > 1$. But the expansion is also the asymptotic expansion of, for example,

$$e^{-n} + f(n).$$

A Taylor (power) series expansion is an asymptotic expansion valid under very wide conditions compared with very restrictive conditions for a convergent power series - because of the ease of obtaining successive terms this technique is very much used, although other techniques (e.g. steepest descent) are available.

In calculations with an asymptotic expansion, the accuracy which can be obtained does not necessarily increase with the order of the expansion: it is determined by the (given) value of n, and in many naturally arising expansions the greatest accuracy attainable follows by stopping at the smallest (without regard to sign) term in the expansion. (A precise analysis is often impracticable because of the complexity of the function expanded.)

A typical application is to find expectations of random variables determined by a sample of size n. Suppose X_n is a discrete variate with probability function $p_x(n)$, $x = x_1, x_2, \ldots$; for example, X_n may be the mean of a random sample from a Poisson(λ) distribution, in which case $x = 0, 1/n, 2/n, \ldots$ and

$$p_x(n) = e^{-n\lambda} \frac{(n\lambda)^{nx}}{(nx)!} .$$

For many cases of practical interest, the mean and variance of X_n can be written as

$$\mathscr{E}(X_n) = \mu_n \text{ , tending to a finite limit as } n\to\infty,$$

$$\mathscr{V}(X_n) = \frac{\sigma_n^2}{n} \text{ where } \sigma_n^2 \text{ tends to a finite and non-zero limit as } n\to\infty .$$

Then the expectation of a function of X_n, $g(X_n)$ say, is examined through the expansion

$$g(X_n) = g(\mu_n) + (X_n - \mu_n)g'(\mu_n) + \frac{1}{2!}(X_n - \mu_n)^2 g''(\mu_n) + R_3$$

where

$$R_3 = \frac{1}{3!}(X_n - \mu_n)^3 g''(\mu_n + \theta(X_n - \mu_n)),\ 0 < \theta < 1,$$

and the expansion is terminated as shown merely for convenience. Taking expectations,

$$\mathscr{E}\{g(X_n)\} = g(\mu_n) + \frac{1}{2}\frac{\sigma_n^2}{n} g''(\mu_n) + \mathscr{E}(R_3);$$

provided

$$\mathscr{E}g(X_n) = g(\mu_n) + \frac{1}{2}\cdot\frac{\sigma_n^2}{n} g''(\mu_n) + 0(1/n^2)$$

(i.e. that $n^2\mathscr{E}(R_3)$ is bounded as $n\to\infty$);

then $g(\mu_n) + \frac{1}{2}\cdot\frac{\sigma_n^2}{n} g''(\mu_n)$ is the first order asymptotic expansion for $g(X_n)$, and $g(\mu_n)$ is the zero order expansion.

However, in a number of cases of interest, $g(X_n)$ does not exist, though the formal process may be carried out. Thus, in the Poisson example already mentioned, choosing $X_n = \overline{X}$, the sample mean, and

$$g(x) = 1/x,$$

formally

$$\mathscr{E}(1/\overline{X}) = 1/\mu_n + 0(1/n)$$

is obtained, though since $\Pr(x = 0)$ is not zero the expectation does not exist. The formal result may still be useful, and an explanation of this is as follows. Consider the variate $\bar{Y}$ with probabilities proportional to those of $\bar{X}$ at the same values except for $\bar{X} = 0$; for all $\bar{Y} = y$ other than $y = 1, 2, 3, \ldots$ the probabilities are zero. Hence

$$\Pr(\bar{Y} = y) = \Pr(\bar{X} = y \mid \bar{X} \neq 0)$$

$$= \frac{p_y(n)}{1 - p_0(n)}, \qquad y = 1/n, 2/n, \ldots;$$

so that in general

$$\mathscr{E}(\bar{Y}) = \frac{\mu_n}{1 - p_0(n)}, \quad \mathscr{V}(\bar{Y}) = \frac{1}{\{1 - p_0(n)\}^2} \cdot \frac{\sigma_n^2}{n}$$

and in this patricular case (for the Poisson)

$$\mu_n = \lambda, \quad \sigma_n^2 = \lambda.$$

Then $\mathscr{E}(1/\bar{Y}) = \dfrac{1 - p_0(n)}{\mu_n} + 0(1/n)$ follows exactly as before, though now (provided $\mu_n > 0$ exists) $\mathscr{E}(1/\bar{Y})$ exists. But if n is large, the distributions of $\bar{X}$ and $\bar{Y}$ are 'similar' and concentrated near $\mu_n > 0$, and so $p_0(n)$ is 'small' - if it is small to order $1/n$ or higher,

$$\mathscr{E}(1/\bar{Y}) = \frac{1}{\mu_n} + 0(1/n).$$

In the Poisson case, $p_0(n) = e^{-n\lambda}$, which is of exponentially small order so $\mathscr{E}(1/\bar{Y}) = \frac{1}{\lambda} + 0(1/n)$ while $\mathscr{E}(1/\bar{X})$ does not exist). In fact, in a situation where this behavior appears, what is wanted practically is

$$\mathscr{E}\left(\frac{1}{\bar{X}} \mid \bar{X} > 0\right) = \mathscr{E}\left(\frac{1}{\bar{Y}}\right);$$

but the conditional distribution is generally troublesome to handle. Thus what is used is the formal expression for the asymptotic expansion of

$$\mathscr{E}(1/\bar{X})$$

to some specified order; and in meaningful cases (with or without investigation) - when the probabilitiy of the points deleted to obtain existence tends to zero to a sufficiently high order - it is also the asymptotic expansion for the conditional expectation. (This is an illustration of the non one-one relation between functions and asymptotic expansions.) The derivatives are unchanged under deletions - it is the expectations which alter.

So, tentatively (if an exact investigation is not practicable), the successive terms of an expansion are evaluated:

$$\mathscr{E}(g(X_n)) = \alpha_0 + \alpha_1/n + \alpha_2/n^2 + \ldots,$$

and, if the first few terms are decreasing numerically, they are used down to the smallest (or until they are small enough) - if they are not decreasing, the expansion is regarded with suspicion. (If the expression is convergent, there is no smallest term, and the approximation can be made arbitrarily good.) A crude working rule could be formulated in terms of α_0 and α_1 if only the zero order term were to be used (or α_0, α_1 and α_2 if the first order term were included); e.g. the zero order term will be satisfactory if $|\alpha_1|/n$ is less than 1% of $|\alpha_0|$, i.e. if $n \geq 100 \left|\frac{\alpha_1}{\alpha_0}\right|$.

These ideas can be illustrated with the Poisson case, for which

$$\mathscr{E}(1/\bar{Y}) = \frac{n}{e^{n\lambda}-1} \sum_{s=1}^{\infty} \frac{1}{s} \cdot \frac{(n\lambda)^s}{s!}, \quad \text{exactly,}$$

The expansion is then

$$\mathscr{E}(1/\bar{Y}) = \frac{1}{\lambda} + \frac{\mu_{2:\bar{Y}}}{\lambda^3} - \frac{\mu_{3:\bar{Y}}}{\lambda^4} + \frac{\mu_{4:\bar{Y}}}{\lambda^5} - \ldots$$

where $\mu_{r:\bar{Y}} = \mathscr{E}\{\bar{Y} - \mathscr{E}(\bar{Y})\}^r$

$= \mathscr{E}\{\bar{X} - \mathscr{E}(\bar{X})\}^r$, to arbitrary order in $1/n$

$= \mu_{r:\bar{X}}$,

and substituting for these moments

$$\mathcal{E}(1/\bar{Y}) = \frac{1}{\lambda} + \frac{1}{n} \cdot \frac{1}{\lambda^2} + \frac{1}{n^2} \cdot \frac{2!}{\lambda^3} + \ldots + \frac{1}{n^5} \cdot \frac{5!}{\lambda^6} + 0(1/n^6)$$

Taking the case $\lambda = 1$ throughout and successive values of $n = 1, 2, 3, 10, 20$ the expansions are

n = 1: $\mathcal{E}(1/\bar{Y} = 1 + 1 + 2 + 6 + 24 + 120 + 0(1/n^6)$.
The successive terms do not inspire confidence in the use of the expansion for numerical purposes.

n = 2: $\mathcal{E}(1/\bar{Y}) = 1 + 0.5 + 0.5 + 0.75 + 1.5 + 3.75 + 0(1/n^6)$
To two terms, the sum is 1.5, with doubtful accuracy for most purposes. (Even a rough answer is often better than none, and especially when it is known to be rough.)

n = 5: $\mathcal{E}(1/\bar{Y}) = 1 + 0.2 + 0.08 + 0.048 + 0.0384 + 0.0384 + 0(1/n^6)$
To five terms, the sum is 1.3664; the first decimal place is suspicious because of the size of the third, fourth and fifth terms.

n = 10: $\mathcal{E}(1/\bar{Y}) = 1 + 0.1 + 0.02 + 0.006 + 0.0024 + 0.0012 + 0(1/n^6)$
To six terms, the sum is 1.1296, with reasonable confidence in the second decimal.

n = 20: $\mathcal{E}(1/\bar{Y}) = 1 + 0.05 + 0.005 + 0.00075 + 0.00015 + 0.0000375 + 0(1/n^6)$
To six terms the sum is 1.0559375 with some confidence in the fourth decimal.

From Grab and Savage (1954), the "exact" values are

n	1	2	5	10	20
$\mathcal{E}(1/\bar{Y})$	0.76699	1.1532	1.2888	1.1302	1.0560

and these are in line with the remarks above. The pleasant behavior of the Poisson distribution may also be noticed: the expansions do very well for remarkably small values of n.

The arbitrariness of the 'rule' regarding numerical accuracy of an approximation can also be illustrated with this example: exactly the same formal expansions are obtained for

$\mathcal{E}(1/\bar{X})$ (does not exist),

$\mathcal{E}(1/\bar{X} | \bar{X} > 0)$ (as above),

and $\mathcal{E}(1/\bar{X} | \bar{X} > 1/n)$.

This last is

$$\frac{n}{e^{n\lambda}-1-n\lambda}\sum_{s=2}^{\infty}\frac{1}{s}\cdot\frac{(n\lambda)^s}{s!}=\frac{e^{n\lambda}-1}{e^{n\lambda}-1-n\lambda}\mathcal{E}(1/\bar{X}|\bar{X}>0)-\frac{n^2\lambda}{e^{n\lambda}-1-n\lambda},$$

confirming the statement made about the validity of the expansion; numerically, for $n=5$ and $\lambda=1$,

$$\mathcal{E}(1/\bar{X}|\bar{X}>1/5) = 1.159 \quad \text{compared with}$$

$$\mathcal{E}(1/\bar{X}|\bar{X}>0) = 1.289 \text{ and } \mathcal{E}(1/\bar{X}|\bar{X}\geq 1) = 0.800.$$

(Reference may also be made to an example in Kendall and Stuart I (1963), pp. 297-8, where for the normal distribution the sampling moments of

$$M_3/M_2^{3/2}$$

are investigated by these methods and where an exact result is also available, and also to comments by D'Agostino and Tietjen (1971) on M_4/M_2^2.)

For the Neyman(μ,ν) distribution and the practically interesting case of the sample mean $\bar{X}$, for which $n\bar{X}$ is Neyman$(n\mu,\nu)$ distributed and

$$\mathcal{E}(1/\bar{X}|\bar{X}>0) = n\mathcal{E}(1/n\bar{X}|n\bar{X}>0),$$

direct evaluation gives

$$\mathcal{E}(1/\bar{X}|\bar{X}>0) = 0.659,$$

to be compared with $1/\mathcal{E}(\bar{X}) = 0.5$ and $1/\mathcal{E}(\bar{X}|\bar{X}>0) = 0.499$. In this example the zero order term is $1/\mu\nu = 0.5$; the first two terms are

$$\frac{1}{\mu\nu}+\mathcal{V}(X)\frac{1}{\mu^3\nu^3}+0(1/n^2)$$

$$=\frac{1}{\mu\nu}+\frac{1}{n}\cdot\frac{1+\nu}{\mu^2\nu^2}+0(1/n^2)$$

i.e. $0.5+0.1=0.6$ numerically. Continuing the expansion gives

$$\frac{1}{\mu\nu}+\frac{1}{n}\cdot\frac{1+\nu}{\mu^2\nu^2}+\frac{1}{n^2}\,\frac{3\mu\nu(1+\nu)^2-(1+3\nu+\nu^2)}{\mu^3\nu^3}+0(1/n^3)$$

i.e. $0.5 + 0.1 + 0.095 = 0.695$ numerically. So from the expansion one might speculate with some mild confidence that

$\mathcal{E}(1/\bar{X}|\bar{X}>0)$ is about 0.6 or 0.7; in fact it is 0.659 as stated above; however $\mathcal{E}(1/\bar{X}|\bar{X} \geq 2) = 0.389$.

2.1 STEEPEST DESCENT

Another standard routine directly applicable to the p.g.f. is the method of steepest descent, for which a convenient mathematical reference is de Bruijn (1961). No account of the theory is here given - the approximation obtained by its application to the p.g.f.

$$g(z) = \sum_{x=0}^{\infty} p_x z^x$$

when only the first term of the asymptotic expansion is retained is

$$p_x \sim \frac{g(z_x)}{\sqrt{(2\pi)}(z_x)^{x+1}\sqrt{\phi''(z_x)}}$$

where $\phi(z) = \ln\{g(z)/z^x\}$ and z_x is such that $\phi'(z_x) = 0$ with $\phi''(z_x) > 0$.

So trivially, if the Poisson $e^{-\lambda+\lambda z}$ is taken for $g(z)$,

$$\phi(z) = -\lambda+\lambda z - x \ln z,$$

$$\phi'(z) = \lambda - \frac{x}{z}, = 0 \quad \text{for} \quad z_x = \frac{x}{\lambda},$$

$$\phi''(z_x) = \frac{\lambda^2}{x} > 0,$$

and thus the first term is

$$e^{-\lambda} \frac{\lambda^x}{\sqrt{(2\pi)}e^{-x}x^{x+\frac{1}{2}}};$$

the expression beneath the vinculum is the (asymptotic first term) Stirling result for x!

Other examples are quoted under the headings of particular distributions.

Appendix 3

CONDITIONAL EXPECTATION

Suppose X and Y have the joint distribution function $F_{X,Y}(x,y) = F(x,y)$, say, i.e., $Pr(X,Y \varepsilon E) = \int\int_E dF(x,y)$ for any (measurable) set E. Then the expectation of $h(X,Y)$ is defined (or derived) as

$$\mathscr{E}\{h(X,Y)\} = \int\int h(x,y)\ dF(x,y)$$

taken over all values of x and y. So, quite generally,

$$\mathscr{E}(X+Y) = \int\int x\ dF(x,y) + \int\int y\ dF(x,y) = \mathscr{E}(X) + \mathscr{E}(Y);$$

but

$$\mathscr{E}(XY) = \int\int xy\ dF(x,y)$$

cannot be reduced without further information about X and Y. In the simplest case, X and Y are independent and the joint distribution factorizes into the product of the marginal distributions:

$$F(x,y) = F_X(x)\ F_Y(y), \quad \text{say,}$$

when

$$\begin{aligned}\mathscr{E}(XY) &= \{\int x\ dF_X(x)\}\{\int y\ dF_Y(y)\} \\ &= \mathscr{E}(X) \cdot \mathscr{E}(Y).\end{aligned}$$

However, when X and Y are dependent, there may be knowledge about the distribution of Y, say, for each fixed value x of X:

$$F(x,y) = F_{Y|X}(x,y)\ F_X(x),$$

corresponding to the formula (or definition of conditional probability)

$$\Pr(AB) = \Pr(A|B)\ \Pr(B).$$

In this case, quite generally, where the existence of the joint expectation is assumed,

$$\mathscr{E}\{h(X,Y)\} = \int \{\int h(x,y)\ dF_{Y|X}(x,y)\}\ dF_X(x),$$

and in some circumstances the second expression may be handled more readily. Using entirely the expectation notation, the result can be written

$$\mathscr{E}\{h(X,Y)\} = \mathscr{E}_X[\mathscr{E}_Y\{h(X,Y)|X\}] \text{ or } \mathscr{E}_X\{\mathscr{E}_{Y|X}h(X,Y)\};$$

the proof is essentially a standard theorem on the reduction of a multiple integral to repeated integrals (Fubini's theorem). In particular,

$$\mathscr{E}_X[\mathscr{E}_Y(Y|X)] = \mathscr{E}(Y).$$

A slightly different point of view may be helpful. Suppose for the bivariate X,Y and a given function $h(X,Y)$ the expectation with respect to Y for a fixed $X=x$ is $h(x)$:

$$\mathscr{E}_Y\{h(x,Y)\} = h(x).$$

Hence for the random variable X the expectation with respect to Y of $h(X,Y)$ is defined to be $h(X)$:

$$\mathscr{E}_Y\{h(X,Y)\} = h(X).$$

It is then possible to consider the expectation of $h(X)$,

$$\mathscr{E}\{h(X)\} = \mathscr{E}_X[\mathscr{E}_Y\{h(X,Y)\}],$$

and the "theorem of total expectation" asserts that

$$\mathscr{E}_X[\mathscr{E}_Y\{h(X,Y)\}] = \mathscr{E}\{h(X,Y)\}$$

where the expectation of the right hand side is over the bivariate distribution and is supposed to exist.

Because so many functionals commonly used in statistics can be interpreted as conditional expectations, the theorem occurs constantly, for example in mixed, whence stopped, distributions, and indeed it can be used for the most general and abstract definition of the concept of probability.

Another common expectation is the variance:

$$\mathscr{V}\{h(X,Y)\} = \mathscr{E}\{h^2(X,Y)\} - [\mathscr{E}\{h(X,Y)\}]^2.$$

Because it is not linear in $\mathscr{E}(\cdot)$, some care is necessary: unlike the "natural" result for the mean, a conditional expression for the variance is more complicated:

$$\mathcal{V}(Y) = \mathcal{E}\{\mathcal{V}(Y|X)\} + \mathcal{V}\{\mathcal{E}(Y|X)\}$$

(i.e. the unconditional variance is the sum of the expectation of the conditional variance and the variance of the conditional expectation).

In the same way, if $(X_1, X_2, \ldots, X_k)$ has a k-variate distribution,

$$\mathcal{C}(X_i, X_j) = \mathcal{E}_Y\{\mathcal{C}(X_i, X_j|Y)\} + \mathcal{C}_Y\{\mathcal{E}(X_i|Y), \mathcal{E}(X_j|Y)\} ,$$

and a sometimes useful special case is

$$\mathcal{C}(X,Y) = \mathcal{C}\{X, \mathcal{E}(Y|X)\}.$$

EXERCISES

1. If X and Y have a joint distribution, prove from the fundamental result

$$\mathcal{E}\{h(X,Y)\} = \mathcal{E}_X[\mathcal{E}_Y\{h(X,Y)|X\}]$$

and the definition of $\mathcal{V}\{h(X,Y)\}$ that

$$\mathcal{V}(Y) = \mathcal{V}\{\mathcal{E}(Y|X)\} + \mathcal{E}\{\mathcal{V}(Y|X)\} .$$

If $S_N = X_1 + X_2 + \ldots + X_N$ where the $X_1, X_2, \ldots$ (typified by X) are independently and identically distributed and N is independently distributed, deduce that

$$\mathcal{E}(S_N) = \mathcal{E}(X) \ \mathcal{E}(N),$$

$$\mathcal{V}(S_N) = \mathcal{V}(X) \ \mathcal{E}(N) + \{\mathcal{E}(X)\}^2 \mathcal{V}(N).$$

$$\left(\mathcal{V}(Z|x) = \mathcal{E}(Z^2|x) - \{\mathcal{E}(Z|x)\}^2 , \text{ whence } \mathcal{E}\{\mathcal{V}(Z|X)\} = \mathcal{E}\mathcal{E}(Z^2|X) - \mathcal{E}[\{\mathcal{E}(Z|X)\}^2].\right)$$

2. For $X > 0$, show that to zero order $\mathcal{C}(X, Y/X) = 0$; to first order that $\mathcal{C}(X,Y/X) = 0$ if $\mathcal{C}(X,Y) = \mu_Y \sigma_X^2/\mu_X$ and that (exactly) $\mathcal{C}(X,Y/X) = \mu_Y \cdot \{1 - \mathcal{E}(X) \ \mathcal{E}(1/X)\}$, whence, since $\mathcal{E}(X) \ \mathcal{E}(1/X) > 1$, $\mathcal{C}(X,Y/X) < 0$ if $\mu_Y > 0$.

Appendix 4
ESTIMATION

4.1 GRAPHICAL DISCRIMINATION BETWEEN FAMILIES OF DISTRIBUTIONS

In a vaguely defined observational situation, counts of some phenomenon are often made in an exploratory fashion, the intention being to produce a frequency distribution and then to look for regularities: is the "shape" of the distribution preserved under certain modifications of the observational situation, and is it modified in some recognizable way with modifications of the situation?

It is legitimate to fit theoretical distributions in these circumstances, not as the only or even main analysis of the data, and to look for stability in the results of the the fitting. But what distribution or distributions? It is now practicable, with the aid of a computer, to fit a large variety of distributions relatively rapidly, though perhaps expensively on any realistic costing of computer time, but it is obviously wasteful systematically to 'fit' distributions which a simple preliminary examination would have shown to be inappropriate - e.g. if the mean exceeds the variance for the data, the fitting of a negative binomial distribution should not be attempted. (There are some further references in the section Goodness of Fit Testing: see also Pereira (1977).)

On the basis of the results for ratios of factorial cumulants (or equivalently, probability ratio cumulants) for several distributions which are often competitors for fitting purposes, Hinz and Gurland (1967) suggest a simple graphical use of $\kappa_{(r+1)}/\kappa_{(r)} = \eta_r$ for successive small values of $r \neq 0$, say

$r = 1,2,3$ small, because of the high sampling variability associated with high order sample cumulants. The following Figure 35 shows the behaviour of these ratios for some common distributions. Of course sample factorial cumulsnts $K_{(r)}$ will have random variation super-imposed on these simple patterns (and these distributions may not be the only ones relevant), but the value of graphical analyses in an exploratory stage is very high, and can at the least often prevent the commission of absurdities.

In a similar way it may be possible to use the sample equivalent of $\Pi_x = (x+1)p_{x+1}/p_x$, the expression which repeatedly turns up in maximum likelihood estimation for stopped distributions. For three common distributions simple recurrence relations obtain: a plot of Π_x against x follows in Figure 36. Again, the plotting of sample proportions P_x as estimates of p_x leads to less distinct patterns than these; and for a more complex distributions such simple recurrence relations do not obtain. For further discussion and references see Johnson and Kotz (1960), Chapter 2.4.

Yet another series of diagrams is suggested by Ord (1970), based on the use of relations between the variance to mean ratio τ and the third cumulant to variance ratio σ (see Figure 37):

$$\tau = \frac{\kappa_2}{\kappa_1}\ , \ \sigma = \frac{\kappa_3}{\kappa_2}\ .$$

For some distributions these relations turn out to be as follows.

Poisson	$\sigma = \tau = 1$
binomial and negative binomial	$\sigma = 2\tau - 1$
Neyman	$\sigma = \tau + 1 - 1/\tau$
Poisson-binomial	$\sigma = \tau + \chi - \chi/\tau$
Poisson-Pascal	$\sigma = \tau + \gamma - \gamma/\tau$
(Polya-Aeppli	$\sigma = \frac{1}{2}(3\tau - 1/\tau)$)

The (τ,σ) diagram is shown, and the Neyman curve divides the region between the lines $\sigma = \tau$ and $\sigma = 2\tau$ so that for any given value of τ the third-cumulant-to-variance ratio for the Poisson-binomial is less than that for the Neyman which in turn is less than that for the Poisson-Pascal. (Further, Poisson-Pascal

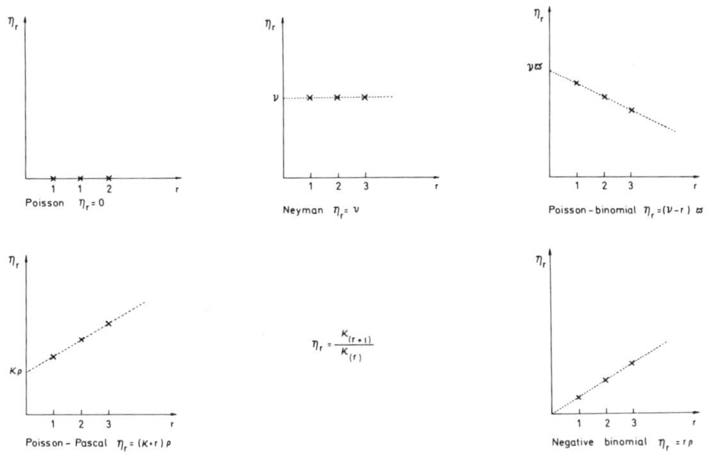

FIGURE 35 Ratios of factorial cumulants for several distributions.

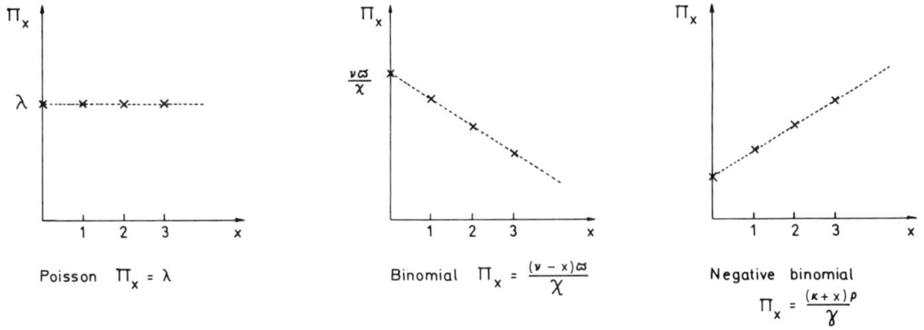

FIGURE 36 Graphs of $\Pi_x = (x+1)p_{x+1}/p_x$ for three distributions.

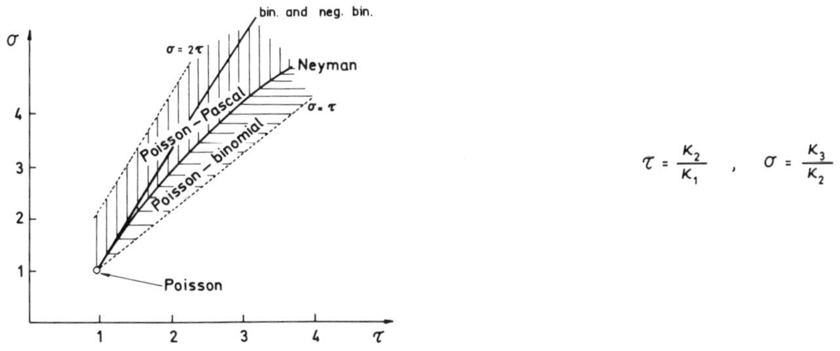

FIGURE 37 The (τ,σ) diagram showing several distributions.

distributions above the binomial line $\sigma = 2\tau-1$ have a half mode.) Plots of these for sample values may again be useful for graphical discriminatory purposes.

Somewhat more complex graphical analyses relating to a number of the common distributions are described in Gart (1970) and Grimm (1970), and in a survey in Holms (1974).

4.2 MOMENT, MAXIMUM LIKELIHOOD, AND MINIMUM CHI-SQUARED

In much scientific work, the precise purpose of an estimation procedure is not sharply defined. The point of view taken here is therefore that, given the parametric form of a probability distribution from which random samples are available, the sampling distributions of various estimators of the parameters can be investigated. From the properties of these sampling distributions the estimators can be compared, directly and with respect to inference procedures based on them.

As far as much of the literature on discrete distributions is concerned, it would almost seem that the major purpose of estimation is to calculate 'expected' (i.e. fitted) frequencies in order to compare these with observed frequencies. In a preliminary stage it is certainly important to be able to recognize the inappropriateness of an anticipated model, but of much greater importance after this stage is the use of estimates for comparative or analytical purposes: if the model does not apply, then to a greater or lesser extent such analyses will be approximate. Further, a 'good' model is likely to have its parameters directly interpretable in terms of the modelled structure: it is changes in the structure which hopefully will be reflected by estimation procedures based on observations (samples) taken on the physical system.

A short account is given of the methods of estimation used for the distributions discussed in this book. It is confined to the results required, with rather more explanation in the case of less familiar techniques. More detail on moment and maximum likelihood methods is to be found in Cramer (1947), Kendall and Stuart II (1965) and, especially, Rao (1965). Because of the brevity of the account and its applied nature, statements such as "if $\hat{\alpha}$ is the maximum likelihood estimator of α, then $f(\hat{\alpha})$ is the m.l. estimator of $f(\alpha)$" are made without qualification.

A comment on the "practicalities" associated with various methods of estimation to be discussed and computational facilities may be ventured in the following terms. Special, ad hoc, methods have mostly been introduced to ease the work of actually calculating estimates: they are therefore often useful for pencil and paper calculations. The method of moments is often suitable for pencil and paper, or hand calculating (electro-mechanical, or desk electronic) facilities. Maximum likelihood often requires iterative procedures and so demands substantial effort unless a programmable computer is available. Minimum chi-squared also requires extensive calculation impracticable without a computer, but since iteration is not required the procedure can be effectively just as fast as for moments given even rather minor computing facilities. Of course it is easy to find exceptions to these broad statements, but for routine use of the procedures in relatively complex cases they given an indication of which procedures ought to be under consideration.

A distinction is maintained throughout between estimator and estimate. An estimator is a function of a random sample: it is a random variable with a probability distribution; an estimate is a (realized) value of an estimator for a particular, numerical, sample. The term statistic is used for a function of a random sample whether or not the function is being thought of as estimating some parameter, the usual circumstance in which estimator is applied.

It is sometimes helpful to keep in mind a distinction made explicit by Kemp (1971) between the parameters appearing in a specification. It will be enough here to exemplify the notions involved. First, there are the model parameters: thus a Poisson variate which is defined as the convolution of two independent Poisson(μ) and Poisson(ν) variates has 2 model parameters. Second, there are the mathematical parameters: in the above example the convolution variate is fully specified by 1 parameter ($\mu+\nu=\lambda$, say). Finally, given observations on the convolution variate there is 1 estimation parameter. Similarly, if ν trials are made of tossing a coin in which the probability of getting a head uppermost at a single trial is π, then there are 2 model parameters, 2 mathematical parameters and, given observations on the number of heads observed in a fixed known number ν of tosses, 1 estimation parameter. It is to be noticed that even if the numbers of model and mathematical parameters are equal the parameters may be different; and that, while it may be mathematically convenient to transform the parameter specification in some way to simplify the estimation procedure, if one is concerned with the model parameters because of their physical interpretation it is the behaviour of their estimators (and not of arbitrarily chosen functions of them) which has to be studied.

A summary "figure of merit" of an estimator T is given by the variance of T, $\mathcal{V}(T)$, or in units of T, by $\sqrt{\mathcal{V}(T)}$; this is often used, e.g., to obtain approximate confidence intervals. (In many applications the use of the figure is complicated by the necessity of using not $\mathscr{V}(T)$ but an estimate of $\mathscr{V}(T)$.) In the multi-estimator case, for the vector $\underset{\sim}{T} = (T_1, T_2, \ldots, T_m)^t$, the corresponding "figure of merit" is the generalized variance (or its 2m-th root):

$$\gamma = \begin{vmatrix} \mathscr{V}(T_1), & \mathscr{C}(T_1,T_2, \ldots, & \mathscr{C}(T_1,T_m) \\ & \ldots & \\ \mathscr{C}(T_m,T_1), & \mathscr{C}(T_m,T_2), \ldots, & \mathscr{V}(T_m) \end{vmatrix} = |\underset{\sim}{\mathscr{V}}(\underset{\sim}{T})|,$$

the determinant of the covariance matrix of $\underset{\sim}{T}$.

Apart from the obvious impossibility of conveying much of the complexity of the behaviour of a multivariate distribution by a single expression such as this, there are warnings which are worth noticing. The simplest relates to dimensions: comparable to the use of the standard deviation rather than the variance for a single estimator, one should examine $\gamma^{1/2m}$ rather than γ. As an extension, since $\underset{\sim}{\mathscr{V}}(\underset{\sim}{T})$ can be written as (Kowal (1971))

$$\underset{\sim}{\mathscr{V}}(\underset{\sim}{T}) = \underset{\sim}{\Delta}\,\underset{\sim}{P}\,\underset{\sim}{\Delta}$$

where $\underset{\sim}{P} = (\mathrm{corr}(T_i, T_j))$, the matrix of correlations (in place of covariances), and

$$\underset{\sim}{\Delta} = \begin{pmatrix} \sqrt{\mathscr{V}(T_1)} & & 0 \\ & \ddots & \\ 0 & & \sqrt{\mathscr{V}(T_m)} \end{pmatrix},$$

the generalized variance can be expressed as

$$\gamma = |P|\nu^m ,$$

where ν is the geometric mean of the variances of $T_1, \ldots, T_m$. In this form it is particularly clear that γ may be small not because any of the variances $\mathscr{V}(T_i)$ are small but because $|\underset{\sim}{P}|$ is small; and indeed it is common to find that estimators used for contagious distributions are very highly correlated, with correspondingly small values of $|\underset{\sim}{P}|$. In practice, the estimation of γ worsens this behaviour, and the comparison of two vector estimators by the ratio of their generalized variances is clearly at best partial, with the added complication that the covariance matrices actually used are likely to be first order asymptotic expansions for these matrices rather than the matrices.

In the Note by Kowal (1971) it is pointed out that in the special case where for two vector estimators multivariate normality and equal correlation matrices are supposed, the ratio of the generalized variances and ratio of the geometric means of the individual variances are obviously equivalent for comparisons when parameter values are known. However, calculations there reported also show that if the covariance matrices are in fact diagonal (though not known to be) it is advantageous to make comparisons through the geometric means of the individual variances rather from the determinants of the covariance matrices (i.e. the generalized variances).

There are also some remarks about transformations below, under Minimum Chi-Squared Estimation, and in the Appendix on Approximations for Means, Variances and Covariances.

Under regular transformations, ratios of generalized variances have an attractive invariance property. Suppose it is proposed to specify a distribution not by the parameter vector $\underset{\sim}{\tau}$ but some regular function $\underset{\sim}{\sigma} = \underset{\sim}{s}(\underset{\sim}{\tau})$ of $\underset{\sim}{\tau}$. (So, e.g., in the Neyman(μ,ν) case, an alternative to $\underset{\sim}{\tau} = (\mu,\nu)^t$ is $\underset{\sim}{\sigma} = (\alpha,\lambda)^t$ where $\alpha = \mu\nu$ and $\lambda = \mu e^{-\nu}$; for the negative-binomial(κ,ρ), $\underset{\sim}{\sigma} = (\kappa\rho,\kappa)^t$ is a sometimes recommended alternative.) Writing $\overset{o}{\underset{\sim}{T}}$ and $\hat{\underset{\sim}{T}}$ for the moment and m.l. (e.g.) estimators of $\underset{\sim}{\tau}$, and $\overset{o}{\underset{\sim}{S}}$ and $\hat{\underset{\sim}{S}}$ correspondingly for $\underset{\sim}{\sigma}$, then to first order in the sample size

$$\underset{\sim}{\mathcal{V}}(\underset{\sim}{S}) = \frac{\partial \underset{\sim}{s}}{\partial \underset{\sim}{\tau}} \underset{\sim}{\mathcal{V}}(\underset{\sim}{T}) \left(\frac{\partial \underset{\sim}{s}}{\partial \underset{\sim}{\tau}}\right)^t$$

both for moment and m.l. estimators. Hence, writing $GV[\underset{\sim}{T}]$ for the determinant of $\underset{\sim}{\mathcal{V}}(\underset{\sim}{T})$,

$$\frac{GV[\hat{\underset{\sim}{S}}]}{GV[\overset{o}{\underset{\sim}{S}}]} = \frac{GV[\hat{\underset{\sim}{T}}]}{GV[\overset{o}{\underset{\sim}{T}}]} ;$$

i.e., as measured by this overall criterion one does no better and no worse by use of such a re-specification. Of course if interest is really not in the vector parameter but only in one component (or several components), it may be possible to choose a suitable function advantageously - e.g., so that a component estimator is (perhaps only nearly) independent of the others and has a small standard error.

Method of Moments

This method (as a formal method due to K. Pearson) proceeds by equating sample distribution moments so that as many equations are obtained as there are parameters to be estimated, starting at the lowest order moment because of the higher sampling variances of higher order moments. For a one parameter distribution with parameter λ the moment equation for the moment estimator $\overset{o}{\lambda}$

$$\left.\begin{aligned} \bar{X} &= \mathscr{E}(X)\Big|_{\lambda=\overset{o}{\lambda}} \\ \text{or} \qquad M_1' &= \mu_1'(\overset{o}{\lambda}) \end{aligned}\right\},$$

where $\mathscr{E}(X^i) = \mu_1'(\lambda) = \Sigma x^i p_x$, $p_x \equiv p_x(\lambda)$, and $M_i' = \frac{1}{n}\,\Sigma x^i F_x$.
Of course this equation need not have a unique solution, nor even solutions, but it often does have a single solution. If the explicit solution is

$$\overset{o}{\lambda} = g(\bar{X}),$$

the sampling variance of $\overset{o}{\lambda}$ is found to first order by using the Taylor expansion

$$\overset{o}{\lambda} = g(\bar{X}) = g(\alpha) + (\bar{X} - \alpha)\,\frac{dg(\alpha)}{d\alpha} + \ldots \quad \text{with} \quad \alpha = \mu_1'(\lambda)$$

for brevity

$$\text{i.e.} \qquad \overset{o}{\lambda} - g(\alpha) = (\bar{X} - \alpha)\,\frac{dg(\alpha)}{d\alpha} + \ldots, \quad \text{and} \quad \mathscr{E}(\overset{o}{\lambda}) = g(\alpha) + \ldots, \\ = \lambda + \ldots$$

whence squaring and taking expectations,

$$\mathscr{V}(\overset{o}{\lambda}) = \left\{\frac{dg(\alpha)}{d\alpha}\right\}^2 \mathscr{V}(\bar{X}) = \left\{\frac{dg(\alpha)}{d\alpha}\right\}^2 \cdot \frac{\mathscr{V}(X)}{n}\,,$$

the familiar first order formula. If no explicit solution can be obtained, the moment equation is expanded directly:

$$M_1' = \mu_1'(\overset{o}{\lambda}) = \mu_1'(\lambda) + (\overset{o}{\lambda} - \lambda)\frac{d\mu_1'(\lambda)}{d\lambda} + \ldots$$

whence

$$M_1' - \mu_1'(\lambda) = (\overset{o}{\lambda} - \lambda)\frac{d\mu_1'(\lambda)}{d\lambda} + \ldots,$$

or $$\overset{o}{\lambda} - \lambda = \frac{1}{d\mu_1'/d\lambda}(M_1' - \mu_1') \quad \text{to first order.} \qquad \text{(A)}$$

This equation can be used both to set up an iterative routine (the Newton-Raphson procedure) for $\overset{o}{\lambda}$ and to obtain the approximate sampling variance of $\overset{o}{\lambda}$. Given a starter value λ_1 for $\overset{o}{\lambda}$, for example obtained by equating a particular observed proportion (e.g. P_0) and the corresponding probability (p_0), a second approximation λ_2 is obtained from

$$\lambda_2 = \lambda_1 + \frac{1}{\left(\frac{d\mu_1'}{d\lambda}\right)_{\lambda=\lambda_1}} \cdot \{M_1' - \mu_1'(\lambda_1)\},$$

and further approximations may be obtained by iteration. (See the Appendix "Iterative Procedures for Solving $F(x) = 0$" for additional remarks.) Similarly, squaring the equation (A) and taking expectations yields directly

$$\mathcal{V}(\overset{o}{\lambda}) = \frac{1}{\left(\frac{d\mu_1'}{d\lambda}\right)^2}\mathcal{V}(M_1') \quad \text{to first order.}$$

(A more detailed examination of the derivation, in this and the more complex examples which follow, shows that the "first order" terms are of first order in $1/n$, n being the sample size. There is further comment in the Appendix on Asymptotic Expansions.)

For a distribution with 2 parameters μ and ν, the moment equations are (with some danger of confusion between the functions μ_1', μ_2' and the parameter μ)

$$M_1' = \mu_1'(\overset{o}{\mu}, \overset{o}{\nu})$$

and $$M_2' = \mu_2(\overset{o}{\mu}, \overset{o}{\nu}),$$

although in many applications with discrete distributions it is algebraically simpler to use sample and distribution factorial cumulants rather than moments. Taylor expansions as for the single parameter case can be used: if explicit solutions can be found, say

$$\overset{o}{\mu} = f(M_1', M_2') \quad \text{and} \quad \overset{o}{\nu} = g(M_1', M_2')$$

where for brevity it is convenient to write

$$M = M_1',\ N = M_2',\ \mathscr{E}(M_1') = \alpha,\ \mathscr{E}(M_2') = \beta \quad \text{and} \quad f_\alpha = \left(\frac{\partial f}{\partial X}\right)_{X=\alpha, Y=\beta}, \text{ etc,}$$

then $\overset{o}{\mu} = f(\alpha,\beta) + (M-\alpha)f_\alpha + (N-\beta)f_\beta + \ldots,$

$$\overset{o}{\nu} = g(\alpha,\beta) + (M-\alpha)g_\alpha + (N-\beta)g_\beta + \ldots.$$

Hence $\mathscr{E}(\overset{o}{\mu}) = f(\alpha,\beta) + \ldots,\quad \mathscr{E}(\overset{o}{\nu}) = g(\alpha,\beta) + \ldots,$

and writing the expansions as

$$\overset{o}{\mu} - f(\alpha,\beta) = (M-\alpha)f_\alpha + (N-\beta)f_\beta + \ldots,$$

$$\overset{o}{\nu} - g(\alpha,\beta) = (M-\alpha)g_\alpha + (N-\beta)g_\beta + \ldots,$$

squaring, or multiplying, and taking expectations gives to first order

$$\mathscr{V}(\overset{o}{\mu}) = f_\alpha^2\ \mathscr{V}(M) + 2f_\alpha f_\beta \mathscr{C}(M,N) + f_\beta^2\ \mathscr{V}(N),$$

$$\mathscr{C}(\overset{o}{\mu},\overset{o}{\nu}) = f_\alpha g_\alpha \mathscr{V}(M) + (f_\alpha g_\beta + f_\beta g_\alpha)\mathscr{C}(M,N) + f_\beta g_\beta \mathscr{V}(N),$$

and $$\mathscr{V}(\overset{o}{\nu}) = g_\alpha^2\ \mathscr{E}(M) + 2g_\alpha g_\beta \mathscr{C}(M,N) + g_\beta^2\ \mathscr{V}(N).$$

Explicit expressions for $\mathscr{V}(M)$, $\mathscr{C}(M,N)$ and $\mathscr{V}(N)$ are readily obtained. Taking a more general case, and reverting to a notation in which the variates $X_1, X_2, \ldots, X_n$ are shown as all distinct,

$$\mathscr{C}(M_i', M_j') = \mathscr{E}\{(M_i' - \mu_i)(M_j' - \mu_j')\}$$

$$= \mathscr{E}\{(\tfrac{1}{n}\ \Sigma\ X_r^i - \mu_i')(\tfrac{1}{n}\ \Sigma X_s^j - \mu_j')\}$$

$$= \frac{1}{n^2}\left\{\mathscr{E}(\Sigma X_r^{i+j}) + \mathscr{E}(\sum_{r\neq s}\Sigma\ X_r^i\ X_s^j)\right.$$

$$\left. - n\mu_j'\ \mathscr{E}(\Sigma X_r^i) - n\mu_i'\ \mathscr{E}(\Sigma X_s^j) + n^2\mu_i'\mu_j'\right\}$$

$$= \frac{1}{n}\ (\mu_{i+j}' - \mu_i'\mu_j'),$$

and this holds for $i=j$ as well - i.e. for a variance. Thus, in terms of the moments $\mu_2, \mu_2', \ldots$ of the original variate

$$\mathcal{V}(M) = \frac{\mu_2}{n}, \quad \mathcal{C}(M,N) = \frac{\mu_3' - \mu_2'\mu_1'}{n} \quad \text{and} \quad \mathcal{V}(N) = \frac{\mu_4' - \mu_2'^2}{n}.$$

If it is not possible to invert explicitly the moment equations

$$M_1' = \mu_1'(\overset{o}{\mu}, \overset{o}{\nu}), \quad M_2' = \mu_2'(\overset{o}{\mu}, \overset{o}{\nu})$$

for $\overset{o}{\mu}$ and $\overset{o}{\nu}$, as it rarely is in the case of truncated distributions for example, an iterative Newton-Raphson procedure can be arranged to give as well the estimated variances $\mathcal{V}(\overset{o}{\mu})$, $\mathcal{V}(\overset{o}{\nu})$ and covariance $\mathcal{C}(\overset{o}{\mu}, \overset{o}{\nu})$ in terms of $\mathcal{V}(M_1')$, $\mathcal{V}(M_2')$ and $\mathcal{C}(M_1', M_2')$. Changing the notation for brevity, in the moment equations immediately above, write

$$\overset{o}{\mu} = X, \quad \overset{o}{\nu} = Y \quad \text{and}$$

$$M_1' = \mu_1'(X,Y), \quad M_2' = \mu_2'(X,Y) \quad \text{as} \quad M = m(X,Y), \quad N = n(X,Y).$$

Then carrying out the usual first order Taylor expansions,

$$\left.\begin{aligned} M - m(\mu,\nu) &= (X-\mu)m_\mu + (Y-\nu)m_\nu \\ N - n(\mu,\nu) &= (X-\mu)n_\mu + (Y-\nu)n_\nu \end{aligned}\right\} \text{ to first order,} \tag{B}$$

where $$m_\mu = \left.\frac{\partial m(X,Y)}{\partial X}\right|_{X=\mu, Y=\nu}, \text{ etc.}$$

Thus, for starter values μ_1, ν_1 (obtained for example for a zero-truncated distribution by assuming a zero class frequency F_0 and using the corresponding equations, if they are simple enough, for the untruncated case), second approximations μ_2, ν_2 are obtained by solving

$$\left.\begin{aligned} M - m_1 &= (\mu_2 - \mu_1)n_{\mu_1} + (\nu_2 - \nu_1)m_{\nu_1} \\ N - n_1 &= (\mu_2 - \mu_1)n_{\mu_1} + (\nu_2 - \nu_1)n_{\nu_1} \end{aligned}\right\}$$

where $m_1 = m(\mu_1, \nu_1)$ and $m_{\mu_1} = \left.\frac{\partial m(X,Y)}{\partial X}\right|_{X=\mu_1, Y=\nu_1}$, etc.,

i.e., as

$$\mu_2 = \mu_1 + \frac{(M-m_1)n_{\nu_1} - (N-n_1)m_{\nu_1}}{m_{\mu_1} n_{\nu_1} - m_{\nu_1} n_{\mu_1}}$$

$$\nu_2 = \nu_1 + \frac{(N-n_1)m_{\mu_1} - (M-m_1)n_{\mu_1}}{m_{\mu_1} n_{\nu_1} - m_{\nu_1} n_{\mu_1}},$$

with iteration if necessary. The variances and covariance follow by squaring the equations (B) and taking expectations, giving (to this order)

$$\mathcal{V}(M) = m_\mu^2 \, \mathcal{V}(X) + 2m_\mu m_\nu \mathcal{C}(X,Y) + m_\nu^2 \, \mathcal{V}(Y)$$

and
$$\mathcal{V}(N) = n_\mu^2 \, \mathcal{V}(X) + 2n_\mu n_\nu \mathcal{C}(X,Y) + n_\nu^2 \, \mathcal{V}(Y),$$

while multiplying them and taking expectations gives

$$\mathcal{C}(M,N) = m_\mu n_\mu \, \mathcal{V}(X) + (m_\mu n_\nu + m_\nu n_\mu) \, \mathcal{C}(X,Y) + m_\nu n_\nu \, \mathcal{V}(Y).$$

The solutions of these three equations can then be given explicitly as follows, where the original notation is partly returned to:

$$\delta^2 \mathcal{V}(\mathring{\mu}) = n_\nu^2 \mathcal{V}(M_1') - 2m_\nu n_\nu \mathcal{C}(M_1', M_2') + m_\nu^2 \, \mathcal{V}(M_2'),$$

$$\delta^2 \mathcal{C}(\mathring{\mu}, \mathring{\nu}) = -n_\nu n_\mu \mathcal{V}(M_1') + (m_\nu n_\mu + m_\mu n_\nu) \mathcal{C}(M_1', M_2') - m_\nu m_\mu \, \mathcal{V}(M_2'),$$

$$\delta^2 \mathcal{V}(\mathring{\nu}) = n_\mu^2 \, \mathcal{V}(M_1') - 2m_\mu n_\mu \, \mathcal{C}(M_1', M_2') + m_\mu^2 \, \mathcal{V}(M_2'),$$

with $\delta = m_\nu n_\mu - m_\mu n_\nu$.

However, if a high speed computer is available it is preferable even for this 2 parameter case to work entirely matrix-wise. In general, for k parameters $\lambda_1, \lambda_2, \ldots, \lambda_k$, writing $(\lambda_1, \lambda_2, \ldots, \lambda_k) = \underset{\sim}{\lambda}^t$ with ^{t} for transpose, the k moment equations are

$$\left.\begin{aligned} M_1' &= \mu_1'(\overset{o}{\underset{\sim}{\lambda}}) \\ &\cdots \\ M_k' &= \mu_k'(\overset{o}{\underset{\sim}{\lambda}}) \end{aligned}\right\} \quad \text{or} \quad \underset{\sim}{M}' = \underset{\sim}{\mu}'(\overset{o}{\underset{\sim}{\lambda}}).$$

These have the first order Taylor expansion (see the Exercises: the argument is the natural and obvious extension of the two variable case worked out in some detail above)

$$\underset{\sim}{\mu}'(\overset{o}{\underset{\sim}{\lambda}}) = \underset{\sim}{\mu}'(\underset{\sim}{\lambda}) + \frac{\partial \underset{\sim}{\mu}}{\partial \underset{\sim}{\lambda}} (\overset{o}{\underset{\sim}{\lambda}} - \underset{\sim}{\lambda}).$$

So $$\overset{o}{\underset{\sim}{\lambda}} = \underset{\sim}{\lambda} - \left(\frac{\partial \underset{\sim}{\mu}'}{\partial \underset{\sim}{\lambda}}\right)^{-1} (\underset{\sim}{\mu}'(\underset{\sim}{\lambda}) - \underset{\sim}{M}') \quad \text{to first order,}$$

whence the iteration formula

$$\underset{\sim}{\lambda}_{r+1} = \underset{\sim}{\lambda}_r - \left(\frac{\partial \underset{\sim}{\mu}'}{\partial \underset{\sim}{\lambda}}\right)^{-1}\Bigg|_{\underset{\sim}{\lambda}_r} (\underset{\sim}{\mu}'(\underset{\sim}{\lambda}_r) - \underset{\sim}{M}'), \quad r = 1, 2, \ldots,$$

can be used, given the starter values $\underset{\sim}{\lambda}_1$. Also, taking the transpose of the Taylor expansion, multiplying the original and transposed equations and taking expectations, leads to

$$\underset{\sim}{\mathcal{V}}(\underset{\sim}{M}') = \frac{\partial \underset{\sim}{\mu}'}{\partial \underset{\sim}{\lambda}} \underset{\sim}{\mathcal{V}}(\overset{o}{\underset{\sim}{\lambda}}) \left(\frac{\partial \underset{\sim}{\mu}'}{\partial \underset{\sim}{\lambda}}\right)^t$$

or

$$\underset{\sim}{\mathcal{V}}(\overset{o}{\underset{\sim}{\lambda}}) = \left(\frac{\partial \underset{\sim}{\mu}'}{\partial \underset{\sim}{\lambda}}\right)^{-1} \underset{\sim}{\mathcal{V}}(\underset{\sim}{M}') \left[\left(\frac{\partial \underset{\sim}{\mu}'}{\partial \underset{\sim}{\lambda}}\right)^{-1}\right]^t$$

to first order, the (i,j)-th element, $\mathcal{C}(M_i', M_j')$, of $\underset{\sim}{\mathcal{V}}(\underset{\sim}{M}')$ having been given explicitly above. All these operations would be carried out within the computer, and it is convenient that the same Jacobian matrix $\partial\underset{\sim}{\mu}'/\partial\underset{\sim}{\lambda}$ appears both in the iteration and covariance formulae; further, $\partial\underset{\sim}{\mu}'/\partial\underset{\sim}{\lambda} = (\partial\underset{\sim}{\mu}'/\partial\underset{\sim}{\kappa}_{(\,)})\cdot(\partial\underset{\sim}{\kappa}_{(\,)}/\partial\underset{\sim}{\lambda})$ is often useful because of the frequent simplicity of factorial cumulant expressions.

Whatever means may be adopted to find $\mathcal{V}(\overset{o}{\lambda})$, although $\mathcal{V}(M')$ involves moments of degree up to 2k, these moments need not to be estimated directly by the relatively highly variable sample moments; instead, the distributiom moments may be expressed in terms of the k parameters, and the estimates of these used to estimate $\mathcal{V}(M')$, in a sense making the estimation more model dependent. If, for example, the factorial cumulants have simple expressions in terms of the parameters, it will be convenient to use them and then use the Jacobian product immediately above - perhaps stored in a computer - to convert to moments.

In practice these expressions are used by supposing the distribution of $\overset{o}{\mu}$, e.g., to be characterized by its (approximate) mean and variance, and then comparing these with the results of similar investigations for other estimators.

Moment estimators are usually probability consistent (i.e., the probability that a moment estimator deviates from its parameter value by any specified amount tends to zero as the sample size n increases), Fisher consistent (replacement of the F_x by their expectations produces identities), usually unbiased to zero order with correction terms of order $1/n$, usually have sampling variances with leading terms of order $1/n$ obtained from expansions carried out as above, and are usually asymptotically normally distributed.

Comparisons with m.l. estimators

For a single parameter λ, if $\hat{\lambda}$ is the m.l. estimator and T some other at least asymptotically unbiased estimator, based on a random sample of size n, with

$$\mathcal{V}(\hat{\lambda}) = \frac{\alpha_1}{n} + \frac{\alpha_2}{n^2} + \ldots, \quad \mathcal{V}(T) = \frac{\beta_1}{n} + \frac{\beta_2}{n^2} + \ldots$$

(asymptotic normality is also usually required), then "the" efficiency - strictly, the first order asymptotic efficiency - of T relative to $\hat{\lambda}$ is conventionally defined as

$$\frac{\alpha_1}{\beta_1} \quad \left(\text{or } 100\,\frac{\alpha_1}{\beta_1} \text{ as a percentage}\right).$$

A second order efficiency may be correspondingly defined as

$$\frac{\alpha_1/n + \alpha_2/n^2}{\beta_1/n + \beta_2/n^2} = \frac{\alpha_1}{\beta_1} \cdot \frac{1 + \dfrac{\alpha_2/\alpha_1}{n}}{1 + \dfrac{\beta_2/\beta_1}{n}} = (\text{"the" effic.}) \times (2^{\text{nd}} \text{ order term}).$$

(The implicit assumptions, that expansions of this form exist, that commonly $\alpha_1 < \beta_1$, that the bias is negligible and that the variance is the appropriate measure, may be noted.) For several parameters, the generalized variance, defined as the determinant of the corresponding variance-covariance matrix, is used: e.g., for

$$\begin{vmatrix} \mathscr{V}(\hat{\mu}), & \mathscr{C}(\hat{\mu},\hat{\nu}) \\ \mathscr{C}(\hat{\mu},\hat{\nu}) & \mathscr{V}(\hat{\nu}) \end{vmatrix} = \frac{\varepsilon_2}{n^2} + \frac{\varepsilon_3}{n^3} + \ldots,$$

$$\begin{vmatrix} \mathscr{V}(\mathring{\mu}) & \mathscr{C}(\mathring{\mu},\mathring{\nu}) \\ \mathscr{C}(\mathring{\mu},\mathring{\nu}) & \mathscr{V}(\mathring{\nu}) \end{vmatrix} = \frac{\phi_2}{n^2} + \frac{\phi_3}{n^3} + \ldots,$$

"the" efficiency (first order asymptotic efficiency) of the moment estimators is conventionally defined to be

$$\frac{\varepsilon_2}{\phi_2},$$

with a corresponding second order expression.

Method of Maximum Likelihood

For the random sample $F_0, F_1, \ldots, F_x, \ldots,$ these being the frequencies corresponding to $x = 0, 1, \ldots, x, \ldots$ respectively, from the distribution with probability function p_x, the likelihood is

$$L = p_0^{F_0} p_1^{F_1} \ldots p_x^{F_x} \ldots$$

In the two parameter case $p_x \equiv p_x(\mu,\nu)$, say, the functions $\hat{\mu}, \hat{\nu}$ which maximize L over variations in μ and ν are called the maximum likelihood, or m.l., estimators of μ, ν (R. A. Fisher). In practice, standard calculus techniques are used, and it is more convenient to deal with

$$\ln L = \Sigma F_x \ln p_x;$$

the m.l. estimators are then (hopefully) the roots $\hat{\mu}, \hat{\nu}$ of the m.l. equations

$$\left.\begin{aligned} \left(\frac{\partial \ln L}{\partial \mu}\right)_{\mu=\hat{\mu},\ \nu=\hat{\nu}} &= 0 \\ \left(\frac{\partial \ln L}{\partial \nu}\right)_{\mu=\hat{\mu},\ \nu=\hat{\nu}} &= 0 \end{aligned}\right\} .$$

(That a maximum is obtained is often not investigated: it may be examined through second order derivatives or by appeal to the general theory (e.g. Cramer (1947)), although since L is a product of probabilities it is positive, bounded above and often a very well behaved function with exactly one root of the likelihood equations.)

Under rather general conditions, m.l. estimators are at most biased to order $1/n$ where $n = \Sigma F_x$, consistent (both probability, and Fisher), sufficient when sufficient estimators exist, and possess relatively simple formulae for first order sampling variances and the covariance. The m.l. estimator of a function of a parameter is that function of the m.l. estimator of the parameter. Further, an asymptotically normal sampling distribution is often obtained. (For this property, particularly, the general theory requires the satisfaction of the m.l. equations: a maximum at which the derivatives do not vanish need not possess an asymptotically normal distribution for the estimators so obtained, e.g.) Asymptotically, too, given regularity conditions no estimator can have a smaller sampling variance than the m.l. estimator (this is not an assertion about samples of any given sample size).

A brief outline of the one parameter case will serve to recall without too much complexity the basic procedures which are also applied in multiparameter circumstances. For the single parameter λ, the m.l. equation is

$$\left.\frac{\partial \ln L}{\partial \lambda}\right|_{\hat{\lambda}} = 0.$$

Only exceptionally can this be solved explicitly, and so iterative procedures must generally be used, as outlined in the moment estimation section, and commented on in the Appendix on Iterative Methods. The results will be exhibited in the notation convenient for the present formulation, but they correspond very closely to those already obtained, the method (of Taylor expansion) being essentially the same. Expanding the left side of the m.l. equation gives

$$\frac{\partial \ln L}{\partial \lambda} + (\hat{\lambda} - \lambda)\frac{\partial^2 \ln L}{\partial \lambda^2} + \ldots = 0$$

or

$$\hat{\lambda} = \lambda - \frac{\partial \ln L/\partial \lambda}{\frac{\partial}{\partial \lambda}(\partial \ln L/\partial \lambda)}$$

on neglecting the higher order terms: for a starter value λ_1

and writing $\partial \ln L/\partial\lambda = H(\lambda)$ the Newton-Raphson iterative scheme is therefore

$$\lambda_{r+1} = \lambda_r - \frac{H(\lambda_r)}{H'(\lambda_r)}, \qquad r = 1, 2, \ldots,$$

where λ_r (usually even for $r = 1$) is a random variable, in spite of the notation. Paying more attention to the details of the Taylor expansion, it can be shown that

$$\mathscr{E}\left(\frac{\partial \ln L}{\partial\lambda}\right) = 0$$

and $$\mathscr{V}(\hat{\lambda}) = \frac{-1}{\mathscr{E}\left(\dfrac{\partial^2 \ln L}{\partial\lambda^2}\right)} = \frac{1}{\mathscr{E}\left(\dfrac{\partial \ln L}{\partial\lambda}\right)^2}, \text{ to } O(1/n),$$

$-(\partial^2 \ln L/\partial\lambda^2)$ being called the "information" of the sample. An alternative iteration procedure, called the "method of scoring", uses

$$\lambda_{r+1} = \lambda_r - \frac{H(\lambda_r)}{\mathscr{E}\{H'(\lambda_r)\}}$$

$$= \lambda_r + \mathscr{V}(\hat{\lambda})_{\lambda=\lambda_r} \cdot H(\lambda_r).$$

For some cautions in the use of such iterative methods in a statistical context, see Barnett (1966).

Returning to the two parameter case, μ and ν, it is then even more exceptional to be able to solve explicitly the m.l. equations. However, it is sometimes possible to eliminate one of the estimators, say $\hat{\mu}$, from the equations to obtain an implicit equation, say $H(\hat{\nu}) = 0$, for the other. This one variable equation can then be treated as above: given a starter value ν_1 the iterature formula is

$$\nu_{r+1} = \nu_r - \frac{H(\nu_r)}{H'(\nu_r)}, \qquad r = 1, 2, \ldots,$$

and again $\mathscr{E}\{H'(\nu)\}$ may be used in place of $H'(\nu)$. In general such elimination is not possible, and, even if it is, with a high speed computer a direct approach is easy to carry out provided convergence difficulties are not encountered. Writing the matrix of negative second derivatives as

$$\underset{\sim}{I} = \begin{bmatrix} -\dfrac{\partial^2 \ln L}{\partial\mu^2}, & -\dfrac{\partial^2 \ln L}{\partial\mu\partial\nu} \\ \\ -\dfrac{\partial^2 \ln L}{\partial\mu\partial\nu}, & -\dfrac{\partial^2 \ln L}{\partial\nu^2} \end{bmatrix}$$

and carrying out first order expansions of the m.l. equations leads to

$$\left.\begin{aligned} -(\hat{\mu}-\mu)\frac{\partial^2 \ln L}{\partial\mu^2} - (\hat{\nu}-\nu)\frac{\partial^2 \ln L}{\partial\mu\partial\nu} &= \frac{\partial \ln L}{\partial\mu} \\ -(\hat{\mu}-\mu)\frac{\partial^2 \ln L}{\partial\mu\partial\nu} - (\hat{\nu}-\nu)\frac{\partial^2 \ln L}{\partial\nu^2} &= \frac{\partial \ln L}{\partial\nu} \end{aligned}\right\}$$

or $$\underset{\sim}{I} \cdot \begin{pmatrix} \hat{\mu}-\mu \\ \hat{\nu}-\nu \end{pmatrix} = \begin{pmatrix} \partial \ln L/\partial\mu \\ \partial \ln L/\partial\nu \end{pmatrix},$$

whence $$\begin{pmatrix} \hat{\mu}-\mu \\ \hat{\nu}-\nu \end{pmatrix} = \underset{\sim}{I}^{-1} \cdot \begin{pmatrix} \partial \ln L/\partial\mu \\ \partial \ln L/\partial\nu \end{pmatrix}.$$

(The generalization to more than 2 parameters is obvious.) Hence, introducing $\underset{\sim}{\lambda}$ for the parameter vector $(\mu,\nu)^t$, and showing explicitly the argument $\underset{\sim}{\lambda}$ for $\underset{\sim}{I}$ and $\partial \ln L/\partial\underset{\sim}{\lambda}$, the iterative scheme is

$$\underset{\sim}{\lambda}_{r+1} = \underset{\sim}{\lambda}_r + \underset{\sim}{I}(\underset{\sim}{\lambda}_r)^{-1} \cdot \left.\frac{\partial \ln L}{\partial\underset{\sim}{\lambda}}\right|_{\underset{\sim}{\lambda}=\underset{\sim}{\lambda}_r}.$$

Since it can be shown that the covariance matrix

$$\underset{\sim}{\mathscr{V}}(\hat{\underset{\sim}{\lambda}}) = \begin{pmatrix} \mathscr{V}(\hat{\mu}) & ,\mathscr{C}(\hat{\mu},\hat{\nu}) \\ \mathscr{C}(\hat{\mu},\hat{\nu}), & \mathscr{V}(\hat{\nu}) \end{pmatrix}$$

is given to $O(1/n)$ by the inverse of the "information matrix" $\underset{\sim}{\mathscr{I}}$, where

$$\underset{\sim}{\mathscr{I}} = \mathscr{E}(\underset{\sim}{I}),$$

i.e.,

$$\begin{pmatrix} \mathscr{V}(\hat{\mu}) & \mathscr{C}(\hat{\mu},\hat{\nu}) \\ \mathscr{C}(\hat{\mu},\hat{\nu}), & \mathscr{V}(\hat{\nu}) \end{pmatrix} = \begin{pmatrix} -\mathscr{E}\left(\frac{\partial^2 \ln L}{\partial\mu^2}\right), & -\mathscr{E}\left(\frac{\partial^2 \ln L}{\partial\mu\partial\nu}\right) \\ -\mathscr{E}\left(\frac{\partial^2 \ln L}{\partial\mu\partial\nu}\right), & -\mathscr{E}\left(\frac{\partial^2 \ln L}{\partial\nu^2}\right) \end{pmatrix}^{-1} \text{ to } O(1/n),$$

the "method of scoring" leads to the iteration formula

$$\underset{\sim}{\lambda}_{r+1} = \underset{\sim}{\lambda}_r + \underset{\sim}{\mathscr{V}}(\hat{\underset{\sim}{\lambda}})\Big|_{\underset{\sim}{\lambda}=\underset{\sim}{\lambda}_r} \cdot \frac{\partial \ln L}{\partial \underset{\sim}{\lambda}}\Big|_{\underset{\sim}{\lambda}=\underset{\sim}{\lambda}_r} .$$

(It is sometimes helpful to recognize alternative expressions for the elements of the information matrix, such as

$$\mathscr{E}\left(\frac{\partial^2 \ln L}{\partial\mu\partial\nu}\right) = -\mathscr{E}\left(\frac{\partial \ln L}{\partial\mu} \cdot \frac{\partial \ln L}{\partial\nu}\right) = n\mathscr{E}\left(\frac{\partial^2 \ln p_x}{\partial\mu\partial\nu}\right)$$

$$= -n\mathscr{E}\left(\frac{\partial \ln p_x}{\partial\mu} \cdot \frac{\partial \ln p_x}{\partial\nu}\right) ;$$

for the background theory and extensions, reference may be made to Rao (1965).) All these calculations are best left in matrix form for computer application, and the convenience that they are expressed in terms of matrices used both for iteration and variance-covariance estimation should be noticed.

Attention should also be drawn to general minimizing/maximizing procedures, with an exemplification by Ross (1970) in which emphasis is laid on the use of transformations which when appropriate may substantially simplify both numerical analyses and their interpretation.

4.3 MINIMUM CHI-SQUARED ESTIMATION (GENERALIZED LEAST SQUARES)

Since most of the desirable properties of the sampling distribution of m.l. estimators are asymptotic (in the sample size), other estimators will also possess these: comparisons for small or moderate sample sizes must be made on other than first order asymptotic results. Since m.l. estimators are often rather complex to find in specific cases, a good deal of attention has been directed to alternative classes of "best asymptotical normal" (BAN) estimators, and one such class is that of minimum chi-squared estimators.

The procedure and background are briefly outlined because of their relative inaccessibility.

Motivation and Basic Formulae

If observed frequencies $F_0, F_1, F_2, \ldots, F_c$ and expected or fitted frequencies $\phi_0, \phi_1, \phi_2, \ldots, \phi_c$ are to be compared, the usual chi-squared goodness of fit expression is

$$\Sigma(F_x - \phi_x)^2/\phi_x = \Sigma(F_x - \phi_x)\ \phi_x^{-1}\ (F_x - \phi_x);$$

this can be written in matrix form as

$$\begin{bmatrix} F_0 - \phi_0 \\ F_1 - \phi_1 \\ \vdots \\ F_c - \phi_c \end{bmatrix}^t \begin{bmatrix} \phi_0 & 0 & \cdots & 0 \\ 0 & \phi_1 & \cdots & 0 \\ \cdots & \cdots & \cdots & \cdots \\ 0 & 0 & \cdots & \phi_c \end{bmatrix}^{-1} \begin{bmatrix} F_0 - \phi_0 \\ F_1 - \phi_1 \\ \vdots \\ F_c - \phi_c \end{bmatrix} \qquad \text{(A)}$$

$$= (\underset{\sim}{F} - \underset{\sim}{\phi})^t\ (\text{diag}\ \underset{\sim}{\phi})^{-1}(\underset{\sim}{F} - \underset{\sim}{\phi}),$$

where ^{t} denotes the transpose. In this, the vector $\underset{\sim}{\phi} = (\phi_0, \phi_1, \ldots, \phi_c)^t$ is the expectation of $\underset{\sim}{F} = (F_0, F_1, \ldots, F_c)^t$ at least asymptotically and the diagonal matrix is essentially one of the variances of the elements of $\underset{\sim}{F}$. (It may be recalled that the early derivations of the chi-squared goodness of fit test contained a step in which the frequencies were supposed Poisson, with the consequence that their variances were equal to their expectations.) It is clearly possible to choose estimators of the parameters in the ϕ_x in such a way as to minimize (whence the name <u>generalized least squares</u>) this usual chi-squared quadratic form, and if chi-squared were actually being used as the criterion of goodness of fit the procedure would be the logical choice (See Neyman (1949), Rao (1957), and an account in Kendall and Stuart II (1967).)

But there is an alternative function, also asymptotically chi-squared distributed, which is used in a similar way. If $\underset{\sim}{T} = (T_1, T_2, \ldots, T_m)^t$ is multivariate normally distributed with mean vector $\underset{\sim}{\mathscr{E}}(\underset{\sim}{T}) = \underset{\sim}{\tau} = (\tau_1, \tau_2, \ldots, \tau_m)^t$ and non-singular covariance matrix

$$\underset{\sim}{\mathscr{V}}(\underset{\sim}{T}) = \underset{\sim}{\mathscr{E}}\{(\underset{\sim}{T} - \underset{\sim}{\tau})(\underset{\sim}{T} - \underset{\sim}{\tau})^t\}$$

$$= \begin{bmatrix} \mathscr{V}(T_1) & \mathscr{C}(T_1,T_2) & \cdots & \mathscr{C}(T_1,T_m) \\ \mathscr{C}(T_2,T_1) & \mathscr{V}(T_2) & \cdots & \mathscr{C}(T_2,T_m) \\ & \cdots\cdots\cdots\cdots & & \\ \mathscr{C}(T_m,T_1) & \mathscr{C}(T_m,T_2) & \cdots & \mathscr{V}(T_m) \end{bmatrix} ,$$

it is also the case that the quadratic form

$$(\underset{\sim}{T} - \underset{\sim}{\tau})^t \; \underset{\sim}{\mathscr{V}}(\underset{\sim}{T})^{-1} \; (\underset{\sim}{T} - \underset{\sim}{\tau}) \qquad \text{(B)}$$

is (exactly) distributed as chi-squared with m degrees of freedom. For many sets of statistics $\underset{\sim}{T}$, such as the sample moments up to order m, the distribution of $\underset{\sim}{T}$ is at least asymptotically normal whatever the distribution being sampled (subject to mild regularity conditions), and the quadratic form is then asymptotically chi-squared distributed. Writing the parameter vector in the parent distribution as $\underset{\sim}{\lambda} = (\lambda_1, \lambda_2, \cdots, \lambda_k)^t$ $k \leq m$, and the relations between $\underset{\sim}{\tau}$ and $\underset{\sim}{\lambda}$ as $\underset{\sim}{\tau} = \underset{\sim}{\tau}(\underset{\sim}{\lambda})$, the quadratic form can be minimized for variations in $\underset{\sim}{\lambda}$, and Barankin and Gurland (1951) have shown that, among regular estimators which are functions of $\underset{\sim}{T}$ only, the functions of $\underset{\sim}{T}$ so found - say $\underset{\sim}{\lambda}^*$ - are asymptotically minimum variance unbiased at least for the exponential family. Such functions may thus be called <u>minimum chi-squared estimators</u>; also, any convenient one-one function (subject to regularity conditions) of $\underset{\sim}{T}$ and any consistent estimator $\underset{\sim}{V}$ or estimate $\underset{\sim}{v}$ of $\underset{\sim}{\mathscr{V}}(\underset{\sim}{T})$ may be used without changing the asymptotic properties. ($\underset{\sim}{v}$ is obtained by replacing values of the parameter $\underset{\sim}{\tau}$ by the values of consistent estimators of them; the estimators of $\underset{\sim}{\lambda}$ should perhaps then be called <u>modified minimum chi-squared estimators</u>, or <u>generalized (weighted) least squares estimators</u>.)

Thus, minimum chi-squared estimators could be described either as those which minimize (A), which is approximately chi-squared distributed, or those which minimize a modification of (B), such as

$$\underset{\sim}{Q} = (\underset{\sim}{T} - \underset{\sim}{\tau})^t \underset{\sim}{v}^{-1} (\underset{\sim}{T} - \underset{\sim}{\tau}) ,$$

also approximately chi-squared. The former generally leads to rather complicated procedures, and it is the latter which is here described in detail: its theoretical background and further details will be found in Barankin and Gurland (1951) and Ferguson (1958), while in the present form it is due to Katti and Gurland (1962) and Hinz and Gurland (1967). The high generality of the method should be noticed, because of the choice available in the selection of $\underset{\sim}{T}$: e.g., in the case where the dimension of $\underset{\sim}{T}$ is the same as that of $\underset{\sim}{\lambda}$, a suitable selection (of sample moments) will lead to the method of moments, or to the 'method of zero frequency and first moment'. (It may be helpful to remark that the application of minimum chi-squared methods to the Neyman distribution, though not independent of the outline which follows these remarks, is described in more detail than for any of the other distributions.) In early applications of these results the equations obtained were non-linear - this made the theory straightforward but the numerical analysis rather intractable, and as a result later work has been directed to exploiting the generality of the method by looking specifically for linear equations.

The Linear Minimum Chi-squared Equation, and Sampling Variances.

Suppose the probability function of $X = 0,1,2,\ldots$ is indexed by the parameter vector $\underset{\sim}{\lambda} = (\lambda_1, \lambda_2, \ldots, \lambda_k)^t$, where this is to be estimated from a random sample with frequencies $F_0, F_1, F_2, \ldots;\ \Sigma F_x = n \geq k$. It will usually be the case that for a set $\underset{\sim}{S}$ of k statistics (e.g., the zero class frequency and the first $k-1$ moments), for which $\mathscr{E}(\underset{\sim}{S}) = \underset{\sim}{\sigma}(\underset{\sim}{\lambda})$, say, a function of $\underset{\sim}{\lambda}$ determined by the choice of $\underset{\sim}{S}$, the minimization of the chi-squared quadratic form

$$(\underset{\sim}{S} - \underset{\sim}{\sigma}(\underset{\sim}{\lambda}))^t \{\text{est. } \mathscr{V}(\underset{\sim}{S})\}^{-1} (\underset{\sim}{S} - \underset{\sim}{\sigma}(\underset{\sim}{\lambda}))$$

with respect to $\underset{\sim}{\lambda}$ will lead to non-linear equations for the estimators $\underset{\sim}{\lambda}^*$. And in the usual cases of complex distributions, k sufficient statistics for $\underset{\sim}{\lambda}$ do not exist.

If possible, it is therefore advantageous to re-parameterize the probability function in terms of $\underset{\sim}{\theta} = (\theta_1, \ldots, \theta_k)^t$ and to choose a corresponding statistic $\underset{\sim}{T} = (t_1, \ldots, t_m)^t$ whose (at least) asymptotic expectation $\underset{\sim}{\tau} = \mathscr{E}(\underset{\sim}{T})$ is linearly related to

$\underset{\sim}{\theta}$: say $\underset{\sim}{\tau} = \underset{\sim}{w}\,\underset{\sim}{\theta}$ $\quad m \geq k$, where $\underset{\sim}{w}$ is a constant matrix. These are thus linear equations, leading (below) correspondingly to linear minimum chi-squared (normal) equations. Supposing the relations between $\underset{\sim}{\lambda}$ and $\underset{\sim}{\theta}$ are written $\underset{\sim}{\lambda} = \underset{\sim}{\lambda}(\underset{\sim}{\theta})$, and the estimators of $\underset{\sim}{\theta}$ are written $\underset{\sim}{\theta}^*$, it may need some care in the original choices of $\underset{\sim}{\theta}$ and $\underset{\sim}{T}$ to make the numerical determination of $\underset{\sim}{\lambda}^*$ from $\underset{\sim}{\lambda}^* = \underset{\sim}{\lambda}(\underset{\sim}{\theta}^*)$ equally straightforward; but examples where this is so are given in appropriate sections, originating with Gurland (1965), Hinz and Gurland (1967) and Rao and Katti (1970).

In general, then, the chi-squared distribution quadratic form to be minimized with respect to $\underset{\sim}{\theta}$ is

$$Q = (\underset{\sim}{T} - \underset{\sim}{w}\underset{\sim}{\theta})^t\ \underset{\sim}{v}^{-1}\ (\underset{\sim}{T} - \underset{\sim}{w}\underset{\sim}{\theta}) \quad \text{or} \quad (\underset{\sim}{T} - \underset{\sim}{w}\underset{\sim}{\theta})^t\ \underset{\sim}{V}^{-1}\ (\underset{\sim}{T} - \underset{\sim}{w}\underset{\sim}{\theta}),$$

where $\underset{\sim}{V}$ is a consistent estimator of $\underset{\sim}{\mathcal{V}}(\underset{\sim}{T})$ and $\underset{\sim}{v}$ an estimate of it. It is convenient to write $\underset{\sim}{V}^{-1} = \underset{\sim}{I}$ and $\underset{\sim}{v}^{-1} = \underset{\sim}{i}$, the notation being suggested by the information matrix. Hence

$$\frac{\partial Q}{\partial \underset{\sim}{\theta}} = \left(\frac{\partial\ \underset{\sim}{w}\underset{\sim}{\theta}}{\partial \underset{\sim}{\theta}}\right)^t \frac{\partial Q}{\partial\ \underset{\sim}{w}\underset{\sim}{\theta}} = -2\underset{\sim}{w}^t\ \underset{\sim}{i}(\underset{\sim}{T} - \underset{\sim}{w}\underset{\sim}{\theta}) \ ,$$

and so the linear minimum chi-squared equations are

$$\underset{\sim}{w}^t\ \underset{\sim}{i}\ \underset{\sim}{T} = \underset{\sim}{w}^t\ \underset{\sim}{i}\ \underset{\sim}{w}\ \underset{\sim}{\theta}^*,$$

i.e., $$\underset{\sim}{\theta}^* = (\underset{\sim}{w}^t\ \underset{\sim}{i}\ \underset{\sim}{w})^{-1}\ \underset{\sim}{w}^t\ \underset{\sim}{i}\ \underset{\sim}{T}, \qquad [1mc(A)]$$

the usual generalized least squares normal equations. (The vector differentiation results needed are given in the Exercises; see these also for some justification of this and the following results when $\underset{\sim}{I}$ is used instead of $\underset{\sim}{i}$.)

The first order covariance matrix is obtained immediately:

$$\underset{\sim}{\mathcal{V}}(\underset{\sim}{\theta}^*) = (\underset{\sim}{w}^t\ \underset{\sim}{\mathcal{V}}(\underset{\sim}{T})^{-1}\ \underset{\sim}{w})^{-1}, \qquad [1mc(B)]$$

and hence, returning to the parameterization $\underset{\sim}{\lambda}$, the first order result

$$\underset{\sim}{\mathcal{V}}(\underset{\sim}{\lambda}^*) = \frac{\partial \underset{\sim}{\lambda}}{\partial \underset{\sim}{\theta}}\ \underset{\sim}{\mathcal{V}}(\underset{\sim}{\theta}^*) \left(\frac{\partial \underset{\sim}{\lambda}}{\partial \underset{\sim}{\theta}}\right)^t \qquad [1mc(C)]$$

follows from the standard Taylor expansion

$$\underset{\sim}{\lambda}^* = \underset{\sim}{\lambda}(\underset{\sim}{\theta}^*) = \underset{\sim}{\lambda}(\underset{\sim}{\theta}) + \frac{\partial \underset{\sim}{\lambda}}{\partial \underset{\sim}{\theta}} (\underset{\sim}{\theta}^* - \underset{\sim}{\theta}) \quad \text{to first order.}$$

Hence, rearranging, and post-multiplying by the transpose,

$$\{\underset{\sim}{\lambda}^* - \underset{\sim}{\lambda}(\underset{\sim}{\theta})\}\{\underset{\sim}{\lambda}^* - \underset{\sim}{\lambda}(\underset{\sim}{\theta})\}^t = \frac{\partial \underset{\sim}{\lambda}}{\partial \underset{\sim}{\theta}} (\underset{\sim}{\theta}^* - \underset{\sim}{\theta})(\underset{\sim}{\theta}^* - \underset{\sim}{\theta})^t \left(\frac{\partial \underset{\sim}{\lambda}}{\partial \underset{\sim}{\theta}}\right)^t$$

and taking expectations gives (ℓmc[C]).

A systematic treatment of $\underset{\sim}{\mathcal{V}}(\underset{\sim}{T})$ is therefore required both here and for $\underset{\sim}{i}$ in (ℓmc[A]). For $\underset{\sim}{M}'$ the vector of sample moments about the origin.

$$\underset{\sim}{M}' = (M_1', M_2', \ldots, M_m')^t,$$

$$\mathcal{C}(M_i', M_j') = \frac{1}{n} (\mu_{i+j}' - \mu_i' \mu_j')$$

where

$$\mathcal{E}(\underset{\sim}{M}') = \underset{\sim}{\mu}' = (\mu_1', \mu_2', \ldots, \mu_m')^t$$

and

$$\underset{\sim}{\tau} = \underset{\sim}{\tau}(\underset{\sim}{\mu}'),$$

the usual expressions apply:

$$\underset{\sim}{\mathcal{V}}(\underset{\sim}{T}) = \frac{\partial \underset{\sim}{\tau}}{\partial \underset{\sim}{\mu}'} \underset{\sim}{\mathcal{V}}(\underset{\sim}{M}') \left(\frac{\partial \underset{\sim}{\tau}}{\partial \underset{\sim}{\mu}'}\right)^t$$

and, if it is convenient, a chain of transformations

$$\frac{\partial \underset{\sim}{\tau}}{\partial \underset{\sim}{\mu}'} = \frac{\partial \underset{\sim}{\tau}}{\partial \underset{\sim}{\sigma}} \cdot \frac{\partial \underset{\sim}{\sigma}}{\partial \underset{\sim}{\mu}'}$$

where $\underset{\sim}{\sigma}$ is an appropriate intermediate function (e.g., involving factorial cumulants).

In applications, the matrices will be estimated by using estimates of the elements, and as usual the first order generalized variance is the determinant of the first order covariance matrix.

To apply these general results, it is convenient to use the fact that factorial cumulants take simple forms for a number of contagious distributions, and their ratios are even simpler. Write then

$$\eta_0 = \kappa_{(1)} \ , \ \eta_r = \frac{\kappa_{(r+1)}}{\kappa_{(r)}} \ , \qquad r = 1,2,\ldots$$

and $\underset{\sim}{\eta} = (\eta_0, \eta_1, \cdots, \eta_{m-1})^t, \equiv \underset{\sim}{\tau}$ in the earlier notation with $\underset{\sim}{H}$ defined analogously in terms of sample factorial cumulants $K_{(r)}$, and $\underset{\sim}{\mathscr{E}}(\underset{\sim}{H}) = \underset{\sim}{\eta}_r$ at least asymptotically. ($\underset{\sim}{\eta}$ and $\underset{\sim}{\theta}$ are chosen so that a constant $\underset{\sim}{w}$ can be found such that $\underset{\sim}{\eta} = \underset{\sim}{w}\,\underset{\sim}{\theta}$.) Using the transformation formulae already written down,

$$\underset{\sim}{\mathscr{V}}(\underset{\sim}{H}) = \frac{\partial \underset{\sim}{\eta}}{\partial \underset{\sim}{\kappa}_{(\ )}} \cdot \frac{\partial \underset{\sim}{\kappa}_{(\ )}}{\partial \underset{\sim}{\mu}'} \underset{\sim}{\mathscr{V}}(\underset{\sim}{M}') \left(\frac{\partial \underset{\sim}{\eta}}{\partial \underset{\sim}{\kappa}_{(\ )}} \ \frac{\partial \underset{\sim}{\kappa}_{(\ )}}{\partial \underset{\sim}{\mu}} \right)^t \qquad [1mc(D)]$$

$$= \underset{\sim}{J}\ \underset{\sim}{\mathscr{V}}(\underset{\sim}{M}')\ \underset{\sim}{J}^t, \quad \text{say,}$$

where the Jacobian $\partial \underset{\sim}{\kappa}_{(\ )}/\partial \underset{\sim}{\mu}$ can be obtained up to 4th order from the relations [1,6] in the Appendix on Moments and Cumulants, and the elements of $\partial \underset{\sim}{\eta}/\partial \underset{\sim}{\kappa}_{(\ )}$ are given by

$$\frac{\partial \eta_0}{\partial \kappa_{(1)}} = 1 \ ,$$

$$\frac{\partial \eta_r}{\partial \kappa_{(r+1)}} = \frac{1}{\kappa_{(r)}} \ , \quad \frac{\partial \eta_r}{\partial \kappa_{(r)}} = \frac{-\kappa_{(r+1)}}{\kappa_{(r)}^2} \ , \qquad r = 1,2,\ldots,m-1,$$

and all other $\dfrac{\partial \eta_r}{\partial \kappa_{(s)}} = 0$. The use of an estimate of $\underset{\sim}{\mathscr{V}}(\underset{\sim}{H})^{-1}$ thus gives, explicitly, estimates of $\underset{\sim}{\theta}^*$, $\underset{\sim}{\lambda}^*$ and $\underset{\sim}{\mathscr{V}}(\underset{\sim}{\lambda}^*)$.

For completeness, it may be noted that if cumulants of higher order than 4 are required, though this should be quite rare because of sampling instability,

$$\frac{\partial \underset{\sim}{\kappa}_{(\ )}}{\partial \underset{\sim}{\mu}} = \frac{\partial \underset{\sim}{\kappa}_{(\ )}}{\partial \underset{\sim}{\kappa}} \cdot \frac{\partial \underset{\sim}{\kappa}}{\partial \underset{\sim}{\mu}}$$

can be used to give quite simple formulae convenient for use with a computer. For

$$\kappa_{(r)} = \sum_{s=1}^{r} S_r^{(s)} \kappa_s$$

or

$$\underset{\sim}{\kappa}_{(\ )} = \underset{\sim}{S}\,\underset{\sim}{\kappa},$$

where $\underset{\sim}{S}$ is the $m \times m$ matrix of Stirling Numbers of the First Kind (see the Appendix) and so $\dfrac{\partial \underset{\sim}{\kappa}_{(\ )}}{\partial \underset{\sim}{\kappa}} = \underset{\sim}{S}$. Similarly (Kendall and Stuart I (1963), p. 69; or see the Exercises)

$$\frac{\partial \mu'_r}{\partial \kappa_s} = \binom{r}{s} \mu'_{r-s} \quad \text{for} \quad r \geq s$$

and 0 otherwise, whence $\partial\underset{\sim}{\kappa}/\partial\underset{\sim}{\mu}$.

However, it often happens that the observed zero class is large, and commonsense and asymptotic efficiency calculations suggest the desirability of incorporating P_0, the observed zero class proportion, in the set of statistics $\underset{\sim}{T}$. Again it is possible to be systematic (Hinz and Gurland (1967)) and express the results in terms of those already obtained for $\underset{\sim}{H}$. A function, say A, of P_0 and $\underset{\sim}{H}$ is chosen so that its (asymptotic) expectation, α, can be included in the linear relation $\underset{\sim}{\tau} = \underset{\sim}{w}\,\underset{\sim}{\theta}$, where

$$\underset{\sim}{T} = (H_0, H_1, \ldots, H_{m-1} \vdots A)^t, \qquad \underset{\sim}{\tau} = (\eta_0, \eta_1, \ldots, \eta_{m-1} \vdots \alpha)^t$$

$$= (\underset{\sim}{H} \vdots A)^t \qquad\qquad = (\underset{\sim}{\eta} \vdots \alpha)^t,$$

$\underset{\sim}{\theta}$ is unchanged, and $\underset{\sim}{w}$ has an extra row.

If such a $\underset{\sim}{w}$ can be found corresponding to α, then the normal equation [lmc(A)] is unchanged in form, but $\underset{\sim}{i}$, an estimate of $\underset{\sim}{\mathcal{V}}(\underset{\sim}{T})^{-1}$, must reflect the new $\underset{\sim}{T}$. In fact for the above case, writing $\underset{\sim}{S} = (\underset{\sim}{H} \vdots P_0)^t$ with $\underset{\sim}{\sigma}$ the parametric analogue,

$$\text{then} \quad \underset{\sim}{\mathcal{V}}(\underset{\sim}{T}) = \frac{\partial \underset{\sim}{\tau}}{\partial \underset{\sim}{\sigma}} \underset{\sim}{\mathcal{V}}(\underset{\sim}{S}) \left(\frac{\partial \underset{\sim}{\tau}}{\partial \underset{\sim}{\sigma}}\right)^t, \qquad \text{[lmc(E)]}$$

and

$$\underset{\sim}{\mathcal{V}}(\underset{\sim}{S}) = \left[\begin{array}{c|c} \underset{\sim}{\mathcal{V}}(\underset{\sim}{H}) & -\frac{p_0}{n}\underset{\sim}{J}\,\underset{\sim}{M} \\ \hline -\frac{p_0}{n}(\underset{\sim}{J}\,\underset{\sim}{M})^t & \frac{p_0(1-p_0)}{n} \end{array}\right] \qquad [1mc(F)]$$

with

$$\underset{(m+1)\times(m+1)}{\frac{\partial\underset{\sim}{\tau}}{\partial\underset{\sim}{\sigma}}} = \left[\begin{array}{ccccc|c} 1 & 0 & \cdots & & 0 & 0 \\ 0 & 1 & \cdots & & 0 & 0 \\ \vdots & \vdots & & & \vdots & \vdots \\ 0 & 0 & \cdots & & 1 & 0 \\ \hline \frac{\partial\alpha}{\partial\eta_0}, & \frac{\partial\alpha}{\partial\eta_1}, & \frac{\partial\alpha}{\partial\eta_2}, & 0,\ldots,0 & & \frac{\partial\alpha}{\partial p_0} \end{array}\right] \qquad [1mc(G)]$$

Illustrations which show that appropriate choices for $\underset{\sim}{w}$ and α can be made are given in the sections relating to specific distributions.

In fact convenient choices for $\underset{\sim}{T}$ turn out to include, as well as those mentioned, factorial cumulants and zero and first class proportions, and probability ratio cumulants: see Hinz and Gurland (1967), and Rao and Katti (1971). These probability ratio cumulants are suggested by the analysis in the section dealing with infinite divisibility of Poisson stopped variates, where it was convenient to write the p.g.f. in the form

$$\begin{aligned} \ln g(z) &= \ln (p_0 + p_1 z + p_2 z^2 + \ldots + p_x z^x + \ldots) \\ &= \ln p_0 + \ln(1 + \rho_1 z + \rho_2 z^2 + \ldots + \rho_x z^x + \ldots) \end{aligned}$$

$$(\text{where } \rho_x = p_x/p_0, \quad x = 1,2,\ldots).$$

It is thus possible to introduce a 'probability ratio cumulant' δ_r (bearing the same relation to $r!\rho_r$ as κ_r does to μ'_r) from

$$\ln g(z) = \ln p_0 + \sum_{r=1}^{\infty} \delta_r \frac{z^r}{r!},$$

and then to consider the ratios of successive δ's,

$$\gamma_r = \frac{\delta_{r+1}}{\delta_r}\ ,\quad \gamma_0 = \kappa_{(1)}\ ,$$

analogous to the ratios of factorial cumulants. For the Neyman(μ,ν) distribution

$$\gamma_0 = \mu\nu,\ \gamma_1 = \gamma_2 = \gamma_3 = \ldots = \nu;$$

the use of sample analogues to the γ's can then be made the basis of a chi-squared estimation procedure. Its first order asymptotic efficiency was found to be not as high as for the method based on the η functions, but Hinz and Gurland remark that in their fitting the estimator so based "performed better than anticipated. This would suggest that a comparison of estimators on the basis of large sample efficiency may be somewhat misleading in judging the performance of estimators based on small or moderately sized samples". It might also be added that 'moderately sized samples' are in some cases much larger than one would have suspected before 1960, and that since the asymptotic generalized variances whose ratio gives the asymptotic efficiency depend critically on the covariance of the estimators the interpretation of "efficiency" for the estimation of more than a single parameter is by no means straightforward.

Since the procedures described above all involve minimizing a quadratic form, reference may be made from here to the Appendix on Goodness of Fit Testing where a test procedure based on this minimized quadratic form is described.

Finally it is repeated that "best asymptotically normal" estimators may be obtained by minimum chi-squared methods, and that these properties are preserved through the kinds of transformation described above. Of course the 'small' sample properties of these estimators are likely all to differ: only large sample properties are preserved in these circumstances, the meaning attaching to 'large' being discussed under Asymptotic Expansions. In particular, $\underset{\sim}{i}$ is usually determined by using moment estimates: estimates based on any consistent estimators would do equally well and, again, for small samples the properties of the (modified) minimum chi-squared estimators would differ. In practice, what has been used without exception is the $\underset{\sim}{i}$ determined by the same observations as are used for the estimation of $\underset{\sim}{\tau}$; in the minimum chi-squared equation any simple (often moment-like) estimate of $\underset{\sim}{i}$ is used, but in the estimate of $\underset{\sim}{\mathcal{V}}(\underset{\sim}{\lambda}^*)$ the values from $\underset{\sim}{\lambda}^*$ itself are commonly substituted.

4.3a EXERCISES

1. a) By differentiating $m(t) = \exp k(t)$ once with respect to κ_s and equating coefficients (the functions are moment and cumulant generating functions), show that

$$\frac{\partial \mu'_r}{\partial \kappa_s} = \binom{r}{s}\mu'_{r-s} \;, \quad r \geq s.$$

   b) Use the relations

$$\kappa_{(r)} = \Sigma\, S_r^{(s)}\, \kappa_s \;, \quad \mu'_{(r)} = \Sigma\, S_r^{(s)}\, \mu'_s$$

   to obtain the Jacobians

$$\frac{\partial \underset{\sim}{\kappa}_{(\,)}}{\partial \underset{\sim}{\kappa}} = \underset{\sim}{S} = \frac{\partial \underset{\sim}{\mu}'_{(\,)}}{\partial \underset{\sim}{\mu}'} \;,$$

   $\underset{\sim}{S}$ being the matrix of Stirling Numbers of the First Kind.

   Show that $\dfrac{\partial \underset{\sim}{\kappa}}{\partial \underset{\sim}{\kappa}_{(\,)}} = \$ \,.$

2. For the Poisson-with-zeroes distribution in the form

$$1 - \theta + \theta e^{-\lambda+\lambda z}$$

   show that the maximum likelihood estimators of λ and θ can be obtained from the following.

i) Recurrence relation for probabilities:

$$p_0 = \theta + (1-\theta)e^{-\lambda} ,$$

$$p_1 = (1-\theta)\lambda e^{-\lambda} ,$$

$$p_{x+1} = \frac{\lambda}{x+1} p_x, \quad x = 1,2,\ldots$$

ii) M.l. equations:

$$\frac{\hat{\lambda}}{1-e^{-\hat{\lambda}}} = \frac{\bar{X}}{1-P_0} ,$$

$$\hat{\theta} = \frac{1-P_0}{1-e^{-\hat{\lambda}}} ,$$

where an iterative solution of the former is found by using as starter value

$$\lambda_1 = \frac{\bar{X}}{1-P_0} , \quad \text{neglecting} \quad e^{-\hat{\lambda}} ,$$

and then

$$\lambda_{i+1} = \bar{X}\,\frac{1-e^{-\lambda_i}}{1-P_0} , \quad i = 1,2,\ldots .$$

iii) Application:

Count	0	1	2	3	4+
Observed frequencies	227	70	21	6	1
Fitted frequencies	227.0	69.1	22.9	5.0	1.0

for which $\hat{\lambda} = 0.662$, $\hat{\theta} = 0.623$.

(Martin and Katti (1965), Katti and Rao (1965).)

3. Suppose a Poisson(λ) distribution applies to the observed frequency distribution:

x	0	1	2	3+
F_x	4	1	2	0

Show that the maximum likelihood estimate of λ is $5/7 = 0.714\ 28$ to 5 decimals. If the observed frequency distribution is

x	0	1	2+
F_x	4	1	2

show that the maximum likelihood estimate $\hat{\lambda}$ is the root of

$$\frac{2\hat{\lambda}e^{-\hat{\lambda}}}{1-e^{-\hat{\lambda}}-\hat{\lambda}e^{-\hat{\lambda}}} + \frac{1}{\hat{\lambda}} - 5 = 0,$$

which is 0.801 29 to 5 decimals. (See the Appendix on Right Tail Censored Data.)

4. If the information matrix for a two parameter distribution is

$$\begin{bmatrix} -\mathscr{E}\left(\dfrac{\partial^2 \ln L}{\partial\mu^2}\right), & -\mathscr{E}\left(\dfrac{\partial^2 \ln L}{\partial\mu\partial\nu}\right) \\[2ex] -\mathscr{E}\left(\dfrac{\partial^2 \ln L}{\partial\mu\partial\nu}\right), & -\mathscr{E}\left(\dfrac{\partial^2 \ln L}{\partial\nu^2}\right) \end{bmatrix} = \begin{bmatrix} \iota_{11} & \iota_{12} \\ \iota_{12} & \iota_{22} \end{bmatrix}$$

where the parameters are μ, ν, show that the first order asymptotic covariance matrix of the m.l. estimators $\hat{\mu}, \hat{\nu}$ can be written as

$$\frac{1}{1-\rho^2}\begin{bmatrix} \dfrac{1}{\iota_{11}} & \dfrac{-\iota_{22}}{\iota_{11}\iota_{22}} \\[2ex] \dfrac{-\iota_{12}}{\iota_{11}\iota_{22}} & \dfrac{1}{\iota_{22}} \end{bmatrix}$$

where $\rho \equiv \rho(\hat{\mu},\hat{\nu}) = \mathscr{C}(\hat{\mu},\hat{\nu})/\surd\{\mathscr{V}(\hat{\mu})\,\mathscr{V}(\hat{\nu})\}$

$$= \iota_{12}/\{\iota_{11}\,\iota_{22}\}.$$

5. In practice, the linear minimum chi-squared equation (preceding (lmc[A]))

$$\underset{\sim}{w}^t \underset{\sim}{i}\ \underset{\sim}{T} = \underset{\sim}{w}^t \underset{\sim}{i}\ \underset{\sim}{w}\ \underset{\sim}{\theta}^*$$

is replaced by

$$\underset{\sim}{w}^t \underset{\sim}{I}\ \underset{\sim}{T} = \underset{\sim}{w}^t \underset{\sim}{I}\ \underset{\sim}{w}\ \underset{\sim}{\theta}^+ \tag{A}$$

where $\underset{\sim}{I} = \underset{\sim}{\iota}(\underset{\sim}{T})$ is a consistent estimator of

$$\underset{\sim}{\iota} = \underset{\sim}{\iota}(\underset{\sim}{\tau}) = \mathcal{V}(\underset{\sim}{T})^{-1} .$$

By carrying out Taylor expansions of this replacement equation, it can be shown to be equivalent to first order to [lmc(A)]. For $\underset{\sim}{\tau}^+ = \underset{\sim}{w}\ \underset{\sim}{\theta}^+$, and

$$\underset{\sim}{T} - \underset{\sim}{\tau}^+ = \underset{\sim}{T} - \underset{\sim}{\tau} - (\underset{\sim}{\tau}^+ - \underset{\sim}{\tau}),$$

$$\underset{\sim}{\iota}(\underset{\sim}{T}) - \underset{\sim}{\iota}(\underset{\sim}{\tau}) = \left(\frac{\partial \iota_{rs}}{\partial \underset{\sim}{\tau}} (\underset{\sim}{T} - \underset{\sim}{\tau})\right) \quad \text{to first order,}$$

so that (A) becomes on retaining only first order terms

$$\underset{\sim}{w}^t \underset{\sim}{\iota}(\underset{\sim}{T} - \underset{\sim}{\tau}) = \underset{\sim}{w}^t \underset{\sim}{\iota}\ \underset{\sim}{w}(\underset{\sim}{\theta}^+ - \underset{\sim}{\theta})$$

$$= \underset{\sim}{\Gamma}(\underset{\sim}{\theta}^+ - \underset{\sim}{\theta}) \text{ , say.}$$

By transposing, multiplying, and taking expectations,

$$\mathcal{V}(\underset{\sim}{\theta}^+) = \underset{\sim}{\Gamma}^{-1} ,$$

the result in the text for θ^*.

6. <u>Non-linear Minimum Chi-squared Estimation</u>. For a sample of n consider the statistic

$$\underset{\sim}{S} = (S_1, \ldots, S_m)^t, \quad m \leq n,$$

where $\underset{\sim}{S}$ is (at least) asymptotically normally distributed with expectation vector

$$\underset{\sim}{\sigma} = (\sigma_1, \ldots, \sigma_m)^t$$

and covariance matrix $\mathcal{V}(\underset{\sim}{S})$, $\underset{\sim}{\sigma}$ being functionally related to the indexing parameter vector $\underset{\sim}{\lambda} = (\lambda_1, \ldots, \lambda_k)^t$, $k \leq m$, of the distribution from which the sample comes. Write $\underset{\sim}{v}$ for an estimate of $\underset{\sim}{\mathcal{V}}(\underset{\sim}{S})$, obtained from an at least consistent estimator, and $\underset{\sim}{i} = \underset{\sim}{v}^{-1}$.

To minimize the (asymptotically) chi-squared distributed quadratic form

$$Q = (\underset{\sim}{S} - \underset{\sim}{\sigma})^t \, \underset{\sim}{i} \, (\underset{\sim}{S} - \underset{\sim}{\sigma}),$$

where $\underset{\sim}{\sigma} = \underset{\sim}{\sigma}(\underset{\sim}{\lambda})$, differentiate with respect to $\underset{\sim}{\lambda}$:

$$\frac{\partial Q}{\partial \underset{\sim}{\lambda}} = \left(\frac{\partial \underset{\sim}{\sigma}}{\partial \underset{\sim}{\lambda}}\right)^t \frac{\partial Q}{\partial \underset{\sim}{\sigma}} = -2 \left(\frac{\partial \underset{\sim}{\sigma}}{\partial \underset{\sim}{\lambda}}\right)^t \underset{\sim}{i} \cdot \{\underset{\sim}{S} - \underset{\sim}{\sigma}(\underset{\sim}{\lambda})\},$$

whence the minimum chi-squared equation for $\underset{\sim}{\lambda}^*$:

$$\left(\frac{\partial \underset{\sim}{\sigma}}{\partial \underset{\sim}{\lambda}}\right)^t_{\underset{\sim}{\lambda}^*} \underset{\sim}{i} \cdot \{\underset{\sim}{S} - \underset{\sim}{\sigma}(\underset{\sim}{\lambda}^*)\} = \underset{\sim}{0},$$

k scalar equations for $\lambda_1^*, \ldots, \lambda_k^*$.

Hence, with a Taylor series expansion of the minimum chi-squared equation, using the procedure of the preceding exercise, show that

$$\underset{\sim}{\mathcal{V}}(\underset{\sim}{\lambda}^*) = \left(\left(\frac{\partial \underset{\sim}{\sigma}}{\partial \underset{\sim}{\lambda}}\right)^t \underset{\sim}{\mathcal{V}}(\underset{\sim}{S})^{-1} \frac{\partial \underset{\sim}{\sigma}}{\partial \underset{\sim}{\lambda}}\right)^{-1}.$$

7. If $\underset{\sim}{T} = (T_1, \ldots, T_p)^t$ is an unbiased estimator of $\underset{\sim}{\tau} = (\tau_1, \ldots, \tau_p)^t$ with covariance matrix $\underset{\sim}{\mathcal{V}}(\underset{\sim}{T})$, and $\underset{\sim}{a} = (a_1, \ldots, a_p)^t$ is a constant, then $\underset{\sim}{a}^t \underset{\sim}{T}$ is unbiased for $\underset{\sim}{a}^t \underset{\sim}{\tau}$ and $\mathcal{V}(\underset{\sim}{a}^t \underset{\sim}{T}) = \underset{\sim}{a}^t \, \underset{\sim}{\mathcal{V}}(\underset{\sim}{T}) \underset{\sim}{a}$. If, further $\underset{\sim}{T}_1$ and $\underset{\sim}{T}_2$ are unbiased for $\underset{\sim}{\tau}$, where $\underset{\sim}{T}_i = (T_{i1}, \ldots, T_{ip})^t$, and $\underset{\sim}{\mathcal{V}}(\underset{\sim}{T}_2) - \underset{\sim}{\mathcal{V}}(\underset{\sim}{T}_1)$ is positive semidefinite, then

i) $\mathcal{V}(T_{1j}) \leq \mathcal{V}(T_{2j})$

i.e. the variance of each component of $\underset{\sim}{T}_1$ is not greater than the variance of the corresponding component of $\underset{\sim}{T}_2$;

ii) $\underset{\sim}{\mathcal{V}}(\underset{\sim}{a}^t \underset{\sim}{T}_1) \le \underset{\sim}{\mathcal{V}}(\underset{\sim}{a}^t \underset{\sim}{T}_2)$

for every fixed $\underset{\sim}{a}$.

4.4 INTERVAL AND REGION ESTIMATION

If T is $N(\lambda,\sigma^2)$, then where u_α is the upper 100% point of a standardized normal variate, $\Pr\{\lambda \in (T \pm u_{\frac{1}{2}\alpha}\sigma)\} = 1-\alpha$, i.e. a $100(1-\alpha)\%$ confidence interval for λ is given by

$$(T \pm u_{\frac{1}{2}\alpha}\sigma).$$

If then s^2 is a large sample estimate of σ^2, $(T \pm u_{\frac{1}{2}\alpha} s)$ will be an approximate $100(1-\alpha)\%$ confidence interval for λ. In the usual applications in this book all that is known is that T is asymptotically $N(\lambda,\sigma^2)$ - the same approximate confidence interval is still used.

For several parameters, say (μ,ν), and estimators, say (S,T), similar remarks apply. If (S,T) is asymptotically bivariate normal with mean (μ,ν) and covariance matrix

$$\begin{pmatrix} \sigma_S^2 & \sigma_{ST} \\ \sigma_{ST} & \sigma_T \end{pmatrix} = \underset{\sim}{\mathcal{V}}\binom{S}{T}$$

(correlation $\sigma_{ST}/\sigma_S\sigma_T$), an approximate confidence region for (μ,ν) is given by the appropriate elliptical contour of the bivariate normal in which the covariance matrix is replaced by a sample estimate - the region thus has the boundary

$$\left(\frac{S-\mu}{s_S}\right)^2 - 2s_{ST}\frac{S-\mu}{s_S^2}\cdot\frac{T-\nu}{s_T^2} + \left(\frac{T-\nu}{s_T}\right)^2 = \text{constant}$$

where s_S, s_T, s_{ST} are the values of consistent estimators of σ_S, σ_T, σ_{ST}. Partly because of computational difficulties, and partly because of difficulties of interpretation, it is common to ignore the correlation between S and T, and to determine separate confidence intervals for μ and ν from their marginal distributions using, e.g., T and s_T for μ as in the preceding paragraph. If the correlation is high, as it often is, the procedure may obviously be misleading; however, its justification may be found in the requirements of a practical situation in which an assertion about only one of μ or ν is required, when the marginal distribution is appropriate. (It might also be added that on different occasions different parameters may be of interest, and for re-use of the data no obviously valid procedure is available.)

An entirely different type of inference may be used in place of these approximate confidence regions, with an interpretation in terms of relative likelihood (see, e.g., Sprott (1970), and the booklength treatment of Edwards (1972)). To illustrate, with an observed sample $\underset{\sim}{x}$ for the single parameter λ, the likelihood $L(\underset{\sim}{x};\lambda)$ is determined at its maximum $L(\underset{\sim}{x};\hat{\lambda})$, and, at least for a continuous likelihood with a single maximum, two values $\hat{\lambda}_1 < \hat{\lambda}_2$ can be found so that at these values $L(\cdot)$ is some convenient fraction of its maximum $L(\underset{\sim}{x};\hat{\lambda})$:

i.e. $$L(\underset{\sim}{x};\hat{\lambda}_1)/L(\underset{\sim}{x};\hat{\lambda}) = L(\underset{\sim}{x};\hat{\lambda}_2)/L(\underset{\sim}{x};\hat{\lambda}) = 0.15, \text{ say.}$$

(The reason for suggesting 15% will appear in a moment. See Hudson (1971) for complications introduced by discreteness.) The interpretation offered for this practice is that values outside the interval $(\hat{\lambda}_1, \hat{\lambda}_2)$ are implausible compared with those within, moving out from the most plausible value $\hat{\lambda}$; a graph illustrating the situation is exhibited in the following figure.

The corresponding diagram for a confidence interval based on the approximating normal density for $\hat{\lambda}$ centered at the m.l. estimate $\hat{\lambda}_0$, in which the central area is (say) 95%, is also illustrated. Quite apart from the earlier interpretation of the likelihood ratio, when these two curves are rescaled with maxima the same they may be compared visually: a lack of reasonable agreement points to the need of caution in the interpretation of the confidence interval, at the least because of poor approximation by the normal distribution. An illustration is afforded by the artificial frequency data

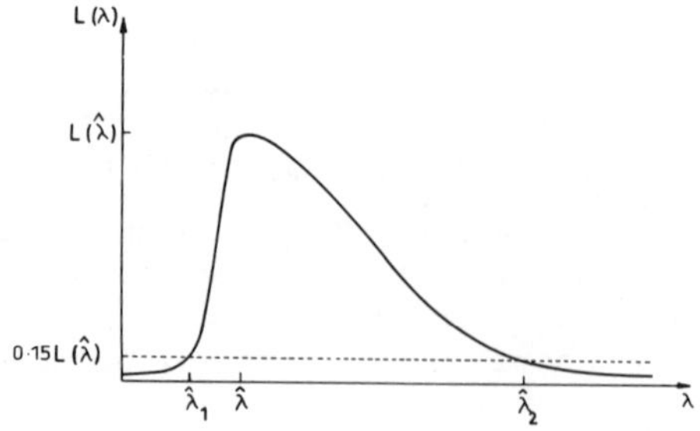

FIGURE 38 Interval determined from likelihood function.

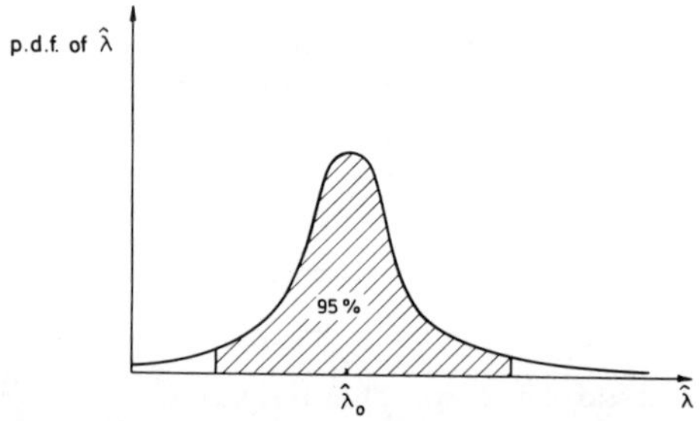

FIGURE 39 Confidence interval based on normal approximation.

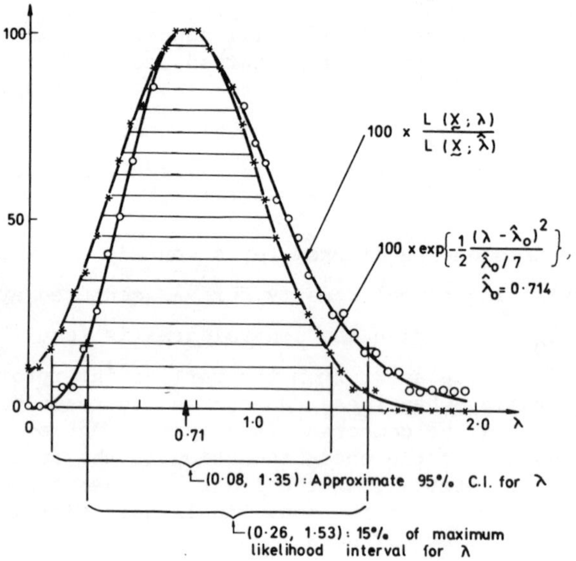

FIGURE 40 Comparison of likelihood and confidence interval.

x	0	1	2	3+
F_x	4	1	2	0

where these are used to provide a visual example of the disagreement possible.

Similar statements are applicable to multiparameter cases, though the computational effort to determine multidimensional contours is often considerable. Although these "plausibility" regions with a likelihood ratio interpretation are not intended to have the frequency interpretation of confidence regions, at least for comparative purposes it is convenient to choose the ratio and confidence coefficient so that they are compatible in some sense. This may be done approximately by appealing to likelihood ratio theory: for $\underset{\sim}{\lambda} = (\lambda_1, \ldots, \lambda_k)^t$ a k component parameter, and $\underset{\sim}{X}$ a "large" sample, under reasonable regularity conditions the distribution of

$$-2 \ln \frac{L(\underset{\sim}{X};\hat{\underset{\sim}{\lambda}})}{L(\underset{\sim}{X};\underset{\sim}{\lambda})} = R, \quad \text{say},$$

is approximately that of chi-squared with k degrees of freedom. So for any given confidence coefficient, the ratio of ordinates to the maximum (likelihood) ordinate can be determined so that the probability content (in the confidence interval sense) of the interval determined by ordinates with this ratio corresponds to that specified. For the single parameter case, reference to chi-squared with 1 d.f. gives

$$0.95 = \Pr(R < 3.84)$$

whence

$$\frac{L(\underset{\sim}{X};\hat{\underset{\sim}{\lambda}})}{L(\underset{\sim}{X};\underset{\sim}{\lambda})} > e^{-3.84/2} = 0.147,$$

leading to the approximate value of 15% of the maximum ordinate for the boundary ordinates which determine the interval to correspond with a 95% confidence interval, as stated previously. Similarly, for two parameters,

$$0.95 = \Pr(R < 5.991)$$

leads to

$$\frac{L(\underset{\sim}{X};\hat{\underset{\sim}{\lambda}})}{L(\underset{\sim}{X};\underset{\sim}{\lambda})} > e^{-3.00} = 0.0498,$$

or, very conveniently, 5%, to correspond with a 95% region. A numerical illustration is given in the Chapter on the Neyman distribution for the case of 2 parameters.

4.5 RIGHT TAIL CENSORED DATA

A practical problem in estimation which arises rather frequently in working over published data is exemplified by the frequency distribution recorded as

(A)

x	0	1	2+
F_x	4	1	2

.

It is impossible to know whether this is

(B)

0	1	2	3+
4	1	2	0

or (C)

0	1	2	3	4+
4	1	1	1	0

,

or indeed which of an infinity of samples.

At the expense of considerable added complexity, maximizing the likelihood, or minimizing the classical chi-squared expression, can deal with this condensation - e.g., for maximum likelihood, what is to be maximized is

$$4 \ln p_0(\theta) + \ln p_1(\theta) + 2 \ln \{1 - p_0(\theta) - p_1(\theta)\}.$$

(See an exercise following Appendix 4.2 for an example.)

For estimation based on the mean, a simple iterative procedure can be applied, though the rate of convergence may not be high. If the data are

x	0	1 ...	k-1	k+
F_x	F_0	F_1 ...	F_{k-1}	F_{k+}

$\Sigma F_x = n$

where a one parameter p.f. $p_x(\theta)$ is to be fitted, with

$$\mu(\theta) = \sum_{x=0}^{\infty} x p_x(0),$$

a starter value a_1 for the mean of the last class is taken to be k (or k+1, etc.). Then the first approximation θ_1 for θ is given by the solution of

$$\mu(\theta_1) = \frac{1}{n} \sum_{x=0}^{k-1} x F_x + a_1 F_{k+} \quad .$$

With this θ_1 the p.f.'s $p_x(\theta_1)$ are evaluated, where

$$p_{k+}(\theta_1) = 1 - \sum_{x=0}^{k-1} p_x(\theta_1),$$

and from

$$\mu(\theta_1) = \sum_{x=0}^{k-1} x p_x(\theta_1) + a_2 p_{k+}(\theta_1),$$

i.e. $$a_2 = \frac{\mu(\theta_1) - \sum_{x=0}^{k-1} x p_x(\theta_1)}{p_{k+}(\theta_1)}$$

a second value a_2 is found; whence θ_2, etc.

For the artificial data (A) above, supposing a Poisson distribution with parameter $\lambda(=\theta$ above, and $\mu(\theta) = \theta)$, taking $a_1 = 2$ leads to $\lambda_1 = 5/7 = 0.714...$, whence to $a_2 = 2.27,...$ and after 5 iterations to $\lambda_5 = 0.801...$ (with $a_5 = 2.30...$). The application and the data are artificial, and there is no special interpretation of the values obtained. The fitted frequencies are:

x	0	1	2	3	4
$\lambda_1 = 0.714$	3.43	2.45	0.87	0.21	0.04
$\lambda_5 = 0.801$	3.14	2.52	1.01	0.27	0.06

As is readily verified, this is also the maximum likelihood fitting.

Taking the flying bomb hit data quoted in Feller I (1968) p. 161, the first three columns as given by Feller are (nearly) as follows. The third column, of fitted frequencies $\phi_x(\hat{\lambda}) = np_x(\hat{\lambda})$, corresponds to the maximum likelihood estimate $\hat{\lambda} = 537/576 = 0.932...$ using the uncondensed data. The fourth

TABLE 22

Flying-Bomb Hits on London

x	F_x	$\phi_x(\hat{\lambda})$	F_x	$\phi_x(\lambda_5)$
0	229	226.74	229	228.86
1	211	211.39	211	211.24
2	93	98.54		97.48
3	35	30.62		29.99
4	7	7.14		6.92
5	0			
6	0		136	
7	1	1.57		1.51
8+	0			

(Feller I, VI Table IV)

column shows a very heavy condensation of the data, and if the frequency of 136 were supposed concentrated at $x = 2$, the estimate of λ is 0.838, much too low. Using $a_1 = 2$ and $\lambda_1 = 0.838$ leads eventually to $\lambda_5 = 0.922\ 99$ (and $a_5 = 2.357\ldots$; the actual sample mean for 2+ is 2.397), and the excellent fit shown in the fifth column. There is obviously an improvement from the simple estimate 0.838 to 0.923 by use of the iterative procedure, and in fact if it had been desired to predict the frequencies at $x = 2,3,\ldots,$ the procedure has done very well.

For references and a more complete discussion, see Carter and Myers (1973) and Dempster et al. (1977).

For several parameters a slight modification can be used. Taking $p_x(\theta,\phi)$, with mean $\mu(\theta,\phi)$ and second moment about the origin $\mu_2'(\theta,\phi)$, then for the starter value a_1 as before,

$$\mu(\theta_1,\phi_1) = \frac{1}{n} \sum_{x=0}^{k-1} x\, F_x + a_1\, F_{k+}$$

and

$$\mu_2'(\theta_1,\phi_1) = \frac{1}{n} \sum_{x=0}^{k-1} x^2\, F_x + a_1^2\, F_{k+}$$

are used to obtain first values $\theta_1, \phi_1,$ the usual moment values for the choice of $a_1 = k$ previously described. Then

$$a_2 = \frac{\mu(\theta_1,\phi_1) - \sum_{x=0}^{k-1} x\, p_x(\theta_1,\phi_1)}{p_{k+}(\theta_1,\phi_1)},$$

with iteration until the required accuracy has been achieved.

Taking the artificial data (A) and the Neyman(μ,ν) distribution one is led (rather slowly) from the initial moment values (for (B)) of

$$\mu_1,\ \nu_1 = 8.33,\ 0.0857$$

to

$$\mu_9,\ \nu_9 - 1.88,\ 0.475$$

with frequencies as shown.

x	0	1	2+
F_x	4	1	2
moment on 2	3.53	2.32	1.15
iterated on 2+	3.44	1.91	1.65

It is perhaps of interest to notice that the maximum likelihood solution for the data (B) (see the exercises on the Neyman distribution) is

$$\hat{\mu},\ \hat{\nu} = 1.38,\ 0.519,$$

very much more like this moment fit to the data (A) than to the rather abrupt (B).

At the best, incompletely recorded data generally require much more work for their analysis, and the above procedures are perhaps rather to be regarded as for emergency rather than routine use. Closely related, but not specifically discussed here, is the case where only 0, 1+ (absence, presence) are recorded, generally because sufficiently larger quantities of data can then be collected to offset the losses incurred. No detailed standard error analyses will be described for these modified procedures.

Appendix 5

GOODNESS OF FIT TESTING

By far the most common "goodness of fit" test of an hypothesis in applications of discrete distributions is the chi-squared goodness of fit test applied to observed and fitted ("expected") frequencies. Although little formal use is made here of such comparisons, a brief account is given of some aspects directly relevant to the fitting of discrete distributions; see also Easterling (1976) and Conover (1972).

5.1 CLASSICAL CHI-SQUARED GOODNESS OF FIT TEST

The classical test and its asymptotic theory were introduced in 1900 by K. Pearson, who continued his 1896 use of the symbol χ^2 for what is here written as Q^2 and for which a common recent notation is the rather overworked X^2.

Suppose a family of distributions, with probability functions

$$p_0, p_1, \ldots, p_x, \ldots$$

where $p_x \equiv p_x(\theta)$ depends on a known (possibly vector) parameter θ, yields a corresponding random sample of fixed size n with observed frequencies

$$F_0^*, F_1^*, \ldots, F_x^*, \ldots, \qquad \Sigma\, F_x^* = n.$$

Then the expected frequencies $n\ p_x(\theta)$, say, corresponding to θ can be examined, and if necessary grouped (condensed) into classes so no class has an expected frequency of less than unity (see below for comment on "unity"), these expected frequencies being written as

$$\phi_1(\theta),\ \phi_2(\theta),\ldots,\phi_c(\theta)\ ,$$

the corresponding observed frequencies being

$$F_1, F_2, \ldots, F_c\ , \qquad \Sigma F_x = n.$$

The sum $\sum_{i=1}^{c} \frac{\{F_i - \phi_i(\theta)\}^2}{\phi_i(\theta)} = Q^2$, say, is a measure of the agreement between the observations and the member of the family specified by θ. On the hypothesis that $F_1, F_2, \ldots, F_c$ constitute a condensation of a random sample from the specified distribution Q^2 is approximately chi-squared distributed with $c-1$ degrees of freedom, a test of the hypothesis at the 100α% level being provided by rejecting if the observed Q^2 exceeds the upper tabular 100α% point, $\chi^2_{\alpha;c-1}$, of the chi-squared distribution. In this, the "exact" chi-squared distribution (to which the tables refer) is that of

$$\sum_{i=1}^{c} \left(\frac{X_i - \mu_i}{\sigma_i}\right)^2$$

where the X_i are independent normal variates with means μ_i and variances σ_i^2 respectively, and $\chi^2_{\alpha;k}$ is the value of chi-squared based on k degrees of freedom which is exceeded in random sampling 100α% of the time. The number "unity" referred to is nominated to try to keep the approximation of Q^2 to a chi-squared variate at a reasonable level - for otherwise Q^2 may be large because some ϕ_i are small rather than because their $|F_i - \phi_i|$ are large - while not losing too much power of the test by over-condensation. It is in the nature of a compromise and obviously ought to be treated as a guide rather than as a rule - in particular, it is wise to look at individual terms in Q^2 if a large Q^2 is found. (See also Yarnold (1970).)

However, the usual situation is that, while the family of distributions is taken as known, θ has to be estimated from the observations. The problem of what happens to the asymptotic distribution of the corresponding Q^2 with estimation, though recognized, was not dealt with until Fisher's work of the 1920's established that the use of minimum Q^2 (or asymptotically equivalent) estimators still led to the chi-squared distribution but with an appropriate reduction in the degrees of freedom, and the use of procedures such as the method of moments, which are asymptotically equivalent only for rather special cases, was shown to lead (in general) to corresponding Q^2 expressions which are not chi-squared distributed and in fact are stochastically too large, thus upsetting nominal significance levels. It was not until the 1950's that detailed studies were made of the effect on the distribution of Q^2 of using estimators based alternatively on the original or the condensed data, when it turned out that an asymptotic chi-squared distribution did not obtain for estimators based on the original data and that sometimes this effect could be large.

If $\tilde{\theta}$ is a "suitable" estimator, such as a multinomial maximum likelihood estimator based on maximizing the multinomial log-likelihood

$$\sum_{i=1}^{c} F_i \ln \phi_i(\theta), \quad \text{at} \quad \theta = \hat{\theta}_m \quad \text{say,}$$

or a multinomial minimum-chi-squared estimator based on minimizing the previously defined Q^2, with minimum $\tilde{Q}_m^2$ at $\theta = \tilde{\theta}$ say, these being asymptotically equivalent, then it is still the case that $\tilde{Q}_m^2$ and the corresponding $\hat{Q}_m^2$ are asymptotically chi-squared distributed but with $k - 1 - p$ degrees of freedom, p being the number of parameters estimated (the number of components of θ).

However the estimators $\hat{\theta}_m$ and $\tilde{\theta}_m$ are not the usual maximum likelihood or minimum chi-squared estimators - e.g., for the Poisson(λ) family the estimator $\hat{\theta}_m$ under condensation will not be the usual sample mean - and if the usual estimators $\hat{\theta}$ or $\tilde{\theta}$ based on the uncondensed observations are used the corresponding $\hat{Q}^2$ and $\tilde{Q}^2$ do not have the chi-squared distribution of the expression

$$\sum_{i=1}^{c-1-p} U_i^2$$

where the U_i are independent standard normal variates, but of

$$\sum_{i=1}^{c-1-p} U_i^2 + \sum_{j=1}^{p} \lambda_j V_i^2$$

where the V_j are also independent standard normal variates and the $\lambda_j(\theta)$ (which tend to zero as the condensation disappears) satisfy $0 < \lambda_j < 1$. It is rather common to use, e.g., the estimator $\hat{\theta}$ (or an equivalent) and to neglect the effect of condensation. For a further discussion of these and related questions, essentially concerned with the goodness of approximations, see Cochran (1952), Chernoff and Lehmann (1954), Watson (1959), Barton (1967), Pahl (1969) and the book by Lancaster (1969).

It may sometimes be desirable to examine the power (in the Neyman Pearson sense) of the chi squared goodness of fit test against certain alternatives. To do so, the distribution of non-central chi-squared is required: with the above notation, $\Sigma X_i^2/\sigma_i^2$ is distributed as non-central chi-squared with c degrees of freedom and non-centrality parameter $\Sigma \mu_i^2/\sigma_i^2$. In the present case, if the hypothesis of a Neyman distribution with expected frequencies $\phi_0, \phi_1, \phi_2, \ldots$ were to be tested against the alternative of a negative binomial distribution with expected frequencies $\psi_0, \psi_1, \psi_2, \ldots,$ the test being defined by:

Reject Neyman hypothesis at level α if

$$Q^2 = \Sigma \frac{(F_i - \phi_i)^2}{\phi_i} > \chi_\alpha^2 ,$$

then the power of this test is given by

$$\Pr(Q^2 > \chi_\alpha^2 \mid \text{negative binomial hypothesis})$$

where Q^2 is now distributed as non-central chi-squared with non-centrality $\Sigma(\phi_i - \psi_i)^2/\phi_i$. Tables of these probabilities are available, e.g., in Owen (1964). The derivation of these results uses a sequence of asymptotic approximations, with no control of the error terms, it might be added. An application of the theory is given under Comparisons with Other Distributions in Chapter IV.

Unfortunately there is no theory to deal with what has been perhaps the most common application of chi-squared tests: the use of these tests to choose the "best fitting family" when a member of each of a variety of families is fitted and the fits compared. Thus Student's 1907 data on yeast cells have been fitted repeatedly in the literature, with Poisson, negative binomial, Neyman,... distributions, and chi-squared tests carried out - many of the fits are excellent, and there the theory rests. Another instance is afforded by the Poisson-binomial(μ,ν,π) distribution in which a succession of values $\nu = 2,3,4,\ldots$ is taken, and a value finally chosen without benefit of (even asymptotic) theory. Some discussion of such questions will be found in Ord (1972, § 6.9), Cox and Hinkley (1974, § 9.3) and Lindsey (1974) with a bibliography in Pereira (1977). In comparing likelihoods under various hypotheses it is particularly interesting to compare any proposed simple model (i.e. with few estimated parameters) with the multinomial model which will reproduce precisely the observations (i.e., with probability parameters the observed proportions in the classes). Such a relative likelihood (Lindsey (1974) - see some following paragraphs) gives, so to speak, an upper limit to the goodness of fit which can be achieved, and hence something more than merely a comparison between arbitrarily selected models. It may be appropriate to mention again here that the purpose of fitting as seen in this book is primarily to estimate (and use) parameters, and that doing so is justified by the meaningful way in which the estimates are interpretable against the way the data were obtained.

5.2 SIMULATED LEVELS OF SIGNIFICANCE

Suppose the theoretical frequencies in the various classes are $\phi_1(\theta),\ldots,\phi_i(\theta),\ldots,\phi_c(\theta)$, and that a realized sample has frequencies $f_1,\ldots,f_c$ which by use of a suitable estimator $\tilde{\theta}$ (e.g. maximum likelihood, or minimum chi-squared) lead to a (realized) estimate θ_0 of θ. To test the hypothesis that the sample comes from the theoretical distribution, the value

$$\sum_{i=1}^{c} \frac{\{f_i - \phi_i(\theta_0)\}^2}{\phi_i(\theta_0)} = q_0^2 \quad , \quad \text{say,}$$

is calculated; the significance level corresponding to this number is sought (or it is to be compared with a tabular value at some significance level).

In place of the usual approximation using tables of the chi-squared distribution, Sprott (1970) suggests a computer simulation to estimate this level. Since $\phi_i(\theta)$ are multinomial frequencies, and $\phi_i(\theta)/n$ the corresponding probabilities, for any chosen θ these can be simulated by the use of rectangular random numbers, leading to frequencies $N_i(\theta)$, say, in a typical run with $\Sigma N_i = \Sigma\phi_i = n$. Applying the estimator $\tilde{\theta}$, above, to the $N_i(\theta)$ leads to the random variable

$$Q^2(\tilde{\theta},\theta) = \Sigma \frac{\{N_i(\theta) - \phi_i(\tilde{\theta})\}^2}{\phi_i(\tilde{\theta})};$$

the distribution of this random variable is thus generated by a sequence of runs for each of which (with the same θ) a set of $N_i(\theta)$ leads to $\tilde{\theta}$ and thus $Q^2(\tilde{\theta},\theta)$. It will be recalled that standard chi-squared theory asserts that this is (nearly) independent of θ for large n. It follows that the proportion of the simulated values of $Q^2(\tilde{\theta},\theta)$ which exceeds q_0^2 is an estimate of

$$\Pr(Q^2(\tilde{\theta},\theta) > q_0^2),$$

i.e. of the significance level of q_0^2. Given good computer facilities such simulations may not be unreasonably lengthy, even if the calculations are repeated over a range of plausible values of θ.

5.3 AN ALTERNATIVE CHI-SQUARED TEST

In conjunction with linear minimum chi-squared (generalized least squares) estimators, Hinz and Gurland (1970) have suggested a direct use of the minimized quadratic form for a much more easily calculated "chi-squared test". It will be recalled that minimum chi-squared estimators (see the Appendix) do not minimize the classical chi-squared goodness of fit criterion discussed earlier in this section.

The notation used is as follows: $\underset{\sim}{T}$ is an m-dimensional and (at least) asymptotically normally distributed statistic, with mean $\underset{\sim}{w}\,\underset{\sim}{\theta}$ ($\underset{\sim}{w}$ being a known constant $m \times p$ matrix, $p \leq m$, and $\underset{\sim}{\theta}$ a p-dimensional parameter of the family of distributions being fitted) and positive definite covariance matrix $\underset{\sim}{\Sigma}$. Then the quadratic form minimized in $\underset{\sim}{\theta}$ is

$$(\underset{\sim}{T} - \underset{\sim}{w}\underset{\sim}{\theta})^t \, \underset{\sim}{\Sigma}^{-1}(\underset{\sim}{T} - \underset{\sim}{w}\underset{\sim}{\theta}) \;,$$

where $\underset{\sim}{\Sigma}$ is regarded as constant: write this minimum as

$$Q_F^* = (\underset{\sim}{T} - \underset{\sim}{w}\underset{\sim}{\theta}^*)^t \, \underset{\sim}{\Sigma}^{-1} \, (\underset{\sim}{T} - \underset{\sim}{w}\underset{\sim}{\theta}^*) \;,$$

where $\underset{\sim}{\theta}^* = (\underset{\sim}{w}^t\underset{\sim}{\Sigma}^{-1}\underset{\sim}{w})^{-1} \; \underset{\sim}{w}^t\underset{\sim}{\Sigma}^{-1} \, \underset{\sim}{T}$

is the minimizing $\underset{\sim}{\theta}$.

In order to obtain the distribution of Q_F^*, write

$$\underset{\sim}{r} = \underset{\sim}{w}(\underset{\sim}{w}^t\underset{\sim}{\Sigma}^{-1}\underset{\sim}{w})^{-1} \; \underset{\sim}{w}^t\underset{\sim}{\Sigma}^{-1} \;,$$

so that

$$\underset{\sim}{w} \, \underset{\sim}{\theta}^* = \underset{\sim}{r} \, \underset{\sim}{T};$$

then

$$Q_F^* = \underset{\sim}{T}^t(\underset{\sim}{1}-\underset{\sim}{r})^t \, \underset{\sim}{\Sigma}^{-1} \, (\underset{\sim}{1} - \underset{\sim}{r}) \, \underset{\sim}{T}$$

$$= \underset{\sim}{T}^t \, \underset{\sim}{b} \, \underset{\sim}{T}$$

for

$$\underset{\sim}{b} = (\underset{\sim}{1}-\underset{\sim}{r})^t \, \underset{\sim}{\Sigma}^{-1} \, (\underset{\sim}{1}-\underset{\sim}{r}) = (\underset{\sim}{1}-\underset{\sim}{r})^t \, \underset{\sim}{\Sigma}^{-1} = \underset{\sim}{\Sigma}^{-1} \, (\underset{\sim}{1}-\underset{\sim}{r}) .$$

However, if $\underset{\sim}{X} : N(\underset{\sim}{\mu},\underset{\sim}{\Sigma})$, it is known that $\underset{\sim}{X}^t \, \underset{\sim}{b} \, \underset{\sim}{X}$ is non-central chi-squared distributed with degrees of freedom $\mathrm{rank}(\underset{\sim}{b})$ and non-centrality $\frac{1}{2} \, \underset{\sim}{\mu}^t\underset{\sim}{b} \, \underset{\sim}{\mu}$ iff $\underset{\sim}{b} \, \underset{\sim}{\Sigma}$ is idempotent (e.g. Graybill (1961)). In the present case $\underset{\sim}{b} \, \underset{\sim}{\Sigma}$ is idempotent, since

$$\underset{\sim}{b} \, \underset{\sim}{\Sigma} \cdot \underset{\sim}{b} \, \underset{\sim}{\Sigma} = (\underset{\sim}{b} \, \underset{\sim}{\Sigma} \, \underset{\sim}{b})\underset{\sim}{\Sigma} = \underset{\sim}{b} \, \underset{\sim}{\Sigma} \;,$$

and $\mathrm{rank}(\underset{\sim}{b}) = \mathrm{rank}(\underset{\sim}{b} \, \underset{\sim}{\Sigma}) = \mathrm{tr}(\underset{\sim}{b} \, \underset{\sim}{\Sigma}) = \mathrm{tr}(\underset{\sim}{1}-\underset{\sim}{r}) = m-p$ since $\mathrm{tr}(\underset{\sim}{r}) = \mathrm{tr}\{(\underset{\sim}{w}^t \, \underset{\sim}{\Sigma}^{-1} \, \underset{\sim}{w})^{-1} \, \underset{\sim}{w}^t \, \underset{\sim}{\Sigma}^{-1} \, \underset{\sim}{w}\} = p$. Further, the non-centrality is

$$\tfrac{1}{2}(\underset{\sim}{w} \, \underset{\sim}{\theta})^t \, \underset{\sim}{b} \, \underset{\sim}{w} \, \underset{\sim}{\theta} = \underset{\sim}{0}$$

since $(\underset{\sim}{1}-\underset{\sim}{r}) \, \underset{\sim}{\theta} = \underset{\sim}{0}$. Thus Q_F^* has an asymptotic (central) chi-squared distribution with $m-p$ degrees of freedom, and an asymptotic test at level α of the hypothesis that the family is that specified is given by using the critical region

$$Q_F^* > \chi^2_{\alpha;m-p} .$$

(If on the other hand the alternative hypothesis gives the (at least asymptotic) expectation of $\underset{\sim}{T}$ as $\underset{\sim}{\tau}$ the non-centrality is $\frac{1}{2} \underset{\sim}{\tau}^t \underset{\sim}{b} \underset{\sim}{\tau}$ (the other parts of the argument go through more or less unchanged), whence the power of the above test can be obtained from tables of the non-central chi-squared distribution. A much fuller account of the test, including its power and favorable comparisons with the chi-squared goodness of fit test, is to be found in Hinz and Gurland (1970).)

In the common case of a two parameter distribution $(p = 2)$ and a three dimensional estimator (e.g. based on the zero class frequency and the first two moments, so $m = 3$), the easily calculated minimum value of the quadratic form, Q_F^*, is referred to the chi-squared table with $3 - 2 = 1$ d.f. to check the adequacy of the fit. This can of course be done without calculating the fitted frequencies (and hence in particular without grouping), and while it is often convenient to do so for the more complex distributions the procedure loses some intuitive appeal because it depends - in the above example - only on the first two moments and the zero class frequency and not directly on the variation of frequencies; on the other hand, it is based directly on the parameter estimates, the primary objects of interest.

An example is given in the Chapter on the Neyman distribution.

5.4 RELATIVE LIKELIHOOD

For observed frequencies F_x and probabilities $p_x(\underset{\sim}{\theta})$, it is usual to consider the likelihood

$$L(\underset{\sim}{F};\underset{\sim}{\theta}) = \prod_x \{p_x(\underset{\sim}{\theta})\}^{F_x} , \quad \underset{\sim}{\theta} = (\theta_1, \ldots, \theta_q)^t$$

or its logarithm

$$\ln L = \sum_x F_x \ln p_x(\underset{\sim}{\theta}) , \quad p_x > 0.$$

But when these are used to examine the plausibility of the hypothesis $\{p_x(\underset{\sim}{\theta})\}$ for $\underset{\sim}{F} = (F_x)^t$, they suffer from the great disadvantage of having no natural scaling: in ordinary cases L

takes very small values, and the more voluminous the data the smaller L is likely to be.

An intuitively appealing procedure is to compare $L(\underset{\sim}{F};\underset{\sim}{\theta})$ for any hypothetical $\underset{\sim}{\theta}$ with $L(\underset{\sim}{F};\hat{\underset{\sim}{\theta}})$ where $\hat{\underset{\sim}{\theta}}$ is chosen so that L is maximized, and indeed this leads to the standard asymptotic likelihood ratio theory. In detail, suppose the p-dimensional parameter space of $\underset{\sim}{\theta} = (\theta_1, \theta_2, \ldots, \theta_q, \theta_{q+1}, \ldots, \theta_p)^t$ is Ω, and that an hypothesis H_0 of interest is $(\theta_1, \theta_2, \ldots, \theta_q) = (\theta_{10}, \theta_{20}, \ldots, \theta_{q0})$, $q \leq p$, the "reduced" parameter space Ω_0 having dimension p-q. Then maximizing L in Ω, and in Ω_0, the ratio

$$R = \frac{\max_{\Omega_0} L(\underset{\sim}{F};\underset{\sim}{\theta})}{\max_{\Omega} L(\underset{\sim}{F};\underset{\sim}{\theta})}$$

can be considered, "small" values of R casting doubt on H_0. (More precisely, sup rather than max may be appropriate.) Under rather general conditions,

$$-2 \ln R$$

is asymptotically chi-squared distributed with q degrees of freedom; with a formal hypothesis testing procedure, H_0 would be rejected if $-2 \ln R$ were "too large", judged by reference to the appropriate upper chi-squared percentage point. But this result requires a rather specific structure for the comparison and, e.g., provides no basis for judging the goodness of a complete maximum likelihood fit, nor of the goodness of fit of (say) a Thomas compared with a Neyman distribution.

However, if $n = \Sigma F_x$ and $P_x = F_x/n$, the log-likelihood given by

$$\Sigma F_x \ln P_x, \quad P_x > 0,$$

corresponds to the best fit possible (the hypothesis is a many-parameter multinomial, and one would usually wish to consider fewer parameters than this), and thus a point of reference is provided for scaling. Hence

$$\Sigma F_x \ln P_x - \Sigma F_x \ln p_x(\underset{\sim}{\theta}), \quad p_x > 0,$$

may be called the relative log-likelihood of the hypothesis or distribution $\{p_x(\underset{\sim}{\theta})\}$ (Lindsey (1974)), and a "large" value of this difference (i.e., the logarithm of the ratio) indicates poor correspondence between the frequencies predicted by the hypothesis and those observed. The use of this over a variety of distributions at least gives an ordering of the distributions although there is no allowance built into the comparison for the different complexities of different hypotheses. (Perhaps one might, speculatively, take twice the relative log-likelihood to have an approximate chi-squared distribution with (large) degrees of freedom given by the number of non-zero frequencies minus one and minus the number of distributional parameters estimated by maximum likelihood.)

So, for the Flying Bomb data already quoted, a Neyman distribution gives twice the log relative likelihood to be 8.872 (and $6 - 3 = 3$ degrees of freedom?), while a Poisson distribution gives 9.263 (with 4 df?).

Appendix 6

ITERATIVE PROCEDURES FOR SOLVING F(x) = 0

When estimation equations are to be solved, only sometimes do these turn out to be such that explicit formulae for their roots can be given - indeed, quite often there is neither a guarantee that there will be roots nor, if there are, that these will be admissible values. The following remarks relate to some of the common procedures employed in these circumstances, and are confined to their simplest aspects - the remarks are intended to provide illumination for the methods used in the body of the book, and hence some assistance when the procedures do not deliver answers of the kind sought; they are not an account of a body of numerical analysis theory. Throughout, the methods are taken as tentative, and proceed on the assumption that the roots sought do exist.

If several "variables" are involved, it is often possible to eliminate all but one of these and thus reduce the equations to be solved to the form $F(\hat{x}) = 0$, the value of x for which $F(x)$ vanishes being written as $\hat{x}$. The procedures described below are iterative, requiring a first guess (a starter value) for x, say $x = x_1$, and they are rather dependent on x_1 being not too far from $\hat{x}$; although some corresponding procedures for more than one variable are used in the body of the book (e.g. the Poisson-Pascal distribution) these tend to be even more heavily dependent on good starter values and - particularly in automatic computing applications - the complexity of their behavior may call for specific advice from computing or numerical analysis specialists. The remarks below therefore relate explicitly only to iteration on a single variable, and even there they are at an intuitive rather than a rigorous level. For a more detailed analysis in a statistical context, see Barnett (1966), particularly with reference to multiple roots and global maxima.

A special case which occurs quite frequently is given by the (non-unique) re-writing of $F(\hat{x}) = 0$ in the form $\hat{x} = f(\hat{x})$. Thus the maximum likelihood equation for $\hat{\lambda}$ in the Poisson(λ) distribution with the zero class truncated is

$$\frac{\hat{\lambda}}{1-e^{-\hat{\lambda}}} = \bar{X}$$

in the usual notation; in the form $F(\hat{x}) = 0$ (writing x for λ to be consistent with the previous notation) this can be written as

$$F(\hat{x}) \equiv \frac{\hat{x}}{1-e^{-\hat{x}}} - \bar{X} = 0 \quad \text{or} \quad F(\hat{x}) \equiv \hat{x} - \bar{X}\,(1 - e^{-\hat{x}}) = 0; \quad .$$

and in the form $\hat{x} = f(\hat{x})$ as

$$\hat{x} = \bar{X}(1-e^{-\hat{x}}).$$

The iterative sequence is very simple: for a starter value x_1 the second value is found from $x_2 = f(x_1)$; and in general

$$x_{r+1} = f(x_r), \quad r = 1,2,\ldots$$

However it turns out that the rate of convergence of the sequence or even whether it does converge at all (if it converges it must converge to a root) is highly dependent on the form of $f(x)$ (in particular, on its first derivative), and the choice of f is substantially unrestricted - e.g. above, trivially

$$\hat{x} = 2\bar{X}(1-e^{-\hat{x}}) - \hat{x},$$

or, "solving" for the $\hat{x}$ in the exponent,

$$\hat{x} = \ln \bar{X} - \ln(\bar{X} - \hat{x}).$$

Nevertheless, because of the procedure's simplicity, it is often a convenient technique; two typical kinds of behavior in the following Figure 41 for different $f(x)$ are

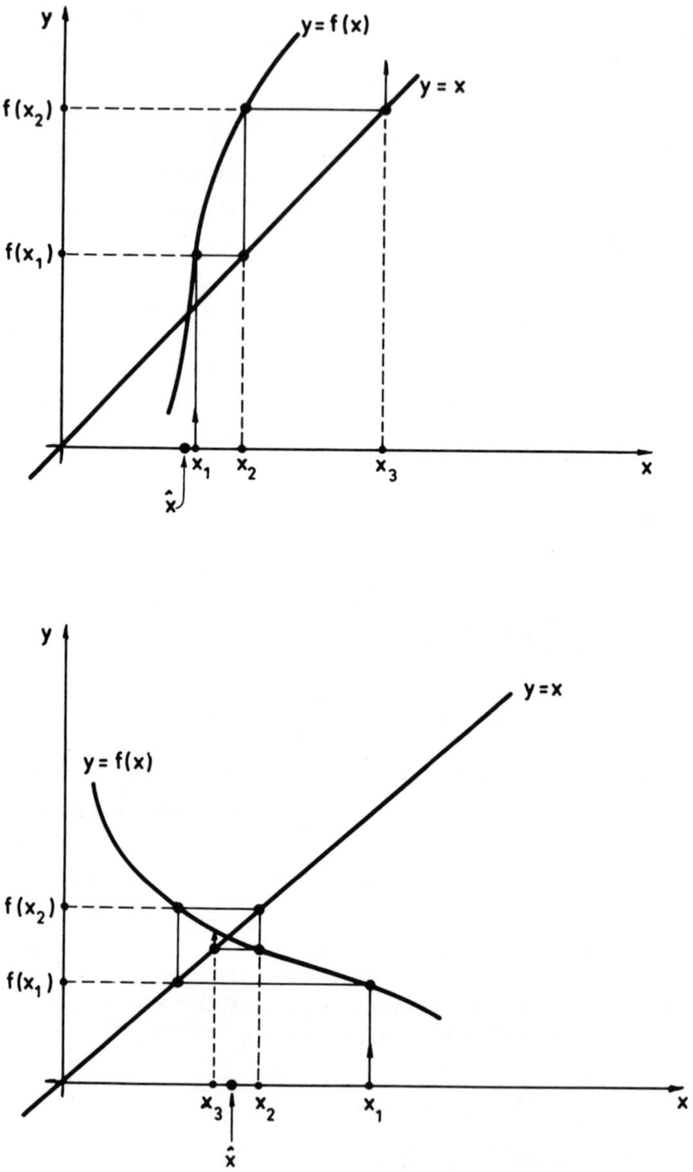

FIGURE 41 Iteration on $x = f(x)$.
(*top*) Divergence.
(*bottom*) Convergence.

illustrated. But it must be remembered that, usually, both the assumptions of the existence of a root and an appropriate starter value are assumptions of hope rather than fact (as is the case too with the following procedures): it is sometimes possible to establish by a preliminary examination that a root exists and a starter value is such that convergence obtains, but this is exceptional in a statistical context.

However, it may be possible (Corbishley (1972)) to write the rearranged form $\hat{x} = f(\hat{x})$ (of $F(\hat{x}) = 0$) in such a way as to produce rapid convergence (and indeed to change divergence to convergence). This can be done when the starter value x_1 is such that $f'(x_1)$ is not close to 1 but is close to $f'(\hat{x})$. Then for m a value near to $f'(x_1)$, and satisfying these two conditions, iteration on the rearranged form

$$\hat{x} = \frac{f(\hat{x}) - m\hat{x}}{1 - m} \equiv g(\hat{x}) \ , \quad \text{say} \ ,$$

with $x_{r+1} = g(x_r)$, $r = 1,2,\ldots,$ leads to rapid convergence. Taking the zero truncated Poisson case above with $\overline{X} = 2$ for illustration, consider the equation

$$\hat{x} = 2(1-e^{-\hat{x}}) . \qquad (\hat{x} = f(\hat{x}))$$

Inspection of a negative exponential table gives a starter value of $x_1 = 1.6$, and iterations run:

x_1	x_2	x_3	x_4	x_5	x_6
1.6	1.596207	1.594672	1.594050	1.593797	1.593695

without reaching stability at this sixth iteration (it does at the twelfth). However,

$$f'(x) = 2e^{-x} \quad \text{and} \quad f'(1.6) = 0.40\ldots \ .$$

So using the above transformation gives

$$\hat{x} = \frac{2(1-e^{-\hat{x}}) - 0.4\ \hat{x}}{0.6} \ , \qquad (\hat{x} = g(\hat{x}))$$

and iterations run:

x_1	x_2	x_3	x_4	x_5
1.6	1.593678	1.593625	1.593624	1.593624

This modification, apart from the possibility of converting divergence to convergence, is of particular value for hand iteration, when the number of iterations may be of substantial importance.

The second procedure described is the Newton-Raphson iteration. In its simplest form, if $x = \hat{x}$ is the root sought of $F(\hat{x}) = 0$, proceeding formally to a mean value or Taylor expansion about the starter value $x = x_1$ gives

$$0 = F(\hat{x}) = F(x_1) + (\hat{x} - x_1)\ F'(x_1) + \tfrac{1}{2}(\hat{x} - x_1)^2\ F''(\bar{x})$$

where $\bar{x}$ lies between $\hat{x}$ and x_1;

there is also the obvious assumption that F possesses well enough behaved derivatives for this expansion to hold. So

$$\hat{x} = x_1 - \frac{F(x_1)}{F'(x_1)} - \tfrac{1}{2}\frac{(\hat{x}-x_1)^2}{F'(x_1)} \cdot F''(\bar{x}),$$

$$F'(x_1) \neq 0.$$

Consequently, if x_1 is "near enough" for the change in F to be sensibly linear, when $F''(x)$ can be neglected, a second approximation is

$$x_2 = x_1 - \frac{F(x_1)}{F'(x_1)},$$

the Newton-Raphson formula. The argument can be refined to obtain conditions which ensure that x_2 is a better approximation, though this is rarely worth doing; and in practice the whole procedure is usually iterated in order to provide practical evidence regarding convergence, termination of the procedure being applied when the difference between successive iterates falls below some preassigned value. The readily obtained graphical representation provides helpful illumination of the procedure, and the following diagrams illustrate (for smooth functions) both its convergence and divergence, and its failure to detect existing roots.

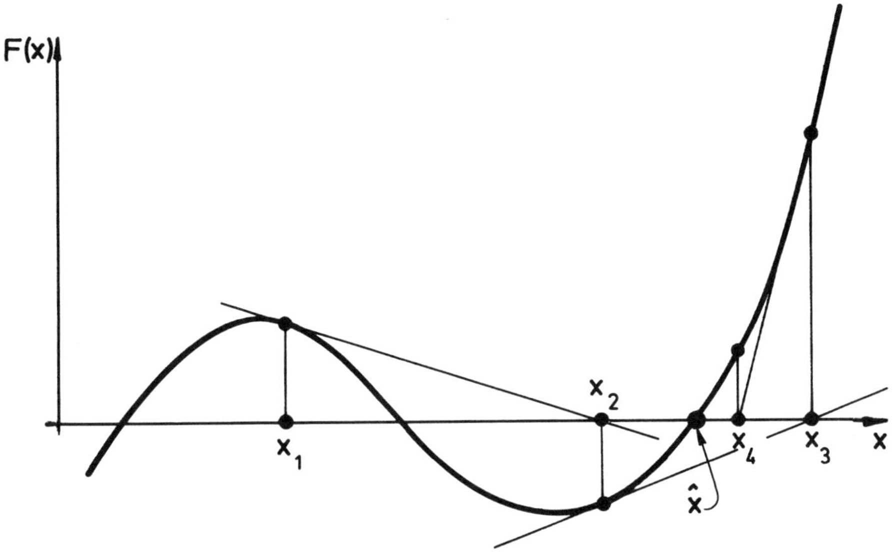

FIGURE 42a Newton-Raphson: convergence, but not to nearest root.

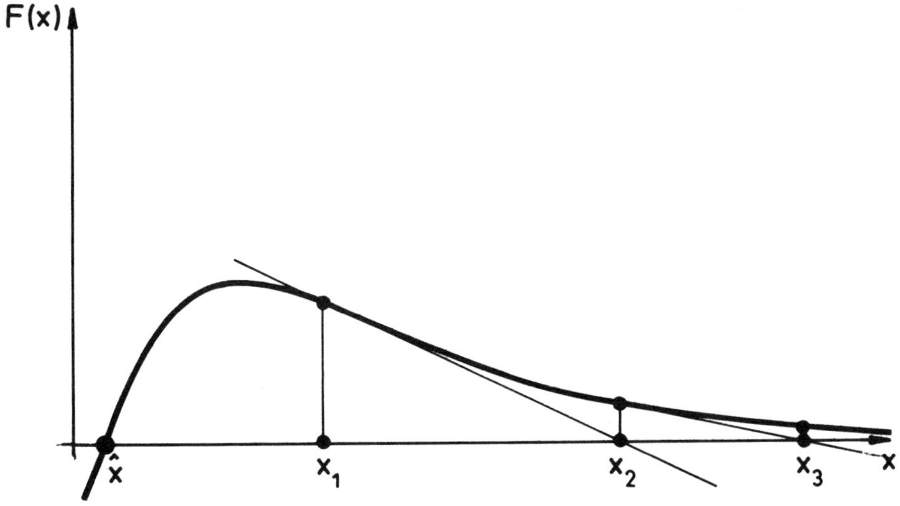

FIGURE 42b Newton-Raphson: divergence.

A third procedure, regula falsi or the rule of false positions, can sometimes be used to supplement or replace the Newton-Raphson. (There are of course very many other rules available: the above are widely used, fairly simple and relatively easy to control procedures.) It has the great advantage of inevitably converging on a root (for continuous F), though not necessarily to the one really required; and it does not require the derivative of F. But for the same accuracy it "generally" takes more iterations than does the Newton-Raphson procedure (a precise statement is rather complicated), and it requires two starter values x_1 and x_2 such that $F(x_1)$ and $F(x_2)$ have opposite signs. An analytical statement will not be given: in geometrical language the graph of $F(x)$ between x_1 and x_2 is replaced by the straight line joining $(x_1, F(x_1))$ and $(x_2, F(x_2))$, the intersection of this line with the x axis determining x_3. The iteration then consists of using whichever of the new pairs (x_1, x_3) and (x_2, x_3) have opposite signs for $F(x)$ in order to determine x_4, and so on. Again a diagram, necessarily representing convergence, illustrates the procedure, in Figure 43.

In all these procedures, the question of when to terminate an iteration sequence has to be answered. It is common to terminate when the absolute difference between successive iterates is "small enough" by some criterion, absolute or relative, and when the function value is "small enough". Failure of the procedure may thus occur when it leads to very small steps where the function is very flat (whether or not it is near a root) - an example is given in Gebski (1975), showing the care necessary to recognize that a root has been obtained. An advantage of a simple iterative procedure is that it is easier to understand its behavior in an anomalous case, and perhaps to modify it appropriately.

In addition to these procedures, there are also available general package programmes for maximization (or minimization) in several variables. Because estimation criteria are often in terms of optimization, e.g., maximizing the likelihood, the packages can sometimes be used directly for estimation though of course not necessarily advantageously for efficiency in calculation. When all else fails, the value of a criterion can be calculated over a grid of plausible parameter values, leading at worst to better starter values; it may also be found, e.g., that the criterion is insensitive to changes in parameter values and that the estimate is thus not well determined, a direct verification of indications often provided by the second derivatives of the function defining the criterion.

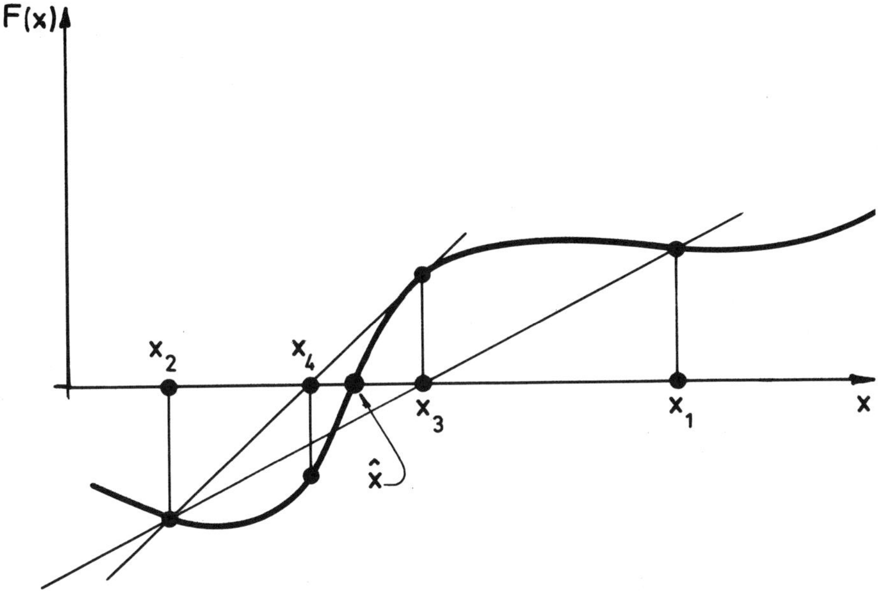

FIGURE 43 Regula Falsi.

Appendix 7

MOMENTS AND CUMULANTS

This Appendix lists, for easy reference, a number of "exact" results (many of which are either definitions or direct deductions from the definitions). They are mostly available in a scattered fashion in the literature, with a variety of notations (especially with respect to moments about the origin and the mean), but those particularly relevant to lattice variates are here collected in the notation of the rest of the book.

For a variate X with distribution function $F(x)$ certain expectation sequences appear repeatedly - these are briefly defined. Recalling that the expectation (where it exists) of a function $h(X)$ is

$$\mathcal{E}\{h(X)\} = \int h(x)\, dF(x),$$

the definitions and notations are as follows, all generating functions being exponential (i.e. the coefficient of $t^r/r!$ is that relevant).

7.1 MOMENTS ABOUT THE ORIGIN

$$\mu'_r = \mathcal{E}(X^r),$$

$$\text{with } \mu'_1 = \mathcal{E}(X) = \mu \text{ for brevity}, \quad r = 0,1,2,\ldots;$$

and moment g.f. $\mathcal{E}(e^{tX}) = \sum_{r=0}^{\infty} \mu'_r \frac{t^r}{r!}$.

7.2 MOMENTS ABOUT THE MEAN

$$\mu_r = \mathscr{E}\{(X-\mu)^r\} ;$$

with g.f. $\mathscr{E}\{e^{t(X-\mu)}\} = e^{-\mu t}\ \mathscr{E}(e^{tX})$.

Moments about an arbitrary value instead of μ are similarly defined.

7.3 FACTORIAL MOMENTS ABOUT THE ORIGIN

$$\mu'_{(r)} = \mathscr{E}\{X(X-1)\ldots(X-r+1)\} = \mathscr{E}\{X^{(r)}\};$$

with g.f. $\mathscr{E}\{(1+t)^X\}$.

7.4 FACTORIAL MOMENTS ABOUT THE MEAN

$$\mu_{(r)} = \mathscr{E}\{(X-\mu)(X-\mu-1)\ldots(X-\mu-r+1)\} = \mathscr{E}\{(X-\mu)^{(r)}\};$$

with g.f. $\mathscr{E}\{(1+t)^{X-\mu}\} = (1+t)^{-\mu}\ \mathscr{E}\{(1+t)^X\}$.

7.5 CUMULANTS

Writing the rth cumulant as κ_r ,

$$\sum_{r=1}^{\infty} \kappa_r \frac{t^r}{r!} = \ln\left(\sum_{r=0}^{\infty} \mu'_r \frac{t^r}{r!}\right) = \ln\{\mathscr{E}(e^{tX})\}.$$

7.6 FACTORIAL CUMULANTS

Similarly, for $\kappa_{(r)}$,

$$\sum_{r=1}^{\infty} \kappa_{(r)} \frac{t^r}{r!} = \ln\left(\sum_{r=0}^{\infty} \mu'_{(r)} \frac{t^r}{r!}\right) = \ln[\mathscr{E}\{1+t)^X\}].$$

Much more rarely, ascending factorial functions are used - e.g.

$$\mu'_{\{r\}} = \mathscr{E}\{X(X+1)(X+2)\ldots(X+r-1)\} = \mathscr{E}(X^{\{r\}})$$

$$= (-1)^r\ r!\ \mathscr{E}\{\binom{-X}{r}\}$$

with generating function

$$\mathscr{E}\{(1-t)^{-X}\}.$$

It is frequently desirable to change from one set of constants to another (particularly in minimum chi-squared estimation procedures) and so a short catalogue follows, giving explicitly all the inter-relations up to order 4. (Most of these can be obtained more expeditiously via g.f.s than directly; for repeated use such relations, or a subset to enable any required transformation to be achieved, should be stored in a computer.) Both sets of relations for a given pair of functions are shown together - e.g. those expressing κ_r in terms of μ_r, and μ_r in terms of κ_r are given under the one heading. In order to locate a particular set, the functions are labelled as follows:

μ'_r	μ_r	$\mu'_{(r)}$	$\mu_{(r)}$	κ_r	$\kappa_{(r)}$
1	2	3	4	5	6

so that the two sets mentioned above are labelled

[2,5] ,

the first member of [i,j] being chosen to be less than the second. The arrangement of the sets is then

[1,2], [1,3], [1,4], [1,5], [1,6]; [2,3], [2,4], [2,5], [2,6];

[3,4],...; [5,6].

μ'_r and μ_r

$$\mu'_1 = \mu$$
$$\mu'_2 = \mu_2 + \mu^2$$
$$\mu'_3 = \mu_3 + 3\mu_2\mu + \mu^3$$
$$\mu'_4 = \mu_4 + 4\mu_3\mu + 6\mu_2\mu^3 + \mu^4$$

[1,2]

$$\mu_1 = 0 \ , \ \mu = \mu'_1$$
$$\mu_2 = \mu'_2 - \mu^2$$
$$\mu_3 = \mu'_3 - 3\mu'_2\mu + 2\mu^3$$
$$\mu_4 = \mu'_4 - 4\mu'_3\mu + 6\mu'_2\mu^2 - 3\mu^4$$

μ'_r	and	$\mu'_{(r)}$

$$\mu'_1 = \mu'_{(1)} = \mu$$

$$\mu'_2 = \mu'_{(2)} + \mu$$

$$\mu'_3 = \mu'_{(3)} + 3\mu'_{(2)} + \mu$$

$$\mu'_4 = \mu'_{(4)} + 6\mu'_{(3)} + 7\mu'_{(2)} + \mu$$

[1,3]

$$\mu'_{(1)} = \mu'_1 = \mu$$

$$\mu'_{(2)} = \mu'_2 - \mu$$

$$\mu'_{(3)} = \mu'_3 - 3\mu'_2 + 2\mu$$

$$\mu'_{(4)} = \mu'_4 - 6\mu'_3 + 11\mu'_2 - 6\mu$$

μ'_r	and	$\mu_{(r)}$

$$\mu'_1 = \mu$$

$$\mu'_2 = \mu_{(2)} + \mu^2$$

$$\mu'_3 = \mu_{(3)} + \mu_{(2)}(3\mu + 3) + \mu^3$$

$$\mu'_4 = \mu_{(4)} + \mu_{(3)}(4\mu+6) + \mu_{(2)}(6\mu^2 + 12\mu + 7) + \mu^4$$

[1,4]

$$\mu_{(1)} = 0, \ \mu = \mu'_1$$

$$\mu_{(2)} = \mu'_2 - \mu^2$$

$$\mu_{(3)} = \mu'_3 + \mu'_2 (-3\mu - 3) + 2\mu^3 + 3\mu^2$$

$$\mu_{(4)} = \mu'_4 + \mu'_3 (-4\mu - 6) + \mu'_2(6\mu^2 + 18\mu + 11) - 3\mu^4 - 12\mu^3 - 11\mu^2$$

μ'_r	and	κ_r

$$\mu'_1 = \kappa_1 = \mu$$
$$\mu'_2 = \kappa_2 + \mu^2$$
$$\mu'_3 = \kappa_3 + 3\kappa_2\mu + \mu^3$$
$$\mu'_4 = \kappa_4 + 4\kappa_3\mu + 3\kappa_2^2 + 6\kappa_2\mu^2 + \mu^4$$

[1,5]

$$\kappa_1 = \mu'_1 = \mu$$
$$\kappa_2 = \mu'_2 - \mu^2$$
$$\kappa_3 = \mu'_3 - 3\mu'_2 + 2\mu^3$$
$$\kappa_4 = \mu'_4 - 4\mu'_3\mu - 3\mu'^2_2 + 12\mu'_2\mu^2 - 6\mu^4$$

μ'_r	and	$\kappa_{(r)}$

$$\mu'_1 = \kappa_{(1)} = \mu$$
$$\mu'_2 = \kappa_{(2)} + \mu^2 + \mu$$
$$\mu'_3 = \kappa_{(3)} + \kappa_{(2)}(3\mu+3) + \mu^3 + 3\mu^2 + \mu$$
$$\mu'_4 = \kappa_{(4)} + \kappa_{(3)}(4\mu+6) + 3\kappa_{(2)}^2 + \kappa_{(2)}(6\mu^2+18\mu+7) + \mu^4 + 6\mu^3 + 7\mu^2 + \mu$$

[1,6]

$$\kappa_{(1)} = \mu'_1 = \mu$$
$$\kappa_{(2)} = \mu'_2 - \mu^2 - \mu$$
$$\kappa_{(3)} = \mu'_3 + \mu'_2(-3\mu-3) + 2\mu^3 + 3\mu^2 + 2\mu$$
$$\kappa_{(4)} = \mu'_4 + \mu'_3(-4\mu-6) - 3\mu'^2_2 + \mu'_2(12\mu^2+18\mu+11) - 6\mu^4 - 12\mu^3 - 11\mu^2 - 6\mu$$

μ_r	and	$\mu'_{(r)}$

$$\mu_1 = 0, \qquad \mu = \mu'_{(1)}$$

$$\mu_2 = \mu'_{(2)} - \mu^2 + \mu$$

$$\mu_3 = \mu'_{(3)} + \mu'_{(2)}(-3\mu + 3) + 2\mu^3 - 3\mu^2 + \mu$$

$$\mu_4 = \mu'_{(4)} + \mu'_{(3)}(-4\mu + 6) + \mu'_{(2)}(6\mu^2 - 12\mu + 7)$$
$$-3\mu^4 + 6\mu^3 - 4\mu^2 + \mu$$

[2,3]

$$\mu'_{(1)} = \mu$$

$$\mu'_{(2)} = \mu_2 + \mu^2 - \mu$$

$$\mu'_{(3)} = \mu_3 + \mu_2(3\mu - 3) + \mu^3 - 3\mu^2 + 2\mu$$

$$\mu'_{(4)} = \mu_4 + \mu_3(4\mu - 6) + \mu_2(6\mu^2 - 18\mu + 11)$$
$$+ \mu^4 - 6\mu^3 + 11\mu^2 - 6\mu$$

μ_r	and	$\mu_{(r)}$

$$\mu_1 = 0$$

$$\mu_2 = \mu_{(2)}$$

$$\mu_3 = \mu_{(3)} + 3\mu_{(2)}$$

$$\mu_4 = \mu_{(4)} + 6\mu_{(3)} + 7\mu_{(2)}$$

[2,4]

$$\mu_{(1)} = 0$$

$$\mu_{(2)} = \mu_2$$

$$\mu_{(3)} = \mu_3 - 3\mu_2$$

$$\mu_{(4)} = \mu_4 - 6\mu_3 + 11\mu_2$$

μ_r	and	κ_r

$$\mu_1 = 0$$
$$\mu_2 = \kappa_2$$
$$\mu_3 = \kappa_3$$
$$\mu_4 = \kappa_4 + 3\kappa_2^2$$

[2,5]

$$\kappa_1 = \mu_1' = \mu$$
$$\kappa_2 = \mu_2$$
$$\kappa_3 = \mu_3$$
$$\kappa_4 = \mu_4 - 3\mu_2^2$$

μ_r	and	$\kappa_{(r)}$

$$\mu_1 = 0 \,, \qquad \mu = \kappa_{(1)}$$
$$\mu_2 = \kappa_{(2)} + \mu$$
$$\mu_3 = \kappa_{(3)} + 3\kappa_{(2)} + \mu$$
$$\mu_4 = \kappa_{(4)} + 6\kappa_{(3)} + 3\kappa_{(2)}^2 + \kappa_{(2)}(6\mu + 7) + 3\mu^2 + \mu$$

[2,6]

$$\kappa_{(1)} = \mu$$
$$\kappa_{(2)} = \mu_2 - \mu$$
$$\kappa_{(3)} = \mu_3 - 3\mu_2 + 2\mu$$
$$\kappa_{(4)} = \mu_4 - 6\mu_3 - 3\mu_2^2 + 11\mu_2 - 6\mu$$

$\mu'_{(r)}$	and	$\mu_{(r)}$

$$\mu'_{(1)} = \mu'_1 = \mu$$

$$\mu'_{(2)} = \mu_{(2)} + \mu^2 - \mu$$

$$\mu'_{(3)} = \mu_{(3)} + 3\mu_{(2)}\mu + \mu^3 - 3\mu^2 + 2\mu$$

$$\mu'_{(4)} = \mu_{(4)} + 4\mu_{(3)}\mu + \mu_{(2)}(6\mu^2 - 6\mu) + \mu^4 - 6\mu^3 + 11\mu^2$$

[3,4]

$$\mu_{(1)} = 0, \quad \mu = \mu'_{(1)}$$

$$\mu_{(2)} = \mu'_{(2)} - \mu^2 + \mu$$

$$\mu_{(3)} = \mu'_{(3)} - 3\mu'_{(2)}\mu + 2\mu^3 - 2\mu$$

$$\mu_{(4)} = \mu'_{(4)} - 4\mu'_{(3)}\mu + \mu'_{(2)}(6\mu^2 + 6\mu) - 3\mu^4 - 6\mu^3 + 3\mu^2 + 6\mu$$

$\mu'_{(r)}$	and	κ_r

$$\mu'_{(1)} = \kappa_1 = \mu$$

$$\mu'_{(2)} = \kappa_2 + \mu^2 - \mu$$

$$\mu'_{(3)} = \kappa_3 + \kappa_2(3\mu - 3) + \mu^3 - 3\mu^2 + 2\mu$$

$$\mu'_{(4)} = \kappa_4 + \kappa_3(4\mu - 6) + 3\kappa_2^2 + \kappa_2(6\mu^2 - 18\mu + 11) + \mu^4 - 6\mu^3 + 11\mu^2 - 6\mu$$

[3,5]

$$\kappa_1 = \mu'_{(1)} = \mu$$

$$\kappa_2 = \mu'_{(2)} - \mu^2 + \mu$$

$$\kappa_3 = \mu'_{(3)} + \mu'_{(2)}(-3\mu + 3) + 2\mu^3 - 3\mu^2 + \mu$$

$$\kappa_4 = \mu'_{(4)} + \mu'_{(3)}(-4\mu + 6) - 3\mu'^2_{(2)} + \mu'_{(2)}(12\mu^2 - 18\mu + 7) - 6\mu^4 + 12\mu^3 - 7\mu^2 + \mu$$

$\mu'_{(r)}$	and	$\kappa_{(r)}$

$$\mu'_{(1)} = \kappa_{(1)} = \mu$$

$$\mu'_{(2)} = \kappa_{(2)} + \mu^2$$

$$\mu'_{(3)} = \kappa_{(3)} + 3\kappa_{(2)}\mu + \mu^3$$

$$\mu'_{(4)} = \kappa_{(4)} + 4\kappa_{(3)}\mu + 3\kappa^2_{(2)} + 6\kappa_2\mu^2 + \mu^4$$

[3,6]

$$\kappa_{(1)} = \mu'_{(1)} = \mu$$

$$\kappa_{(2)} = \mu'_{(2)} - \mu^2$$

$$\kappa_{(3)} = \mu'_{(3)} - 3\mu'_{(2)}\mu + 2\mu^3$$

$$\kappa_{(4)} = \mu'_{(4)} - 4\mu'_{(3)}\mu - 3\mu'^2_{(2)} + 12\mu'_{(2)}\mu^2 - 6\mu^4$$

$\mu_{(r)}$	and	κ_r

$$\mu_{(1)} = 0$$

$$\mu_{(2)} = \kappa_2$$

$$\mu_{(3)} = \kappa_3 - 3\kappa_2$$

$$\mu_{(4)} = \kappa_4 - 6\kappa_3 + 3\kappa_2^2 + 11\kappa_2$$

[4,5]

$$\kappa_1 = \mu$$

$$\kappa_2 = \mu_{(2)}$$

$$\kappa_3 = \mu_{(3)} + 3\mu_{(2)}$$

$$\kappa_4 = \mu_{(4)} + 6\mu_{(3)} - 3\mu^2_{(2)} + 7\mu_{(2)}$$

$\mu_{(r)}$	and	$\kappa_{(r)}$

$$\mu_{(1)} = \kappa_{(1)} = \mu$$
$$\mu_{(2)} = \kappa_{(2)} + \mu$$
$$\mu_{(3)} = \kappa_{(3)} - 2\mu$$
$$\mu_{(4)} = \kappa_{(4)} + 3\kappa_{(2)}^2 + 6\kappa_{(2)}\mu + 3\mu^2 + 6\mu$$

[4,6]

$$\kappa_{(1)} = \mu_{(1)} = \mu$$
$$\kappa_{(2)} = \mu_{(2)} - \mu$$
$$\kappa_{(3)} = \mu_{(3)} + 2\mu$$
$$\kappa_{(4)} = \mu_{(4)} - 3\mu_{(2)}^2 - 6\mu$$

κ_r	and	$\kappa_{(r)}$

$$\kappa_1 = \kappa_{(1)} = \mu$$
$$\kappa_2 = \kappa_{(2)} + \mu$$
$$\kappa_3 = \kappa_{(3)} + 3\kappa_{(2)} + \mu$$
$$\kappa_4 = \kappa_{(4)} + 6\kappa_{(3)} + 7\kappa_{(2)} + \mu$$

[5,6]

$$\kappa_{(1)} = \kappa_1 = \mu$$
$$\kappa_{(2)} = \kappa_2 - \mu$$
$$\kappa_{(3)} = \kappa_3 - 3\kappa_2 + 2\mu$$
$$\kappa_{(4)} = \kappa_4 - 6\kappa_3 + 11\kappa_2 - 6\mu$$

For multivariate distributions, similar functions are defined - for example, in the bivariate case of X, Y with distribution function F(x,y), so

$$\mathscr{E}\{h(X,Y)\} = \int\int h(x,y)\ dF(x,y)\ ,$$

moments about the mean are defined as

$$\mu_{rs} = \mathscr{E}\{(X-\mu_{10})^r\ (Y-\mu_{01})^s\}$$

where

$$\mu_{10} = \mathscr{E}(X),\quad \mu_{01} = \mathscr{E}(Y)\ .$$

In particular the variances are

$$\mu_{20} = \mathscr{V}(X),\quad \mu_{02} = \mathscr{V}(Y),$$

and the covariance $\mathscr{C}(X,Y)$ is

$$\mathscr{C}(X,Y) = \mu_{11} = \mathscr{E}\{(X-\mu_{10})(Y-\mu_{01})\}\ ;$$

the correlation between X and Y is the standardized (dimensionless) function

$$\frac{\mathscr{C}(X,Y)}{\sqrt{\{\mathscr{V}(X)\ \mathscr{V}(Y)\}}}\ .$$

In the general case, for the variate $\underset{\sim}{X} = (X_1,X_2,\ldots,X_m)^t$ and dummy variable $\underset{\sim}{t} = (t_1,t_2,\ldots,t_m)^t$, the moment g.f. is

$$\mathscr{E}\left(e^{\Sigma t_i S_i}\right) = \mathscr{E}\left(e^{\underset{\sim}{t}^t \underset{\sim}{X}}\right)\ ;$$

other g.f.'s are similarly defined

7.7 RECURRENCE RELATIONS FOR CUMULANTS

For a number of common distributions there are useful recurrence relations for cumulants, of the form

$$\kappa_{r+1} = \frac{1}{\beta}\ \frac{\partial\kappa_r}{\partial\theta}\ ,\qquad r = 1,2,\ldots \qquad (*)$$

where θ is a parameter of the distribution and β some function of θ. For example, the Poisson(λ) distribution has

$$\kappa_{r+1} = \lambda \frac{\partial \kappa_r}{\partial \lambda}, \quad \kappa_1 = \lambda.$$

Following Qvale (1932), since the cumulant g.f. is

$$k(t;\theta) = \Sigma\, \kappa_r \frac{t^r}{r!},$$

$$\frac{\partial k}{\partial \theta} = \Sigma \frac{\partial \kappa_r}{\partial \theta} \cdot \frac{t^r}{r!} \quad \text{and} \quad \frac{\partial k}{\partial t} = \Sigma\, \kappa_{r+1} \frac{t^r}{r!}.$$

Hence

$$\frac{\partial k}{\partial \theta} - \beta \frac{\partial k}{\partial t} = -\beta \kappa_1 = \alpha, \quad \text{say},$$

if the relation (*) holds, α not being a function of t, and this partial differential equation therefore gives a condition for the existence of such a recurrence relation.

In terms of the p.g.f.

$$g(z;\theta) = \Sigma\, g_x z^x = \exp\{k(\ln z;\, \theta)\},$$

writing $z = e^t$ gives

$$\frac{\partial \ln g}{\partial \theta} - \beta z \frac{\partial \ln g}{\partial z} = \alpha.$$

Equating coefficients of z^x, if

$$\frac{\partial \ln g_x}{\partial \theta} = \alpha + \beta x,$$

then

$$\kappa_{r+1} = \frac{1}{\beta} \cdot \frac{\partial \kappa_r}{\partial \theta}.$$

Thus, for the Poisson(λ) distribution

$$\frac{\partial k}{\partial \lambda} - \beta \frac{\partial k}{\partial t} = 1 + e^t - \beta \lambda e^t$$

and the right hand side is constant for $\beta = 1/\lambda$; similarly $\dfrac{\partial \ln g_x}{\partial \lambda} = -1 + \dfrac{x}{\lambda}$ leads to $\beta = 1/\lambda$. Further examples are given in the Exercises.

On the other hand, for the Neyman(μ,ν) distribution

$$\frac{\partial k}{\partial \mu} - \beta \frac{\partial k}{\partial t} = - \beta\kappa_1$$

leads to a condition which cannot be satisfied: no relation of the above form exists. However, a natural generalization is

$$\kappa_{r+1} = \frac{1}{b}\frac{\partial \kappa_r}{\partial \theta} - \frac{a}{b}\kappa_r$$

corresponding to

$$\frac{\partial \kappa_r}{\partial \theta} = a\kappa_r + b\kappa_{r+1} ,$$

a and b being functions of θ. Exactly as before,

$$\frac{\partial k}{\partial \theta} - ak - b\frac{\partial k}{\partial t} = b\kappa_1$$

(though the introduction of explicit probabilities is not now helpful). Thus, taking the Neyman distribution with

$$k = -\mu + \mu e^{-\theta+\theta e^t}$$

leads to a left hand side of the differential equation of

$$\mu\{a - (1+a)\varepsilon + \varepsilon e^t(1-b\theta)\}, \quad \text{for} \quad \varepsilon = \varepsilon^{-\theta+\theta e^t} .$$

This is independent of t if

$$a = -1 \quad \text{and} \quad b = 1/\theta$$

whence in the usual notation $(\theta = \nu)$

$$\kappa_{r+1} = \nu\left(\frac{\partial \kappa_r}{\partial \nu} + \kappa_r\right) .$$

Of somewhat less interest is the form

$$k = -\theta + \theta e^{-\nu+\nu e^t} ,$$

leading to

$$\kappa_r = \mu\frac{\partial \kappa_r}{\partial \mu} , \quad (\theta = \mu) ,$$

from which it follows that

$$\kappa_r = \mu \times \text{function of } \nu .$$

Similar results for the Poisson-binomial and Poisson-Pascal distributions are given in the exercises.

7.8. SAMPLE FUNCTIONS

Corresponding to the moments and cumulants for a distribution there are functions for a random sample $X_1, X_2, \ldots, X_n$; these are defined exactly as for a distribution. Thus, for example, the rth sample moment about the origin is

$$M'_r = \frac{1}{n} \sum_{i=1}^{n} X_i^r \text{ , corresponding to } \mu'_r \text{ ,}$$

and similarly

$$M_r,\ M'_{(r)},\ M_{(r)},\ K_r,\ K_{(r)}$$

correspond exactly to

$$\mu_r,\ \mu'_{(r)},\ \mu_{(r)},\ \kappa_r,\ \kappa_{(r)} \ .$$

The relations interconnecting these sample functions are precisely those for the corresponding distribution functions.

However, the relations between sample and distribution functions need more precise examination: they become important immediately estimation problems arise, for example. Thus, while for moments about the origin

$$\mathscr{E}(M'_r) = \mu'_r \ ,$$

for moments about the mean

$$\mathscr{E}(M_r) \neq \mu_r \quad \text{for} \quad r > 1 \ -$$

e.g. $$\mathscr{E}(M_2) = \frac{n}{n-1}\mu_2 \ , \quad \text{and} \quad \mathscr{E}(K_2) = \frac{n}{n-1}\kappa_2 \ .$$

Hence k-statistics are introduced (Kendall and Stuart I (1963), e.g.) defined so that k_r, of degree not greater than r in the sample members, satisfies

$$\mathscr{E}(k_r) = \kappa_r ;$$

and $$\mathscr{E}(k_{(r)}) = \kappa_{(r)} \ .$$

(This notation clashes with usage elsewhere in this book, and is restricted to this section.)

Writing $S_r = \Sigma X_i^r$, the first four are:

$$k_1 = \frac{1}{n} S_1$$

$$k_2 = \frac{1}{n^{(2)}} (nS_2 - S_1^2)$$

$$k_3 = \frac{1}{n^{(3)}} (n^2 S_3 - 3nS_2 S_1 + 2S_1^3)$$

$$k_4 = \frac{1}{n^{(4)}} ((n^3 + n^2)S_4 - 4(n^2 + n)S_3 S_1 - 3(n^2 - n)S_2^2 + 12nS_2 S_1^2 - 6S_1^4),$$

or

$$k_1 = M_1' = \overline{X}$$

$$k_2 = \frac{n}{n-1} M_2$$

$$k_3 = \frac{n^2}{(n-1)(n-2)} M_3$$

$$k_4 = \frac{n^3}{(n-1)(n-2)(n-3)} \{(n+1)M_4 - 3(n-1)M_2^2\} .$$

$$k_{(1)} = k_1$$

$$k_{(2)} = k_2 - k_1$$

$$k_{(3)} = k_3 - 3k_2 + 2k_1$$

$$k_{(4)} = k_4 - 6k_3 + 11k_2 - 6k_1$$

(These are the same as the relations between $\kappa_{(r)}$ and κ_r , since they are linear.)

There is some variety of usage attached to 'sample variance': it is used both for

$$M_2 = \frac{1}{n} \Sigma(X_i - \overline{X})^2$$

and for

$$k_2 = \frac{1}{n-1} \Sigma(X_i - \overline{X})^2 .$$

Here it will be used exclusively for the unbiased estimator k_2 of $\mu_2 = \kappa_2$, and will mostly be written S^2, M_2 being called the sample second moment (about the mean).

Tables are available for finding the sampling variances of sample moments and unbiased estimators of these sampling variances. E.g.,

$$\mathcal{V}(M_2) = \left(\frac{n-1}{n}\right)^2 \left(\frac{\kappa_4}{n} + \frac{2\kappa_2^2}{n-1}\right),$$

and an unbiased estimator of (M_2) is

$$\frac{(n-1)^2}{n^2(n+1)} \left(\frac{n-1}{n} k_4 + 2k_2^2\right);$$

see Kendall and Stuart I (1963). In applications in this book, "exact" results are not usually relevant, and the usual "large sample approximations" are given in the Appendix <u>Approximate Means, Variances and Covariances</u>.

7.9 EXERCISES

1. Show that the Qvale relation

$$\frac{\partial \ln g_x}{\partial \theta} = \alpha + \beta x$$

leads to the following recurrences.

binomial(ν,θ): $$\kappa_{r+1} = \theta(1-\theta) \frac{\partial \kappa_r}{\partial \theta} .$$

negative binomial(κ,θ): $$\kappa_{r+1} = \theta(1+\theta) \frac{\partial \kappa_r}{\partial \theta} .$$

logarithmic(θ): $$\kappa_{r+1} = \theta \frac{\partial \kappa_r}{\partial \theta} .$$

2. Show that the extended Qvale relation

$$\frac{\partial K}{\partial \theta} - aK - b \frac{\partial K}{\partial t} = -b\kappa_1$$

leads to the following recurrence relations.

Poisson-binomial(μ,ν,π) , $\pi=\theta$,

$$\kappa_{r+1} = \pi(1-\pi)\frac{\partial\kappa_r}{\partial\pi} + \nu\pi\kappa_r .$$

Poisson-Pascal(μ,κ,ρ) , $\rho=\theta$,

$$\kappa_{r+1} = \rho(1+\rho)\frac{\partial\kappa_r}{\partial\rho} + \kappa\rho\kappa_r .$$

(Polya-Aeppli: $\kappa_{r+1} = \rho(1+\rho)\frac{\partial\kappa_r}{\partial\rho} + \rho\kappa_r$)

(E.g. Poisson-binomial: left hand side is for $\varepsilon = 1-\theta+\theta e^t$

$$a\mu + \mu\varepsilon^{\nu-1}\{-\nu - a + a\theta + e^t(\nu - a\theta - \nu b\theta)\} ,$$

which is independent of t for

$$a = -\nu/\chi , \quad b = 1/\pi\chi .)$$

3. For use when estimators expressed in terms of the sample moments are relevant, it is sometimes convenient to have expressions in terms of cumulants for

$$\mathscr{C}(M'_r, M'_s) = \mu'_{r+s} - \mu'_r\mu'_s .$$

Show that:

$$\mu'_2 - \mu'^2_1 = \kappa_2 = \kappa_{(2)} + \mu, \quad \mu = \mu'_1$$

$$\mu'_3 - \mu'_2\mu'_1 = \kappa_3 + 2\kappa_2\mu$$

$$= \kappa_{(3)} + \kappa_{(2)}(2\mu+3) + 3\mu^2$$

$$\mu'_4 - \mu'^2_2 = \kappa_4 + 4\kappa_3\mu + 2\kappa_2^2 + 4\kappa_2\mu^2$$

$$= \kappa_{(4)} + \kappa_{(3)}(4\mu+6) + 2\kappa^2_{(2)} + \kappa_{(2)}(4\mu^2+16\mu+7)$$

$$+ 4\mu^3 + 6\mu^2 + \mu .$$

Appendix 8

SERIES MANIPULATIONS

8.1 CAUCHY PRODUCT OF SERIES

Given the series

$$\alpha_1 + \alpha_2 + \dots + \alpha_i + \dots \quad \text{and} \quad \beta_1 + \beta_2 + \dots + \beta_j + \dots ,$$

a 'natural' meaning to give to the indicated product

$$(\alpha_1 + \alpha_2 + \dots)(\beta_1 + \beta_2 + \dots)$$

is

$$\alpha_1\beta_1 + (\alpha_1\beta_2 + \alpha_2\beta_1) + (\alpha_1\beta_3 + \alpha_2\beta_2 + \alpha_3\beta_1) + \dots .$$

This is especially so if the series are power series:

$$a_0 + a_1 z + \dots + a_i z^i + \dots$$

and

$$b_0 + b_1 z + \dots + b_j z^j + \dots ,$$

for this grouping is then that of successive powers of z. So the Cauchy product of these series is defined to be

$$a_0 b_0 + (a_0 b_1 + a_1 b_0) z + (a_0 b_2 + a_1 b_1 + a_2 b_0) z^2 + \dots$$
$$+ p_x z^x + \dots, \quad \text{say,}$$

where $p_x = \sum\sum_{i+j=x} a_i b_j$, all $i,j = 0,1,2,\ldots,x$

such that $i+j = x$

$$= \sum_{j=0}^{x} a_i b_{x-i} .$$

For power series, sufficient conditions for the convergence of the product series are that each of the component series is absolutely convergent: convergence is however not particularly relevant to generating functions.

This product is especially important to convolutions: if $a_0 + a_1 z + \ldots$ and $b_0 + b_1 z + \ldots$ are the p.g.f.'s of independent variates then the coefficient of z^x in their Cauchy product is the probability that the convolution variate takes the value x; similarly, the moments and cumulants for a convolution variate are given by coefficients in Cauchy products of moment and cumulant g.f.'s.

8.2 DOUBLE SUMS: INTERCHANGE OF ORDER OF SUMMATION

The (absolutely convergent, or finite) triangular array

$$\begin{array}{l} a_{00} + a_{01} + a_{02} + \ldots + a_{0n} + \ldots \\ \quad + a_{11} + a_{12} + \ldots + a_{1n} + \ldots \\ \qquad + a_{22} + \ldots + a_{2n} + \ldots \\ \qquad\qquad \ldots \\ \qquad\qquad + a_{mn} + \ldots \\ \qquad\qquad\qquad \ldots \end{array}$$

can be written either as

$$\sum_{n=0}^{\infty} a_{0n} + \sum_{n=1}^{\infty} a_{1n} \sum_{n=2}^{\infty} a_{2n} + \ldots + \sum_{n=m}^{\infty} a_{mn} + \ldots$$

$$= \sum_{m=0}^{\infty} \sum_{n=m}^{\infty} a_{mn} , \text{ summing first by rows,}$$

or as

$$\sum_{m=0}^{0} a_{m0} + \sum_{m=0}^{1} a_{m1} + \sum_{m=0}^{2} a_{m2} + \ldots + \sum_{m=0}^{n} a_{mn} + \ldots$$

$$= \sum_{n=0}^{\infty} \sum_{m=0}^{n} a_{mn}\ , \text{ summing first by columns.}$$

This result,

$$\sum_{m=0}^{\infty} \left(\sum_{n=m}^{\infty} a_{mn} \right) = \sum_{n=0}^{\infty} \left(\sum_{m=0}^{n} a_{mn} \right) ,$$

is of frequent use - see for example the expression of moments in terms of factorial moments.

There are various special forms in which the result appears:

$$\sum_{m=\alpha}^{\beta} \sum_{n=m}^{\beta} a_{mn} = \sum_{n=\alpha}^{\beta} \sum_{m=\alpha}^{n} a_{mn} ;$$

for $a_{mn} = b_m c_n$ the sums can be dealt with separately; and for power series

$$\sum_{m=0}^{\infty} \left(\sum_{n=m}^{\infty} a_{mn} z^n \right) = \sum_{x=0}^{\infty} z^x \left(\sum_{s=0}^{x} a_{sx} \right)$$

so that the coefficient of z^x is the finite sum in parentheses - this device is used extensively in operations with p.g.f.'s in order to identify the p.f. p_x. A further special case is

$$\sum_{m=0}^{\infty} z^m \sum_{n=0}^{\infty} a_{mn} z^n = \sum_{x=0}^{\infty} z^x \sum_{s=0}^{x} a_{s,x-s} ,$$

used, e.g., in the discussion of the Polya-Aeppli distribution.

There is also the integral form

$$\int_0^{\infty} \left\{ \int_0^{y} f(x,y)dx \right\} dy = \int_0^{\infty} \left\{ \int_x^{\infty} f(x,y)dy \right\} dx.$$

Appendix 9

STIRLING NUMBERS

For distributions in which the variate values are integers, the factorial powers

$$x^{(m)} = x(x-1)(x-2)\dots(x-m+1)$$

very frequently lead to simpler manipulation than do the ordinary powers. The Stirling Numbers are the coefficients in the linear relations which connect these powers: unfortunately there is no standard notation for them. What is used here is a minor adaptation of that used in Jordan (1965) and Abramowitz and Stegun (1968) with cursive writing and typescript in mind, so that the German or script S for the Second Kind of Number is replaced by a slashed S, $\$$. As well, the placement of the suffixes has to be noted. What is here and in Abramowitz and Stegun written $S_m^{(n)}$, is written S_n^m by Richardson (1954) and, following him, Gurland in a number of places.

9.1 STIRLING NUMBERS OF THE FIRST KIND

Early factorial powers written out in full as powers of x are:

$$x^{(1)} = \qquad = x,$$

$$x^{(2)} = x(x-1) = -x + x^2$$

$$x^{(3)} = x(x-1)(x-2) = 2x - 3x^2 + x^3 ,$$

$$x^{(4)} \qquad = -6x + 11x^2 - 6x^3 + x^4 .$$

Stirling Numbers of the First Kind are defined as the coefficients of powers of x in such expansions:

$$x^{(m)} = S_m^{(1)}\ x + S_m^{(2)}\ x^2 + \ldots + S_m^{(m)}\ x^m ,$$

where the subscript on S corresponds to the index of the factorial power of x. A short Table is given below.

Short Table of Stirling Numbers $S_m^{(n)}$ and $\mathscr{S}_m^{(n)}$

$S_m^{(n)} : x^{(m)} = x(x-1)(x-2)\ldots(x-m+1) = S_m^{(1)}x + S_m^{(2)}x^2 + S_m^{(3)}x^3 + \ldots + S_m^{(n)}x^m$

m \ n	1	2	3	4	5	6	7	8
1	1							
2	-1	1						
3	2	-3	1					
4	-6	11	-6	1				
5	24	-50	35	-10	1			
6	-120	274	-225	85	-15	1		
7	720	-1764	1624	-735	175	-21	1	
8	-5040	13068	-13132	6769	-1960	322	-28	1

$\mathscr{S}_m^{(n)} : x^m = \mathscr{S}_m^{(1)}x^{(1)} + \mathscr{S}_m^{(2)}x^{(2)} + \mathscr{S}_m^{(3)}x^{(3)} + \ldots + \mathscr{S}_m^{(m)}x^{(m)}$

m \ n	1	2	3	4	5	6	7	8
1	1							
2	1	1						
3	1	3	1					
4	1	7	6	1				
5	1	15	25	10	1			
6	1	31	90	65	15	1		
7	1	63	301	350	140	21	1	
8	1	127	966	1701	1050	266	28	1

Since $x^{(m+1)} = x^{(m)}(x-m)$, multiplying the defining relation by $x-m$ and equating coefficients of x^n on each side leads to the recurrence relation

$$S_{m+1}^{(n)} = S_m^{(n-1)} - m\, S_m^{(n)}$$

where

$$S_m^{(m)} = 1,\ S_m^{(1)} = (-1)^{m-1}\,(m-1)! \quad \text{and} \quad S_m^{(0)} = 0.$$

Writing $x = 1$ in the defining relation gives ("row sums are zero")

$$0 = S_m^{(1)} + S_m^{(2)} + \ldots + S_m^{(m)}\ ;$$

$x = -1$ gives

$$(-1)(-2)\ldots(-m) = (-1)^1 S_m^{(1)} + (-1)^2 S_m^{(2)} + \ldots + (-1)^m S_m^{(m)}$$

or

$$m! = \sum_{n=1}^{m} (-1)^{m+n}\, S_m^{(n)} = \Sigma |S_m^{(n)}|\ ,$$

the terms on the right being all positive.

Taking expectations of the definite relation expressed in terms of random variables yields immediately

$$\mu'_{(m)} = \sum_{n=1}^{m} S_m^{(n)}\, \mu'_n\ ;$$

and there are exactly corresponding relations between cumulants and factorial cumulants.

Since a binomial coefficient is simply expressed in terms of a factorial power a corresponding expansion follows:

$$\binom{x}{m} = \frac{x^{(m)}}{m!} = \frac{1}{m!}\, \Sigma\, S_m^{(n)}\, x^n\ .$$

Further, the "ascending" factorial power

$$\begin{aligned} x^{\{m\}} &= x(x+1)(x+2)\ldots(x+m-1) = (-1)^m(-x)(-x-1)\ldots(-x-m+1) \\ &= (x+m-1)^{(m)} \qquad\qquad = (-1)^m(-x)^{(m)}; \end{aligned}$$

hence

$$x^{\{m\}} = \Sigma(-1)^{m+n}\, S_m^{(n)} x^n \qquad = \Sigma|S_m^{(n)}|\ x^n\ .$$

9.2 STIRLING NUMBERS OF THE SECOND KIND

It is equally the case that 'ordinary' powers can be expanded in terms of factorial powers:

e.g. $$x = x^{(1)},$$
$$x^2 = x + x(x-1) = x^{(1)} + x^{(2)},$$
$$x^3 = x + x(x-1) + x(x-1)(x-2) = x^{(1)} + 3x^{(2)} + x^{(3)} ,$$
$$x^4 = x^{(1)} + 7x^{(2)} + 6x^{(3)} + x^{(4)};$$

thus Stirling Numbers of the Second Kind are defined by

$$x^m = \mathcal{S}_m^{(1)} x^{(1)} + \mathcal{S}_m^{(2)} x^{(2)} + \ldots + \mathcal{S}_m^{(m)} x^{(m)} .$$

It is convenient for both Kinds to extend the definitions with

$$S_m^{(0)} = \mathcal{S}_m^{(0)} = \delta_{m0}$$

where δ_{mn} is the Kronecker delta (which is 0 for $m \neq n$ and 1 for $m = n$).

Since $x = (x-n) + n$, and $(x-n)x^{(n)} = x^{(n+1)}$, multiplying the defining relation by x and equating coefficients of $x^{(n)}$ gives

$$\mathcal{S}_{m+1}^{(n)} = \mathcal{S}_m^{(n-1)} + n\, \mathcal{S}_m^{(n)}$$

with

$$\mathcal{S}_m^{(m)} = 1 , \mathcal{S}_m^{(1)} = 1.$$

Writing $x = -1$ gives

$$(-1)^m = \Sigma(-1)^n\, n!\, \mathcal{S}_m^{(n)};$$

taking expectations of X^m gives,

$$\mu'_m = \Sigma\, \mathcal{S}_m^{(n)}\, \mu'_{(n)}$$

with corresponding results for cumulants.

In terms of ascending factorials $x^{\{n\}}$, the expansion

$$x^m = \Sigma(-1)^{m+n}\, \mathcal{S}_m^{(n)}\, x^{\{n\}}$$

follows immediately.

Tables of both Kinds of Stirling Numbers are to be found in Abramowitz and Stegun (1968); $n!S_m^{(n)}$ are often called "the differences of zero", $\Delta^n 0^m = \{\Delta^n x^m\}_{x=0}$, and are tabulated in a number of places (see later; and the short tabulation in Section 9.1).

9.3 ORTHOGONAL PROPERTIES

The two kinds of Stirling Numbers have orthogonality properties. Since $x^m = \Sigma S_m^{(n)} x^{(n)}$ and $x^{(n)} = \Sigma \mathcal{S}_n^{(p)} x^p$, replacement of $x^{(n)}$ in the first equation gives

$$x^m = \sum_n S_m^{(n)} \sum_p \mathcal{S}_n^{(p)} x^p$$
$$= \Sigma \Sigma S_m^{(n)} \mathcal{S}_n^{(p)} x^p .$$

So equating coefficients of powers of x,

$$\sum_n S_m^{(n)} \mathcal{S}_n^{(p)} = 1 \quad \text{if} \quad p = m$$
$$= 0 \quad \text{otherwise,}$$

or, using the Kronecker delta δ_{mp} (1 for $p = m$, 0 otherwise),

$$\sum_n S_m^{(n)} \mathcal{S}_n^{(p)} = \delta_{mp}$$
$$= \sum_n S_n^{(p)} \mathcal{S}_m^{(n)} ,$$

where the second relation follows similarly.

The basic relations can also be written in matrix notation. Define

$$\underset{\sim}{x} = (x, x^2, \ldots, x^m)^t, \quad \underset{\sim}{x}^F = (x^{(1)}, x^{(2)}, \ldots, x^{(m)})^t,$$

$$\underset{\sim}{S} = \begin{bmatrix} S_1^{(1)} & 0 & & \ldots & 0 \\ S_2^{(1)} & S_2^{(2)} & 0 & \ldots & 0 \\ S_3^{(1)} & S_3^{(2)} & S_3^{(3)} & \ldots & 0 \\ & \cdots & \cdots & & \\ S_m^{(1)} & S_m^{(2)} & S_m^{(3)} & \ldots & S_m^{(m)} \end{bmatrix}$$

and $\underset{\sim}{\$}$ as for $\underset{\sim}{S}$ with $\$_i^{(j)}$ in place of $S_i^{(j)}$. Then the defining equations for $\underset{\sim}{S}$ and $\underset{\sim}{\$}$ are

$$\underset{\sim}{x}^F = \underset{\sim}{S}\,\underset{\sim}{x} \quad \text{and} \quad \underset{\sim}{x} = \underset{\sim}{\$}\,\underset{\sim}{x}^F ,$$

and the orthogonal relations follow immediately:

$$\underset{\sim}{S}\,\underset{\sim}{\$} = \underset{\sim}{\$}\,\underset{\sim}{S} = \underset{\sim}{1}.$$

9.4 DIFFERENCES AND DERIVATIVES OF ZERO

Since
$$\begin{aligned}\Delta\, x^{(m)} &= (x+1)^{(m)} - x^{(m)} \\ &= (x+1)\; x(x-1)\ldots(x+1-\overline{m-1}) - x(x-1)\ldots(x-\overline{m-1}) \\ &= x(x-1)\ldots(x-\overline{m-2})(x+1-x+m-1) \\ &= m\; x^{(m-1)} ,\end{aligned}$$

then
$$\begin{aligned}\Delta^n\; x^{(m)} &= m(m-1)\ldots(m-n+1)x^{(m-n)}, \quad n \le m \\ &= m^{(n)}\; x^{(m-n)} .\end{aligned}$$

Hence
$$\begin{aligned}\Delta^n\; x^m &= \Delta^n\; \Sigma\; \$_m^{(p)}\; x^{(p)} \\ &= \Sigma\; \$_m^{(p)}\; p^{(n)}\; x^{(p-n)} , \quad n \le p\end{aligned}$$

with zero terms otherwise.

Writing $x = 0$, and noting $n^{(n)} = n!$,

$$\Delta^n\; 0^m = n!\; \$_m^{(n)} , \quad \text{the "differences of zero"},$$

i.e.,
$$\$_m^{(n)} = \frac{\Delta^n\; 0^m}{x!} .$$

These are tabulated, e.g., in Fisher and Yates (1967).

Exactly in the same way, for $D = d/dx$,

$$Dx^m = mx^{m-1}$$

and
$$D^n\; x^m = m^{(n)}\; x^{m-1} ,$$

whence
$$D^n\; x^{(m)} = \Sigma\; S_m^{(p)}\; p^{(n)}\; x^{p-n} .$$

Writing $x = 0$ thus gives

$$S_m^{(n)} = \frac{D^n \, 0^{(m)}}{n!} \quad .$$

Alternative arguments use Taylor and Newton expansions. The (polynomial) Taylor expansion for $x^{(m)}$ about $x = 0$ is

$$x^{(m)} = \sum_{n=1}^{m} \frac{x^n}{n!} D^n \, 0^{(m)}$$

$$= \sum_{n=1}^{m} S_m^{(n)} \, x^n$$

by the definition of the Stirling numbers; equating coefficients gives the result for derivatives of zero. Similarly, the Newton expansion for x^m is

$$x^m = \sum_{n=1}^{m} \frac{x^{(n)}}{n!} \Delta^n \, 0^m \qquad \left(\frac{x^{(n)}}{n!} = \binom{x}{n} \right)$$

$$= \sum_{n=1}^{m} \mathfrak{S}_m^{(n)} \, x^{(n)} \, ,$$

whence the result for differences.

For many statistical purposes the notation of differences and derivatives of zero is particularly clumsy and tends to obscure the simplicity of their applications.

9.5 THE FAA DI BRUNO FORMULA

The 'function of a function' rule for the nth derivative of a 'composite' function is needed frequently, and can be expressed in a variety of ways.

Suppose $H = g(f)$ and $f = f(z)$; what is sought is an expression for

$$\frac{d^m H}{dz^m} = H^{(m)}(z)$$

in terms of derivatives of g and f.

a) Direct differentiation gives, somewhat tediously, the first few derivatives. Writing

$$g' = dg/df, \quad f' = df/dz, \text{ etc.}$$

then

$$H' = g'f',$$

$$H'' = g''(f')^2 + g'f'',$$

$$H''' = g'''(f')^3 + 3g''f' + g'f''',$$

$$H^{iv} = g^{iv}(f')^4 + 6g'''f''(f')^2 + g''(4f'''f' + 3(f'')^2) + g'f^{iv} ,$$

and, writing $\{f^{(m)}\}^p = f_m^p$ in these next two formulae only,

$$H_5 = g_5f_1^5 + 10g_4f_2f_1^3 + g_3(10f_3\ f_1^2 + 15f_2^2\ f_1)$$
$$+ g_2(5f_4f_1 + 10f_3f_2) + g_1f_5 ,$$
$$H_6 = g_6f_1^6 + 15g_5f_2f_1^4 + g_4(20\ f_3f_1^3 + 45f_2^2f_1^2)$$
$$+ g_3(15f_4f_1^2 + 60f_3f_2f_1 + 15f_2^3)$$
$$+ g_2(6f_5f_1 + 15f_4f_2 + 10f_3^2) + g_1f_6 .$$

A computer algorithm is given by Klimko (1973).

b) To obtain the general formula, repeated Taylor expansions can be used. For

$$f(z) = \sum_{s=0}^{\infty} \frac{(z-z_0)^s}{s!} f^{(s)}(z_0)$$

and

$$H(z) = \sum_{n=0}^{\infty} \frac{(f-f_0)^n}{n!} g^{(n)}(f_0) \quad \text{where} \quad f_0 = f(z_0)$$

$$= \sum_{n=0}^{\infty} \frac{1}{n!} \left\{ \sum_{s=1}^{\infty} \frac{(z-z_0)^s}{s!} f^{(s)}(z_0) \right\}^n g^{(n)}(f_0) .$$

The repeated Cauchy product of $\Sigma\, a_s z^s/s!$ gives

$$\left\{ \sum_{s=1}^{\infty} a_s \frac{z^s}{s!} \right\}^n = \sum_{m=n}^{\infty} \frac{z^m}{m!} \sum \frac{m!}{n_1!n_2!\dots n_m!} a_1^{n_1} \left(\frac{a_2}{2!}\right)^{n_2} \dots \left(\frac{a_m}{m!}\right)^{n_m}$$

where the second sum is over non-negative integers n_i such that

$$n_1 + n_2 + \ldots + n_m = n$$

$$n_1 + 2n_2 + \ldots + mn_m = m,$$

and hence the coefficient of $(z - z_0)^m/m!$ in the expansion for $H(z)$ is

$$\sum_{n=0}^{m} g^{(n)}(f_0) \sum \frac{m!}{n_1!n_2!\ldots n_m!} (f_0')^{n_1} \left(\frac{f_0^{(2)}}{2!}\right)^{n_2} \ldots \left(\frac{f_0^{(m)}}{m!}\right)^{n_m}$$

$$= H^{(m)}(z_0)$$

where $f_0^{(m)} = f^{(m)}(z_0)$, the Faa di Bruno Formula, in which the subscript 0 can now be dropped. Two special cases are, for non-negative integers $n_1, n_2, \ldots, n_m$:

$$\frac{d^m e^{k(t)}}{dt^m} = m!\ e^{k(t)} \sum \frac{1}{n_1!n_2!\ldots n_m!} \left(\frac{k'}{1!}\right)^{n_1} \left(\frac{k^{(2)}}{2!}\right)^{n_2} \ldots \left(\frac{k^{(m)}}{m!}\right)^{n_m}$$

summed over $1.n_1 + \ldots m\, n_m = m.$

and

$$\frac{d^m \ln \mathscr{m}(t)}{dt^m} = m! \sum_{n=1}^{m} (-1)^{m-1} (n-1)! \sum \frac{1}{n_1!\ldots n_m!} \left(\frac{\mathscr{m}'}{\mathscr{m}.1!}\right)^{n_1} \ldots \left(\frac{\mathscr{m}^{(m)}}{\mathscr{m}.m!}\right)^{n_m}$$

summed over $n_1 + \ldots + n_m = n,$

$$1.m_1 + \ldots + m\, n_m = m\ .$$

If $k(t)$ is interpreted as the cumulant g.f. and $\mathscr{m}(t)$ as the moment g.f., so

$$k(t) = \sum_{r=1}^{\infty} \kappa_r \frac{t^r}{r!}\ ,\quad \mathscr{m}(t) = \sum_{r=0}^{\infty} \mu_r' \frac{t^r}{r!}$$

and

$$k(t) = \ln \mathscr{m}(t)\ ,$$

the usual relations between κ_r and μ_r' are obtained.

Other applications are given in the section relating to Stopped Distributions, and Lukacs (1955) has additional material.

c) An expression can also be given in terms of differences: the details may be found in Jordan (1965), para. 12, (8). Writing

$$\Delta_h f(z) = f(z+h) - f(z)$$

and $$H(z) = g(f(z)),$$

$$H^{(m)}(z_0) = \sum_{n=1}^{m} \frac{g^{(n)}(f)}{n!} \left[\frac{d^m}{dh^m} \left\{ \Delta_h f(z) \right\}^n \right]_{h=0}.$$

An example, used in the section on Mixed Distribution given by taking $f(z) = e^z$. Then

$$\Delta_h e^z = e^z(e^h - 1)$$

and $$\{\Delta e^z\}^n = e^{nz}(e^h-1)^n = e^{nz}\ n!\ \Sigma \frac{h^m}{m!} \mathfrak{S}_m^{(n)}$$

Differentiating this p times with respect to h

$$e^{nz}\ n!\ \Sigma \frac{m^{(p)}\ h^{m-p} \mathfrak{S}_m^{(n)}}{m!} \ ;$$

writing $h = 0$ gives 0 except for $p = m$ when it is

$$e^{nz}\ n!\ \mathfrak{S}_m^{(n)}$$

and so

$$\frac{d^m g(e^z)}{dz^m} = \Sigma \frac{d^n g}{df^n} e^{nz} \mathfrak{S}_m^{(n)} \quad \text{for} \quad f = e^z .$$

A second application is for $f(z) = \ln z$; then

$$\frac{d^m\ g(\ln z)}{dz^m} = \Sigma \frac{d^n g}{df^n} \frac{1}{z^n} {}'S_m^{(n)} \quad \text{for} \quad f = \ln z.$$

9.6 THE OPERATOR $x \frac{d}{dx}$

It is common to meet repeated applications of the process in which differentiation is followed by multiplication by the independent variable:

$$x \frac{df(x)}{dx} = x\ D\ f(x) = T\ f(x), \text{ say, with } T = x\ D.$$

Clearly $T^2 = (xD)(xD) = x\ D + x^2 D^2$,

$$T^3 = x\ D + 3x^2 D^2 + x^3 D^3 ,$$

and in general

$$T^m = a_1 xD + a_2 x^2 D^2 + \ldots + a_m x^m D^m$$

where $a_1, a_2, \ldots$ are constants. The simplest function to use to determine these is $f(x) = x^q$, since

$$D^n\ x^q = q^{(p)}\ x^{q-n}$$

and $$T^m\ x^q = q^m\ x^q .$$

Hence

$$q^m\ x^q = \Sigma\ a_n\ q^{(n)}\ x^q$$

i.e. $$q^m = \Sigma\ a_n\ q^{(n)} .$$

But the factorial expansion of q^m is by definition $\Sigma\ \mathcal{S}_m^{(n)}\ q^{(n)}$, and so

$$T^m = \Sigma\ \mathcal{S}_m^{(n)}\ x^n\ D^n .$$

Applying this to the p.g.f. $g(z)$ of a discrete variate X, so

$$g(z) = \Sigma\ p_x\ z^x ,$$

then

$$T^r\ g(z) = \Sigma\ x^r\ p_x\ z^x ,$$

and writing $z = 1$ gives the rth moment μ_r' :

$$T^r\ g(z)\big|_{z=1} = \mu_r' , \text{ or } T^r g(1) = \mu_r' .$$

Thus, for the binomial distribution in which $g(z) = (\chi + \pi z)^\nu$,

$$T^r\ g(z) = \Sigma\ \mathcal{S}_r^{(p)}\ z^p\ D^p\ (\chi + \pi z)^\nu$$
$$= \Sigma\ \mathcal{S}_r^{(p)}\ \nu^{(p)}\ \pi^p\ (\chi + \pi z)^{\nu - p} ,$$

and on writing $z = 1$,

$$\mu_r' = \Sigma\ \mathcal{S}_r^{(p)}\ \nu^{(p)}\ \pi^p .$$

By use of the orthogonal relations, D^m can be expressed in terms of $T, T^2, \ldots, T^m$. For

$$T^n = \Sigma\ \mathcal{S}_n^{(q)}\ x^q\ D^q\ ;$$

multiplying by $S_m^{(n)}$ and summing over n gives

$$\Sigma\ S_m^{(n)}\ T^n = \Sigma\ (\Sigma\ S_m^{(n)}\ \mathcal{S}_n^{(q)}\ x^q\ D^q$$
$$= x^m D^m$$

i.e. $$D^m = \frac{1}{x^m}\ \Sigma\ S_m^{(n)}\ T^n .$$

9.7 THE OPERATOR $x\Delta$.

As it was convenient to study $x\,\frac{d}{dx}$, it is also useful to consider

$$x\ \Delta = U\ , \text{ say,}$$

with $$U^2 = (x\ \Delta)(x\ \Delta) = x\ \Delta + x(x+1)\ \Delta^2$$
$$= x\ \Delta + x^{\{2\}}\ \Delta^2$$
$$U^3 = x\ \Delta + 3x^{\{2\}}\ \Delta^2 + x^{\{3\}}\ \Delta^3\ , \text{ etc.}$$

Precisely as before, using $\Delta^n x^{\{q\}} = q^{(n)}(x+n)^{\{q-n\}}$,

$$U^m = \Sigma\ \mathcal{S}_m^{(n)}\ x^{\{n\}}\ \Delta^n$$
$$= \Sigma\ \mathcal{S}_m^{(n)}(x+n-1)^{(n)}\ \Delta^n\ ;$$

the inverse relation is

$$\Delta^m = \frac{1}{x^{\{m\}}}\ \Sigma\ S_m^{(n)}\ U^n\ .$$

9.8 RELATIONS BETWEEN DERIVATIVES AND DIFFERENCES

Using a difference with arbitrary interval h, relations between derivatives and differences can be obtained in terms of Stirling Numbers. From the Taylor expansion

$$f(x) = f(a) + (x-a)\ D\ f(a) + \ldots + \frac{(x-a)^m}{m!}\ D^m\ f(a) + \ldots$$

But Newton's expansion is

$$f(x) = f(a) + (x-a)_h\ \frac{\Delta f(a)}{h} + \ldots + \binom{x-a}{m}_h\ \frac{\Delta^m f(a)}{h^m} + \ldots,$$

where

$$\binom{x-a}{m}_h = \frac{(x-a)(x-a-h)\ldots(x-a-mh+h)}{m!}\ ,\ \Delta = \Delta_h\ ,$$

and

$$\Delta^n\ \binom{x-d}{m} = h^n \binom{x-a}{m-n}_h\ .$$

Since

$$\binom{x-a}{m}_h = \frac{1}{m!}\ \Sigma\ S_m^{(p)}\ h^{m-p}(x-a)^p\ ,$$

its nth derivative is

$$D^n \binom{x-a}{m}_h = \frac{1}{m!}\ \Sigma\ S_m^{(p)}\ h^{m-p}\ p^n (x-a)^{p-n}$$

which at $x = a$ reduces to

$$\frac{n!}{m!}\ h^{m-n}\ S_m^{(n)}\ .$$

Hence

$$D^s\ f(x) = \sum_{n=s}^{\infty} \frac{(x-a)^{n-s}}{(n-s)!}\ D^n\ f(a)$$

whence $$D^n\ f(a) = \sum_{m=n}^{\infty} \frac{n!}{m!}\ \frac{\Delta^m f(a)}{h^n}\ S_m^{(n)}\ .$$

The inverse result, giving differences in terms of derivatives follows by multiplying the above result by

$$h^n\ \mathscr{S}_n^{(k)}/n!$$

and summing over n from k onwards, using the orthogonal relations: thus, finally, replacing n by m, and k by n,

$$\Delta^n\ f(a) = \sum_{m=n}^{\infty} \frac{n!}{m!}\ h^m\ D^m\ f(a)\ \mathscr{S}_m^{(n)}\ .$$

9.9 GENERATING FUNCTIONS

The definition of the Stirling Number of the First Kind gives a (power) generating function for it in terms of its superscript:

$$x^{(m)} = \Sigma \mathscr{S}_m^{(n)} x^n .$$

Generating functions in terms of the subscript are widely useful: they are of exponential type:

$$\frac{\{\ln (1+z)\}^n}{n!} = \sum_{m=n}^{\infty} S_m^{(n)} \frac{z^m}{m!} ,$$

$$\frac{(e^z - 1)^n}{n!} = \sum_{m=n}^{\infty} \mathscr{S}_m^{(n)} \frac{z^m}{m!} .$$

Easy proofs follow from use of moment generating functions. For the moment g.f. of a variate X is

$$\mathscr{E}(e^{zX}) = \sum_m \mu'_m \frac{z^m}{m!} ;$$

the factorial moment g.f. is

$$\mathscr{E}\{(1+t)^X\} = \sum_m \mu'_{(m)} \frac{t^m}{m!} , \text{ say.}$$

These g.f.'s are the same if $1+t=e^z$; replacing

$$t \text{ by } e^z - 1$$

thus leads to

$$\Sigma \mu'_{(n)} \frac{(e^z - 1)^n}{n!} = \Sigma \mu'_m \frac{z^m}{m!} .$$

But

$$\mu'_m = \sum_{n=0}^{m} \mathscr{S}_m^{(n)} \mu'_{(n)} , \text{ including } m=0 \text{ with } \mathscr{S}_m^{(0)} = \delta_{m0} ,$$

so $$\Sigma \mu'_m \frac{z^m}{m!} = \sum_{m=0}^{\infty} \sum_{n=0}^{m} \mathscr{S}_m^{(n)} \mu'_{(n)} \frac{z^m}{m!}$$

$$= \sum_{n=0}^{\infty} \mu'_{(n)} \left(\sum_{m=n}^{\infty} \mathscr{S}_m^{(n)} \frac{z^m}{m!} \right),$$

i.e. $$\frac{(e^z - 1)^n}{n!} = \sum_{m=n}^{\infty} \mathcal{S}_m^{(n)} \frac{z^m}{m!} .$$

Replacing z by $\ln(1+t)$ similarly leads to the g.f. for $S_m^{(n)}$.

Since

$$e^t - 1 = \sum_{m_i=1}^{\infty} \frac{t^{m_i}}{m_i!} ,$$

considering the Cauchy product corresponding to

$$(e^t - 1)^n$$

leads to the (combinatorial) formula

$$\mathcal{S}_m^{(n)} = \frac{m!}{n!} \sum \frac{1}{m_1! m_2! \ldots m_n!}$$

where the sum is taken over the $m_i > 0$ such that $\Sigma m_i = m$. In precisely the same way

$$S_m^{(n)} = (-1)^{m+n} \frac{m!}{n!} \sum \frac{1}{m_1 m_2 \ldots m_n} .$$

An important application of the g.f. for the Second Kind of Stirling Number is to the moments of the Poisson distribution: because of the fundamental role the Poisson distribution plays in discrete distribution theory, these appear in a wide variety of contexts. The Poisson moment g.f. is

$$e^{-\lambda+\lambda e^t} = e^{\lambda(e^t-1)}$$

$$= \sum_n \frac{\lambda^n (e^t-1)^n}{n!}$$

$$= \sum_n \lambda^n \sum_{m=n}^{\infty} \mathcal{S}_m^{(n)} \frac{t^m}{m!}$$

$$= \sum_{m=0}^{\infty} \frac{t^m}{m!} \sum_{n=0}^{m} \lambda^n \mathcal{S}_m^{(n)} ,$$

i.e. the mth moment μ'_m is given by

$$\mu'_m = \sum_{n=0}^{m} \mathcal{S}_m^{(n)} \lambda^n .$$

The most comprehensive and accessible reference on Stirling Numbers is probably Jordan (1965). A variety of statistical applications is given in Douglas (1971), where an effort is made to demonstrate how little more than the definitions are needed to unify a number of otherwise apparently unrelated results. In particular, while an expression like

$$\left(x \frac{d}{dx}\right)^m = \sum \mathcal{S}_m^{(n)} x^n \left(\frac{d}{dx}\right)^n$$

can be used to obtain m^{th} moments in many cases, these can be obtained directly and naturally from the appropriate generating functions without such a special device.

9.10 EXERCISE

Define the rth power and factorial moments μ'_r and $\mu'_{(r)}$ about the origin for a variate X with p.f. $\phi(x)$. Introducing the Stirling Numbers, obtain general formulae which connect μ'_r and $\mu'_{(r)}, \mu'_{(r-1)}, \ldots,$ and $\mu'_{(r)}$ and $\mu'_r, \mu'_{r-1}, \ldots$.

If X is a Poisson variate with mean λ, obtain general formulae for $\mu'_{(r)}$ and μ'_r.

If Y is a truncated-at-zero Poisson variate with

$$\Pr(Y=y) = \frac{e^{-\lambda}}{1-e^{-\lambda}} \frac{\lambda^y}{y!}, \quad y = 1,2,3,\ldots \quad \text{and} \quad T = Y_1 + Y_2 + \ldots + Y_n$$

where each Y_i is independent with the above distribution of Y, find an explicit expression for $\Pr(T=t)$.

$$\left(\mu'_{(r)} = \lambda^r ; \ \Pr(T=t) = \frac{n!}{(e^{\lambda}-1)^n} \mathcal{S}_t^{(n)} \frac{\lambda^t}{t!}\right).$$

```
      ∇R←STIRL1 M
[1]    →(R←1=M)/0
[2]    R←STIRL1 M-1
[3]    R←(0,R)+(R×-M-1),0
      ∇
```

```
      ∇R←STIRL2 M
[1]    →(R←1=M)/0
[2]    R←STIRL2 M-1
[3]    R←(0,R)+(R×⍳M-1),0
      ∇
```

FIGURE 44 APL functions for calculating Stirling numbers of order M.

Appendix 10

VECTOR AND MATRIX RESULTS

The following results are collected for reference: they are used repeatedly in various sections. (The transpose of a vector or matrix is indicated by a raised ^{t}.)

10.1 THE CHAIN RULE: JACOBIANS

Consider the n component vector $\underset{\sim}{x} = (x_1,\ldots,x_n)^t$, the scalar function $h = f(\underset{\sim}{x})$, the vector function $\underset{\sim}{\theta} = \underset{\sim}{\phi}(\underset{\sim}{x}) = (\phi_1(\underset{\sim}{x}),\ldots,\phi_p(\underset{\sim}{x}))^t$, and the functional connection of $\underset{\sim}{y} = (y_1,\ldots,y_q)^t$ with $\underset{\sim}{x}$ denoted by $\underset{\sim}{y} = \underset{\sim}{u}(\underset{\sim}{x})$, so that h and $\underset{\sim}{\theta}$ are also functions of $\underset{\sim}{y}$, say $h = g(\underset{\sim}{y})$ and $\underset{\sim}{\theta} = \underset{\sim}{\psi}(\underset{\sim}{y})$.

The vector derivative of f with respect to $\underset{\sim}{x}$ is defined as usual:

$$dh = \frac{\partial f}{\partial x_1}\,dx_1 + \ldots + \frac{\partial f}{\partial x_n}\,dx_n = \left(\frac{\partial f}{\partial x_1},\ldots,\frac{\partial f}{\partial x_n}\right)\cdot(dx_1,\ldots,dx_n)^t$$

$$= \left(\frac{\partial f}{\partial \underset{\sim}{x}}\right)^t \cdot d\underset{\sim}{x}$$

where $\frac{\partial f}{\partial \underset{\sim}{x}} = \begin{pmatrix} \partial f/\partial x_1 \\ \vdots \\ \partial f/\partial x_n \end{pmatrix}$, $d\underset{\sim}{x} = \begin{pmatrix} dx_1 \\ \vdots \\ dx_n \end{pmatrix}$.

Equally,

$$dh = \left(\frac{\partial g}{\partial \underset{\sim}{y}}\right)^t \cdot d\underset{\sim}{y} ,$$

and $$dy_1 = \left(\frac{\partial \mu_1}{\partial \underset{\sim}{x}}\right)^t \cdot d\underset{\sim}{x}, \ldots, dy_q = \left(\frac{\partial u_q}{\partial \underset{\sim}{x}}\right)^t \cdot d\underset{\sim}{x} ,$$

i.e. $$d\underset{\sim}{y} = \begin{pmatrix} \partial u_1/\partial x_1, \ldots, \partial u_1/\partial x_n \\ \ldots \\ \partial u_q/\partial x_1, \ldots, \partial u_q/\partial x_n \end{pmatrix} \cdot d\underset{\sim}{x}$$

$$= \frac{\partial \underset{\sim}{u}}{\partial \underset{\sim}{x}} \cdot d\underset{\sim}{x}$$

where $\partial \underset{\sim}{u}/\partial \underset{\sim}{x}$ is called the Jacobian matrix of the transformation. Equating the coefficients of $d\underset{\sim}{x}$ in the two expressions for dh,

$$\left(\frac{\partial f}{\partial \underset{\sim}{x}}\right)^t = \left(\frac{\partial g}{\partial \underset{\sim}{y}}\right)^t \cdot \frac{\partial \underset{\sim}{u}}{\partial \underset{\sim}{x}}$$

i.e. $$\frac{\partial f}{\partial \underset{\sim}{x}} = \left(\frac{\partial \underset{\sim}{u}}{\partial \underset{\sim}{x}}\right)^t \cdot \frac{\partial g}{\partial \underset{\sim}{y}} \quad \left(\text{or } \frac{\partial}{\partial \underset{\sim}{x}} = \left(\frac{\partial \underset{\sim}{y}}{\partial \underset{\sim}{x}}\right)^t \cdot \frac{\partial}{\partial \underset{\sim}{y}}\right) \tag{1}$$

Again

$$d\underset{\sim}{\theta} = \frac{\partial \underset{\sim}{\phi}}{\partial \underset{\sim}{x}} \cdot d\underset{\sim}{x}$$

$$= \frac{\partial \underset{\sim}{\psi}}{\partial \underset{\sim}{y}} \cdot d\underset{\sim}{y} = \frac{\partial \underset{\sim}{\psi}}{\partial \underset{\sim}{y}} \cdot \frac{\partial \underset{\sim}{u}}{\partial \underset{\sim}{x}} \cdot d\underset{\sim}{x} ,$$

whence

$$\frac{\partial \underset{\sim}{\phi}}{\partial \underset{\sim}{x}} = \frac{\partial \underset{\sim}{\psi}}{\partial \underset{\sim}{y}} \cdot \frac{\partial \underset{\sim}{u}}{\partial \underset{\sim}{x}} \quad \left(\text{or } \frac{\partial}{\partial \underset{\sim}{x}} = \frac{\partial}{\partial \underset{\sim}{y}} \cdot \frac{\partial \underset{\sim}{y}'}{\partial \underset{\sim}{x}}\right) \tag{2}$$

(Notice the difference between (1) and (2): $\partial/\partial \underset{\sim}{x}$ is ambiguous.)

10.2 DERIVATIVES OF LINEAR AND QUADRATIC FORMS

Write $\underset{\sim}{v} = (v_1,\ldots,v_n)^t$ and

$$\underset{\sim}{\alpha} = \begin{pmatrix} \alpha_{11},\ldots,\alpha_{1n} \\ \cdots \\ \alpha_{n1},\ldots,\alpha_{nn} \end{pmatrix}, \quad \underset{\sim}{\beta} = \begin{pmatrix} \beta_{11},\ldots,\beta_{1n} \\ \\ \beta_{m1},\ldots,\beta_{mn} \end{pmatrix}.$$

By direct examination of the expressions involved, for example

$$\underset{\sim}{v}^t \underset{\sim}{x} = \underset{\sim}{x}^t \underset{\sim}{v} = v_1 x_1 + \ldots + v_n x_n = \frac{\partial}{\partial \underset{\sim}{x}} (\underset{\sim}{x}^t \underset{\sim}{v}) ,$$

$$\frac{\partial}{\partial \underset{\sim}{x}} (\underset{\sim}{v}^t \underset{\sim}{x}) = \begin{pmatrix} v_1 \\ \vdots \\ v_n \end{pmatrix} = \underset{\sim}{v} = \frac{\partial}{\partial \underset{\sim}{x}} (\underset{\sim}{x}^t \underset{\sim}{\beta}^t) ,$$

$$\frac{\partial}{\partial \underset{\sim}{x}} (\underset{\sim}{\beta} \underset{\sim}{x}) = \underset{\sim}{\beta} ,$$

$$\frac{\partial}{\partial \underset{\sim}{x}} (\underset{\sim}{x}^t \underset{\sim}{x}) = 2\underset{\sim}{x},$$

$$\frac{\partial}{\partial \underset{\sim}{x}} (\underset{\sim}{x}^t \underset{\sim}{\alpha} \underset{\sim}{x}) = \underset{\sim}{\alpha} \underset{\sim}{x} + \underset{\sim}{\alpha}^t \underset{\sim}{x}$$

$$= 2\underset{\sim}{\alpha} \underset{\sim}{x} \quad \text{if } \underset{\sim}{\alpha} \text{ is symmetric.}$$

10.3 EXPECTATION VECTOR AND COVARIANCE MATRIX OF LINEAR FUNCTIONS

If $\underset{\sim}{X}$ is a random variable, with mean $\mathscr{E}(\underset{\sim}{X})$ and covariance $\underset{\sim}{\mathscr{V}}(\underset{\sim}{X})$ with $\underset{\sim}{\beta}$ the constant matrix specified above, then

$$\mathscr{E}(\underset{\sim}{\beta} \underset{\sim}{X}) = \underset{\sim}{\beta} \, \mathscr{E}(\underset{\sim}{X})$$

and

$$\underset{\sim}{\mathscr{V}}(\underset{\sim}{\beta} \underset{\sim}{X}) = \underset{\sim}{\beta} \, \underset{\sim}{\mathscr{V}}(\underset{\sim}{X}) \, \underset{\sim}{\beta}^t .$$

GLOSSARY

The following descriptions of abbreviations and notations are of the briefest kind, with particular emphasis on symbols used in a number of different ways. They are intended primarily to serve as a reminder of things known but temporarily forgotten, rather than as definitions or complete statements.

No rational ordering of the entries seems practicable.

estimate	:	a (numerical) value of an estimator
estimator	:	a random variable used for purposes of estimation
g.f.	:	generating function
iff	:	if and only if (a necessary and sufficient condition)
p.f.	:	probability function
p.g.f.	:	probability g.f.
sup	:	supremum or least upper bound (lub)
variate	:	synonomous with random variable

$\mathscr{C}$ $\quad \mathscr{C}(X,Y), = \mathscr{E}[\{X- \mathscr{E}(X)\}\{Y- \mathscr{E}(Y)\}]$

(Covariance)

$\mathscr{E}$ $\quad \mathscr{E}(m(X)) = \int_{\text{all } x} m(x)\, dF(x)$

(expected value of, or expectation of, m(X))

$\underset{\sim}{\mathscr{E}}$ $\quad \underset{\sim}{\mathscr{E}}(\underset{\sim}{X})$ or $\mathscr{E}(\underset{\sim}{X}) = (\mathscr{E}(X_1), \ldots, \mathscr{E}(X_m))^t$

k_r k-statistics : used only in Appendix 7.

K_r sample cumulants

$K_{(r)}$ sample factorial cumulants

M'_r, M_r sample moments about origin, mean

$M'_{(r)}, M_{(r)}$ sample factorial moments

Pr Pr(A): The probability of an event A

$S_m^{(n)}$, $\$_m^{(n)}$ Stirling Numbers of First, Second Kinds

u_α upper 100% point of a standard normal variate.

$\mathscr{V}$ $\mathscr{V}(X) = \mathscr{E}[\{X - \mathscr{E}(X)\}^2] = \mathscr{E}(X^2) - \{\mathscr{E}(X)\}^2$

(Variance)

$\underset{\sim}{\mathscr{V}}$ $\underset{\sim}{\mathscr{V}}(\underset{\sim}{X}) = (\mathscr{C}(X_i, X_j))$, $i,j = 1,2,\ldots,m$

(Covariance matrix of $\underset{\sim}{X} = (X_1, X_2, \ldots, X_m)^t$)

B $B(x,y) = \int_0^1 t^{x-1} (1-t)^{y-1} dt, \quad x,y > 0$

(beta x,y)

Γ $\Gamma(x) = \int_0^\infty e^{-t} t^{x-1} dt, \quad x > 0$

(gamma x) = (x-1)!

δ_{mn} (Kronecker delta) For integral m,n δ_{mn} is 0 if $m \neq n$ and is 1 if $m = n$.

ε as in $x \varepsilon A$: x is an element of the set A.

$\chi^2_{\alpha;p}$ upper 100α% point of a chi-squared variate with p degrees of freedom

! as in $x! = x(x-1)\ldots3.2.1$ for x a positive integer (if not, $(x-1)! = \Gamma(x)$)

(factorial x, x shriek)

$\sim$ as in $x \sim y$: x is asymptotically equal to (often read as 'twiddles')

as in $\underset{\sim}{X}$: $\underset{\sim}{X}$ is a (column) vector or a matrix

as in $\tilde{\mu}$: $\tilde{\mu}$ is an estimator of a parameter μ

: as in X : Poisson(λ), meaning X is Poisson distributed with parameter λ

as in $\underset{\sim}{X} : N(\underset{\sim}{\mu}, \underset{\sim}{\Sigma})$, meaning $\underset{\sim}{X}$ is normally distributed with mean $\underset{\sim}{\mu}$ and covariance matrix $\underset{\sim}{\Sigma}$.

as in $X : e^{-\lambda+\lambda z}$, meaning X has the indicated (Poisson(λ)) p.g.f.

frequently to be read as "such that".

(m) as in $x^{(m)}$: factorial power $x^{(m)} = x(x-1)\ldots(x-m+1)$

as in $f^{(m)}(x)$: m^{th} derivative

(note special use with Stirling Numbers)

{m} as in $x^{\{m\}}$: ascending factorial power $x^{\{m\}} = x(x+1)\ldots(x+m+1)$

' as in $f'(x)$: first derivative

as in $\mu'_{(r)}$: moment about the origin (as opposed to about the mean)

| as in $\Pr(A|B)$: the probability of A given B

as in $f(x)|_{x=x_1}$: the value of $f(x)$ at $x = x_1$, $f(x_1)$.

t as in $\underset{\sim}{\alpha}^t$: transpose of a vector or matrix (interchange rows and columns)

$|x|$: the numerical value of x taken as positive (read as "mod x"; $|3| = |-3| = 3$)

^ as in $\hat{\mu}$: maximum likelihood estimator of μ

o as in $\overset{o}{\mu}$: moment estimator of μ

* as in μ^*: minimum chi-squared (or some other specified) estimator of μ

REFERENCES

Abramowitz, M. and Stegun, I. (ed.) (1966). *Handbook of Mathematical Functions*, U. S. Nat. Bur. Stand., App. Math. Ser. 55.

Ahmed, M. S. (1961). On a locally most powerful boundary randomized similar test for the independence of two Poisson variables. *Annals of Mathematical Statistics*, 32, 809-827.

Aitken, A. C. (1947). *Statistical Mathematics* (5th ed.) Oliver and Boyd.

Anscombe, F. J. (1950). Sampling theory of the negative binomial and logarithmic series distributions. *Biometrika*, 37, 358-382.

Arbous, A. G. and Kerrich, J. E. (1951). Accident statistics and the concept of accident-proneness. *Biometrics*, 7, 340-342.

Archibald, E. E. A. (1948). Plant populations - a new application of Neyman's contagious distributions. *Annals of Botany*, 12, 221-235.

Barankin, E. W. and Gurland, J. (1951). On asymptotically normal, efficient estimators. *University of California Publications in Statistics*, 1, 89-129.

Bardwell, G. E. and Crow, E. L. (1964). A two-parameter family of hyper-Poisson distributions. *Journal of the American Statistical Association*, 59, 133-141.

Barnes, H. and Marshall, S. M. (1951). On the variability of replicate plankton samples and some applications of contagious series to the statistical distribution of catches over restricted periods. *Journal of the Marine Biology Association, U.K.* 30, 233-263.

Barnes, H. and Stanbury, F. A. (1951). A statistical study of plant distribution during the colonization and early development of vegetation on china clay residues. *Journal of Ecology*, 39, 171-181.

Barnett, V. D. (1966). Evaluation of the maximum likelihood estimator where the likelihood equation has multiple roots. *Biometrika*, 53, 151-165.

Bartholomew, D. J. (1967). *Stochastic Models for Social Processes*. Wiley.

Bartko, J. J., Greenhouse, S. W. and Patlak, C. S. (1968). On expectations of some functions of Poisson variates. *Biometrics*, 24, 97-102.

Bartlett, M. S. (1952). The statistical significance of odd bits of information. *Biometrika*, 39, 228-246.

Barton, D. E. (1957). The modality of Neyman's contagious distribution of type A. *Trabajos Estadistica*, 8, 13-22.

Barton, D. E. (1967). Completed runs of length k above and below median (Query 25). *Technometrics*, 9, 682-694.

Barton, D. E., David, F. N. and Merrington, M. (1963). Tables for the solution of the exponential equation $\exp(b) - b/(1-p) = 1$. *Biometrika*, 50, 169-176.

Bateman, G. I. (1950). The power of the χ^2 index of dispersion test when Neyman's contagious distribution is the alternate hypothesis. *Biometrika*, 37, 59-63.

Beall, G. (1940). The fit and significance of contagious distributions when applied to observations on larval insects. *Ecology*, 21, 460-474.

Beall, G. and Rescia, R. R. (1953). A generalization of Neyman's contagious distributions. *Biometrics*, 9, 344-386.

Best, D. J. (1974). The variance of the inverse binomial estimator. *Biometrika*, 61, 385-386.

Best, D. J. (1975). The difference between two Poisson expectations. *Australian Journal of Statistics*, 17, 29-35.

Blackman, G. E. (1935). A study by statistical methods of the distribution of species in grassland associations. *Annals of Botany*, 49, 749-777.

Blackman, G. E.(1942). Statistical and ecological studies on the distribution of species in plant communities (i). *Annals of Botany*, 6, 351-366.

Berg, S. (1975). A note on the connection between factorial series distribution and zero-truncated power series distribution. *Scandanavian Actuarial Journal*, 233-237.

Berman, M. M. (1975). *Poisson-stopped distributions*. (Unpublished M. Stat. project, U.N.S.W.).

Beyer, W. H. (ed.) (1968). *Handbook of Tables for Probability and Statistics*, 2nd ed. Chemical Rubber Co.

Blischke, W. R. (1965). Mixtures of discrete distributions. *Classical and Contagious Discrete Distributions*, (ed. G. P. Patil). Statistical Publications Society, Calcutta, 351-372.

Bliss, C. I. (1953). Fitting the negative binomial distribution to biological data. *Biometrics*, 9, 176-196.

Bliss, C. I. and Owen, A. R. G. (1958). Negative binomial distributions with a common k. *Biometrika*, 45, 37-58.

Blumenthal, S., Dahiya, R. C., and Gross, A. J. (1978). Estimating the complete sample size from an incomplete Poisson sample. *Journal of the American Statistical Association*, 73, 182-187.

Bowman, K. O. and Shenton, L .R. (1967). *Remarks on estimation problems for the parameters of the Neyman type A distribution.* Oak Ridge National Laboratory, ORNL-4102/UC-32.

Bulmer, M. G. (1974). On fitting the Poisson lognormal distribution to species-abundance data. *Biometrics*, 30, 101-110.

Cain, S. A. and Evans, F. C. (1952). The distribution patterns of three plant species in an old-field community in southeastern Michigan. *Contributions from the Laboratory of Vertebrate Biology, University of Michigan*, 52, 1-11.

Carter, W. H. and Myers, R. H. (1973). Maximum likelihood estimation from linear combinations of discrete probability functions. *Journal of the American Statistical Association*, 68, 203-206.

Cassie, R. M. (1952). Frequency distribution models in the ecology of plankton and other organisms. *Journal of Animal Ecology*, 31, 65-92.

Cernuschi, F. and Castagnetto, L. (1946). Chains of rare events. *Annals of Mathematical Statistics*, 17, 53-61.

Chernoff, H. and Lehmann, E. L. (1954). The use of maximum likelihood estimates in χ^2 tests of goodness of fit. *Annals of Mathematical Statistics*, 25, 579-586.

Choi, K. and Bulgren, W. G. (1968). An estimation procedure for mixtures of distributions. *Journal of the Royal Statistical Society, (B)*, 30, 444-460.

Cliff, A. D. and Ord, J. K. (1975). Model building and the analysis of spatial pattern in human geography (with discussion). *Journal of the Royal Statistical Society, B,* 37, 297-348.

Clapham, A. R. (1936). Over dispersion in grassland communities and the use of statistical methods in plant ecology. *Journal of Ecology,* 24, 232-251.

Cochran, W. G. (1950). Estimation of bacterial densities by means of the "most probable number." *Biometrics,* 6, 105-116.

Cochran, W. G. (1952). The χ^2 test of goodness-of-fit. *Annals of Mathematical Statistics,* 23, 315-345.

Cohen, A. C. (1961). On a class of pseudo-contagious distributions. *Technical Report* 11, *Institute of Statistics, University of Georgia.*

Cole, L. C. (1946). A theory for analyzing contagiously distributed populations. *Ecology,* 27, 329-341.

Conover, W. J. (1972). A Kolmogorov goodness-of-fit test for discontinuous distributions. *Journal of the American Statistical Association,* 67, 591-596.

Cook, R. D. and Martin, F. B. (1974). A model for quadrat sampling with "visibility bias." *Journal of the American Statistical Association,* 69, 345-349.

Corbishley, H. R. (1972). Improving direct iteration. *Mathematical Gazette,* 56, 110-113.

Cox, D. R. and Hinkley, D. V. (1974). *Theoretical Statistics,* Chapman and Hall.

Cramer, H. (1947). *Mathematical Methods of Statistics,* Princeton U. P.

Cresswell, W. L. and Froggatt, P. (1963). *The Causation of Bus Driver Accidents.* Oxford U.P.

Curtiss, J. H. (1943). On transformations used in the analysis of variance. *Annals of Mathematical Statistics,* 14, 107-122.

D'Agostino, R. B. and Tietjen, G. L. (1971). Simulation probability points of b_2 for small samples. *Biometrika,* 58, 669-672.

Dandekar, V. M. (1955). Certain modified forms of binomial and Poisson distributions, *Sankhya*, 15, 237-259.

Daniels, H. E. (1954). Saddlepoint approximations in statistics. *Annals of Mathematical Statistics*, 25, 631-650.

Darwin, J. H. (1960). An ecological distribution akin to Fisher's logarithmic distribution. *Biometrics*, 16, 51-60.

David, F. N. and Johnson, N. L. (1952). The truncated Poisson. *Biometrics*, 8, 275-285.

David, F. N., Kendall, M. G. and Barton, D. E. (1966). *Symmetric Functions and Allied Tables*, C.U.P.

David, F. N. and Moore, P. (1954). Notes on contagious distributions in plant populations. *Annals of Botany*, 18, 47-53.

De Bruijn, N. G. (1961). *Asymptotic Methods in Analysis*. 2nd ed. North Holland Publication Company.

Dempster, A. P., Laird, N. M. and Rubin, D. B. (1977). Maximum likelihood from incomplete data via the EM algorithm. *Journal of the Royal Statistical Society, B*, 39, 1-38.

Dickerson, O. D., Katti, S. K. and Hofflander, A. E. (1961). Loss distributions in non-life insurance. *Journal of Insurance*, 28, 45-54.

Douglas, J. B. (1955). Fitting the Neyman type A (two parameter) contagious distribution. *Biometrics*, 11, 149-173.

Douglas, J. B. (1965). Asymptotic expansions for some contagious distributions. *Classical and Contagious Discrete Distributions*, ed. G. P. Patil, Statistical Publications Society, Calcutta, 291-302.

Douglas, J. B. (1970). Statistical models in discrete distributions. *Random Counts in Scientific Work*, ed. G. P. Patil, Penn State University Press, 203-232.

Douglas, J. B. (1971). Stirling numbers in discrete distributions. *Statistical Ecology*, Vol. I, ed. G. P. Patil, E. C. Pielou, and W. E. Waters, Penn State University Press.

Douglas, J. B. (1975). Clustering and aggregation. *Sankhya B*, 37, 398-417.

Dubey, S. D. (1970). Compound Gamma, Beta and F distributions. *Metrika*, 16, 27-31.

Easterling, R. G. (1976). Goodness of fit and parameter estimation. *Technometrics*, 18, 1-10.

Edwards, A. W. F. (1972). *Likelihood.* Cambridge, U.P.

Engen, S. (1974). On species frequency models. *Biometrika*, 61, 263-270.

Evans, D. A. (1953). Experimental evidence concerning contagious distributions in ecology. *Biometrika*, 40, 186-211.

Evans, F. C. (1952). The influence of size of quadrat on the distributional pattern of plant populations. *Contributions from the Laboratory of Vertebrate Biology, University of Michigan*, No. 54.

Faa Di Bruno, F. (1855). *Tortolini Annali di science mathematische e fisiche*, 6, 479-480.

Feller, W. (1943). On a general class of "contagious" distributions. *Annals of Mathematical Statistics*, 14, 389-400.

Feller, W. (1967). *An Introduction to Probability Theory and its Applications*, Vol. I, Wiley.

Feller, W. (1966). *An Introduction to Probability Theory and its Applications*, Vol. II, Wiley.

Ferguson, T. S. (1958). A method of generating best asymptotically normal estimates with application to the estimation of bacterial densities. *Annals of Mathematical Statistics*, 29, 1046-1062.

Fisher, R. A. (1931). Properties of the Functions. (Part of the Introduction to *British Association Mathematical Tables*, 1), British Association for the Advancement of Science, London.

Fisher, R. A. (1943). Fisher, R. A., Corbet, A. S. and Williams, C. B. The relation between the number of species and the number of individuals in a random sample from an animal populations. *Journal of Animal Ecology*, 12, 42-58.

Fisher, R. A. and Yates, F. (1967). *Statistical Tables*, 6th ed. Oliver and Boyd.

Fisz, M. (1963). *Probability Theory and Mathematical Statistics*, 3rd ed., Wiley.

Fraser, D. A. S. (1976). *Probability and Statistics. Theory and Applications.* Duxbury.

Frishman, F. (1975). On the arithmetic means and variances of products and ratios of random variables. *Statistical Distributions in Scientific Work I*. D. Reidel Publishing Co.

Froggatt, P. (1970). Application of discrete distribution theory to the study of non-communicable events in medical epidemiology. *Random Counts in Scientific Work, Vol. 2*, G. P. Patil, ed. Penn State University Press.

Gail, M. (1974). Power computations for designing comparative Poisson trials. *Biometrics*, 30, 231-237.

Galliher, H. P., Morse, P. M. and Simond, M. (1959). Dynamics of two classes of continuous-review inventory systems. *Operations Research*, 7, 362-383.

Gart, J. J. (1970). Some simple graphically oriented statistical methods for discrete data. *Random Counts in Scientific Work, Vol. I.*, G. P. Patil, ed. Penn State University Press, 171-191.

Gates, C. E. and Ethridge, F. G. (1972). A generalized set of discrete frequency distributions with FORTRAN program. *Mathematical Geology*, 4, 1-24.

Gebski, V. J. (1975). Solving the maximum likelihood equations of the Poisson-Pascal distribution using inverse interpretation. *Skandanavian Actuarial Journal*, 203-206.

Gillings, D. B. (1974). Some further results for bivariate generalizations of the Neyman Type A distribution. *Biometrics*, 30, 619-628.

Gleeson, A. C. and Douglas, J. B. (1975). Quadrat sampling and the estimation of Neyman Type A and Thomas distributional parameters. *Australian Journal of Statistics*, 17, 103-113.

Godambe, A. V. and Patil, G. P. (1969). Infinite divisibility and additivity of certain probability distributions with an application to mixtures and randomly stopped sums. *Department of Statistics, The Pennsylvania State University Technical Reports and Preprints, No*. 20.

Graybill, F. A. (1961). *An Introduction to Linear Statistical Models*, Vol. 1. McGraw-Hill.

Grab, E. L. and Savage, I. R. (1954). Tables of the expected value of 1/X for positive Bernoulli and Poisson variables. *Journal of the American Statistical Association*, 49, 169-177.

Greig-Smith, P. (1952). The use of random and contiguous quadrats in the study of the structure of plant communities. *Annals of Botany*, 16, 293-316.

Greig-Smith, P. (1964). *Quantitative Plant Ecology*. 2nd ed. Butterworths.

Greenwood, M. and Yule, G. U. (1920). An inquiry into the nature of frequency distributions representative of multiple happenings with particular reference to the occurrence of multiple attacks of disease or of repeated accidents. *Journal of the Royal Statistical Association*, 83, 255-279.

Griffiths, D. A. (1977). Avoidance-modified generalized distributions and their application to studies of super-parisitism. *Biometrics*, 33, 103-112.

Grimm, H. (1964). Tafeln der Neyman-Verteilung Type A. *Biometrische Zeitschrift*, 6, 10-23.

Grimm, H. (1970). Graphical methods for the determination of type and parameters of some discrete distributions. *Random Counts in Scientific Work*, Vol. I, G. P. Patil, ed. Penn State University Press, 193-206.

Gupta, R. C. (1974). Modified power series distribution and some of its applications. *Sankhya Ser. B*, 36, 288-298.

Gupta, R. C. (1975). Maximum likelihood estimation of a modified power series distribution and some of its applications. *Communications in Statistics*, 4, 689-697.

Gurland, J. (1948). *Best asymptotically normal estimates*. Unpublished Ph.D. thesis, University of California, Berkeley.

Gurland, J. (1957). Some interrelations among compound and generalized distributions. *Biometrika*, 44, 265-268.

Gurland, J. (1958). A generalized class of contagious distributions. *Biometrics*, 14, 229-249.

Gurland, J. (1954). Some applications of the negative binomial and other contagious distributions. *American Journal of Public Health*, 49, 1388-1399.

Gurland, J. and Edwards, C. B. (1961). A class of distributions applicable to accidents. *Journal of the American Statistical Association*, 56, 503-517.

Gurland, J. (1965). A method of estimation for some generalized Poisson distributions. *Classical and Contagious Discrete Distributions*. G. P. Patil, ed. Statistical Publications Society, Calcutta, 141-158.

Gurland, J. and Hinz, P. (1971). Estimating parameters, testing fit, and analyzing untransformed data pertaining to the negative binomial and other distributions. *Statistical Ecology, Vol. 1 Spatial Patterns and Statistical Distributions*. Penn State University Press, G. P. Patil, E. C. Pielou, and W. E. Waters, ed., 143-178.

Haight, F. A. (1959). The generalized Poisson distribution. *Annals of the Institute of Statistical Mathematics, Tokyo*, 11, 101-105.

Haight, F. A. (1967). *Handbook of the Poisson Distribution*. Wiley.

Haldane, J. B. S. (1942). The fitting of binomial distributions. *Annals of Eugenics*, 11, 179-181.

Hamden, M. (1975). Correlation between the numbers of two types of children when the family size distribution is zero-truncated negative binomial. *Biometrics*, 31, 765-766.

Harvey, D. W. (1966). Geographical processes and the analysis of point patterns. *Transactions and Papers, Institute of British Geographers*, 40, 81-95.

Heyde, C. C. and Seneta, E. (1972). The simple branching process, a turning point test and a fundamental inequality: A historical note on I. J. Bienayme. *Biometrika*, 59, 680-683.

Hill, B. M. (1975). Aberrant behaviour of the likelihood function in discrete cases. *Journal of the American Statistical Association*, 70, 717-719.

Hinz, P. and Gurland, J. (1967). Simplified techniques for estimating parameters of some generalized Poisson distributions. *Biometrika*, 54, 555-566.

Hinz, P. and Gurland, J. (1970). Test of fit for the negative binomial and other contagious distributions. *Journal of the American Statistical Association*, 65, 887-903.

Hodges, J. L. and Le Cam, L. (1960). The Poisson approximation to the Poisson binomial distribution. *Annals of Mathematical Statistics*, 31, 737-740.

Hoel, P. G. (1943). On indices of dispersion. *Annals of Mathematical Statistics*, 14, 155-162.

Hogg, R. V., Uthoff, V. A., Randalls, R. H. and Davenport, A. S. (1972). On the selection of the underlying distribution and adaptive estimation. *Journal of the American Statistical Association*, 67, 597-600.

Holgate, P. (1964). Estimation for the bivariate Poisson distribution. *Biometrika*, 51, 241-245.

Holgate, P. (1966). Bivariate generalizations of Neyman's Type A distribution. *Biometrika*, 53, 241-244.

Holgate, P. (1970). The modality of some compound Poisson distributions. *Biometrika*, 57, 666-667.

Holmes, D. I. (1974). Graphical methods for the analysis of discrete data. *The Statistician*, 23, 129-134.

Hudson, D. J. (1971). Interval estimation from the likelihood function. *Journal of the Royal Statistical Society, B*, 33, 256-262.

Huzurbazar, V. S. (1956-7). Sufficient statistics and orthogonal parameters. *Sankhya*, 17, 217-220.

Irwin, J. O. (1964). The personal factor in accidents. *Journal of the Royal Statistical Society, Ser. A.*, 127, 438-451.

Irwin, J. O. (1965). Inverse factorial series as frequency distributions. *Classical and Contagious Discrete Distributions*, G. P. Patil, ed. Statistical Publishing Society, Calcutta, 159-174.

Irwin, J. O. (1968). The generalized Waring distribution, applied to accident theory. *Journal of the Royal Statistical Society, Ser. A*, 131, 205-225.

Irwin, J. O. (1975). The generalized Waring distribution II. *Journal of the Royal Statistical Society, A*, 138, 204-227.

Irwin, J. O. (1975). The generalized Waring distribution III. *Journal of the Royal Statistical Society, A*, 138, 374-384.

Iwao, S. and Kuno, E. (1971). An approach to the analysis of aggregation pattern in biological populations. *Statistical Ecology, Vol. 1: Spatial Patterns and Statistical Distributions*, G. P. Patil, E. C. Pielou, and W. E. Waters, Penn State University Press, 461-513.

Jeffreys, H. (1961). *The Theory of Probability*, 3rd ed. Cambridge, U. P.

Johnson, N .L. and Kotz, S. (1964). *Distributions in Statistics: Discrete Distributions*. Houghton Mifflin.

Johnson, N. L. and Kotz, S. (1977). *Urn models and their applications: an approach to modern discrete probability theory*. Wiley.

Jordan, C. (1965). *Calculus of Finite Differences*. Chelsea.

Joshi, S. W. (1974). Integral expressions for the tail probabilities of the power series distributions. *Sankhya, B*, 36, 462-465.

Kabir, A. B. M. L. (1968). Estimation of parameters of a finite mixture of distributions. *Journal of the Royal Statistical Society, Ser. B.*, 30, 472-482.

Kathirgamatamby, N. (1953). Note on the Poisson index of dispersion. *Biometrika*, 40, 225-228.

Katti, S. K. (1966). Interrelations among generalized distributions and their components. *Biometrics*, 22, 44-52.

Katti, S. K. (1967). Infinite divisibility of integer valued random variables. *Annals of Mathematical Statistics*, 38, 1306-1308.

Katti, S. K. and Gurland, J. (1961). The Poisson Pascal distribution. *Biometrics*, 17, 527-38.

Katti, S. K. and Gurland, J. (1962a). Some methods of estimation for the Poisson binomial distribution. *Biometrics*, 18, 42-51.

Katti, S. K. and Gurland, J. (1962b). Efficiency of certain methods of estimation for the negative binomial and the Neyman Type A distributions. *Biometrika*, 49, 215-226.

Katti, S. K. and Rao, A. V. (1965). *The Log-Zero-Poisson Distribution*. Florida State University Statistical Report M106.

Katti, S. K. and Rao, A. V. (1970). The log-zero Poisson distribution. *Biometrics*, 26, 501-513.

Katti, S. K. and Sly, L. E. (1965). Analysis of contagious data through behavioristic models. *Classical and Contagious Discrete Distributions*, G. P. Patil, ed., 303-319.

Keilson, J. and Gerber, H. (1971). Some results for discrete modality. *Journal of the American Statistical Association*, 66, 386-389.

Kemp, A. W. (1974). *Towards a unification of the theory of counts*. Statistics Reports and Preprints, University of Bradford, 15.

Kemp, A. W. and Kemp, C. D. (1966). An alternative derivation of the Hermite distribution. *Biometrika*, 53, 627-628.

Kemp, A. W. and Kemp, C. D. (1968). On a distribution associated with a certain stochastic process. *Journal of the Royal Statistical Society, Series B*, 30, 160-163.

Kemp, C. D. (1967a). 'Stuttering-Poisson' distributions. *Journal of Statistical and Social Inquiry Society of Ireland*, 21, 151-157.

Kemp, C. D. (1967b). On a contagious distribution suggested for accident data. *Biometrics*, 23, 241-255.

Kemp, C. D. (1970). "Accident proneness" and discrete distribution theory. *Random Counts in Scientific Work Vol. II.*, G. P. Patil, ed., Penn State University Press, 41-66.

Kemp, C. D. (1971). Properties of some discrete ecological distributions. *Statistical Ecology, Vol. 1: Spatial Patterns and Statistical Distributions*. G. P. Patil, E. C. Pielou, and W. E. Waters, ed., Penn State University Press, 1-22.

Kemp, C. D. and Kemp, A. W. (1965). Some properties of the 'Hermite' distribution. *Biometrika*, 52, 381-394.

Kemp, C. D. and Kemp, A. W. (1967). A special case of Fisher's 'modified Poisson series'. *Sankhya*, Ser. A, 29, 103-104.

Kendall, M. G. and Stuart, A. (1963). *The Advanced Theory of Statistics*, Volume I (2nd ed.), Griffin.

Kendall, M. G. and Stuart, A. (1967). *The Advanced Theory of Statistics*, Volume II, (2nd ed.), Griffin.

Khatri, C. G. (1959). On certain properties of power series distributions. *Biometrika*, 46, 486-490.

Khatri, C. G. (1963). A fitting procedure for a generalized binomial distribution. *Annals of the Institute of Statistical Mathematics, Tokyo*, 14, 133-141.

Khatri, C. G. and Patel, I. R. (1961). Three classes of univariate discrete distributions. *Biometrics*, 17, 567-75.

Klimko, E. M. (1973). An algorithm for calculating indices in Faa di Bruno's formula. *Bit*, 13, 38-49.

Koerts, J. and Abrahamse, A. P. J. (1969). *On the Theory and Application of the General Linear Model.* Rotterdam U.P.

Kowal, R. R. (1971). Disadvantages of the generalized variance as a measure of variability. *Biometrics*, 27, 213-216.

Ladd, D. W. (1975). An algorithm for the binomial distribution with dependent trials. *Journal of the American Statistical Association*, 70, 333-340.

Lancaster, H. O. (1960). *The Chi-Squared Distribution.* Wiley.

Leiter, R. E. and Hamdan, M. A. (1973). Some bivariate probability models applicable to traffic accidents and fatalities. *International Statistical Review*, 41, 87-100.

Lind, P. (1974). Properties of and estimation in the Thomas distribution. (Unpublished M.Sc. Thesis, Univ. N.S.W.).

Linnik, Yu.V. (1964). *Decomposition of Probability Distributions* (English translation of Russian 1960 edition). Oliver and Boyd.

Lindsey, J. K. (1974). Comparison of probability distributions. *Journal of the Royal Statistical Association, B*. 36, 38-47.

Lloyd, M. (1967). Mean crowding. *Journal of Animal Ecology*, 36, 1-30.

Loeve, M. (1963). Probability Theory, 3rd ed., Princeton.

Luders, R. (1934). Die statistic der seltenen ereignisse. *Biometrika*, 26, 108-128.

Lukacs, E. (1955). Applications of Faa di Bruno's formula in mathematical statistics. *American Mathematical Monthly*, 62, 340-348.

Lukacs, E. (1960). *Characteristic Functions.* Griffin.

Maceda, E. C. (1948). On the compound and generalized Poisson distributions. *Annals of Mathematical Statistics*, 19, 414-416.

Magistad, J. G. (1961). Some discrete distributions associated with life testing. *Proceedings 7th National Symposium on Reliability and Quality Control*. 1-11.

Maritz, J. S. (1952). Note on a certain family of discrete distributions. *Biometrika*, 39, 196-198.

Martin, D. C. and Katti, S. K. (1962). Approximations to the Neyman Type A distribution for practical problems. *Biometrics*, 18, 354-364.

Martin, D. C. and Katti, S. K. (1965). Fitting of certain contagious distributions to some available data by the maximum likelihood method. *Biometrics*, 21, 34-48.

McGilchrist, C. A. (1969). Discrete distribution estimators from the recurrence relation for probabilities. *Journal of the American Statistical Association*, 64, 602-609.

McGuire, J. V., Brindley, T. A. and Bancroft, T. A. (1957). The distribution of European cornborer larvae Pyrausta Nubilalis (hbn) in field corn. *Biometrics*, 13, 65-78, and corrections *Biometrics*, 14, (1958), 432-434.

Mead, R. (1974). A test for spatial pattern at several scales using data from a grid of contiguous quadrats. *Biometrics*, 30, 295-307.

Miles, R. E. (1959). The complete amalgamation into blocks, by weighted means, of a finite set of real numbers. *Biometrika*, 46, 317-327.

Molenaar, W. (1965). *Some remarks on mixtures of distributions*. Stichting Mathematisch Centrum, Amsterdam, Report S343.

Molenaar, W. and Van Zwet, W. R. (1966). On mixtures of distributions. *Annals of Mathematical Statistics*, 37, 281-283.

Molenaar, W. (1970). *Approximations to the Poisson, Binomial and Hypergeometric Distribution Functions*. Math. Centre Tracts 31, Mathematisch Centrum Amsterdam.

Montmort, P.R. (1708). *Essi d'analyse sur les jeux de hasard*. 1st ed. Paris (1708), 2nd ed., Paris (1714).

Moran, P. A. P. (1968). *An Introduction to Probability Theory*. Oxford U.P.

Nelson, D. L. (1975). Some remarks on generalizations of the negative binomial and Poisson distributions. *Technometrics*, 17, 135-136.

Newbold, E. M. (1927). Practical applications of the statistics of repeated events, particularly to industrial accidents. *Journal of the Royal Statistical Society, A*, 90, 487-547.

Neyman, J. (1939). On a new class of "contagious" distributions applicable in entomology and bacteriology. *Annals of Mathematical Statistics*, 10, 35-57.

Neyman, J. (1949). Contribution to the theory of the χ^2 test. *Proceedings (First) Berkeley Sumposium on Mathematical Statistics and Probability*, University of California Press, 239-273.

Neyman, J. (1965). Certain chance mechanisms involving discrete distributions. *Classical and contagious discrete distributions*. G. P. Patil, ed. Statistical Publishing Society, Calcutta, 4-13.

Niven, I. (1969). Formal power series. *American Mathematical Monthly*, 76, 871-889.

Noack, A. (1950). A class of random variables with discrete distributions. *Annals of Mathematical Statistics*, 21, 127-132.

Ord, J. K. (1970). The negative binomial model and quadrat sampling. *Random Counts in Scientific Work, Vol. II*, G. P. Patil, ed. Penn State University Press. 151-164.

Ord, J. K. (1972). *Families of Frequency Distributions*. Griffin.

Owen, D. B. (1962). *Handbook of Statistical Tables*. Addison Wesley.

Pahl, P. J. (1969). On testing for goodness of fit of the negative binomial distribution when expectations are small. *Biometrics*, 25, 143-151.

Patel, Y. C. (1971). Some problems in estimation for the parameters of the Hermite Distribution. Unpublished Ph.D. thesis, University of Georgia.

Patel, Y. C. (1976). Estimation of the parameters of the triple and quadruple stuttering-Poisson distributions. *Technometrics*, 18, 67-74.

Patel, Y. C., Shenton, L. R. and Bowman, K. O. (1974). Maximum likelihood estimation for the parameters of the Hermite distribution. *Sankhya, B,* 36, 54-162.

Patil, G. P. (1961). Asymptotic bias and variance of ratio estimates in generalized power series distributions and certain applications. *Sankhya, Series A,* 23, 269-280.

Patil, G. P. (1962). Certain properties of the generalized power series distribution. *Annals of the Institute of Statistical Mathematics, Tokyo,* 14, 179-182.

Patil, G. P. (1962b). On homogeneity and combined estimation for the generalized power series distribution and certain applications. *Biometrics,* 18, 365-374.

Patil, G. P. (1962c). Maximum likelihood estimation for generalized power series distributions and its application to a truncated binomial distribution. *Biometrika,* 49, 227-238.

Patil, G. P. (1962d). Estimation by two-moments method for generalized power series distribution and certain applications. *Sankhya, Series A,* 24, 201-214.

Patil, G. P. (1963). Minimum variance unbiased estimation and certain problems of additive number theory. *Annals of Mathematical Statistics,* 34, 1050-1056.

Patil, G. P. and Bildikar, S. (1966). On minimum variance unbiased estimation for the logarithmic series distribution. *Sankhya, Series A,* 28, 239-50.

Patil, G. P. and Bildikar, S. (1967). Multivariate logarithmic series distribution as a probability model in population and community ecology and some of its statistical properties. *Journal of the American Statistical Association,* 62, 655-674.

Patil, G. P. and Wani, J. K. (1965). On certain structural properties of the logarithmic series distribution and the first type Stirling distribution. *Sankhya, Series A,* 27, 271-280.

Patil, G. P., Kotz, S. and Ord, J. K. (ed.) (1975). *Statistical Distributions in Scientific Work,* Vol. 1: Models and structures, Vol. 2: Model building and model selection, Vol. 3: Characterizations and applications. D. Reidel Publishing Company.

Patil, G. P. and Rao, C. R. (1978). Weighted distributions and size biased sampling with applications to wildlife populations and human families. *Biometrics*, 34, 179-190.

Pearson, E. S. and Hartley, H. O. (ed.) (1966). *Biometrika Tables for Statisticians*, Vol. I., 3rd Edition Biometrika Trustees, C.U.P.

Pearson, K. (1918-1919). Peccavimus! *Biometrika*, 12, 266-281. esp. p. 270.

Peizer, D. B. and Pratt, J. W. (1968). A normal approximation for binomial, F, beta and other common related tail probabilities I. *Journal of the American Statistical Association*, 63, 1416-1456.

Pereira, B. de B. (1977). Discriminating among separate models - a bibliography. *International Statistical Review*, 45, 163-172.

Philipson, C. (1960). The theory of confluent hypergeometric functions and its application to compound Poisson processes. *Skandinaviske Aktuarietidsskrift*, 43, 136-162.

Philpot, J. W. (1964). *Orthogonal Parameters for Two Parameter Distributions*. M.Sc. thesis, V.P.I.

Pielou, E. C. (1957). The effect of quadrat size on the estimation of the parameters of Neyman's and Thomas' distributions. *Journal of Ecology*, 45, 31-47.

Pielou, E. C. (1969). *An Introduction to mathematical ecology*. Wiley.

Pólya, G. (1930/31). Sur quelques points de la théorie des probabilités. *Ann. Inst. Henri Poincare*, 1, 117-161.

Pratt, J. W. (1959). On a general concept of "in probability." *Annals of Mathematical Statistics*, 30, 549-558.

Quenouille, M. M. (1948). Statistical Note. *Journal of General Microbiology*, 2, 65-66.

Quenouille, M. M. (1949). A relation between the logarithmic, Poisson and negative binomial series. *Biometrics*, 5, 162-164.

Qvale, P. (1932). Remarks on semi-invariants and incomplete moments. *Skandinaviske Aktuarietidsskrift*, 15, 196-210.

Rade, L. (1972). On the use of generating functions and Laplace transforms in applied probability theory. *International Journal of Mathematical Education in Science and Engineering*, 3, 25-33.

Rao, A. V. (1969). *The Log-Zero-Poisson Distribution*. Ph.D. thesis, Florida State University.

Rao, A. and Katti, S. K. (1969). *Estimation of Parameters of the Log-Zero Poisson Distribution*. Florida State University Statistics Report M165.

Rao, A. V. and Katti, S. K. (1970). Minimum chi-squared estimation for the log-zero-Poisson distribution. *Random Counts in Scientific Work Vol. 2*, G. P. Patil, ed., Penn State University Press, 165-188.

Rao, C. R. (1957). Theory of the method of estimation by minimum chi-square. *Bulletin of the International Statistical Institute*, 35, (2), 25-

Rao, C. R. (1965). *Linear Statistical Inference and its Applications*. Wiley.

Rao, C. R. (1965). On discrete distributions arising out of methods of ascertainment. *Classical and contagious discrete distributions*. G. P. Patil, ed., Statistical Publishing Society, Calcutta, 320-332.

Rao, C. R., Mitra, S. K. and Matthai, A. (1966). *Formulae and Tables for Statistical Work*. Statistical Publishing Society, Calcutta.

Richardson, C. H. (1954). *An Introduction to the Calculus of Finite Differences*. Van Nostrand, N. Y.

Rider, P. R. (1962). Expected values and standard deviations of the reciprocal of a variable from a decapitated negative binomial distribution. *Journal of the American Statistical Association*, 57, 439-445.

Riordan, J. (1937). Moment recurrence relations for binomial, Poisson and hypergeometric frequency distributions. *Annals of Mathematical Statistics*, 8, 103-111.

Riordan, J. (1958). *An Introduction to Combinatorial Analysis*. Wiley.

Roach, S. A. (1968). *The Theory of Random Clumping*. Methuen.

Robertson, C. A. and Fryer, J. G. (1970). The bias and accuracy of moment estimators. *Biometrika*, 57, 57-65.

Rogers, A. (1969). Quadrat analysis of urban dispersion. 1. Theoretical techniques, pp. 47-80. 2. Case studies of urban retail systems, pp. 155-171. *Environment and Planning*, 1.

Rogers, A. and Gomar, N. G. (1969). Statistical inference in quadrat analysis. *Geographical Analysis*, 1, 370-384.

Rogers, A. (1972). Statistical analysis of spatial dispersion: the quadrat method. *The Technological Institute, Northwestern University*, Illinois.

Ross, G. J. S. (1970). The efficient use of function minimization in non-linear maximum-likelihood estimation. *Journal of the Royal Statistical Society C*, 19, 205-221.

Roy, J. and Mitra, S. K. (1957). Unbiased minimum variance estimators in a class of discrete distributions. *Sankhya*, 18, 371-378.

Russell, K. G. (1978). Estimation of the parameters of the Thomas distribution. *Biometrics*, 34, 95-99.

Sankaran, M. (1970). The discrete Poisson-Lindley distribution. *Biometrics*, 26, 145-149.

Satterthwaite, F. E. (1942). Generalized Poisson distribution. *Annals of Mathematical Statistics*, 13, 410-417.

Seal, H. L. (1949). The historical development of the use of generating functions in probability theory. *Bulletin de l'Association des Actuaires Suises*, 49, 209-228.

Shah, S.M. (1961). The asymptotic variances of method of moments estimates of the parameters of the truncated binomial and negative binomial distributions. *Journal of the American Statistical Association*, 56, 990-994.

Shapiro, S. S., Wilk, M. B. and Chen, J. H. (1968). A comparative study of various tests for normality. *Journal of the American Statistical Association*, 63, 1343-1372.

Shenton, L. R. (1949). On the efficiency of the method of moments and Neyman's Type A distribution. *Biometrika*, 36, 450-454.

Shenton, L. R. (1950). Maximum likelihood and the efficiency of the method of moments. *Biometrika*, 37, 111-116.

Shenton, L. R. and Bowman, K. O. (1967). Remarks on large sample estimators for some discrete distributions. *Technometrics*, 9, 587-598.

Shenton, L. R. and Bowman, K. O. (1977). *Maximum Likelihood Estimation in Small Samples*. Griffin.

Shumway, R. and Gurland, J. (1960a). A fitting procedure for some generalized Poisson distributions. *Skandinairsk Aktuarietidsskrift*, 43, 87-108.

Shumway, R. and Gurland, J. (1960b). Fitting the Poisson Bionmial distribution. *Biometrics*, 16, 522-533.

Sichel, H. S. (1975). On a distribution law for word frequencies. *Journal of the American Statistical Association*, 70, 542-547.

Skellam, J. G. (1952). Studies in statistical ecology, I. Spatial pattern. *Biometrika*, 39, 346-362.

Skellam, J. G. (1958). On the derivation and applicability of Neyman's Type A distribution. *Biometrika*, 45, 32-36.

Slakter, M. J. (1966). Comparative validity of the chi-square and two modified chi-square goodness-of-fit tests for small but equal frequencies. *Biometrika*, 53, 619-622.

Sprott, D. A. (1958). The method of maximum-likelihood applied to the Poisson binomial distribution. *Biometrics*, 14, 97-106.

Sprott, D. A. (1965). A class of contagious distributions and maximum likelihood estimation. *Classical and Contagious Discrete Distributions*. G. P. Patil, ed., Statistical Publishing Society, Calcutta, 337-350.

Sprott, D. A. (1970). Some exact methods of inference applied to discrete distributions. *Random Counts in Models and Structures*. Vol. 1 of Random Counts in Scientific Work, G. P. Patil, ed., Penn State University Press, 207-232.

Stark, A. E. (1975). Some estimators of the integer-valued parameter of a Poisson Variate. *Journal of the American Statistical Association*, 70, 685-689.

Steyn, H. S. (1975). On extensions of the binomial distributions of Bernouille and Poisson. *South African Statistical Journal*, 9, 163-

"Student" (1907). On the error of counting with a haemacytometer. *Biometrika*, 5, 351-355.

"Student" (1919). An explanation of deviations from Poisson's law in practice. *Biometrika*, 12, 211-215.

Subrahmaniam, K. (1966). On a general class of contagious distributions: the Pascal-Poisson distribution. *Trabajos de Estadistica*, 17, 109-128.

Tate, R. F. and Goen, R. L. (1958). Minimum variance unbiased estimation for the truncated Poisson distribution. *Annals of Mathematical Statistics*, 29, 755-765.

Taylor, L. R. (1971). Aggregation as a species characteristic. *Statistical Ecology, Vol. 1: Spatial Patterns and Statistics Distributions*, G. P. Patil, E. C. Pielou, and W. E. Waters, ed., Penn State University Press, 357-377.

Teicher, H. (1960). On the mixture of distributions. *Annals of Mathematical Statistics*, 31, 55-73.

Teicher, H. (1961). Identifiability of mixtures. *Annals of Mathematical Statistics*, 32, 244-248.

Teicher, H. (1963). Identifiability of finite mixtures. *Annals of Mathematical Statistics*, 34, 1265-1269.

Thomas, M. (1949). A generalization of Poisson's binomial limit for use in ecology. *Biometrika*, 36, 18-25.

Thomas, M. (1951). Some tests for randomness in plant populations. *Biometrika*, 38, 102-111.

Thomson, G. W. (1952). *Measures of Plant Aggregation Based on Contagious Distribution*. Contributions from the Laboratory of Vertebrate Biology, University of Michigan. No. 53.

Thompson, H. R. (1954). A note on contagious distributions. *Biometrika*, 41, 268-271.

Tucker, H. G. (1963). An estimate of the compounding distribution of a compound Poisson distribution. *Theoria Veroyatnostyei i ee Primenenia*, 8, 211-216.

Van Heerden, D. F. I. and Gonin, H. T. (1966). The orthogonal polynomials of power series probability distributions and their uses. *Biometrika,* 53, 121-128.

Wallace, D. L. (1958). Asymptotic approximations to distributions. *Annals of Mathematical Statistics,* 29, 635-654.

Wani, J. K. and Lo, H. P. (1975). Clopper-Pearson system of confidence intervals for the logarithmic series distribution. *Biometrics,* 31, 771-775.

Warren, W. G. (1971). The center-satellite concept as a basis for ecological sampling. *Sampling and Modelling Biological Populations and Population Dynamics.* Statistical Ecology, Vol. 2. G. P. Patil, E. C. Pielou, and W. E. Waters, ed., Penn State University Press.

Watson, G. S. (1959). Some recent results in chi-square goodness-of-fit tests. *Biometrics,* 15, 440-468.

Watson, H. W. (1873). Solution to Problem 4001 (proposed by Francis Galton). *The Educational Times,* August 1, 115-116.

Westman, W. E. (1971). Mathematical models of contagion and their relation to density and basal area sampling techniques. *Statistical Ecology, Vol. 1: Spatial Patterns and Statistical Distribution.* G. P. Patil, E. C. Pielou, and W. E. Waters, ed. Penn State University Press, 515-536.

Williams, D. G. (1972). Ecological studies on shrub-steppe of the western Riverina, N.S.W. Ph.D. thesis, Australian National University.

Yarnold, J. K. (1970). The minimum expectation in X^2 goodness of fit tests. *Journal of the American Statistical Association,* 65, 564-886.

AUTHOR INDEX

SUBJECT INDEX